A General View of the Agriculture of Aberdeenshire
1811

EXPLANATION.

Clay ....
Good Loam ....
Gravelly Loam ....
Inferior various .....
Heath & barren ....

Scale of British Statute Miles.
1 2 3 4 5 6 7 8 9 10 11 12 13 14 15 16 17 18 19 20

INVERNESS SH.

BANFFSHI[RE]

PERTH SH.   FORFAR or ANGUS SH.

Braeriach H. 4220 f.
Cairntoul H. 4220 f.
Hawtoul H.
Beaulomul H.
Benmuca H.
Sannock M.
Ho. of Braemar
Lochnagar
Mont Keen
Mont Battock
Craitua
Glenmuick
Glentanar
Tullich
Aboyne
Tarland
Coldstone
Migvie
Morven
Towie
Cushny
Kildrummy
Forbes
Strathdon
Glenbucket
Auchindoir
Kearn
Rhynie
Noath
Cabrach
Cairnie

A GENERAL VIEW
OF THE
AGRICULTURE
OF
ABERDEENSHIRE;
DRAWN UP UNDER THE DIRECTION OF
THE BOARD OF AGRICULTURE;
AND ILLUSTRATED WITH PLATES

by
GEORGE SKENE KEITH, D.D.
Minister of Keith-Hall and Kinkell.

1811

The Grimsay Press
2005

The Grimsay Press
an imprint of
Zeticula
57 St Vincent Crescent
Glasgow
G3 8NQ

http://www.thegrimsaypress.co.uk
admin@thegrimsaypress.co.uk

Transferred to digital printing in 2005

Copyright © Zeticula 2005

First published in Great Britain in 1811
in Aberdeen

ISBN 0 902664 38 7

Reproduced from the copy in the Library of
the Scottish Agricultural College, Ayr Campus,
Scotland

All rights reserved. No part of this publication
may be reproduced, stored in a retrieval
system, or transmitted in any form or by any
means, electronic, mechanical, photocopying,
recording or otherwise, without the prior
permission of the publishers.

ADVERTISEMENT.

THE desire that has been generally expressed, to have the AGRICULTURAL SURVEYS of the KINGDOM reprinted, with the additional Communications which have been received since the ORIGINAL REPORTS were circulated, has induced the BOARD of AGRICULTURE to come to a resolution to reprint such as appear on the whole fit for publication.

IT is proper at the same time to add, that the Board does not consider itself responsible for every statement contained in the Reports thus reprinted, and that it will thankfully acknowledge any additional information which may still be communicated.

N. B. *Letters to the Board may be addressed to Sir* JOHN SINCLAIR, *Bart. M. P. the President, No. 32, Sackville-street, Piccadilly, London.*

DIRECTIONS TO THE BINDER.

1. MAP OF THE SOILS....................FACING THE TITLE.
2. ABBEY OF DEER..PAGE 98
3. HADDO-HOUSE..101.
4. KEITH-HALL...103.
5. WESTER FINTRAY..134
6. PITFOUR's ORCHARD.......................................366
7. MOUNTAIN OF LOCHNAGAR..............................643

Page 407, line 24. for *night-soil* read *soiling*.

PLAN

FOR REPRINTING THE
AGRICULTURAL SURVEYS.

BY THE PRESIDENT OF THE BOARD OF AGRICULTURE.

A Board established for the purpose of making every essential inquiry into the agricultural state, and the means of promoting the internal improvement of a powerful empire, will necessarily have it in view to examine the sources of public prosperity, in regard to various important particulars. Perhaps the following is the most natural order for carrying on such important investigations; namely, to ascertain,

1. The riches to be obtained from the surface of the national territory.
2. The mineral or subterraneous treasures of which the country is possessed.
3. The wealth to be derived from its streams, rivers, canals, inland navigations, coasts, and fisheries. And,
4. The means of promoting the improvement of the people, in regard to their health, their industry, and morals, founded on a *statistical* survey, or a minute and careful inquiry into the actual state of every parochial district in the kingdom, and the circumstances of its inhabitants.

Under one or other of these heads, every point of real importance, that can tend to promote the general happiness of a great nation, seems to be included.

Investigations of so extensive and so complicated a nature, must require, it is evident, a considerable space of time before they can be completed. Differing indeed in many respects from each other, it is better, perhaps, that they should be undertaken at different periods, and separately considered. Under that impression the Board of Agriculture has hitherto directed its attention to the first point only, namely, the cultivation of the surface, and the resources to be derived from it.

That the facts essential for such an investigation might be collected with more celerity and advantage, a number of intelligent and respectable individuals were appointed to furnish the Board with accounts of the state of husbandry, and the means of improving the different districts of the kingdom. The returns they sent were printed, and circulated by every means the Board of Agriculture could devise, in the

the districts to which they respectively related; and in consequence of that circulation, a great mass of additional valuable information has been obtained. For the purpose of communicating that information to the public in general, but more especially to those counties the most interested therein, the Board has resolved to reprint the Survey of each County, as soon as it seemed to be fit for publication; and among severals equally advanced, the counties of Norfolk, Lancaster, and Mid-Lothian, were pitched upon, for the commencement of the proposed publication. When all these Surveys shall have been thus re-printed, it will be attended with little difficulty to draw up an abstract of the whole, (which will not probably exceed two or three volumes quarto) to be laid before His Majesty and both Houses of Parliament; and afterwards, a general Report, on the present state of the country, and the means of its improvement, may be systematically arranged according to the various subjects connected with agriculture. Thus every individual in the kingdom may have,

1. An account of the husbandry of his own particular county; or,
2. A general view of the agricultural state of the kingdom at large, according to the counties, or districts, into which it is divided; or,
3. An arranged system of information on agricultural subjects, whether accumulated by the Board since its establishment, or previously known:

And thus information respecting the state of the kingdom, and Agricultural knowledge in general, will be attainable with every possible advantage.

In reprinting these Reports, it was judged necessary, that they should be drawn up according to one uniform model; and after fully considering the subject, the following form was pitched upon, as one that would include in it all the particulars which it was necessary to notice in an Agricultural Survey. As the other Reports will be reprinted in the same manner, the reader will thus be enabled to find out, at once, where any point is treated of, to which he may wish to direct his attention.

PLAN

PLAN OF THE RE-PRINTED REPORTS.

Preliminary Observations.

CHAP.
I. Geographical State and Circumstances.
SECT. 1.—Situation and Extent.
2.—Divisions.
3.—Climate.
4.—Soil and Surface.
5.—Minerals.
6.—Waters.

II. State of Property.
SECT. 1.—Estates and their Management.
2.—Tenures.

III. Buildings.
SECT. 1.—Houses of Proprietors.
2.—Farm Houses and Offices; and Repairs.
3.—Cottages.

IV. Mode of Occupation.
SECT. 1.—Size of Farms—Character of the Farmers.
2.—Rent—in money—in kind—in services.
3.—Tythes.
4.—Poor Rates.
5.—Leases.
6.—Expence and Profit.

CHAP.
V. Implements.
VI. Inclosing.—Fences.—Gates.
VII. Arable Land.
   Sect. 1.—Tillage.
        2.—Fallowing.
        3.—Rotation of Crops.
        4.—Crops commonly cultivated.
        5.—Crops not commonly cultivated.

VIII. Grass.
   Sect. 1.—Natural Meadow and Pasture.
        2.—Artificial Grasses.
        3.—Hay Harvest.
        4.—Feeding.

IX. Gardens and Orchards.
X. Woods and Plantations.
XI. Wastes.
XII. Improvements.
   Sect. 1.—Draining.
        2.—Paring and Burning.
        3.—Manuring.
        4.—Weeding.
        5.—Watering.

XIII. Live Stock.
   Sect. 1.—Cattle.
        2.—Sheep.
        3.—Horses.
        4.—Hogs.
        5.—Rabbits.
        6.—Poultry.
        7. Pigeons.

CHAP.
>SECT. 7.—Pigeons.
>8.—Bees.

XIV. Rural Economy.
>SECT. 1.—Labour—Servants—Labourers—Hours of labour.
>2.—Provisions.
>3.—Fuel.
>4.—Expence and profit.

XV. Political Economy.
>SECT 1.—Roads.
>2.—Canals.
>3.—Fairs.
>4.—Weekly Markets.
>5.—Commerce.
>6.—Manufactures.
>7.—Fisheries.
>8.—Poor.
>9.—Population.

XVI. Obstacles to improvement; including general observations on agricultural legislation and police.

XVII. Miscellaneous Observations.
>SECT. 1.—Agricultural Societies.
>2.—Weights and Measures.

CONCLUSION.—Means of improvement, and the measures calculated for that purpose.

CONTENTS.

	PAGE.
CHAP. I. GEOGRAPHICAL STATE & CIRCUMSTANCES,	1—75
Sect. 1. Situation and Extent,	1
2. Divisions, Ancient and Modern,	5
3. Climate,	20
4. Soil and Surface,	41
5. Minerals,	53
6. Water,	62
CHAP. II. STATE of PROPERTY,	75—93
Sect. 1. Estates and their Management,	75
2. Tenures,	89
CHAP. III. BUILDINGS,	93—142
Sect. 1. Houses of Proprietors,	93
2. Farm Houses, Offices, and Repairs,	129
3. Cottages,	138
CHAP. IV. MODE of OCCUPATION,	142—211
Sect. 1. Size of Farms, and Character of Farmers,	142
2. Distribution of Crop,	161
3. Rent,	169
4. Tythes,	175
5. Poor Rates,	177
6. Leases,	180
7. Expence and Profit,	202
CHAP. V. IMPLEMENTS,	211
CHAP. VI. ENCLOSING,	221
CHAP. VII. ARABLE LANDS,	225—323
Sect. 1. Tillage,	225
2. Fallowing,	229
3. Rotation of Crops,	231
4. Crops commonly cultivated,	243
5. Crops not commonly cultivated,	292

CHAP.

CONTENTS.

CHAP. VIII. GRASS LANDS,	Page	323—360
Sect. 1. Meadow and Natural Grass,		323
2. The Artificial Grasses,		331
CHAP. IX. GARDENS and ORCHARDS,		360—369
CHAP. X. WOODS and PLANTATIONS,		369—385
WASTES,		385—411
Sect. 1. Moors—Extent,		385
2. Mountains,		387
3. Bogs,		891
4. Fens and Marshes,		393
5. Forests,		395
6. Heaths and Downs,		396
CHAP. XII. IMPROVEMENTS.		411—457
Sect. 1. Trenching,		411
2. Draining,		423
3. Paring and Burning,		429
4. Manure,		430
5. Weeding,		442
6. Watering, or Irrigation,		444
CHAP. XIII. EMBANKMENTS,		454—457
CHAP. XIV. LIVE STOCK,		457—513
Sect. 1. Black Cattle—History of the Cattle Trade,		457
2. Sheep,		493
3. Goats,		497
4. Horses,		497
5. Asses. 6. Mules. 7. Hogs,		505
8. Rabbits,		506
9. Poultry,		507
10. Pigeons. 11. Bees,		508
12. Game,		509
CHAP. XV. RURAL ECONOMY,		513—535
Sect. 1. Labour,		513
2. Price of Provisions at different Periods,		525
CHAP. XVI. POLITICAL ECONOMY,		535—609
Sect. 1. Roads,		535
2. Streets and Harbour of Aberdeen,		540
3. Canal from Inverury to Aberdeen,		542
4. Fairs,		546
5. Market Towns,		551
		6. Weights

6. Weights and Measures,		PAGE 551
7. Price of Products compared with Expence of raising them,		561
8. Manufactures—History of these,		577
9. Fisheries,		589
10. Commerce,		591
11. The Poor,—taxed illegally,		597
12. Population,		604

CHAP. XVII. OBSTACLES to IMPROVEMENT, 609—616
Sect. 1. Want of Capital, - 609
 2. Prices. 3. Expences, - 610
 4. 5. 6. Want of power to Inclose, &c. - 612
 7. Want of Disseminated Knowledge, - 612
 8. Enemies of the Farmer, Vermin, &c. - 613

CHAP. XVIII. MISCELLANEOUS ARTICLES, 616—617
Sect. 1. Agricultural Societies established, - 616
 2. Others Wanted, - 616
 3. Mill-Multures, - 617

CONCLUSION.—MEANS of IMPROVEMENT, or the Measures Calculated for that Purpose, - 618—625
 1. Address to the Farmers, - 618
 2. To the Landholders, - 620
 3. To the Legislature, - 621
 Corrections, Additions, and Explanations, - 625
 Classical Tables, - 629

APPENDIX.

1. Observations on British Grasses, - 635
2. Account of Two Journies for ascertaining the Elevation of the principal Mountains of Marr, 641
3. Survey of the Lands near Aberdeen, - 653
4. Rules for equalizing or proportioning the Taxes on Corn, Malt, and Sugar Spirits. - 657
5. Answers to Questions respecting Barley, Malt, and different kinds of Ardent Spirits, - 660

PRELIMINARY OBSERVATIONS.

BEFORE perusing the following Report, the reader will be pleased to attend to a few Preliminary Remarks. They regard,

I. The original Report, or General View of the Agriculture of the County of Aberdeen, which was drawn up by Dr. James ANDERSON, in 1793, and printed in 1794.

II. The Lessons in Agriculture, which the County of Aberdeen can give to other counties of the United Kingdom.

III. The principal errors or defects, in the practice of this county, which we ought to correct or supply, from the knowledge which we now have of the better practice of other counties.

I. As to the original Survey of Aberdeenshire:—The first general view of the Agriculture of the county of Aberdeen was prepared by Dr. JAMES ANDERSON; was printed by the Board of Agriculture with a broad margin for corrections and additions; and was very generally circulated, for that purpose, without any intention of publishing it in its original shape. Owing to the Doctor's not being a native of the county,—to his having removed to Edinburgh some years before he drew up the Report,—and, in particular, to his not being acquainted with its geography, and other matters of detail, there were some inaccuracies in his work, which otherwise had great merit, and shews a general knowledge both of the principles and practice of agriculture. His Report, however, was then, and still will be, read to great advantage, by those who can overlook minute circumstances, and make allowance for those mistakes, in the work of a man of learning and genius, which in his situation were really unavoidable.

As Dr. ANDERSON's Report is in many hands, and contains much valuable information, wherever any important matter of fact is related by him, the Report will be quoted in his own words for the most part, or abridged, where owing to the plan of drawing up the corrected Reports, it cannot be quoted. Where any incorrect statement occurs in the original Report, if this regard a matter of fact, or may tend to mislead, the correction will be made in the proper place. Where the Doctor's opinion or his reasoning cannot be introduced,

duced, any quotations would be improper, and even allusions will be unnecessary. But all the most valuable parts of his work will be inserted in this Survey.

Several gentlemen have made their remarks on the margin of Dr. ANDERSON's General View of the Agriculture of Aberdeenshire. These have been considered with attention; and are either quoted, or otherwise made use of, where they were well founded. It would be improper here to mention the names of gentlemen, who are now alive, and who have made useful corrections, but do not choose to be known to the public. But it is only a tribute of justice to the memory of two gentlemen, the late WILLIAM FORBES LEITH, Esq. of Whitehaugh, and the late DAVID MORRICE, Esq. Sheriff-Substitute of Aberdeen, to state that very important remarks were made by them both; and that the former sent his observations to Dr ANDERSON, who, in the most candid manner acknowledged his error, where he found that Mr LEITH's remarks were well-founded, and sent the corrections to Sir JOHN SINCLAIR, along with Mr LEITH's observations.

Dr ANDERSON having declined to take the trouble of drawing up the corrected Report, according to the plan adopted by the Board, the Writer of this Paper, with the Doctor's approbation, undertook the office: And in May 1804, being in London, he stated to Lord SHEFFIELD, Sir JOHN SINCLAIR, and other Members of the Board of Agriculture, his intention of introducing the two following Sections, as Preliminary Observations; as there was no chapter set apart for them in the body of the Report. The idea met with the approbation of the Board; and they are now submitted to the Reader.

II. WHAT, then, are the Lessons in Agriculture, which the County of Aberdeen can give to the other Counties of the United Kingdom?

It cannot be expected, that a County which is situated in the North of Scotland, and into which an improved system of husbandry has been but lately introduced, can have many practices that deserve imitation in the southern and more cultivated districts. Yet it may be proper to select one useful lesson, that may be learned by the more intelligent farmers, in the neighbourhood of great cities—a lesson, which necessity first taught the proprietors and small farmers, who cultivate the lands in the immediate vicinity of Aberdeen—This is the mixture of the plough and spade husbandry.

From the peculiar situation of the cities of New and Old Aberdeen, on a small neck of land between the rivers Dee and Don, near their entrance into the sea, it became necessary

sary to cultivate all the ground in their neighbourhood. Fifty years ago, there were about 12,000 persons in these two towns, with only a few hundred acres of arable land, (and even this of very indifferent quality), and bounded by a moor covered with heath and stones. Grass for the cowfeeders, garden-roots, and other articles of provision, which could not be brought from the country, were much wanted. And, as the population rapidly increased, and is now 28,000, or more than double of its amount in 1755, it became necessary to improve the old lands as far as it was possible, and also to trench with the spade and mattock a considerable quantity of very rough land, extending about three miles from Aberdeen, and, as before mentioned, either covered with heath, and filled with stones almost to the surface, or interspersed with patches of grass, and large masses of granite lying above ground. In the course of forty years, at least three thousand acres of land have been brought into cultivation, at an expence probably unknown, certainly not experienced, in any other part of the island. The expence indeed could not have been borne in many cases, if the first crop (for so it might be called, as it covered the whole soil), that was raised by the spade and mattock, had not produced from L.30 to L.50 per acre. This was a crop of granite stones, which was sold for paving the streets of London. But, after all, the ground that was thus gained to the community, would not have been able to recompense the cultivator, if a mixture of the spade and plough husbandry had not been introduced. The rent of land in the immediate vicinity of Aberdeen, is extremely high; being now on a lease for years from L.5 to L.10 per acre; and, in a few cases, not less than L.18; nay, when let for a single crop, sometimes as high as L.20. Yet all this is necessary to remunerate the improver, who trenched, dunged, limed, and cultivated this thin soil, which must be frequently manured. It would have yielded too little produce, if tilled only by the plough; and would have been cultivated at too great an expence, if the soil had been constantly digged by the spade. A medium between these two, viz. either the alternate use of the plough and spade, or at least a mixture of plough and spade husbandry, was thus introduced by necessity, and has been attended with the happiest effects.

The gardeners and cowfeeders, who pay very high rents, are by this means enabled to raise two crops in one year, or three crops in two years. Their rotations of cropping are very quick; yet one year of clover cut for soiling, or made into hay, and the pasturing of this grass next year till after midsummer, and then breaking it up for turnips or coleworts,

keeps

keeps this thin land in good condition, and tends to give more tenacity to the soil, which is naturally light, and of a loose texture. It is only in particular situations that this mixture of plough and spade husbandry can be imitated; but where the subsoil is intermixed with masses of granite, consequently where the trench-plough could not be introduced, it is absolutely necessary; and where dung can be purchased, and where rents are extremely high, this practice of the gardeners, cowfeeders, or small farmers, in the immediate vicinity of Aberdeen, is very worthy of imitation.

So much for the lessons in agriculture which this county can give to other districts of the island.

III. WHAT lessons have the farmers in Aberdeenshire to learn from the practice of other counties?

While the Reporter believes it may be useful to point out one article in which Aberdeenshire can give instruction to others, it will be no less useful to call the attention of the landed proprietors and farmers in this county to many excellent lessons which they may learn from the practice of other counties, where an improved system of husbandry has long been established, and has been promoted with ardour, steadiness, and success. Several of these will be incidentally mentioned in the course of this Report; but a few of the leading ones may deserve a place in this section of preliminary observations.

1. One great defect in the farming economy of Aberdeenshire, is the want of attention to the collecting of manure and preparing of dunghills, or compost mixens. As manure is the moving power of the machinery of the agriculturist, the collecting and preparing it ought to be particularly attended to. It no doubt must require a considerable time before the farmers in Aberdeenshire acquire the correct practice of the agriculturists in Norfolk, who are in nothing more eminent than in the collecting and preparing of manure. In that county twelve tumbrils of dung are sufficient for an English acre of turnips, or, fifteen loads to the Scotch acre. In Aberdeenshire thrice as much of manure, both in bulk and weight, is found necessary, because the same attention is not paid to the collecting and preparing it. Few farmers in this county have straw yards; and it is no unusual thing to see horses dung lying in heaps till it is fire-fanged. Too often both peat-ashes and the dung of black cattle are thrown out carelessly, left for a long time uncovered, and allowed to be washed away by the rains. Too often also, thick turfs (provincially termed *muck-feal*), are mixed with the dung without being rotted for some months previously to their mixture.

ture. Hence they are unfit for the making of a compost; and where no attention is paid to the fermentation of a dunghill, the manure cannot be good. It is impossible for a man who has observed the care that is bestowed in collecting and preparing of manure in the southern counties of Scotland, not to see that many of the farmers in this county are in this respect very deficient when compared with their neighbours in the more cultivated parts of the kingdom.

Another defect in the agricultural labour of this county, is the neglecting to employ women in the hoeing and cleaning of land. Necessity, which has often stimulated human industry, will probably in a little time supply this defect. Men cannot be had to act as day labourers; and the women can find very little employment at the staple manufacture, in which they had been employed, viz. the knitting of stockings. Now, when this is become so unprofitable, that a woman cannot earn fourpence a day by the employment; when the money price of farm-labour is so high; and when, even at very high wages, men cannot be found to act as day-labourers, it is both humane and politic to employ women in the lighter parts of agriculture; such as picking off stones, weeding lint, making hay, and hoeing potatoes and turnips. There were times, and there still are places, in which the women had the more servile parts of labour; while their lordly masters filled their creels, or *kessies*, (i. e. baskets made of straw) with the dung, which the women carried on their backs to the field. These practices it is hoped will never be introduced into Aberdeenshire. What can be done by *horses* should never be done by a *human being*. But while men dig with the spade, or manage the plough, the lighter parts of agricultural labour could very properly be done by our women. It is pleasant to see a field, in which a number of women are hoeing turnips or weeding lint; and this pleasure is often found in the southern counties, though it is very rarely enjoyed in Aberdeenshire. Yet this county in general abounds in soil particularly adapted to the turnip husbandry; and double the quantity of turnips that at present are cultivated might be easily raised, if the dung were properly prepared, and women employed in hoeing them. Therefore, both the landed proprietors, and farmers of all descriptions, ought, by giving premiums to those who deserve them, and by offering higher wages to those women who will engage to assist in hoeing turnips and potatoes, to induce as many as they can, to join the farm-servants and day-labourers in cultivating crops which are so valuable, when properly attended to in the season of hoeing and weeding. In the southern counties of Scotland, those young women who agree to work along

along with the hinds, or farm-servants, are allowed one-third of additional wages; and it is not to be doubted that the same encouragement would produce the same effect in Aberdeenshire.

A third error, which prevails too generally in this county, is the limiting the tenants or farmers, who hold of the proprietor, to a certain number of subtenants; but not restricting them as to the quantity of ground which they may let to each person. The original cause of this limitation was the scarcity of peat-moss, to which the landed proprietor affixed too much value. But the way to save the peat-moss, was to have only day labourers with very small crofts, which the farmer ploughed and harrowed; and not to permit large *white crofts*, as they are termed, which the subtenant scratched or ploughed imperfectly with a horse and a cow. For a small farmer, or a subtenant, who has a horse and cart, will drive home as much fuel to his house, as would serve three day-labourers; who, having no horses of their own, are obliged to hire carts to carry home their fuel, and therefore are contented with having a small quantity of it.— It is very seldom that these large crofts are properly cultivated. The farmer ought always to be the only man that yokes a plough; and all the subtenants ought to be married servants or day-labourers attached to the farm, or such artificers and handicrafts as the farmer has occasion to employ. In the southern counties of Scotland, the subtenants, or cottagers, are either the farmers' married servants, or they are lot-men, (as they are termed), who thresh his corn in winter, and act as ditchers and day-labourers during the rest of the year. This practice with regard to subtenants has already been adopted by some of the more intelligent landholders, who are acquainted with the practice of other and better cultivated counties: but as yet it is by no means general; though certainly deserving of imitation.

No fault is here meant to be found with these landed proprietors, who let detached pieces of broken land (i. e. of irregular and rough grounds) which do not ly contiguous to larger farms. What is reprehensible, is the limiting the tenant to a certain number of *fires*, as it is called, or subtenants crofts, instead of leaving him free as to the number of cottages, but limiting the quantity of ground which he may subset to each of them. There cannot be too many day-labourers, nor too few large crofters, who hold their grounds of the farmers. Artificers or handicrafts ought also to be restricted to a moderate quantity of ground, that they may chiefly depend upon their employments, as blacksmiths, masons, plough and cart-wrights, &c. A farmer who subsets

such

such large crofts to persons of this description, is a land-jobber, who wishes to have the land in his own possession as cheap as possible; while neither he nor his subtenants cultivate their ground properly; the former having too little rent as a stimulus for exertion, and the latter depending on the profits of their different occupations to enable them to pay the rack-rent to the farmer.

It is hoped that the pointing out in this place the only important general lesson which the county of Aberdeen can give to more cultivated districts, and confessing candidly the three general errors or defects of the practice of Aberdeenshire, can give offence to no person, and may be of use to many readers of this report; and that, when the Agricultural Reports of the Kingdom are published, and come to be arranged in a *General View of British Agriculture*, this humble attempt may give rise to the introduction of two chapters; the 1st—regarding the practices of every county or district, which merit general imitation; and the 2d—respecting the defects which ought to be supplied, or the errors which ought to be avoided, in any particular county; the farmers in which may derive important information from the practice of other districts.

The beacon that is placed on a rock warns the mariner of his danger; and the light which is placed near the entrance of a harbour serves to direct him how to get safely into port. Both are useful in different ways. And it would be of no small service to the British farmer to know both the beacons or errors which he should avoid, and the good practices or lights by which he should be directed.

It is only farther necessary to state, that by some unaccountable neglect, the corrected Plan of the Board of Agriculture, drawn up by their Secretary in 1806, was not seen by the writer of this Report, till December, 1809, after the greater part of the work was ready for the press. Owing to the insertion of new matter, and a difference of arrangement between the two Plans of the Board, a few repetitions were really unavoidable; and, it is hoped, will be compensated by the additional information.

ERRATA.

ERRATA.

Besides a few trifling Errata, the Reader will please to Correct the following errors, which affect the sense.

Page 25. l. 22. for *wither*, r. *either*.
  64. l. 18. for *a thousand*, r. *four thousand*.
  79. l. 37. for L.08,509, r. 108,509.
    l. 38. for 126,482, r. 26,482.
  83. l. 1. for 258, r. 228.
  117. l. 26. for *well-wooded*, r. *plantations of wood*.
  136. l. 28. for *ears*, r. *court*.
  212. l. 14. for *given* r. *even*.
  218. l. 21. for *litty*, r. *tilly*.
  222. l. 13. for *there*, r. *fences*.
  240. l. 16. for *successively*, r. *successfully*.
  242. l. 6. for *bell*, r. *bushel*.
  283. l. 20. for *rationally*, r. *irrationally*.
  289. l. ult. for *proportion*, r. *preparation*.
  295. l. 11. for *turnips*, r. *yellow-turnips*.
  383. l. 2. for *Rhubarb leaves towards the end*, r. *Rhubarb stalks towards the beginning of summer*.
  403. l. 5. for 60, r. 140.
  405. l. 30. for *two-fifths*, r. *five-twelfths*.
  406. l. 9. for *two-fifths*, r. *five-twelfths*.
  495. l. 8. for *superior*, r. *inferior*.
  497. l. 1. for *of Scotch*, r. *of six Scotch*.
  625. l. 3. for *standard*, r. *of labour*.
  641. l. 27. for 163, r. 183.
  643. l. 7. after *mountain-top*, add, *of Lochnagar*.
  651. l. 34. for *from*, r. *lower than*.
  656. l. 46. for *three inches*, r. *ten inches*.
  665. l. 11. for *is 34*, r. *is proposed to be 34*.

And in page 408, from line 14, take the following Correction for the six-shift course, as a wrong Table was inserted by mistake.

One year, Turnips, 25,000 acres, food for two persons, p. acre,	50,000
Ditto, Potatoes, 25,000 acres, food for 8 persons, per acre,	200,000
Two years Oats, with Grass Seeds, 50,000 acres, food for two persons, per acre,	100,000
Hay and Grass for three years given to the Cattle,	000,000
Oats after lea, 50,000 acres,	100,000
On this six-shift course, 10 acres maintain fifteen persons,	450,000

CHAPTER I.

GEOGRAPHICAL STATE AND CIRCUMSTANCES.

SECT. I.—SITUATION AND EXTENT.

ABERDEENSHIRE may be called a Maritime County, the north-east corner of which is also the north-east extremity of Great Britain. Yet the county of Aberdeen extends so far across the island, that the south-west point of it is considerably nearer to the Atlantic Ocean, in various parts of the west of Scotland, than to the German Sea at Peterhead or Frasersburgh.

It is situated between 56° 52′, and 57° 42′ of north latitude; and between 1° 49′ and 3° 48′ of longitude west of Greenwich.

It is one of the most extensive counties of Scotland. Its greatest length, from Cairn-eilar to Cairnbulg, is 85 miles; and its greatest breadth, from Drumoak to Cairney, or from Aberdeen to King-Edward, is 40 miles. Its circumference, following all its boundaries with other places, is 280 miles; of which 60 are on the sea coast.

Compared to the other counties of South and North Britain, Aberdeenshire, which (by a careful measurement of Arrowsmith's map, abridged from that of General Roy) was found to contain 1950 English square miles, stands inferior to the following counties; viz. Yorkshire, Devonshire, and Lincoln, and perhaps to Northumberland, in the southern; and to Inverness, Ross, Argyle and Perthshires, in the northern division of the island. To all the other counties of Great Britain it is superior in point of extent.

Taking a more comprehensive view of its relative value, its extent is very nearly ~~one sixteenth~~, and its population *one thirteenth* part of Scotland; including the islands; and its area ~~is one fiftieth, and~~ its population ~~one eighty-fifth~~ part of Great Britain. In both these views, therefore, it is to be regarded as one of the most considerable counties of the united kingdom.

Number of Acres in the county.—Aberdeenshire has never been accurately measured by one employed for that sole purpose; but a landmeasurer has undertaken to do so, and to publish a map of the county. TEMPLEMAN, in his Survey of the Globe, states the extent to be only 1170 square miles, or 718,800 English acres; and Dr ANDERSON in his original Report of the Agriculture of Aberdeenshire, seems to adopt this as the whole area of the county; though it is only about three fifths of its extent. One cause why this was so much under-rated is, that both these gentlemen had overlooked a district containing 280 square miles, which lies on the south bank of the Dee, and between it and the Grampian mountains. According to Mr AINSLIE, whose map approaches nearer to the truth, Aberdeenshire contains 1891 miles; and that lately published by ARROWSMITH, makes its area 1950 miles; or 1,248,000 English, or 982,102 Scotch acres. Even this perhaps is a little below the truth. And from the extent of the united parishes of Crathie and Braemar, as stated by the Reverend CHARLES M'HARDY in his very able account of these parishes, it is probable that when an accurate measurement is taken of the extent of the very high mountains which separate Aberdeen from Perth and Inverness shires, it will be found to contain very nearly a million of Scotch, or 1,270,744 English acres; or nearly 1980 square English miles.

Pro-

Proportion of English to Scotch Acres.—It is proper here to state, that 48 statute Scotch are almost exactly equal to 61 English acres; being only one acre more in fourteen thousand. The reader will please remember this proportion of Scotch to English acres. The common proportion of 3 English being equal to 4 Scotch is very inaccurate; as it occasions an error of 5 acres in every 303, or of an acre in 61 English statute acres.

Boundaries of the County.—It is bounded on the N. E. by the German Ocean, for the space of 60 miles; on the S. E. and S. by the county of Kincardine for above 40 miles, or as far as that county extends to the west; on the S. W. by the county of Forfar for 52 miles farther west; afterwards, by the county of Perth for 25 miles, till it meet the county of Inverness at the most westerly point of Aberdeenshire. From this point where these three counties meet, it is bounded on the N. W. by Inverness, till it reach the S. W. corner of Banffshire, for the space of 20 miles; and, lastly, it is bounded on the N. W. and N. by the county of Banff, for above 100 miles, or the whole length of that county. Besides this length of boundary, the insulated parish of St Fergus, which is locally on the sea-coast of Buchan, and county of Aberdeen, is legally in that of Banff.

The most easterly part of Aberdeenshire is near Peterhead, and is called the Buchan-ness. This is also the most easterly point of a large triangle, which juts far out into the German Ocean. This triangle is circumscribed by lines running between Edinburgh, Inverness, and Peterhead; and contains the whole of the counties of Nairn, Banff, Aberdeen, Kincardine and Forfar; also the greatest part of the counties of Moray and Fife, and part of the shires of Perth and Kinross.

The most westerly part is at the head of Braemar, near the point above mentioned, where the counties of Aberdeen, Perth,

Perth, and Inverness meet. This point is distant from Aberdeen, the county town, 65 miles, from Inverness only 42, from Perth 38, from Edinburgh 70, from Glasgow 75, from Berwick on Tweed 103, and from John-o-Groats or the most northern part of Scotland 122. A straight line drawn from Edinburgh to Inverness would touch this point, which is from 8 to 14 miles nearer to various places on the west coast of Scotland, than to Peterhead, Cairnbulg, and Frasersburgh, all belonging to the county of Aberdeen, and situated on the east coast. It may give some idea to a stranger of the situation of this point, which is nearly in the middle of Scotland, to mention that the same length of line, which extends on General Roy's map from this centre to Cairnbulg, would touch the Ord of Caithness on the northeast; the island of St Martin in Loch Broom on the northwest; would include the island of Scalpa, reach to the middle of Eig, to the entrance of Loch Gilp, and the island of Bute, all in the Atlantic Ocean; and after reaching the boundaries, and cutting off a small part of Ayrshire, would extend 12 miles beyond Glasgow; and moved across the island, would reach the German Sea at Deansburn, in Cockburnspath, on the boundary between Haddington and Berwick shires. The circle thus described would include above three-fourths of Scotland; its diameter being 170 miles, and its radius 85, or the length of the county of Aberdeen from S. W. to N. E. The reader from this (perhaps rather minute) description, will be satisfied of the great extent of Aberdeenshire; and of its oblique direction across the island. He will also see that there must be a great difference in point both of fertility and climate between those parts of the county which project as a great promontory into the German Sea, and those mountainous districts which reach across the middle of the island, and extend more than halfway to the Atlantic Ocean.

SECT.

SECT. II.—DIVISIONS.

ABERDEENSHIRE at a remote period seems to have composed two distinct counties or earldoms, viz. Marr and Buchan; the former comprehending the divisions of Marr proper, Garioch and Strathboggie.; and the latter including the thanedoms of Formartin and Belhelvie, which were united in a political connexion with the territory, and subject to the jurisdiction of the earls of Buchan. But when the feudal system was generally established, and when it became expedient to unite several earldoms, under the jurisdiction of one sheriff, or judge, appointed by the sovereign; all the divisions were included in the general name of the County of Aberdeen. From that period Aberdeenshire has been considered as composed of 5 divisions, viz. Marr, Formartin, Buchan, Garioch, and Strathboggie.

The division of *Marr* was by far the largest, containing nearly one half, or *seven fifteenths* of the whole county. It gave the title of Earl to its powerful chieftain, ever since that became a title of nobility in Scotland. It comprehends the whole tract of country lying between the rivers Dee and Don (except perhaps that part of Kincardineshire, which is on the north bank of the Dee.) It also comprehends a very large district on the south side of that river, and between it and the Grampian mountains, which divide Aberdeenshire from the counties of Kincardine, Forfar and Perth. This district, which has been overlooked by Dr ANDERSON and Mr TEMPLEMAN contains 280 square miles; and the whole division of Marr contains at least 900, perhaps 920 miles.

Though Marr be the largest and most southerly, it is one of the most barren divisions of Aberdeenshire. It is in general hilly, even within a few miles of the sea-coast, of which it contains only the neck of land, not two miles in length, between the rivers Dee and Don, near the mouths of which both New and Old Aberdeen are situated; and towards the southern and western boundaries it is in many places mountainous. Few of the hills in the lower part of Marr are less than 500 feet above the level of the sea; and many of the mountains in Braemar are more than 3000 feet above that level. It is divided into a number of districts.

The upper part of it is called Braemar, (i. e. the *highest grounds of Marr*). It is very extensive, and includes a number both of high mountains and of glens or vallies, through which the Dee and inferior rivulets flow. It contains the forests of Braemar, Strathdee, Invercauld, Glentanar, and Birse. On the north-west side of this district lies Strathdon, a very extensive tract of country on both sides of the river, from which it derives its name. This uppermost, and mountainous, subdivision, extending from Birse, and the lower parts of Morven, to Strathdon, is very thinly inhabited. In 9 parishes, which are now under the care of 5 ministers its population in 1801 consisted only of 7,315 persons in 490 miles, or 15 inhabitants to the square mile. Its area is a trifle more than one-fourth of the surface, and its population is only one-seventeenth of that of the whole county. But there are no inhabitants in the highest parts of this mountainous district, except for a few of the summer months.

Below Braemar lies Cromar, a fine valley surrounded by hills, and containing the greatest part of 5 parishes. Between it and the south lies the small district of Inchmarnoch. On the opposite side, or on the north-east, ly Towie, Kildrummie, and the district of Alford, on both sides of the Don, and touching the divisions both of the Garioch and Strathbogie. In Cromar, and in the parishes of Towie, Kildrummie and Alford, are many fertile tracts; and grain of the finest quality is generally raised. This middle division is

by

by far the most fertile part of Marr. Only toward the lower part of it, in the parishes of Kincardine, Midmar, Lumphanan and Cluny, there is a considerable proportion of hilly and barren land. It contains 250 square miles, and 20 parishes, the inhabitants of which in 1801 amounted to 13,586, or 54 persons to the mile; and it sends a part of its produce to Aberdeen, and also to the upper or mountainous district, whose lands are often insufficient for subsisting their thinly scattered inhabitants. The distance from market, and the seacoast, has prevented this district from advancing rapidly in the improvement of a soil which is naturally fertile, and in many places well sheltered by its hills and plantations.

The lowest part of Marr, which lies next the sea, includes the ancient forest of Stocket, (a considerable part of which has been trenched and cultivated at a great expence by the citizens of Aberdeen,) the forest of Drum, and the forest of Kintore: But all these three have long been stripped of their natural woods; and are in many places well cultivated by the inhabitants. Though only a small proportion of wood is raised in this district, this subdivision of Marr 500 years ago was one vast forest, extending from Drumoak, on the north bank of the Dee, to Kintore on the south bank of the Don, a distance of 14 miles from north to south, and an equal distance from the sea-coast.

Dr Anderson observes very properly of this part of Marr, that " it was one of the most rugged and naturally barren
" regions that is to be met with in any of the low parts of
" Scotland; but that by the industry of the inhabitants, a
" considerable part of these inhospitable wastes has been
" converted into fertile fields, at an expence that is perhaps
" unequalled in any other part of Europe." The area of this lower subdivision of Marr is 160 square miles, which includes Old and New Aberdeen, and part of 7 country parishes;

parishes; the population of which is in all 32,393, or 202 to the square mile. But the towns are supported by the other districts; for the lands in this district are insufficient to supply the numerous inhabitants.

The whole division of Marr contains 39 parishes; and its population amounted in 1801 to 53,300 persons. On a general view, the highest subdivision contains a number of high mountains, and extensive forests, with few inhabitants; the middle division is naturally the most fertile, with a moderate population; and the lower parts of it, from the industry of the inhabitants of New and Old Aberdeen, though the soil is barren by nature, have been highly cultivated, and are now become both productive and populous.

The second division of the county is called *Formartin*. It anciently consisted of two thanedoms, Formartin, properly so called, and Belhelvie. The latter extended from the Don to the Ythan, about 12 miles along the coast, and 5 or 6 miles inland. The former reached from the boundaries of Strathboggie to Belhelvie; and was more than thrice as extensive as the latter. Both were subject to the earl of Buchan, till Formartin, including Belhelvie, became a principal division of the county. Separated from Marr by the Don, and from Buchan by the Ythan, for the first 10 miles from the sea-coast, it then crosses the Ythan, and extends to the banks of the Doveran by Turriff, where it is also separated both from the division of Buchan and the county of Banff. Towards the northwest it reaches the division of Strathboggie; and on the west and south-west it is separated from the division of Garioch by a number of bounding hills, which reach from Fintray, within 10 miles of the sea-coast, to the upper part of the parish of Forgue, about 24 miles into the interior part of the county. This division, including the thanages of Formartin and Belhelvie, covers and

forms

forms an area of 280 square miles, includes the greatest part of 16 parishes, and contains 16,769 inhabitants, or 60 persons to the square mile.

Towards the sea-coast of Formartin, the ground is low, and in general of a good quality, and in a high degree of cultivation. In the inland districts it is full of hills and peat-mosses, with a good deal of moorish ground on the west. Yet in the parishes of Fyvie, Auchterless, and Forgue, there are many excellent farmers, who were the earliest, and still are among the most distinguished improvers in the county. In general they have extensive farms, which are well managed; although considerable tracts of moor still remain uncultivated. Formartin gives the title of Viscount to the earl of Aberdeen. It contains only one village of any consequence, viz. Newburgh, which is a sea-port; though of very inconsiderable trade, as its harbour is very bad. It was formerly a fishing village; but the frequent shifting of its bar occasioned it to be deserted by the fishers.

The third division of the county (and that which by an old charter of K. Robert Bruce appears to have been once a county by itself) is *Buchan*. Next to Marr it is the largest, or most extensive; and since the introduction of the new husbandry, it has become the most valuable. The increase of its relative value, compared to the other districts, has been occasioned partly by its vicinity to the German Sea, (for it contains above 40 miles of sea-coast), partly from its flat surface, which renders it more easily cultivated than the other districts, partly from the immense quantities of shelly sand, the value of which was only lately understood, and in no small degree by the liberality of the landholders and the industry of the farmers.

This division was anciently an earldom; and during the prevalence of the feudal system its earls were very powerful.

They

They kept their courts at Ellon, as being most centrical both for Buchan and Formartin. A charter of Fergus earl of Buchan, in 1211, (a copy of which is in the reporter's custody) contains this clause—*Faciendo per annum tres sectas capitales curiæ meæ de Ellon*, i. e. attending yearly three diets of my head court at Ellon. These earls possessed very great estates both in Buchan and Formartin; and were too powerful subjects in a small kingdom; till the forfeiture of the Cummines occasioned their estates to be partitioned among the adherents of King Robert Bruce in 1309.

Buchan, in general, is a flat and level country, containing only one considerable hill, (viz. Mormond), excepting on the north-west, where it bounds with Formartin; and where there are several hills covered with heath. The sea-coast near Formartin is pretty flat, and so much exposed to storms from the east, that the greatest part of the parish of Forvie was completely covered with sand above 200 years ago. Near to the Buchan-ness the coast is more bold and rocky; and the bullers of Buchan are very formidable. Yet the maritime parts of Buchan, besides the towns of Peterhead and Frasersburgh, are indented with a number of creeks and harbours for fishing boats, or for small vessels loaded with lime or coals. At Peterhead the harbour frequently affords protection for ships, when driven past the harbour of Aberdeen.

In the interior parts of Buchan the immense quantities of peat-moss renders the importation of coals unnecessary; and a canal, extending 6 miles in length, from the shelly sand in St Fergus, and made out by JAMES FERGUSON, esq. of Pitfour, member of Parliament for the county, is very beneficial for carrying into the interior parts of his extensive property, considerable quantities of calcareous matter; which, for the purposes of agriculture, supplies the place of lime.

This division of the county contains 450 square miles, (besides those parts of the ancient earldom of Buchan which are situated in Banffshire), and 36,172 inhabitants, or 80 persons to the square mile; who reside in 21 parishes. It was long neglected, and consequently was unproductive; but it has for a considerable time past been cultivated with ardour and perseverance, and now raises excellent crops. In many districts of it, the turnip husbandry is cultivated to great advantage. Other districts are better adapted to the raising of beans; particularly the whole parish of St Fergus; which though a detached part of the county of Banff, is locally situated in the middle of the sea-coast of Aberdeenshire.

Buchan, besides an extensive tract of country, contains several sea-port towns, and both fishing villages on the coast, and other villages in its interior parts. Of the first, the most remarkable is Peterhead, a very thriving town, containing 3000, and Frasersburgh, containing above 1000 inhabitants. Of the last the most populous is New Pitsligo, containing 600, and Turriff, containing 700 persons. And the inferior villages of both descriptions contain at least 3000 persons more. So that 9000 of the inhabitants of Buchan live in towns or villages; and of these nearly 2000 reside in small fish-towns on the sea-coast.

The fourth division of the county of Aberdeen is called the *Garioch*, anciently termed Garviauch. It was formerly an earldom. The first earl of whom we have any certain information, was David earl of Huntingdon and Garioch, youngest son to King David *the First*; of whom King John Baliol, King Robert Bruce, and the present royal family are descended. Other three persons nearly allied to the crown, (the last of whom, Alexander Duke of Albany, was forfeited in 1483,) were successively earls of Garioch. Afterwards on the annexation of the titles of Marr and Garioch, Marr as being the most extensive, gave the title of *earl*, and Garioch that of *lord*, to the noble family of Erskine, before their attainder 1715.

The Garioch is bounded by Marr on the south and west; by Formartin on the east and north-east; and by Strathbogie on the west and north-west. This division, before the introduction of the turnip husbandry, was esteemed by far the most fruitful in the county; and from the great quantity of meal sent into that city, was termed the granary of Aberdeen. But since lime came to be generally used, those places in Marr, Formartin and Buchan, which are nearer to the sea-coast, have advanced more rapidly in agricultural improvement, than the Garioch did, till very lately. Yet the making of two excellent turnpike roads from Aberdeen to Huntly, and to Old Meldrum, has proved very beneficial to this inland division. And the cutting of a canal, along the Don, from Aberdeen to Inverury, which was completed near three years ago, will virtually render the Garioch nearer, because more accessible to Aberdeen; and will soon make a great change both on the aspect, and on the produce of a district which is naturally fertile.

What tends much to the fertility of the Garioch is the situation of its bounding hills. The high hill of Foudland, nealy 900 feet above the level of the sea, and the hill of Culsamond, which is about 600 feet above that level, shelter it on the north. And a range of hills near Old Meldrum, from 400 to 500 feet of elevation, shelter it on the north-east and east. The mountain of Benochie, nearly 1440 feet high, and several other bounding hills, separate it from Marr, on the west and south-west. So that it is both protected in a great measure from the severe blasts of winter, and in no small degree from the blights and mildews of autumn, so

fatal

fatal to other places nearer the sea-coast. Hence some of the finest fields, bearing the weightiest crops of corn and grass, are to be seen in the head of the Garioch, at 30 miles distance from Aberdeen.

The Garioch extends about 24 miles in length, and including one half of the base of its bounding hills, from 3 to 10 miles in breadth. Its area is nearly 150 square miles; and in 1801, the number of its inhabitants, contained in 18 parishes, under the care of 16 clergymen, was 10,447, or 70 inhabitants to the square mile.

This division includes three towns, viz. Old Meldrum, Kintore, and Inverury. The first of these is only a burgh of barony, in which the inhabitants hold their houses and gardens from JAMES URQUHART, esq. of Meldrum, sheriff-depute of Banffshire. The second is a royal burgh, which possesses very fine land, but has little trade. The last is also a royal burgh, and more populous than Kintore, though a younger burgh. Old Meldrum contained in 1801 more inhabitants than both the other, viz. 783. Kintore has only 250; and Inverury 450, persons; but it is increasing rapidly since the canal was opened, or within the last 3 years. And it is highly probable that this burgh will soon become a considerable trading town, from being placed at the head of a navigable canal, in the opening of a rich and extensive valley, for so the Garioch really is.

It needs only be added to the account of the Garioch, that its two royal burghs, Kintore and Inverury, along with Elgin, Cullen, and Banff, elect a member of Parliament for this district of burghs.

The last division of the county of Aberdeen is *Strathbogie*. The Isla and the Dovern separate it on the north and east from Banffshire. In other directions its bounding hills divide it from that county, and from Marr, Garioch and Formartin.

Formartin. It does not extend so far east, as to touch the division of Buchan. But the barony of Gartly is an insulated part of Banffshire, situated in the division of Strathboggie, like the parish of St Fergus in Buchan; and pays the land-tax in the county of Banff. Several of the parishes in this division are partly in that county, and partly in Aberdeenshire. The cause of this arrangement probably was, that the division of Strathboggie was formerly a lordship, like those of Lorn, Badenoch, and Galloway; and that those lords of regality had too great power, as they could repledge a criminal from the king's judges. On the forfeiture of David de Strathbolgie, earl of Athol, and lord of this division, by abolishing that privilege, and dividing the lordship between different counties, this rebellious spirit was checked in the most effectual manner.

The greatest part of this division was given by King Robert Bruce in 1309, under the inferior title of the barony of Strathboggie, to Adam de Gordon, ancestor of the Duke of Gordon, for his services against the English, in maintaining the independence of Scotland; and the whole afterwards became the property of the noble family of GORDON, and was again erected into a lordship by King Robert III.

The whole division consists of 48 daughs, (*davate terræ*;) and each daugh should contain 32 oxen gates of 13 Scotch acres each, or 416 acres; consequently the lordship should contain 19,968 acres of arable land. This however is not one third part of the extent of the whole division, which contains about 150 square miles, in 10 parishes under the charge of 6 clergymen, with only 6,403 inhabitants, or 43 persons to every square mile. But a great part of this country is hilly, especially on the boundary adjoining to Marr, where the extensive parish of Cabrach contains a great deal of moor, peat-mosses, and hilly grounds. The Buck, a mountain

tain in that parish, above 2200 feet high, the hill of Noath, above 1700 feet, and that of Foudland already mentioned, cover a considerable part of this division. The farmers, owing to their distance from the sea-coast, did not begin to improve their land so early as those in the lower parts of the county. But from the opening of the country by different turnpikes, and a number of commutation roads, and from the command of excellent lime at Ardonnel, there is every probability that this division, though late in beginning its improvements, will soon be well cultivated by its native farmers.

The only town in this division is Huntly, formerly in a very flourishing, but now in a declining condition. Its staple manufacture was flax; and during the life of one enterprising merchant, Mr Hugh M'Veagh, the linen manufacture was carried on with great success. But from the excessively high price both of flax and of flax-seed, its trade was greatly hurt; and by an injudicious use of paper credit, those persons who had any capital, were either ruined or deeply injured by those who had neither capital nor principle.

Mr M'Pherson, in his very curious Index, seems to think that Strathbogie was only a part of the earldom of Garioch; and Dr Anderson has omitted it altogether, except calling it one of the lesser divisions, or narrow vales on the side of a rivulet; but though it be a hilly country, lying on both sides of the Boggie, it was always esteemed one of the principal divisions of Aberdeenshire.

It is proper here to observe that the ancient divisions of countries were hills. Nothing but a very large river, which stopped the progress of a barbarian, would be regarded by him as a boundary. The tops of the hills which bounded his view he would regard only in that light. And hence
in

in the highest parts of Aberdeenshire, whatever is within the visible horison is called *a country*. The great contention among our ancestors, in their private wars, was about having the sky of the hill. On that account the Grampian mountains made better boundaries between the counties of Aberdeen and Kincardine, than the river Dee could have made, except near its entrance into the ocean. Hence also the high mountain of Benochie made a better boundary between the Garioch and the district of Marr called Alford, than the Don could have made; and the other bounding hills between the Garioch and either Strathboggie or Formartin, were better boundaries than the smaller rivers, the Boggie, the Ury, or the Ythan, could have been, especially near their sources, and where their streams were inconsiderable.

There are however a few circumstances, in the allocating of certain districts to other counties than that in which they are situated, which deserve to be pointed out in this Report of Aberdeenshire.

It has already been mentioned, that the whole parish of St Fergus, which borders on Peterhead, is a detached part of the county of Banff. It may now be added, that the small parish of Fetterangus, which in 1618 was annexed to Old Deer, and is in the very heart of Buchan, is legally in Banffshire. Nay, the barony of Straloch in the parish of New Machar, though situated within 12 miles of Aberdeen, and in the centre of the division of Formartin, belongs to the same county. And what is yet more remarkable, the site of Marischal-street in the city of Aberdeen, 45 miles distant from Banff, though situated in an ancient royal burgh, and in the county town, is legally in Banffshire. The reason of this probably was (as stated by an able antiquarian, the Rev. JOHN CRAIGIE, late minister of St Fergus, now of Deer) that the Cheynes of Inverugie, the

ancient

ancient proprietors of all these estates, which afterwards came by marriage into the family of Marischal, were heritable sheriffs of Banffshire. On the other hand, nine ploughgates in the parish of Drumoak, and the whole parish of Banchory-Ternan, though situated on the north side of the river Dee, and in the division of Marr, belong now to the county of Kincardine. The reason of this is said to have been, that those lands belonged to the Frasers, thanes of Cowie, who also got the thanedom of Durris, and were nearly connected with King Robert Bruce. The daughter and heiress of one of them was married to ROBERT DE KEITH, Great Marischal of Scotland, who was heritable sheriff of the county of Kincardine. Alexander Fraser, thane of Cowie, was married to Mary Bruce, sister to King Robert Bruce, and his grand-daughter, who inherited the thanedom of Durris, was married to the Great Marischal above mentioned, whose son was married to a daughter of K. Robert II. Whatever was the cause of the improper divisions of counties at a remote period, when a great landed proprietor acquired both power and riches by being a sheriff, and by levying whatever fines he pleased, they ought all now to be set aside; and every county ought to comprehend whatever is situated within its own boundaries.

So much for the ancient political divisions of Aberdeenshire.

The present divisions of this extensive county are different, according as we consider them, with respect to the administration of justice, the various ecclesiastical jurisdictions, or the regulations of the militia.

As to the first of these, the whole county is subject to the jurisdiction of the Sheriff Depute and his Substitutes: And the three counties of Aberdeen, Banff, and Kincardine are classed together in the administration of criminal justice, by

B   the

the Court of Justiciary. The circuit courts are held at Aberdeen twice a-year, generally in April and September; and it is deserving of notice, that though the number of inhabitants of these counties amounts to nearly 200,000, or one-eighth part of the population of Scotland; yet in the space of 40 years, only nine criminals have suffered death.

The Commissary Court, a species of consistorial judicature, first erected by King James VI. and now about to be suppressed by Act of Parliament, has its jurisdiction limited by the ancient boundaries of the Bishop's see; and does not extend to six parishes in the division of Strathbogie, viz. Huntly, Gartly, Rhynie and Essie, Glass and Cairnie; but it includes twelve parishes of Banffshire, viz. Alva, Gamery, Forglen, Inverkeithny, Mortlach, Fordyce, Banff, Cullen, Deskford, Rathven, Boyndie, and Ordequhill, and 4 parishes in the county of Kincardine, viz. Strachan, Upper Banchory, Nether Banchory, and Maryculter.

Both in the Sheriff and Commissary Courts, the judges sit in the county town, Aberdeen; where the Magistrates of the city also hold their Baillie Courts once a week.

But the most useful Court, for determining petty disputes, and questions of small debts, is that of the Justices of the Peace, not in the Quarter Sessions, where they meet at Aberdeen, but in the different districts, bounded by the eight presbyteries, into which the county is divided. All the landed proprietors, who are included in the Commission of the Peace, are judges in these Courts; in which, at a very trifling expence, and without the intervention of lawyers, or the tedious forms of other Courts, they decide in all claims not exceeding L.5 of value, and settle a multitude of petty disputes. It would be a criminal omission in this Report of the Agriculture of Aberdeenshire, not to state that these Courts stand very high in the opinion of the country people, that they are

at-

attended by the most respectable gentlemen of the county, and that every landed proprietor, who has an interest in the prosperity of the inferior classes, by whose labour he is supported, owes it to himself, and to society, to attend these Courts, and impartially to discharge his duty, as a judge, in the district in which his property is situated.

The ecclesiastical divisions of the county include 8 presbyteries, viz. those of Aberdeen, Kincardine O'Neil, Alford, Garioch, Turriff, Ellon, Deer, and Strathboggie. The first seven of these are subject to the Synod of Aberdeen, which also includes the Presbytery of Fordyce; but the last, viz. Strathboggie belongs to the Synod of Moray. The names of the parishes in each presbytery will be afterwards mentioned, when the statistical estimates and population of each are considered. It is unnecessary to be more particular, when treating only of the divisions of the county.

The Lord Lieutenant and Deputy Lieutenants of the county have divided Aberdeenshire into ten districts, in order to execute the Militia Acts in the way that may occasion the least trouble to the people, who are obliged to attend the Courts of Lieutenancy. These divisions, however, being altered at the pleasure of the Deputy Lieutenants, are unnecessary to be here inserted.

It may be briefly remarked, that the Aberdeenshire regiment of Militia consists of 640 men; and that there are 5 regiments of Local Militia. The former is moved over all parts of the kingdom; the latter are stationary in the county, and are trained only a few weeks annually, in the different towns of the county, viz. 2 at Aberdeen, 1 at Ellon, 1 at Peterhead, and 1 at Fraserburgh.

SECT. III.—CLIMATE.

Dr. Anderson's account of the climate of Aberdeenshire, deserves here to be quoted at length. "From the high lati-
"tude of this district, and the general opinion that is enter-
"tained of the inhospitable nature of these northern regions,
"most persons are inclined to believe, that a much greater
"degree of cold here takes place, than is ever experienced.
"Being washed by the sea on two sides, the county of Aber-
"deen experiences a mildness of temperature in winter, even
"greater than most parts of the island. Snow, in the lower
"parts of the county, seldom lies long; and it may be con-
"sidered as a pretty general rule, that when the snow is one
"foot deep at Aberdeen, it is nearly two feet deep at New-
"castle upon Tyne. In the year 1762, when the frost was
"so severe in England, that the Thames at London, was
"frozen over for many weeks together, the weather was so
"mild in Aberdeenshire, as scarcely to interrupt the ordina-
"ry operations of agriculture; and though a little snow lay
"for some weeks on the surface of the ground, there was not
"a day during the whole season, that a plough could not
"have gone. Nor is this an uncommon case. For I have
"reason to believe, that the frost is seldom so intense in the
"lower parts of Aberdeenshire as at London. But if the
"winter's cold be less severe than, in many of the southern
"districts of the island, the summer heats are here, perhaps,
"still less intense. In short, there is a smaller variation be-
"tween the heat and cold at different seasons; and of course,
"there are many crops that may be brought to maturity in
"the South of Britain, which are seldom found to ripen here.
"Grapes there are none, without artificial heat; and French
"beans can scarcely be brought to ripen their seeds, in the
"best sheltered garden, unless in a very favourable season.
This

"This circumstance marks the degree of summer heat with
" great precision.

" The great disadvantage attending the climate of Aber-
" deenshire, and of Scotland in general, when compared
" with that of the southern part of the island, is the lateness
" of the spring, owing to the prevalence of eastern winds,
" and the too frequent fogs and rains at that season, which
" often render the seed time both late and ungenial; and the
" stormy winter, in autumn, frequently accompanied with
" heavy dashes of rain, which require a degree of exertion
" and skill in the husbandman, to protect his crop from da-
" mage, to which the English farmer is totally a stranger.—
" These inconveniencies are all felt in Aberdeenshire. But
" the crops are so seldom hurt by frosts in summer or autumn,
" in the lower parts of the county at least, that in the course
" of 30 years nearly, I never saw a single instance of frosted
" corn, except in the singularly intemperate season of 1782;
" nor were any of the people who have resided there all their
" lives, better acquainted with that phenomenon. The spring
" frosts seldom hurt any thing but the blossom of fruit trees
" occasionally. As a proof that the climate is not uncommon-
" ly backward, I may observe, that one season I had a dish
" of pease gathered from the open field, cultivated by the
" plough, on the King's birth day, the 4th of June. This
" was a very early season; but green pease are commonly
" ripe in the garden, not long after that period."

The Reporter thought it his duty to introduce this long quotation, being the whole that Dr. Anderson has written on the
subject of the climate of Aberdeenshire. The climate of the
lower part of Aberdeenshire is certainly moderate in this respect. That it is not nearly so warm in summer, nor so cold in
winter, as that of the county of Middlesex. But from the account that has been given in the preceding section, of the

great

great length of Aberdeenshire, as extending from the northeast extremity on the sea-coast, in an oblique direction, beyond the middle of the island, it is obvious that there must be a wide difference between the climate of Peterhead, where the coast of Buchan projects so far into the German Ocean, and that of the mountainous districts in the south-west point of Marr, nearly 100 miles from Cairnbulg-head, and both at a high elevation from the level of the sea, and at nearly an equal distance from the east and west coasts. Dr. Anderson's farm, at Monkshill, where he resided 30 years, was within 5 miles of the sea, in the division of Formartin; And what is above quoted, is a pretty correct account of the state of the climate on the coast both of Formartin and Buchan.— But *there* the climate is peculiarly moderate in the winter-months; and the snow seldom lies long; nor are the ploughs much impeded by the frost; being seldom idle above two or three weeks, even in a severe winter. But in the mountainous districts, the highland farmers are frequently unable to plough their lands for two, and sometimes for three months. Between these two extremes, of the elevated lands on the S. W. and the low-lying ground on the sea coast, there must be a very great variety of climate. And it is this important difference with respect to being able to plough almost constantly in winter, which, since the introduction of the new husbandry into this county, has enabled the inhabitants near the sea coast to advance rapidly in the improvement of their soil; while those in the districts of Garioch, Strathbogie, and the higher parts of Marr, have often been unable to put a plough in the ground for a great part of the winter months.

The Writer of this report, in the course of 32 years, had his plough laid idle twice for 13 weeks, once for 10 weeks, and at an average 5 weeks yearly, though he resides only 13 miles from the sea coast. In the higher parts of the county, the

the highland farmer, at a medium of seasons, experiences two months either of hard frost or of stormy weather, during which he is unable to plough his lands. On the other hand, on the south sides of hills, and in sheltered places in the higher districts, the heat is frequently much greater in summer, than upon the sea coast. And betwixt the greatest heat in summer, and the greatest cold in winter, there is a difference of 7 or 8 degrees more in the inland than in the maritime districts. So that it is not accurate to speak of the climate of Aberdeenshire in general, and without distinguishing between these two situations.

The great defects of the climate of this county are the following :—

On the sea coast, where the land is low, and along the haughs, or meadows on the banks of rivers, near their entrances into the sea, frosty fogs in the end of summer, frequently destroy the potatoe crop, or at least injure them considerably. And where a frost at night succeeds rain in the evening, the rays of the morning sun frequently blight the crops of corn, when these happen to be *in flower*. This destroys the germ, and renders it unfit for the *purpose of seed*, although it may fill in the ear, and be useful for food. (If the corns be either in an earlier or more advanced stage, they are in less danger of being injured in the above respect.) The corn in the higher or more elevated grounds, is not so much liable to injury by the frosts or fogs of summer. But in the end of autumn, and beginning of winter, when the snow begins to fall, or the hard frosts commence, the cold is much more severe, and the frost more intense, in lands that are highly elevated above the sea, than on lands on the coast, which both from their low elevation, and their vicinity to the ocean, are naturally warmer. Hence the late harvests, with their unpleasant attendants, to the crop that is exposed to such dan-

dangers; and hence the early and the long winters, which prevent ploughing, and protract the season of sowing.—When the farmer is laid idle, as he frequently is in January or February, and not seldom in March, he has his grounds first to plough, and then to sow as soon as they are in a proper condition for receiving the seed. If the storms, which in this county often attend the vernal equinox, come on before his seed time begins, this throws him back in the spring labour, and he is seldom able to escape the more fatal equinoctial storms in September.

The only exception to this general rule deserves to be mentioned, when comparing the climate of the northern, with that of the southern parts of the island. It is the case of a very late harvest season, when the equinoctial rains come on before the barley of England is harvested. In this case, while the English farmer has his barley crop much injured, that of Aberdeenshire escapes this danger, merely because it was not come to maturity, till after the equinox was over, with its unpleasant attendants. Thus in 1804, the barley of England was deeply injured during the period of harvesting; while that of Scotland was preserved, and of an uncommonly fine quality, because it met with an excellent season for filling the ear, and because the bad weather was over, before it was fit for being cut down, owing to its being late in ripening. This, however, is a very uncommon case.—For the most part, the English farmer has his harvest concluded before the equinox, while the crop of Aberdeenshire is often injured by rains at that season. For example, in 1802, while England, and the south of Scotland had an abundant crop of barley, of excellent quality, the bigg in Aberdeenshire was deficient both in quality and in quantity. In short, from the lateness of the seed time, the farmers of this county are often exposed to all the inconveniences of a late harvest; and

with

with the single exception of 1784, late harvests have occasioned much loss, and late crops have always been found unproductive, and have frequently been much injured.

There is also a considerable difference in the temperature, or heat of different seasons. Thus Mr. MARSHALL informs us, that the foliation of the oak, which is the rule for sowing barley in Norfolk, took place in Surrey on the 20th of April, 1779, and on the 20th of May, 1780; and that in Norfolk the oak foliated May 17th, 1781, and not till June 4th, 1782. In Aberdeenshire, Dr. ANDERSON eat *ripe pease*, i. e. pease fit to be used, June 4th, 1779, which were raised in the field, within a few miles of the sea-coast. But on an average of years, and in Aberdeenshire in general, the pease raised in the field are not ready to be cut down so early as the 4th of August.

The only accurate way of calculating the comparative heat of two places, is by ascertaining what was the heat, at a given time, as ascertained by a Fahrenheit's, or any correct thermometer. In 1802, this heat was taken at the Brewery of Gilcomston, near Aberdeen. The thermometer stood without in the shade; and the following account was given in to the Writer of this Report, by Mr. JOHN SIM, late clerk to that company. Its heat, from May to September, was as follows, viz.—

May.	June.	July.	August.	Four Months.
41.8.	51.28	50.33.	54.20.	49.22.

This was a very low heat during the four summer months, on which vegetation so much depends; but the heat of September and October fell so much short of this, that the medium heat of 6 months was only 46' 27, and that of the spring months only 37' 38; the average heat of these 9 months, during which there was either vegetation, or ripening of corn, was only 42° 23 at Aberdeen. During the three winter months

months it was at a medium 31° 41', or a trifle below the freezing point. And the average heat of the whole year was 12 degrees of Fahrenheit's lower than that at the Royal Observatory of Greenwich. This sufficiently accounted for the late harvest, and both the light crop, and the inferior quality of the bigg of Aberdeenshire in 1802.

But as the season of 1802 was peculiarly unfavourable to the north of Scotland, and as the climate of Aberdeenshire should be judged from the medium heat of different years, the Writer of this Report has procured from the Rev. Dr. WILLIAM LAING, minister of the Episcopal Church at Peterhead, the following Tables, which being taken from the accurate observations of a very ingenious man, who resides nearly at the north-east corner of Great Britain, deserve a place in this Account of the Agriculture of Aberdeenshire.

These Tables show, that in the winter months the frost is often less intense at Peterhead, than in the neighbourhood of London; but that in the spring, summer, and autumn seasons, we have so much less heat in Aberdeenshire, compared with the southern parts of the island, as fully accounts for our slow vegetation and late harvests.

As Peterhead is situated at the N. E. point of Great Britain, it was judged proper to insert these Tables at large, rather than merely to note the average heat of every month. This, however, is also marked at the foot of each page, at 8 A. M. And the depth of rain is also inserted below, at the quantity for every month.

STATE OF THE THERMOMETER, AND REGISTER OF THE WINDS.

1808.

	MAY. Ther. 8 A.M.	NIGHT.	WINDS. 8 A.M.	JUNE. Ther. 8 A.M.	NIGHT.	WINDS 8 A.M.	JULY. Ther. 8 A.M.	NIGHT.	WINDS. 8 A.M.
1	47		W.	52	48	W.	52	48	S. E.
2	47		S. by W.	50	44	S.	53	51	N.
3	45	42	S. by W.	49	47	S.	54	48	N. N. W.
4	45	45	S.	52	47	S.	53	34	N. W.
5	50	43	S. by W.	49	48	S. E.	56	39	W. by S.
6	45	44	S. by W.	49	46		56	52	W. by S.
7	47	45	S. by E.	52	48	S. E.	57	52	S. S. W.
8	50	45	S. E.	50	47	N.	55	52	W. S. W.
9	47	45	S. E.	50	47	N. E.	55	43	S. W.
10	45	42	S. E.	48	46	N. W.	55	52	S. S. W.
11	46	43	S. S. E.	50	47	W.	56	52	S. by W.
12	48	44	S.	56	44	N.	60	50	S. W.
13	46	44	S.	52	48	S. E.	58	54	S.
14	54			54	48	S.	60	57	N. W.
15	53	48	S. by W.	52	48	N. W.	58	55	S. by W.
16	50	45	S. S. W.	50	44	N. by W.	58	53	S.
17	51	48	W. by S.	54	50	S. W.	58	56	S. by E.
18	45	42	N. W.	52	48	S.	59	52	S.
19	48	43	W. by S.	57	48	W. N. W.	59	57	S.
20	54	43	W. by N.	55	51	S.	58	57	S. E.
21	47	47	S. E.	57	50	S. by E.	57	55	S. E.
22	46	44	S. E.	52	49	S. S. E.	60	55	S. E.
23	47	43	S. E.	55	49	S. S. E.	58	55	S. E.
24	47	45	S. by E.	53	52	N. W.	56	54	N. by W.
25	49	44	S. E.	53	52	S.	61	54	
26	50	45	S. by E.	53	52	S.	54	54	N.
27	49	44	S. E.	53	52	S.	58	54	N.
28	47	43	S. E.	55	45	S.	59	56	S. E.
29	49	45	S. ½ by E.	55	53		57	51	E. S. E.
30	50	45	S.	54	44		58	51	S. E.
31	51	47	S. by W.				60	57	S.

Average—May, 48° 14'—June, 52° 26'—July, 57° 6'.

Heat of 3 Months, 52° 55'.

Rain—Inches, May, 1.058—June, 0.98—July, 3.30.

Rain in 3 Months, Inches, 5.338.

1808.

AUGUST.			SEPTEMBER.			OCTOBER.		
Ther.			Ther.			Ther.		
8 A.M.	NIGHT	WINDS. 8 A. M.	8 A.M.	NIGHT	WINDS. 8 A. M.	8 A.M.	NIGHT	WINDS. 8 A. M.
1	59	56		56	54	S. E.	35	31	W.
2	60	56	N. by W.	56	48	W.	41	32	W. S. W.
3	58	53	N. W.	56	52	S. by W.	48	40	S. W.
4	56	54	S.	53	51	S. W.	45	39	W. by S.
5	57	54	S. E.	56	52	S.	44	38	S. W.
6	58	54	S. W.	55	51	N. E.	47	41	S. W.
7	55	53	N. W.	56	52	N. W.	47	38	S. W.
8	57	52	S.	54	53	S.	49	39	N. by W.
9	59	55	E.	52	47	S.	42	38	N. W.
10	57	49	N. W.	53	46	E. N. E.	48	41	N. W.
11	58	48	N. W.	54	50	N.	46	37	S. W.
12	58	55	S.	53	49	N. by W.	40	34	N. W.
13	57	54	S. S. E.	53	48	N.	37	32	N. W.
14	58	51	S. W.	54	50		38	31	S. E.
15	57	49	W.	55	50	N.	45	33	N. E.
16	55	49	W.	55	51	S.	41	33	W.
17	55	51	N. W. by W.	56	54	S.	40	40	N. W.
18	59	50	W.	57	52	S.	38	35	W.
19	58	54	S. E.	55	50	S. E.	42	34	N. E.
20	62	54	S. W.	53	42	W. by N.	41	34	S.
21	61	58		54	47	S. W.	39	34	W.
22	57	51	S.	55	47	S. W.	34	28	S. W.
23	61	47	S. W.	48	44	N W.	34	30	W. S. W.
24	60	54	W. N. W.	42	38	N. W.	41	38	S. by W.
25	60	56	S.	52	44	N. W.	42	37	S. W.
26	50	48	N. W.	54	47	S. S. W.	47	38	S. by W.
27	48	38	S. W.	41	35	W. N. W.	45	39	S. by W.
28	52	47	S. S. W.	33	28	W.	46	43	S.
29	54	47	S. by W.	41	33	N. W.	44	38	S. W.
30	58	42	S. E.	46	33	N. by E.	38	32	S. W.
31	57	55	S. E. by S.				43	34	S. W.

Average—August, 57° 8—Sept. 51° 36′—October, 42° 10′.
———Heat of 3 Months, 50° 25′.

Rain—Inches, August, 2.055—Sept. 3.561—October, 5.200,
Rain in 3 Months, Inches, 10.816.

CLIMATE.

1808.   1809.

	NOVEMBER.			DECEMBER.			JANUARY.	
	Ther.			Ther.			Ther.	
	8 A.M.	NIGHT.	WINDS. 8 A.M.	8 A.M.	NIGHT.	WINDS. 8 A.M.	8 A.M.	NIGHT.	WINDS. 8 A.M.
1	46	40	S. W.	30	27	S. W.	36	34	S. E
2	46	43	S. by E.	41	31	S. E.	35	34	S. E
3	46	44	S. by E.	41	29	E.	33	33	S. E
4	44	42	S. E.	32	27	N. by W.	31	31	S. E.
5	42	39	S. E.	43	29	S. W.	30	29	S. E.
6	41	38	S. W.		31		35	33	S. S. E
7	38	32	S.	38		N. by W.	40	35	S.
8	41		S. W.	42		N. N. W.	38	36	S.
9	45	38	S.	36		N. by E.	38	34	S. by W.
10	44	40		40		N. W.	40	38	S.
11	43	39	N. N. E.	35		W.	38	36	S. by E.
12	42	37	N. E.	42		W. S. W.	34	32	N.
13	34	27	S. W.	42	35	W.	31	27	N. W.
14	36	29	S. S. W	41		N. W.	29	27	E.
15	47	32	S. W.		30		28	26	E.
16	52	45	S. W.	34			32	26	S. E.
17	48	45	S. by W.				32	26	S.
18	44	33	S. W.		20		34	26	S.
19	38	33	N. by E.				37	32	S.
20	32	27	N. W.				28	26	W.
21	38	29	S. W.				21	20	N. W.
22	38	29	S. W				23	14	W. S. W.
23	40	34	W.				28	17	
24	34	30	S. W.				19	12	W.
25	36	30	S. W.			E.	20	10	N.
26	34	31	S. W.			E.	24	10	E.
27	41	31	W.			E.	32		S. E.
28	37	31	N.	38		E.	40	33	S. by E.
29	30	27	S.	39		S. E.	42	39	S. by W.
30	38	27	S. E.	38	38	S. E.	37	36	S. W.
31				35	35	S. E.	32	29	W. by S.
						S. E.			

Average—Nov. 37° 10′—Dec. 38° 10′—Jan. 30° 33′.

———— Heat of 3 Months, 35° 18′.

Rain—Inches, Nov. 3.066—Dec. 3.442—Jan. 2.92.

Rain in 3 Months, Inches, 9.428.

1809.

	FEBRUARY.			MARCH.			APRIL.		
	Ther.			Ther.			Ther.		
	8 A.M.	NIGHT.	WINDS. 8 A.M.	8 A.M.	NIGHT.	WINDS. 8 A.M.	8 A.M.	NIGHT.	WINDS. 8 A.M.
1	36	32	S. E.	40	36	W.	37	35	N.
2	39	35	W. by S.	39	36	S.	36	36	N. W.
3	36	34	E. N. E.	39	36	S. S. W.	35	30	N. by W.
4	32	31	E.	38	34	S. by W.	35	32	N.
5	34	32	E.	36	34	W.	35	31	W.
6	34	31	S. E.	38	34	S.	44		N. W.
7	27	27	S.	38	36	S. W. by S.	43	39	N. W.
8	31	26	S. E.	45	36	S. W.	48	41	S. W.
9	33	26	S. S. E.	43	43	N. W.	42	39	S. W.
10				42	39	S.	38	34	N.
11	37	31	S. W.	43	42	N. W.	39	33	W.
12	37	37	S. E.	42	42	N.	40	32	S. W.
13	37	36	S. E.	42	38	N. W.	42	32	S. E.
14	36	35	N. W.	40	39	N. N. W.	42	33	S. E.
15	39	33	S. by W.	42	36	S. W.	40	33	S.
16	43	38	S. W.	46	43	W.	35	32	N. E.
17	41	40	S. S. W.	42	37	W.	37	32	N. E.
18	40	39	S.	41	36	W.	32	30	N.
19	39	36	S. W.	39	34	W.	35	28	N. W.
20	45	39	S. W.	42	38	S. by E.	36	29	N. W.
21	30	25	N. W.	40	39	S. by E.	37	29	W. by N.
22	33	29	S. S. W.	41	41	S.	40	34	S. E.
23	37	35	N. W.	42	41	S.	41	32	N. E.
24	39	36	S. W.	41	40	S. by E.	43	33	S.
25	40	38	W. by N.	40	35	S. W.	43	33	S. W.
26	44	36	W. by S.	36	34	S. E.	44	39	N.
27	43	41	W. by N.	35	33	S. E.	41	38	N. E.
28	48	46	W. S. W.	35	34	E.	40	32	N. N. W.
29				39	38	N. E.	38	34	N.
30				37	35	N. E.	43	36	S. W.
31				37	35	W. by N.			

Average—Feb. 37° 24′—March, 40° —April 39° 26′.
————Heat of 3 Months, 38° 57′.
Rain—Inches, Feb. 2.45—March, 1.11—April, 3.02.
Rain in 3 Months, Inches, 6.58.

CLIMATE.

1809.

	MAY.			JUNE.			JULY.		
	Ther.			*Ther.*			*Ther.*		
	8 A.M.	NIGHT.	WINDS. 8 A.M.	8 A.M.	NIGHT.	WINDS. 8 A.M.	8 A.M.	NIGHT.	WINDS. 8 A.M.
1	44	35	W.	43	37	N. W.	53	50	S.
2	39	37	N. W.	46	40	E.	50	49	
3	40	32	N. W.	46	41	W. S. W.	47	44	N. W.
4	43	35	S. W.	48	44	S. E.	51	44	N. E.
5	39	32	N. W.	49	44	S. by E.	52	49	N. by E.
6	37	35	S. W.	47	45	S. E.	50	50	N. W.
7	43	42		46	46	S. E.	52	48	S. W.
8	47	43	S. W.	51	45	S. by W.	52	49	S.
9	52	48	S. W.	52	46	S. by W.	52	49	N. W.
10	52	44	S. W.	46	45	N. W.	52	49	N. N. W.
11	46	43	S.	49	45	N. W.	57	48	W.
12	52	43	N. W.	50	40	S. by W.	54	53	N. W.
13	45	41	S.	54	47	W.	51	46	N. W.
14	50	45	S.	54	48	N.	54	48	S. W.
15	48	45	S.	53	39	N. W.	54	46	W.
16	52	45		49	46	W. by S.	56	50	W.
17	50	46	S.	48	44	S. W.	50	46	N. by W.
18	49	44	S.	48	45	N. W.	45	41	N. W.
19	54	48	S.	52	46	N. W.	48	41	W. by N.
20	51	47	S. E.	55	50	N. W.	58	48	N. W.
21	50	47	S.	52	48		52	50	W.
22	48	45	S.	60	50	W.		51	S.
23	58	46	S. W.	62	51	W.	49	49	S. E. or E.
24	54	52	S. E.	54	54	N. W.	56	50	S. by E.
25	47	47	N. W.	52	49	N. W.	54	50	N. W.
26	51	47	S. E.	51	41	N.	54	53	N.
27	47	46	S. by E.	54	54	N. W.	54	52	
28	48	46		51	50	N. W.	58	52	
29	44	41	N.	52	48	N. W.	53	52	W.
30	41	30		52	50	S.	57	51	S.
31	38	32	W.				54	50	S.

Average—May, 47° 4′—June, 50° 52′—July, 52° 38′.
————Heat of 3 Months, 50° 11′.
Rain—Inches, May, 1.66—June, 2.88—July, 4.14.
Rain in 3 Months, Inches, 8.66.

CLIMATE.

1809.

	AUGUST.			SEPTEMBER.			OCTOBER.		
	Ther.		WINDS. 8 A. M.	Ther.		WINDS. 8 A. M.	Ther.		WINDS. 8 A. M.
	8 A. M.	NIGHT.		8 A. M.	NIGHT.		8 A. M.	NIGHT.	
1	56	50	W.	54	50	S. W.	45	39	W.
2	50	49	S. E.	52	51	S.	55	44	S. W.
3	57	49	S. E.	54	52	S. E.	53	52	S.
4	55	48		54	54	E. S. E.	49		N.
5	54	53	N. W.	55	53	E. S. E.	47	46	S. E.
6	52	52	N. W.	54	52	S.	48	47	S.
7	53	50	N. W.	53	51	S. S. E.	48	48	S.
8	53	51	S. W.	54	52	S. E.	48	48	S.
9	56	52	S. S. W.	53	49	S. E.	50	49	S.
10	56	53	S. S. W.	57	52	E. S. E.	47	46	S. by E.
11	55	54	S. E.	54	50	S. E.	43	43	S. E.
12	56	55	S. E.	51	50	N. W.	47	45	S. E.
13	55	52	S. E.	54	53	N. W.	47	47	S. S. E.
14	55	52	S. E.	52	49	S. E.	48	47	S. by E.
15	56	52	S. W.	52	50	S.	47	46	S. S. W.
16	54	52	N. W.	55	45	S. W.	48	48	S. S. W.
17	53	51	N. W.	53	43	W. S. W.	50		S. W.
18	53	51	N. W.	49	47	S. W.	48	48	N. N. W.
19	53	52	N. W.	51	51	N. N. W.	48	48	S.
20	55	47	W.	51	47	S.	50	46	S. W.
21	52	52	S.	49	49	N. W.	51	50	S.
22	54	51	S. W.	52	47	S.	52	51	S. S. W.
23	52	50	S.	51	34	S.	48	48	S.
24	52	50	S.	38	39	S.	49	49	S. S. W.
25	50	50	W. by S.	39	38	N. W.	51	49	S.
26	53	48	S. by E.	43		S. W.	50	47	S.
27	55	52	S. E.	42	37	N.	49	49	S. W.
28	53	48	N. N. W.	43	42	N. W.	49	49	S. W.
29	56	52	S. S. W.	43	37	N. W.	50	50	S. by W.
30	54	53	S.	45		S. by W.	53	51	S.
31	52	44	S.				54	52	W. by S.

Average—August, 52° 52'—Sept. 50° 14'—October, 49° 10'
Heat of 3 Months, 51° 5'.
Rain—Inches, August, 2.845—Sept. 3.455—October, 1.385.
Rain in 3 Months, Inches, 7.685.

1809.   1810.

	NOVEMBER.			DECEMBER.			JANUARY.	
	Ther.			Ther.			Ther.	
	8 A. M.	NIGHT.	WINDS. 8 A. M.	8 A. M.	NIGHT.	WINDS. 8 A. M.	8 A. M.	NIGHT.	WINDS. 8 A. M.
1	43	43	N. by W.	32	31	W.	38	38	W. S. W.
2	37	36	W.	34	31	W.	42	38	S. W.
3	44	33	S. by E.	40	30	S. S. W.	36	36	W. S. W.
4	46	44	S. E.	41	31	S. by W.	44	41	S. S. W.
5	47	46	E.	32	32	S. W.	43	41	S. by W.
6	48	45	S. E.	43	31	S. by W.	43	43	S. by W.
7	44	44	S.	44	43	S. W.	46	36	W.
8	47	43	S. W.	35	34	S. W. by W.	43	35	S. S. W.
9	46	45	W. by N.	83	55	S. W.	41	41	S. S. E.
10	37	37	N. W.	38	38	W. S. W.	42	38	S.
11	47	43	S.	31	31	W. S. W.	42	41	S.
12	44	45	S.	36	32	S. W	39	39	S. E. by S.
13	46	36	S. E.	51	31	S. W.	37	37	S. E.
14	39	33	W.	27	26	S. W.	34	33	S. E.
15	38	32	N. E.	39	27	S. E.	28	26	S. E. by S.
16	32	31	N. W.	38	37	S. E.	31	28	E. S. E.
17	35	32	N.	38	35	E. by S.	30	26	S. W. by W.
18	35	22	N. by E.	39	39	E. N. E.	32	28	W. S. W.
19	27	33	W.	41	39	N. E.	32	28	S. W.
20	38	33	W. by S.	39	33	S.	26	26	S. W.
21	35	33	W. by S.	35	82	S. W.	29	23	W.
22	43	34	N. W.	36	32	S. W.	34	28	N. E.
23	37	34	S. by W.	32	32	W.	32	27	N. W.
24	42	32	N. by E.	39	31	W.	36	28	S.
25	38	39	S. S. W.	37	34	W. N. W.	40	37	W. by S.
26	42	37	S. E.	38	32	S.	39	37	S. W.
27	37	37	W. by S.	32	32	W.	38	38	S.
28	37	34	S.	36	32	S. by W.	33	31	W.
29	34	35	S. W.	37	37	N. W.	37	33	S. W.
30	42		S. S. W.	31	26	W. by S.	36	32	W. by S.
31				38	23	W. by S.	41	36	S ½ W.

Average—Nov. 40° 14′—Dec. 36° 19′—Jan. 36° 39′.
———Heat of 3 Months, 37° 44′.

Rain—Inches, Nov. 2.965—Dec. 2.75—Jan. 1.635.

Rain in 3 Months, Inches, 7.35.

C

1810.

	FEBRUARY.			MARCH.			APRIL.		
	Ther. 8 A.M.	Ther. NIGHT.	WINDS 8 A.M.	Ther. 8 A.M.	Ther. NIGHT.	WINDS 8 A.M.	Ther. 8 A.M.	Ther. NIGHT.	WINDS 8 A.M.
1	45	41	S. W.	46	36	W.	39	37	S. by E.
2	38	38	W. N. W.	32	28	W. ½ W.	37	36	S. E.
3	32	32	W.	35	38	S. E.	41	38	S. by E.
4	32	30	W. by S.	37	35		39	39	N. N. W.
5	39	31	S. W.	39	36	E.	39	38	E.
6	44	40	S. & W.	32	31	N. W.	40	38	S. by E.
7	41	40	N. E. by E.	35	32	S. E. by E.	40	39	S. E. by E.
8	43	42	S. W.	37	34	S. S. E.	39	38	S. E.
9	43	42	S.	39	34	E. by S.	39	36	S. E. by E.
10	37	35	N. W.	38	37	N. E.	40	35	E. by N.
11	32	30	N. W.	35	31	N. N. W.	38	34	E. N. E.
12	30	29	N. W.	37	28	E.	38	33	N W. by W.
13	33	29	E.	35	31	W. S. W.	38	33	N.
14	32	31	E. by N.	32	29	N. by W.	41	34	N W. by W.
15	24	20	W. ½ S.	32	29	W. by N.	41	38	S.
16	24	22	W. by N.	32	30	W. by N.	42	40	S.
17	27	24	N. W. by W.	26	22	W.	43	39	S. S. E.
18	32	24	N. W.	36	22	W. by N.	44	41	S.
19	32	28	N. E.	36	29	W. by S.	44	42	S. S. W.
20	32	30	N. E.	41	34	W. by S.	43	40	S. W. by S.
21	31	21	S. S. W.	39	38	N.	46	40	W. by S.
22	37	32	S. W.	33	28	N. W.	45	44	S. W.
23	39	34	S. S. W.	38	31	W. S. W.	48	44	S. S. W.
24	37	34	S. W.	32	28	N. E.	46	43	N. N. W.
25	31	30	W. S. W.	35	30	S. by E.	45	43	S. E.
26	35	32	W.	38	34	S. S. E.	44	39	S. ½ E.
27	42	38	S. W. by W.	40	37	S. by E.	43	39	S. by W.
28	37	32	S. by W.	38	38	S. E.	43	41	S.
29				41	33	W.	48	39	S. by W.
30				37	32	S. E. by S.	45	41	N. N. W.
31				42	37	S. E.			

Average—Feb. 35° 8′—March, 36° 18′—April, 41° 36′.
————Heat of 3 Months, 37° 39′.
Rain—Inches, Feb. 1.54—March, 2.50—April, 1.83.
Rain in 3 Months, Inches, 5.87.

CLIMATE.

From the above Tables, it appears that the heat of the thermometer, at Peterhead, on an average of two years, taken at eight o'clock in the morning, is during the three spring months ....38° 18′.
That during the three summer months it is only .51° 33′.
That during the three autumnal months it is 50° 45′.
And that during the three winter months it is 36° 31′.

It here deserves to be remarked, that the average heat of the first nine months, during which there is either sowing, vegetation, ripening or harvesting of the different kinds of corn, which are commonly raised in this county, (for the quantity of wheat that is sown in Aberdeenshire is very inconsiderable) is only 46° 52′.

While that of the three winter months, was 36° 31′.

And that of the average of the two years is 44° 17′.

This, however, holds good only of the sea-coast, or lower parts of the county. In the higher districts, in warm seasons, and on the sandy soil near the river Dee, the heat, in summer, for a few weeks, is often very great, but in these elevated situations, the cold, in winter, is equally intense.— Jan. 26th and 27th, 1809, the mercury, in Dr Laing's thermometer, at Peterhead, as appears from the table, was at no time lower than 10° of heat, or 22 below the freezing point. But at Crathie, in Braemar, it was observed by the Rev. Charles McHardy one night, at 4° of heat, or 28 below the freezing point. And his maid-servants reported to him next morning, that a little after midnight, the mercury was all within the ball. In the Reporter's house, 300 feet above the level of the sea, and 13 miles from the coast, it was at 8° on the 26th, and on the 27th at 9°.

It would be both a curious and an useful inquiry, to compare the heat of the thermometer in all the various districts of Great Britain. And a few examples, which may call attention

tention to this inquiry, may not be unacceptable to the reader.

By Mr. VANCOUVER's Report of the Agriculture of Devonshire, it appears that the average heat of the thermometer, at Ilfracombe, within 50 feet of the level of the sea, in the year 1806, was as follows:—

In February, March, and April,	52° 40′.
In May, June, and July,	64° 15′.
In August, September, and October,	63° 20′.
And in November, December, and January,	55° 50′.
Average of the whole year 1806, is	59° 1′.

By Mr. ROBERTSON's Survey of Kincardineshire, the average heat in lat. 56° 58′, at 500 yards from the sea, and 150 feet above its level, is 44° 40′.

This is less than that at Ilfracombe, by 14° 21′.

The Writer of this Report was favoured, by the ingenious Gentleman who gave Mr. ROBERTSON the various heats of the thermometer, mentioned in that Survey, with a particular account of the heat of the different years; and he subjoins those of 1806, classed into the spring, summer, autumn and winter seasons.

February, March, and April,	34° 40′.
May, June, and July,	51° 20′.
August, September, and October,	51° 40′.
November, December, and January,	37° 40′.

The heat of the particular year here selected was rather *above the average* in this county, and was 2 degrees warmer than that of the year 1807.

By the same ingenious Gentleman he has been favoured with the state of the thermometer at the same place from May, 1808 to May, 1810.—It is as follows at 8 in the morning.

1808.

                            1808.   1809.
In May, June, and July,   55° 20'.—51° 10'.
In Aug. Sept. and October, 50° 10'.—52° 10'.
In Nov. Dec. and January, 35°   —36° 20'.
In Feb. March, and April, 39°   —36° 50'.
And the average of the two years is 44' 30'.

In order to ascertain the progress of heat from Peterhead, southward, along the German Ocean, he also applied to JOHN TAYLOR, Esq. of Kirktonhill, who keeps an accurate register of the thermometer, at his house of Kirktonhill, situated 36 miles south of Aberdeen, (in lat. 56° 40) 200 feet above the level of the sea, in the southern extremity of the county of Kincardine, and near the confines of that of Forfar or Angus. By this Gentleman he has also been favoured with the Table subjoined in the next page, *the general average of which*, at morning, noon, and evening, for these seasons, is as follows:—

                              1808.   1809.
May, June, and July,   57° 50'.—57° 10'.
August, Sept. and October, 53 45.—54° 55'.
Nov. Dec. and Jan.   36° 15'.—37° 19',
Feb. March, and April,   37° 30'—39° 16.'

This average is probably a little different from the heat at 8 in the morning; and therefore it is most satisfactory to insert the whole Table, after remarking, that Kirktonhill is about 6 miles inland from the mouth of the North Esk, which runs into the sea, near Montrose, in the county of Angus, nearly in the same parallel of latitude with the most southerly part of Aberdeenshire, consequently must, on many accounts, be of a different temperature from Peterhead, which is at the termination of the great promontory that projects far into the German Ocean, by which it is chiefly surrounded.

MONTHLY ACCOUNT OF THE AVERAGE HEIGHT OF THE THERMOMETER AT KIRKTONHILL,

FROM 1ST MAY, 1808, TO 1ST MAY, 1810.

	Mor. & Even. at 9 A.M. and P.M.	12 Noon to 2 P.M.	Morn. Noon, & Evening.
1808.			
May	50° 15′	59°	53°
June	55	62 30	57 30
July	61 40	68	63
August	60 40	65 40	62 15
September	52 38	59 30	55
October	42 40	47 30	44
November	39 15	44	40 45
December	35	38	38
1809.			
January	30 50	34	32
February	30 40	41	36 9
March	40 30	46 20	42 20
April	39 40	46 30	42
May	51 30	60 10	54 30
June	55 7	62 15	57 30
July	56 40	65	59 30
August	57 15	63 40	59 15
September	52	59 20	54 30
October	49 15	54 30	51
November	38 12	42 20	39 38
December	35	38	36
1810.			
January	35	39	36 20
February	34	39 20	35 40
March	34 40	41	37
April	43 12	49	45 7

Average for 12 months,

Nine morning and evening, 45° 10′.

Twelve noon, to two P. M. 51° 30′.

Morning, noon, and night, 47° 15′.

Rain, 146 days,

Snow, 34 days,

By comparing with these the valuable Report of Mid Lothian, also written by Mr. ROBERTSON, it appears, that on an average

average of 18 years, the heat in the immediate vicinity of Edinburgh was from 1785 to 1792 inclusive

In the 3 spring months, 44° 50′.
In the 3 summer months, 61° 30′.
In the 3 autumn months, 59°.
In the 3 winter months, 40° 20′.
Average of the whole 18 years, 51° 10′.

Taking a general view of the Climate of Aberdeenshire, with respect to the degree of heat, as indicated by the thermometer, it will be found, that it is from 4 to 6 degrees of less heat in the spring, summer, and autumnal months, than the county of Edinburgh, from 9 to 12 degrees less than that of the Royal Observatory at Greenwich, and from 12 to 16 degrees inferior to the most cultivated places in N. Devon.— (The higher grounds in that county are not taken into this account, because from their elevation they are much colder than the lands which are only 50 feet above the level of the sea; and Cornwall is both a maritime county, and so unequal in various respects, that it is also omitted.)

The reasons for examining this point more minutely than is done in the other Reports which this Writer has seen, are, that the laws of climate ought to be attended to both by the Legislature, and the Agriculturist. The former should proportion taxes to the quality of corn that can be raised in the different districts of the kingdom. The latter should not only sow that species of corn, which will answer in the climate in which he lives, but also if the climate be cold and backward, he should pay particular attention to the *time of sowing.*— For example, If wheat is ever generally raised in the county of Aberdeen, it should be sown as soon as possible after the autumnal equinox, that it may be firmly rooted before winter, and able to endure the alternate frosts and thaws of the spring.

With regard to rain, as affecting the climate, it may be observed, that the whole east coast of Scotland, owing to the comparatively small body of water in the German Sea, that washes it, is visited with much less rain, than what is wafted across the Atlantic Ocean to the western coast of Scotland. Hence it is fitter for raising corn; but not so well adapted for raising grass, or those roots which require abundance of water for their nourishment. But in so extensive a county as that of Aberdeen, the quantity of rain that falls in a year, varies from 25 to 38 inches.

Another circumstance, not commonly attended to, occasions a considerable difference in the climate of Aberdeenshire, comparing one year with another. When there is a great quantity of snow, lying undissolved, on the high mountains of Braemar, the south-west wind, in spring, is not a warm and mild zephyr, but is often attended with piercing cold. When the snow either has not fallen, or has been dissolved, then it is very pleasant. The north wind is always, and the north-west wind, for the most part, attended with cold. And at Peterhead, though otherwise temperate in so high a latitude, when the winds blow from the N. E. over the frozen sea, the cold is much more severe than when it blows overland, in the winter months. The south, and south east are the most genial winds to Aberdeenshire, till the snows are melted in the south-west. Even the east wind, of which Dr. Anderson complains, (as he was much exposed to it at Monkshill) though it conveys rain and fogs, and is particularly injurious to the harvesting of corn, does not bring so much cold as generally accompanies the N. and N. E. winds.

It needs only be added on this article, that from our situation as an island, the weather is more variable than on the continent, and in the maritime parts of Aberdeenshire, than in in the mountainous districts which reach to an equal distance between the Atlantic and the German Ocean. But while there is the greatest irregularity in the *fall of rain*, an accurate observer will often see by the *fall of snow*, (on the *tops and ridges* of the highest mountains, *a little lower than the tops* of others, and only on *the very summits of the lower ones*) their *comparative elevation above the level* of the sea ; a matter that should always be attended to, when speaking of the climate of any place. For at a certain distance from that level, the cold of the polar circle is to be found under the Equator. On that account, it would be very inaccurate to speak of the climate of Aberdeenshire, without distinguishing between the elevated lands on the S. W. and the lower parts of the county that border on the German Ocean.

SECT. IV.—SOIL AND SURFACE.

Dr. Anderson's Account of the Soil of Aberdeenshire shall here, as in the former article, be first quoted from the original report ; and then such facts as require to be mentioned, or observations that occur to this writer, shall be subjoined.

" In the lower parts of Aberdeenshire it consists in general
" of a mellow clay, which under proper culture is capable of
" being made to yield very abundant crops. But though this
" be the general nature of the soil, there is a considerable di-
" versity, in regard to the degree of stiffness especially. In
" most of the places that have not been thoroughly cultivated,
" there is a thin stratum of moory soil upon the surface, which
" is of inferior quality ; which however soon disappears when
" under

" under proper management. The sub-soil is almost univer-
" sally a retentive clay. Springs, or what are called spout-
" ing grounds, are by no means so frequent as in many other
" places that have been under such general bad management;
" and most of the fields could be laid thoroughly dry and
" sweet at a very small expence.

" In the higher districts, especially the Garioch, the soil
" is in general a sharp loam, with a considerable portion of
" sand; and the same may be said of most of the arable vales
" in the higher districts. These soils are kindly to cultivate,
" and afford fine grain, and early crops, considering their
" elevated situation. At Kildrummy, near the banks of the
" Don, more than thirty miles from its source, is produced
" bear of so fine a quality, as often to exceed in weight even
" the finest barley from England. Lime on all these soils
" produces a powerful effect in rendering them more produc-
" tive, and altering the kinds of their natural productions.

" Few things are more difficult, than to give such a des-
" cription of soils as to convey any accurate ideas to those for
" whom it was intended. I feel this on the present occasion.
" Our language does not furnish words capable of discriminat-
" ing those nice shades that are of great practical importance.
" When clay or sand, or loam are mentioned, every person
" recalls the idea of the soils of the same denomination he has
" seen; so that if a hundred persons were present, perhaps
" the word would not convey precisely the same idea to any
" two of them. One who had seen tough wirey clays, would
" figure such as these were before him—another, has seen
" clay mixed with a great proportion of sand or gravel.—One
" has seen clay soils of the poorest and most barren aspect—
" another understood by that phrase, the most fertile fields he
" had ever beheld. In short, the diversities are infinite, and
                                                        " the

" the ideas that different persons annex to the same words, are
" nearly as much diversified. The clays of Aberdeenshire
" are in some respects different from any others I have seen;
" so that wishing not to mislead, it behoves me to be some-
" what more particular respecting them.

" There are two opposite extremes of clayey soil; one is
" of a tough gluey coherent nature, which in the extreme of
" this sort is called *Till:* The other is of a loose light friable
" texture, the extreme of which is *Fuller's earth.* The clays
" of Aberdeenshire run between these extremes, and though
" some of them, particularly near Slains, and the lower parts
" of Buchan, are stiff, and approach a little towards the Till,
" yet through the country at large they bear a much nearer
" affinity to the Fuller's earth. In general, the clays are
" light and friable, very easily cut by the plough, and are
" neither apt to stick to it very much, nor to assume that
" stony hardness in dry weather, so common to some clays.
" These clays are excessively pure, and remarkably free
" from any mixture of sand or other matter, so that when
" they are moistened with water, they feel soft and unctous
" to the touch. When a lump of this clay is exposed to the
" weather, the slightest frost makes it crumble down into a
" great number of small pieces, nearly like lime slaking;
" and, even without frost, the very dews make it fall down
" by degrees into a soft powdery state. This has so much
" the appearance of marle, that many persons would give it
" that name, though it contains not the smallest particle of
" calcareous matter in its composition. It is of various tints
" of red, brown, yellow, and white; and is so pure that I
" have used it as a paint with oil, with success; but all the
" kinds contain such a portion of iron, that they become red
" when burnt in the fire. When impregnated with manures,
" and kept under a good state of culture, it soon assumes the
" ap-

"appearance and properties of a light friable loam. But its
"qualities are so different in different states, and they are so
"much varied by particular modes of culture and manures,
"that it becomes necessary here to specify a few of the most
"remarkable of these peculiarities.

"When this kind of soil has been long kept under a course
"of tillage, without being ever laid down to grass, and dung-
"ed every third year, as is the practice with regard to what
"is here called *in-town* land,—it acquires a dark colour, as-
"sumes a light, soft, puffy feel to the feet, and is more open
"and spungy than most other soils. On ploughing, it is
"nearly as incoherent as sand; but has more the consistence
"of meal. There are very few clods upon its surface; and in
"harrowing it falls down into a fine powdery state. No-
"thing can be more favourable for the germinating of small
"seeds, than a soil in this state, and accordingly it naturally
"produces such a prodigious abundance of the weed called
"*Spurrey, Spergula*, provincially *Yarr*; that in some seasons
"not a stalk of grain can be allowed to spring up among it.
"In a bad state of culture, likewise, it is extremely favourable
"to the growth of knot-grass [*avena elatior,*] provincially
"swines arnots, wormwood; [*artemesia vulgaris,*] dead-net-
"tle, [*Lamium*] and wild oats. When such a field as this is
"left out to grass, if previously freed from the root weeds
"above specified, it naturally produces a tolerable crop of
"meadow soft grass, [*Holcus Lanutus,*] but in this state it
"neither produces wheat, nor pease, nor clover; nor does
"rye grass flourish upon it. Bear and oats are the only
"kinds of grain that thrive upon it. But it is admirably a-
"dapted to the culture of turnips.

"When this soil, however, is cleaned from root weeds,
"and manured with *lime*, and sown down with grass seeds, it
"produces the most luxuriant crops, of rye-grass especially,
I have

"I have ever seen; the meadow soft grass now entirely disap-
"pears, and after the ground has been rested some time, it
"acquires a firmness and solidity that adapts it to the rearing
"of corn crops of every kind; the spurrey disappears, and
"wheat or pease thrive upon it as well as on most soils, tho'
"it never is so well adapted to the rearing of broad clover as
"more weighty soils.

"When it has been alternately put under tillage, and left
"in grass, it produces no other corn crop except oats; and
"these, after the first or second crop, but of a poor quality.
"Its weeds are still spurrey; but in greater quantities corn
"marygold [*Crysanthemum segetum*] provincially, *guild*, and
"when it is much exhausted by tillage, the earth nut [*Bu-
"nium bulbocastanum*] in great quantities. When left out to
"grass, it yields a few stalks of the crested dogs-tail, but
"none of the Holcus, unless it has been previously dunged—
"some of the grass leaved plantain [*Plantango maritima*] and
"if over-ploughed, the small leaved sorrel, [*Rumex acetosella*]
"but the principal crop is of the small bent grass [*Agrostis
"tenuissima*]. Neither rye-grass nor clover will live on this
"soil, while in this state; but after it has been limed, its
"nature is entirely changed. Pease, and every other sort
"of grain can be reared on it then to perfection, under a pro-
"per mode of culture; the small bent grass and dogs-tail
"disappear, and in their stead, the poa grasses spontaneously
"come up:—rye grass thrives well, and clover, both white
"and red, especially the white, may be reared up to great
"perfection. It is only after being limed, that the wild
"mustard comes to be a prevalent weed on these soils."

After thus quoting, at length, all that Dr. ANDERSON has written on the subject of Soil, the Reporter shall state what he considers as necessary to be known on this important article.

1. It

1. It is certainly true, that it is often difficult to give a correct definition of any particular soil, and that our language does not always convey clear ideas on this subject.—But the best method of discriminating soils, is to distinguish them by the name of that particular kind of matter, which is found to be the principal ingredient of the soil, when chemically analyzed. Thus, where a great proportion of *sand* is found, the land may be said to be of a *sandy* soil; where *clays* predominate, it may called a *clay* soil; and where there is found a *rich unctuous black earth*, occasioned by *frequent dunging* and *long culture*, the term *loamy* soil, though *not so definite*, may be used.

2. As to the Doctor's remark, that " our language does " not furnish words, capable of discriminating those nice " shades, that are of great practical importance, it deserves to be attended to, that wherever men have *clear ideas*, they will generally have *perspicuous* language; and that where these ideas are not only *clear*, but *just* and *correct*, their language will be distinguished by its *precision*. But as our farmers have more generally exerted their *active powers*, than cultivated ther *intellectual faculties*, they are more apt to *go wrong* when they use *learned words*, than when they express themselves in the *common language* to which they have been *accustomed*.—The terms, used in the plan of the Reports, *clay, loam, sand, chalk,* and *peat*, are generally understood, or may be easily explained; and unless our farmers were all regularly educated, and studied chemistry as a science, they would not add to their knowledge, by acquiring a few *technical* expressions to which they did not affix any *clear* or *precise* ideas.

3. It may be doubted whether the soil, which is described at so great length by Dr. ANDERSON, in the above passage, be a light friable clay. It will be found, on examination, to contain only a very small proportion of argillaceous earth.

But

But we have various kinds of clay in this county, from the hard and tenacious clay, which is found more generally in the division of Buchan, to that friable clay, which is to be met with in the other districts of Aberdeenshire.

To these general remarks, the statement of a few facts may be subjoined.

In the very extensive division of *Marr*, particularly on the banks of the Dee, a sandy soil predominates. Clay is found in a few places only, and where it is found, is generally mixed with small stones, which are sometimes pieces of granite, nearly decomposed, and sometimes of a harder substance.— The old croft land, near the farm-steads, which has been long in culture, and is provincially called *infield*, from the quantity of animal manure, or rotted vegetables, has acquired a black colour, and may be called a sandy loam, in which the sand is more or less apparent, according as the land has been well or ill manured, or has been gently or severely cropped. The *outfields*, as they are called, on which black cattle were folded, and of which the soil was occasionally manured by the dung of these animals, have not acquired the black colour to the eye, nor the unctuous feel to the touch, which distinguish the kind that has been long under culture. Therefore a greater proportion of sand appears in them. And even since the introduction of lime, their colour is not so black as that of the old croft land. This soil is frequently called *hazelly*; though there does not seem to be any good reason for the appellation. The term *sandy* is commonly given to those poor lands, where the sand is very apparent, and where till of late, neither dung nor lime were applied. Excepting the two kinds of peat that are interspersed through this division, (the one that is found on the tops of hills, and the other in low confined places, where the water does not easily find an outlet) the greatest part of the soil of Marr

is

is sandy originally, whatever alteration has been made upon its appearance by the application of manure. Perhaps clay is found in one acre out of fifty. In the higher parts of the county, peat-moss is found on many of the hills; and is carried several miles by the farmers, when shaped and dried like bricks. What is found in the marshes is of inferior quality, but is useful both as fuel, and when mixed with animal dung, as a compost. Chalk is not found in this division. But lime stone of various qualities has been discovered, and applied artificially, to stimulate the power, and as far as it can, to alter the nature of the soil.

In *Formartin*, especially near the sea-coast, a greater proportion of clay is found. But a still greater proportion of peat,—from thin moorish soil of 8 or 10 *inches*, to the deep mossy grounds of 8 or 10 *feet* deep, is interspersed through the more inland parts of this division. Loam, which is no original soil, but the combination of animal dung, and of rotten animals or vegetables, with the soil on which these are laid, abounds in all the croft, or old cultivated lands. But the loamy soil of Formartin is heavier and stronger than that on the banks of the Dee. A considerable portion of sand is found prevalent in the soil of this division, which also contains several quarries of excellent lime-stone, though very little of this is so near the surface, as to be within the reach of the plough. The sides of the hills, and in general the barren lands abound with sand and small stones, near the surface.— Perhaps one-twentieth part of this division is of clay, and one-fifteenth part of peat earth of various qualities.

In *Buchan* also, on the sea-coast, there is a great proportion of clay, some of which is as tenacious as any soil in the Lothians. From extensive cultivation, there is a great proportion of loam in all the old croft lands: But that of sand is in general very small. Of lime-stone there is great abun-

dance

dance in multitudes of loose stones, of different qualities, which in many places are mixed with the soil, and burnt as lime by the farmers; though the introduction of shelly sand has occasioned this coarse lime to be less used than formerly. One-tenth part of the surface of Buchan consists of peat moss, generally of an excellent quality, in the more inland parts of the division.

In the Garioch, which has been long under cultivation, and is naturally very fertile, there is a greater quantity of loam, than in any of the other divisions of the county, in proportion to its extent. Yet even here there are several peat mosses on the bounding hills, and on the north-east side of the valley; and a considerable portion of sand is found in various places. The soil of the old cultivated lands is either a mixture of loam, sand, and clay, or of loam, with only one of these two. Perhaps the basis of one-fifth part of the Garioch is clay, which, by repeated manure, has become a heavy loam; but above two-fifths of this extensive valley may more properly be called a light loam, being a compound of sand, manure, or rotten vegetables, with a small proportion of peat moss. The remainder of the surface is composed of moorish ground, in the different bounding hills, or of peat moss of various depths, a considerable part of which is exhausted, by being used as fuel by the inhabitants. No chalk is found in the Garioch, nor has any limestone been discovered in this division.

In Strathboggie, where there is a less proportion of the soil under cultivation, there is a much less proportion of loam, than where the country has been longer cultivated, and is better peopled. Sand, moorish soil, and peat moss are generally prevalent in the hills—but clay is found in many places, sometimes in a pure state, but for the most part, mixed with small stones, sand, or gravel, that is with granite,

nite, in the different stages of decomposition. Near the banks of the Deveron, the Boggie, and the Isla, where the lands have been longer under culture, a light and fertile loam, the basis of which is sand, mixed with the richer soil washed down from the hills, generally prevails; although a heavy loam, the basis of which is clay, is found in many places. The discovery of the finest limestone at Ardonnel, which is now wrought in very considerable quantities, and is at least equal to any lime from Sunderland, will make, in a few years, a great alteration on the surface of this division.

The qualities of the subsoil have not hitherto been mentioned, yet are deserving of notice; for the fertility of land depends not only upon the quality of the soil, but also on the nature of the subsoil.

On the banks of the Dee, the soil is not only light and sandy, but in general it rests on a gravelly bottom. By frequent ploughing, the *particles of gravel* are diminished, and by combination with dung, or manure, of different kinds, the sand becomes a light loam. But in the subsoil, these particles are larger, not being worn down by the plough.— This *dry* and *thirsty* land, even when *highly cultivated,* stands in need of *frequent rains;* and the crop which it bears is apt to be *parched* or *stinted* in dry seasons. But after a rainy summer, it is more weighty and luxuriant, in harvest, than could be thought possible by a stranger, who saw the plough turning over the light and shallow soil, incumbent on the bed of gravel, which is discovered by every furrow. On the banks of the Don there is a much deeper staple of soil, which generally reaches farther than the plough; and even in a dry season the crop of corn is good, whether the loam be light or heavy. For the light loam of the *haughs,* or meadows along the Don, is much weightier than that which is found near the Dee. In the valley of Garioch, the subsoil, whether

ther clay or sand, is found to be more open, and after exposure to the air, more fertile than that of the bounding hills. In Strathboggie, and in part of Formartin, a kind of *hard till* frequently distinguishes the subsoil, which it is very difficult to pierce. In Buchan, where the clay is extremely tenacious in some places (for example in St. Fergus) the subsoil is most commonly a mixture of clay and small stones. In some places near the sea coast of Belhelvie, a stratum of clay, of three or four feet deep, is incumbent on another of sand, and sometimes of peat-moss.

In the district of Kildrummy, where corn of the finest quality is raised, the soil partly consists of decomposed freestone, and is incumbent upon freestone, which is remarkably fair and durable; as appears by the ruins of the Castle of Kildrummy, which will be noticed in its proper place.

These are only general outlines; for in so extensive a county, there is a great diversity both of soil and subsoil.

With respect to surface, Aberdeenshire in general, is rather hilly than mountainous. In Strathboggie, however, the hills are very high; and in the higher parts of Marr there are several mountains of very great altitude. Three of these, Lochnagar, Cairntowl, and Benabourd, all situated in Braemar, are probably the highest mountains of Great Britain.—Their precise heights above the level of the sea, at Aberdeen, have not as yet been ascertained, owing to the delay in getting a map of the county. But the Writer of this Report shall endeavour if possible, (though it does not ly immediately in his department) to supply this defect; and the elevations, above the level of the sea, of the principal mountains in Aberdeenshire, will be marked in the Appendix. It may be sufficient to mention in this place, that in general, the mountains and hills of this county rise gradually from their bases, and increase in altitude as they recede from the

German Sea. They do not spring from the plains in the abrupt manner of the Pentland hills, near Edinburgh. The only exceptions from this general rule, which merit particular notice, are the beautiful conical hill of Dun-o-Deer, situated in the upper part of the valley of Garioch, and the much higher conical mountain of Noath: on the top of both are constructed vitrified forts, of great antiquity. These hills are also supposed, at a remote period, to have been volcanos.

It may be farther remarked, that the soil of all these hills and mountains, is uniformly deeper on the north than on the south sides. And indeed the same thing holds true of all the north-lying fields, compared to those which have a southern exposure. The variations from thawing to freezing, which in the spring months are so common in Aberdeenshire, are much greater on the south than on the north sides of hills; and on fields which have a southerly, than on those which have a northerly exposure. And in summer, the genial influence of the sun, though it favour early vegetation, in March, does not produce so heavy a crop of grass in July, as is found on the deeper soils which ly to the N. and N. W. Consequently, corn of the best quality, ready for the sickle in August, is generally raised in south lying exposures, while the heaviest loads of grass and forage, accompanied by late harvests, characterize those fields which ly to the north.— Perhaps the powerful action of the sun's rays carries off the more volatile parts of the soil on all south-lying exposures; while those that ly to the north, with many other disadvantages, are less exposed to evaporation, and are generally of a deeper staple.

SECT. V.—MINERALS.

Dr. Anderson observes, that, "in regard to mineral "productions, Aberdeenshire has little to boast. No metals, "no coal, and scarcely any lime-stone." This is not quite correct. Considerable quantities of excellent lime-stone have been found in the county of Aberdeen; and since the Doctor's original Report was printed, very good lime-stone has been discovered, and is now wrought in the parish of Udny, within three miles of the Doctor's own farm at Monkshill.— Lime is also found in the higher parts of Marr, in Cabrach, in Forgue, and in many places of Buchan and Formartin;— But in several of those places, the difficulty of finding fuel to burn the lime-stone is so great, that a much smaller quantity is burnt than what is wanted in the immediate neighbourhood of the lime quarries. It has been already mentioned, that at Ardonnel, in the division of Strathboggie, the lime-stone is as pure, and contains as great a proportion of calcareous matter, as that of any place in Great Britain. And by the opening of excellent turnpike, and other roads, it will prove highly beneficial to the agriculture of that part of the county. It is so valuable for cement and plaister, that it has been carried above 30 miles into the parish of Strathdon, and is preferred to any lime from Sunderland.

In the division of Garioch, although no lime-stone is found, quarries of excellent slate have been opened, sufficient to supply the north of Scotland with that valuable covering for houses; and now that there is an excellent turnpike road to the foot of the hill of Foudland, where they are situated, the slates can be carried 12 miles to Inverury, to the head of the Canal, and from thence to Aberdeen, by this inland navigation.

In Buchan, besides the great quantity of shelly sand, and of limestone of different qualities, very rich carbonate of lime is found at the *dropping* Cove of Slains, which is a great natural curiosity.

Fine grey oxyd of manganese is found at Grandholm, in the division of Formartin, on the left or north bank of the Don, about 6 miles from Aberdeen. This oxyd is accompanied by

1st. Vein stone of sulphate of barytes.

2d. A rich vein of hæmatitic iron ore.

Also black oxyd of manganese, as good as from Cumberland, is found diffused through loose blocks of granite, at Gilcomstone, in the suburbs of Aberdeen.

Dolomite has also been found in Tyrebagger, in the lower division of Marr, within 7 miles of Aberdeen. It seems to be a variety that has not hitherto been described. At first it was mistaken for phosphate of lime, till it was examined, and found to contain no phosphoric acid. When heated, it is luminous in the dark.

No agates are known in this county, and the nature of the rock, viz. granite, which is prevalent in Aberdeenshire, is incompatible with agate.

Granite is the most valuable mineral in this county, and has brought gold into Aberdeen. Complete granite, according to Kirwan, contains, 1. quartz, 2. mica, 3. feltspar. In this state it is found at Peterhead, where the feltspar predominates. Near Aberdeen, the granite is imperfect; for it has little or no feltspar. But great quantities of horn blend, an excellent flux for iron ore, is frequently found in masses, and not seldom as a component part of granite. And at the mouth of the Don, after floods, magnetic iron sand is usually deposited. Iron stone has also been found in the bounding hills of the parish of Fintray, between the Garioch and Formartin.

The

The granite in the neighbourhood of Aberdeen deserves to be more particularly mentioned.

It is not only used for building houses and inclosures, and as metalling for turnpike roads, (of which there are many excellent ones in the county) but it has also been exported to a considerable annual amount, partly in the shape of long stones, and lintels, for pillars, doors, and windows, but chiefly in great quantities of small causeway stones, for paving the streets of London. These are admirably calculated for this purpose; and though the expence of quarrying and dressing them, joined to that of carriage by sea and land, is very great, yet the demand has rather increased than diminished. It is of the utmost consequence, that an article which brings so much money to the county, and is so useful to the metropolis, should be properly manufactured. The Writer of this Report feels it his duty to point out an error that he has seen sometimes committed by the inferior workmen, in preparing these smaller sized stones for the London market. This consists in dressing them too slight, and with too much taper.—The workmen are paid by the measure, not by the weight.—And if their masters, who sell by the ton, do not look sharply after them, there is some risk, that though the materials are excellent, the demand may decrease, because the manufacture is becoming too slight, owing to the stones being shaped too much like a wedge, instead of having only a very little taper.

After making these remarks, Dr. ANDERSON'S account of the Aberdeen granite, and of the method of manufacturing it, deserves to be inserted.

" Granite is the chief mineral production of this county;
" and of this it has inexhaustible stores. It is found in many
" places in quarries, but more universally it is scattered over
" the whole face of the county, in large irregular lumps,
" which

" which sometimes cover the surface of the ground, so as
" scarcely to leave the appearance of the soil. This stone,
" called in the country language *pacey whin*, affords the best
" material for building I have any where seen ; and is manag-
" ed by the masons of that country with surprising adroit-
" ness. It is so hard, as to resist the finest tempered edged
" tool; yet they know how to split it into blocks with asto-
" nishing facility, and to cut it into the size and form they
" wish for. The practice of these illiterate artists, tends to
" refute a very prevailing opinion among philosophers re-
" specting granite; for they all maintain, that, like the blue
" basalt, or whin stone, it discovers no tendency to a regular
" structure, but may be broken with equal facility in all di-
" rections. The Aberdeen masons, however, know by ex-
" perience, that this is not the case; for should they attempt
" to split the stone in any other direction than that of its natu-
" ral greet, as they call it, they never would succeed; so that
" the first thing they do, is to discover the ly of that greet,
" which they do with much facility, and with such certainty,
" that if you were to take twenty persons to examine the
" same block separately, they would all concur in pointing
" out the same direction, and proceed to cut it up in the same
" way. This stone is so hard, that no edged tools, com-
" monly used by masons, can make any impression upon it.
" When they mean to split it, they begin by drawing a
" straight line along the stone, in the direction of its greet;
" they then dig a row of little oblong groves along that line,
" by means of a weighty tool like a hammer drawn to a blunt
" point at both ends, and highly tempered at the point. This
" they call a pick ; it being of the same nature with the tool
" employed by millers for picking their mill-stones. These
" grooves are placed at the distance of a few inches only from
" each other. Into each groove they fix a wedge, the point

" of

" of which is cut over square, so as to leave a triangular ca-
" vity below it. They then strike the wedges successively
" with a very weighty hammer, one after the other, along
" the whole-line, (which makes the wedge, that is formed of
" the best steel, and hardened as much as possible, press
" upon the edges of the grooves,) which, acting with conti-
" nued and increasing force, gradually makes the stone split
" asunder; the fissure going straight to the bottom of the stone,
" in the direction of the line first marked, cleaving it in two
" parts with a fissure nearly as straight, though not so smooth
" as if it had been made with a saw. This operation is re-
" peated as often as necessary, till the whole stone be cut in-
" to slabs as thin as are wanted. They are then split, in the
" same manner into lengths, in the same direction of the
" stone, but at right angles to the former cut, exactly as logs
" are sown into battens; but only one at a time. These also
" are cut into the dimensions wanted for the purpose in hand;
" and afterwards each of these is divided across into the
" dimensions wanted at the time. In this way stones for or-
" dinary building are shaped into the form of bricks, about
" two feet in length, and one foot in other dimensions. These
" stones are called *litter* stones, because, before the roads
" were formed, they used to be carried in a litter to the buil-
" ders, and were sold at fourpence each, delivered at the
" foot of the wall. It is in this manner the long stones are
" shaped that are sent to London for the edges of the foot
" pavements; these are all one foot in breadth, nine inches
" in depth, and as long as the stones out of which they are
" formed will admit. I have seen a good many of these
" stones, nine feet in length; and I once saw about thirty of
" these stones, nearly nine feet each in length, which had
" been cut out of one block. In this way of shaping, no
" waste

"waste is made, but of the irregular pieces at the end and
"sides of the original block.

"For ordinary mason work, the stones are used without
"any farther dressing; but for the fronts of houses, and
"finer works, they are now usually smoothed, so as to form
"what they call aslar work. This is done by picking their
"surface, exactly as a miller does his mill-stones, and then
"smoothing them by a tool, in shape like a small hatchet,
"and thus reducing any little heights to the level of the
"lower cavities, which is a work of much less labour than
"could be conceived by those who have not seen it; for so
"straight and smooth are they cut, that the settled price for
"thus dressing these stones has long been sixpence the square
"foot. When thus dressed, these stones are perfectly
"straight, and truly squared; and they join in building with
"such nicety, that the point of a knife could scarcely find
"access into the joining in some cases. The stones thus dress-
"ed, assume a clear white greyish appearance, which is, to
"my eyes, the most beautiful for masonry of any stone I ever
"saw; and it has the singular quality of retaining that neat
"clean look for ages. Smoke, which sullies freestone so
"soon, makes scarcely any impression upon it, so that a
"building of it that has stood in a town for a hundred years,
"will look more clean and neat than one of the best free-
"stone that has stood only five years. Add to this, that it
"never fails:—no weather having the smallest impression u-
"pon it; and it will be allowed, I think, to be the very
"best building stone that has ever been discovered, and a-
"mong the cheapest also.

"One other excellence that results from the use of this
"stone too, and which contributes much to the neatness and
"elegance of the buildings where it is employed, is the plain-
"ness that every where prevails in regard to ornaments,
"which

" which produces a chasteness in the general stile of the archi-
" tecture, that never fails to please; or at least, the most
" fastidious critic, if he will not commend, can never find
" any thing that disgusts. This is a necessary consequence
" of the nature of the stone; for though it be very easy to
" smooth it, it becomes a matter of great difficulty to dress it
" into small mouldings; and to cut it into dentils, modillions,
" and foliage, or other carved work, is nearly impossible: for
" which reason, these taudry ornaments, so often attempted
" upon softer stone by rude artists, must here be entirely
" avoided. A few plain mouldings are all that it admits of
" at a moderate expence; and these are sufficient to give to
" the buildings a finished appearance, that is all that neatness
" can require, or that grandeur will admit of. On account
" of the great expence of polishing it, no attempt of that
" kind is made by the people at large, though it is well known
" that the stone is very susceptible of it."

Though Dr. ANDERSON's Account of the Aberdeen Granite is very full, and particular, as to the mode of dressing it, it was a piece of justice to his memory to give it entire. The species of granite in the neighbourhhod of Aberdeen, of which so much is sent to the city of London, as already mentioned, is not a perfect granite, for it has little or no feltspar in its composition. But it is impossible to look at the brick houses of London, or even the freestone buildings of Edinburgh, where the taudry ornaments, on a building that has lost its colour, look like dirty ruffles on a coarse shirt, without seeing that they are inferior in point of beauty, as well as durability, to the plain but elegant walls built of granite, which distinguish the mansions of many of the landed proprietors in the county, and most of the houses that have been built along the new streets which have been lately opened in the city of Aberdeen. Though it swells out this article, it is proper

per to add, that the parish church of Cruden, which is fit to accommodate eight hundred hearers, was built within these thirty years, out of a single large stone, or mass of granite, on which Hallow-even fires were formerly made annually on the 12th of November, by the children in the neighbourhood.

Besides granite in all parts of the county, and slates in the Garioch, quartz is found in different places, particularly in Buchan, in large masses, to the extent of several tons each; and has frequently been sent to Newcastle, to be employed in the glass-works and potteries. Asbestos, (a species of stone that is so called because it is not hurt by fire,) is found in the parish of Auchindoir, and head of the Garioch. An excellent bed of freestone also extends all the way from Auchindoir to Kildrummy. That ancient castle stands on a base of freestone, which on the opposite side of the gully, is of a quality, that both for beauty and durability, is not excelled, if it be equalled by any stone of that kind, in any part of the island. One of the towers of the Castle of Kildrumny, which was built from the stone on the opposite side of the hollow, from its remarkably white colour, was called the *Snow Tower*.

The high mountains in Braemar contain stones of a very different kind from granite, or any of the above-mentioned kinds of stone. These are precious stones of various denominations, which among the country people, are indiscriminately termed *Cairngorums*. They are now so much sought after, that a number of the inhabitants, not only of Aberdeenshire, but of the counties of Perth and Inverness, flock to these mountains, in whole families, during the summer season, in quest of gems; and purchasers from London, who are well acquainted with their value, come frequently from that city to buy the precious stones from these poor people. The profits of the finders or miners (for so they may be called) are

ex-

extremely variable: For they sometimes discover very valuable stones, which they sell at high prices; and at other times they are not so fortunate. But in their search after these jewels, they have already trenched above twenty Scotch, or twenty-five English acres; to the depth of from five to six feet: and the Writer of this Report was informed by the Rev. CHARLES McHARDY, minister of Crathie and Braemar, who has the charge of the estate of Invercauld, belonging to Mrs. FARQUHARSON, and by Mr. STEWART, factor to Earl FIFE, that these Cairngorum miners paid no quit rent, either to his Lordship, or to Mrs. FARQUHARSON, for the privilege of searching and digging in the mountains; yet they continue their occupation with a zeal and perseverance, which indicate that the employment is often very lucrative. The Reporter saw some valuable precious stones, when surveying that district, which had been found in these mountains, particularly a *beryl*, in the Cabinet at the house of Invercauld, which an eminent mineralogist estimated at a thousand pounds.

In the upper part of Cromar, and in the hills in Strath-Dee, belonging to the Earl of ABOYNE, and to Mr. FARQUHARSON of Monaltry, there are several indications of lead ore; and several specimens have been found containing a considerable proportion of that metal; but it has not as yet been ascertained whether *the quantity* was so considerable, as to warrant the proprietors to expend any considerable sum of money in a pursuit, for which they are not qualified. Perhaps an offer to a company of miners, giving free access to these hills, for 20 or 25 years, might be attended with advantage to the proprietors at the end of that period.

These are the principal minerals belonging to this county, that deserve to be mentioned.

SECT.

SECT. VI.—WATER.

The county of Aberdeen derives many advantages from having nearly 60 miles of sea coast; where, besides the harbour of Aberdeen, which has been greatly improved within these forty years, and will now, probably, soon be much farther improved, those of Peterhead, Fraserburgh, Newburgh, and other places not known in general commerce, are useful in importing lime and coals to all the maritime districts.

Independently of the advantages, which result from its harbours and sea coast, (the latter of which is very productive of various kinds of fish for the use of the inhabitants) the two principal rivers, the Dee and Don, with their subordinate rivulets, and also the small rivers, the Ythan and Ugie, within the county, and the Deveron and Boggie on its N. and N. W. boundaries, have all been found beneficial in different respects. The Dee and Don in a particular manner, are valuable for their salmon fishings, in which respect the Ythan and Ugie have also been of some use, though their streams are smaller: And, with the exception of the Dee, all of them very useful in turning corn-mills, or the water-wheels of other machinery, near Aberdeen, and in different places thro' the county.

Of all these rivers, the Dee is by far the largest. It takes its rise on *the north side of Cairntowl*, one of the highest mountains in Scotland, and nearly in the centre of North Britain. It runs at first for *three or four* miles in a direction nearly south, through *Glen-garchary*, from which it derives the name of the *burn*, or brook of *Garchary*, at the lower extremity of which it is joined by another small stream, called the Guisachan, when the rivulet first assumes the name of *Dee*. This united stream continues to run in the same direction, nearly south for six miles farther, when it is joined by the Geauly,

(in

(in General Roy's map, improperly termed *Galdie*) which flows from the mountain of Cairn-eiler, at the vertex of the county: but is a shorter and much less considerable stream. After the union of these three rivulets, the Dee becomes a considerable highland river, and flows, not in its original direction (nearly south) but, as the openings between the mountains allow it to run, in the direction of its smaller branch, the Geauly, nearly east, with little variation for about sixty miles, till it reach the sea at Aberdeen. The whole course of the river, *following all its windings*, from Cairntowl, is about 80 or 81 miles; and an *ideal straight line*, drawn from its source to its mouth, or influx into the ocean, would be about 65 miles.

For the *first ten* miles there is no human habitation near the banks of the Dee, but the small farm of the Doubrach is situated at its confluence with the Geauly; and is perhaps the highest inhabited land not only in Aberdeenshire, but in Great Britain. About seven miles below this farm, or seventeen from its source, the river is suddenly contracted by two rocks (one on each side of it) at a place called the *Linn*, so that a person can easily jump across it; and a small plank of wood is laid as a bridge for the timorous. The fall of the current is at this place very considerable; and it has been observed by the inhabitants, that no salmon pass through this rapid and narrow strait.

The river continues to flow through the county of Aberdeen for nearly 24 miles farther, till it enter the county of Kincardine, (or rather to that part of the ancient division of Marr, which formerly belonged to the thanedom of Durris, and was by King Robert Bruce joined to Kincardineshire) through which it runs for about 9 miles, through the parish of Upper Banchory, when it becomes a boundary between the counties of Aberdeen on the North, and Kincardine on the

the South, for about 15 miles, till it enter the German Ocean about a mile below Aberdeen.

Perhaps there is no place in the island in which the air is more pure, the scenery more picturesque, than on the banks of the Dee, particularly in the higher districts. Birch, aller, and oak, are found scattered near its banks. And the Scotch fir, a species of pine that is very difficult to raise in the lower parts of the county, here grows spontaneously in great abundance. It is only necessary to inclose a field: the winds or the crows will supply it with seed in a few years.

During the whole of its course it is distinguished by its rapidity, by its broad and capacious channel, and by the limpid clearness of its waters. Its source is elevated more than 4 thousand feet above the level of the sea; and the fall of the river is pretty uniformly rapid. From the number of its mountain torrents, and the great quantity of water poured from them in a rainy season, the channel is larger than appears to be necessary at other times, when the river is not swoln by floods; and it is not confined by rocks on its banks, except at the *Linn of Dee* above mentioned, and a narrow strait within seven miles of Aberdeen, called the *Throat* of Kingcausie. Yet from the quantity of water in the Dee, and from the rapidity of its current, its channel is deeper, and its meadows or haughs are narrower than those of the Don. From the purity of the gravel on its banks, and the small proportion of peat moss near this river, or its tributary streams, the waters of the Dee are more limpid than those of any river in Scotland, the Tweed perhaps excepted.

The Dee conveys to the ocean the waters of about 900 square miles. For it has as much surface in Kincardineshire, in the parishes of Strachan, Upper Banchory, Durris, Maryculter, Nether Banchory, and Nigg, whose superabundant waters decline into the Dee, as it wants in that part of the division

vision of Marr, where the hills decline and run into the Don. Consequently, the Dee holds the same proportion to the other rivers that flow within the county, that the division of Marr bears to the other divisions of Aberdeenshire, or is nearly equal to all the rest united. Yet from the rapid fall of its waters, it is navigable only about a mile from its mouth, to the city of Aberdeen, except in stream tides, when it is navigable for nearly three miles. It is peculiar to this river, that it does not turn the wheels of any corn mill, and very little of any other machinery, within thirty miles of its entrance into sea.

The Don is next to the Dee, in point both of magnitude of current, and length of course. It rises in the mountains, between Aberdeen and Banffshire, about three miles above Curgarf. It runs through that district in a pretty rapid course for nine or ten miles before it receive any considerable stream; and then flowing through the strath or valley of Strathdon, receives the tribute of a number of rivulets, provincially termed Waters, viz. the Ernan, the Conry, the Nochty, the Deskry, the Bucket, and the Kindy, in the district which is called *Strathdon*; and which is about ten miles in length, and from 500 yards to half an English mile in breadth. Its banks then press closer on the river, for nearly other ten miles, where the district is termed *Donside*, and where the river is augmented by the water of Leochel, and a few inferior streamlets. At the termination of this course, it passes through the fertile district of Kildrummy, and within a few hundred yards of the ancient Castle of that name.— It next enters the rich valley of Alford, passing through the estates of Breda, antiently *broad-haugh*, or *broad-meadow*; Whitehaugh and Haughton—all names derived from their situation in this fertile vale; which is about 5 English miles long. Being afterwards for four or five miles hemmed in by

the

the mountains, it reaches the plains of Monymusk, where rich fields in the valley are skirted by extensive woods on the hills, which recede at a considerable distance from the river. Passing through this district, and the adjacent parish of Kemnay, (where stells are set in the river for catching salmon,) it runs along the boundary of the parish of Inverury, to its confluence with the river Ury, about ten miles below Monymusk, and sixteen from its influx into the sea. Immediately below an elegant bridge over the Don, a portion of the waters of this river is cut off, and carried away in a pretty deep channel, in order to feed the navigable Canal, which is carried 18 miles, following its windings, to the harbour of Aberdeen.

Before the confluence of the Don with the Ury, the former runs nearly in an eastern direction from Curgarf; and with considerable rapidity, a few places excepted, during its whole course. But as the Dee, after its union with the Geauly, runs in the direction of the inferior stream; so the Don, after its junction with the Ury, runs nearly in the line of the latter river, or in a south-east course, though with many windings, in its way to the ocean. The country is so flat, and the meadows or haughs are so rich and extensive, that though the river, at the head of the Canal, is nearly 168 feet above the level of the sea, yet for the first half of its course from Inverury, the united streams of the Don and Ury do not fall above 40 feet; and 14 of 17 locks of the Canal are placed within the last two miles of the harbour of Aberdeen. Owing to this very small declivity, the floods of this river are found to be very destructive to the crops that are raised on the rich meadows, which ly on both sides of the river; tho' the raising of embankments has of late tended to lessen that evil, by preserving the fields which have by this means been separated from the river. For the last eight miles of its course, sides, the Don runs with great rapidity, for four miles indeed more rapidly than the Dee. But on the whole, the difference between the length of the river, following all its turnings and windings, and amounting to 61 miles, and an ideal straight line from its source to its mouth, (which is only 42 miles,) shews that this river flows much slower than the Dee. Its waters too, though they are pretty clear, have not the pure limpid colour of that river; both because there is a greater proportion of peat-moss, whose waters flow into the Don, or its tributary streams, and because the richer soil on their banks contains less gravel than the Dee. There is no comparison, however, between these two rivers, with respect to the fertility of the soil in their neighbourhood. Hence the proverbial rhyme of the country people, in which the soil near the two rivers is contrasted:

"A foot of Don's worth two of Dee,
"Except it be for fish and tree."

The Ury takes its rise in the division of Strathbogie; and after a course of about 24 miles, following its windings, or in an *ideal* straight line of 16 miles, during which it receives the Gady, the Shevock, and the Lochter, it falls, as abovementioned, into the Don, at Inverury. It flows through the greater part of the rich valley of the Garioch, in a direction nearly south-east, though with many beautiful windings; and its companion, the Gady, and the inferior streamlets, the Shevock and Lochter, turn a number of corn-mills in this valuable division of the county. No salmon frequent the Ury, except occasionally in the spawning season.

On the united streams of the Don and Ury, with their subordinate rivulets, much more corn is raised than on the Dee, with all the streams which fall into that river. Yet while the waters of the Dee are collected from a surface of nearly 900 square

square miles, those of the Don, Ury, and their inferior rivulets, (which are supplied with spring or rain water, in the north-west division of Marr, in the south-east parts of Strathboggie, and in the whole of the valley of Garioch, and its bounding hills,) do not belong to a surface of more than 400, perhaps only to 360 square miles. So much more fertile, or more generally cultivated are the lands connected with the Don, than those which belong to the Dee. The Don also, though inferior in size, is not so much inferior in this respect as might be supposed. For the Dee, from its gravelly bed, is believed to lose a considerable proportion of its waters from percolation or filtration; and during the last 18 miles of its course, is not supposed to increase in magnitude. The value of the salmon fishings on the two rivers is not in proportion to the size of their streams. That on the Dee, valued at the rate at which the fishings out of lease have been lately rented, amounts to nearly twenty thousand pounds; while that on the Don, at the same rate, would, if rented, amount to fifteen thousand pounds annually. But the comparative value of the salmon fishings on the Don is increased by their being occasion to employ but few fishers. For there are only two fishings on the river, one near its mouth, and another about two miles from the sea, where cruives have been for a long time erected. These prevent the salmon from getting up, except when the river is flooded. The proprietors of land above these cruives have a right to get them open on Sunday; but by a temporary arrangement, they at present accept of a stipulated annual payment from the owners of the cruives, by which means these are shut during the whole season of fishing.

The Ythan is a much smaller stream than either of these, and in the lower part of its course, runs very slow; but in the higher parts of the county, it is pretty rapid. It rises
near

near the borders of the division of Strathboggie, within a few miles of the Ury, and in the upper part of Formartin. It flows at first in a direction nearly east, through the parishes of Auchterless and Turriff, for about twelve miles. It then receives the *little water* of Gight, and flows in the direction of that stream nearly other twelve miles, in a south-easterly course. Its whole length, following its various windings, is about 30 miles. An excellent scalp of mussels, which are never hurtful, and a few cockles, occasionally are found near the mouth of this river; also a considerable number of pearl mussels have also been found in its lower extremity. One of the jewels of the ancient crown of Scotland, a valuable pearl, was said to have been found by a fisher of the name of Jamieson, and was called by his name. And about 60 years ago, one Mr. Tower, a merchant in Aberdeen, got at one time an hundred pounds* for a quantity of pearls, which were taken out of the mussels that were found in the Ythan. The small town of Newburgh, formerly a fishing village, but now deserted by the fishers, in consequence of the many accidents that happened from the shifting of its bar, is situated near the mouth of this river; and there is a valuable salmon fishing, partly in the river, and partly on the sea coast adjoining.

* A fortunate misunderstanding of terms (which has not always the same happy effect) occasioned Mr. Tower to get the full value of his pearls. He asked from a Jeweller in London, an hundred pounds, as the price of them, meaning only *Scotch money*, or L.8 6s. 8d. sterling.—The Jeweller offered him 80 pounds, which he declined taking, declaring that he paid that sum to the fishers of Ythan, from whom he bought them. The Jeweller replied that they were dear, but that they were excellent pearls, and laid the L.100 on his counter. Mr. Tower saw that he had got *English money*, which he pocketed, concealing his ignorance; but he afterwards knew what price to ask for his pearls.

joining. Only ships of small burden can enter into this harbour, except at stream tides.

The Ugie is composed of two branches, termed North and South Ugie. The former rises in the parish of New Deer, in the inland part of Buchan; and the latter in the parish of Tyrie, nearer to the sea-coast. After traversing a considerable part of this division, and turning the wheels of a number of cornmills, in a fertile district, they unite about six miles from the German Ocean; and at the mouth of the river, thus united, a salmon fishing is rented for a few hundred pounds annually. The whole course of either of the branches to the sea, is rather below 20 miles.

The Boggie is a rapid rivulet, which rises in the lower part of the hills of Cabrach, and runs through that division of the county called from it Strathboggie. After a course of 16 miles, following its windings, or 12 in a straight line, it falls into the Deveron, near Huntly. (Both these rivers will probably be noticed in the Report of the County of Banff.) The Deveron itself, though it is for several miles a bounding river, takes its rise in the upper part of the parish of Cabrach; and after a course of 48 miles, following its windings, and 33 miles in a straight line, it falls into the sea near Banff. It receives the Boggie about twenty miles from its source, near Huntly, and the Islay, about 5 miles from Huntly, at Rothiemay. Near its source it is for 15 miles wholly in the county of Aberdeen, and for two miles next the sea it is wholly in Banffshire. The Islay, which takes its rise from a small lake called Loch Park, near the boundaries betwixt Mortlach and Botriphnie, after a course of 14 miles, falls into the Deveron, at Rothiemay, as above-mentioned. When these rivers are united, the Deveron is about two-thirds of the size of the Don; and the waters, which are collected from a surface of near 300 square miles, are emptied into the German

man Ocean a little to the east of Banff. There is a most valuable salmon fishing on the Deveron, but it is situated in the county of Banff. The water of all these three is less limpid than that of the Don, owing to the greater quantity of peatmoss, which is situated near their banks. But it may be noticed, that the banks of the Deveron have been, like those of the Don, celebrated for their fertility. Hence the old uncouth rhyme, which compares the rivers of the north of Scotland, in respect to the produce of the soil upon their banks.

" Don and Davorn for grass and corn—
" Spey and Dee for fish and tree."

Besides these rivers, which run into the German Ocean, either within the county, or on its borders, there is a small stream in Buchan, now called the *Loch of Strabeg*, nearly 10 miles north of Peterhead; which about 160 years ago had its entrance into the sea entirely choked up by a violent easterly storm, and has continued shut up ever since. In consequence of its being thus pent up by the beach, a lake of considerable magnitude has been formed. It now covers about 550 Scotch, or 690 English acres, or above a square mile; but it has not increased in magnitude, except occasionally, for these last 40 years. The evaporation from its surface is now nearly equal to the quantity of water that flows into the lake; and when the stream is swelled by the land floods, it probably oozes, or filtres through the sand, and thus finds an entrance into the ocean.

This singular lake was formerly called the *burn*, or *water* of Rattray, and is the *ratra amnis* of Buchanan. About ten years ago, Mr. WILLIAM SELLAR, a most ingenious man, attempted to drain it, on the encouragement of a long lease, free of rent, of all the land which he could reclaim; but after

ter a considerable expence was laid out in the attempt, his death, which was a public loss to the country, put an end to the work; and it has not since been resumed.

There are several lakes, of various dimensions, in different places of the county, viz.

1. Loch Muick, in the parish of Glenmuick, is two miles long by two broad, and contains 1500 Scotch, or near 1900 English acres.

2. Loch Kanders, in Marr, is three English miles long, and three-fourths of a mile broad. It contains about 700 Scotch, or 880 English acres, besides a small island in the middle, in which King Malcolm Kenmure had a castle.

3. Loch Builg, two miles, by three-fourths in breadth, or 700 Scotch, or 880 English acres. The same dimensions with Loch Kanders.

4. Loch Callader, one and one-half mile, by three-fourths, or 600 Scotch, or 760 English acres.

5. Loch Candlaten.

6. Loch Vrittachan, and

7. Loch Dawin—small lakes, from 50 to 70 acres, in the head of Marr.

Besides these, near the vertex of the county, Loch Le Nean, in which there are excellent char.

In the lower parts of Marr, are the Lochs of Leys and Drum, about 300 acres each.

All these lakes are stored with fish of various kinds. *Pike, trout,* and *eels,* are the most prevalent, and *char* in a few of them, particularly Loch Le Nean.

Within 10 miles of Aberdeen, is *the Loch of Skene,* belonging to GEORGE SKENE, Esq. of Skene. It contains nearly 500 acres, and abounds in all the above kinds of fish. But the most remarkable thing about it is, that three rivulets, or burns enter into it, one from the east, one from the north, and

and a third from the west. The only declivity from it is a valley in the south, whence its superabundant waters are carried into the Dee, in a southerly direction: and it is also to be remarked, that the burn of the Ord runs into it about two miles after it leaves the lake from the east, and the burn of Gormac, about half a mile farther, falls into it from the west. After this union, it is called the burn of Culter, and falls into the Dee about 7 miles from its exit from the lake, and 8 from Aberdeen.

In the division of Buchan, the Loch of Slains, containing 54 acres, is remarkable for its depth, which varies from 25 to 50 feet, and for containing, besides pike and eels, a number of excellent perches, which are not known ever to have been carried to it.

In the division of Formartin, within 6 miles of Aberdeen, is a small lake called the Bishop's Loch, in which the Bishop of Aberdeen had his summer residence. One cannot suppose that *he* had need to reside there as a place of safety;— yet it had a bridge, which, like those round the moats of the ancient castles, could be drawn up at pleasure.

There are also several pieces of water, which have been artificially made. One of these is at Fyvie Castle, the property of the Honourable General GORDON of Fyvie. One at Delgaty, now belonging to the Trustees of the late Earl FIFE. One at Pitfour, of about 13 acres, belonging to JAMES FERGUSON, Esq. Representative of the county in this and five preceding Parliaments, which is stored with *tench, perch*, and *carp*. And another, at St. Fergus, belonging to the same Gentleman, of above 30 acres, to which the same variety of fishes has been carried, and which was made out for a more public spirited purpose, namely, as a reservoir to feed a canal, 6 miles in length, which Mr. FERGUSON has made out

for

for carrying the shelly sand from the sea-coast to the interior parts of his estate.

There is also a small piece of water made out, near Auchry, by the late spirited improver, Mr. CUMMINE, who drained his lands in the most effectual manner, and united ornament with utility.

Of MINERAL WATERS, or SPRINGS.—There is a most powerful chalybeate mineral well at Peterhead, which has been much resorted to. It has been analyzed by the Rev. Dr. LAING. There is also a chalybeate spring at Frasersburgh; and at both places both hot and cold baths.—At Aberdeen also there are two springs, one a chalybeate, which is called the Well of Spaw; and another near Old Aberdeen, containing a small proportion of sulphur, along with a greater of steel.—And in the upper part of Mayr, near Ballater, the wells of Pananich, another chalybeate, was fitted up by the late worthy Mr. FARQUHARSON of Monaltery, and is much resorted to by the country people, and by several persons of the middle ranks of life.

CHAPTER II.

STATE OF PROPERTY.

SECT. I.—ESTATES, AND THEIR MANAGEMENT.

IN order to give both a just and a comprehensive view of this subject, it is necessary to consider,

1st. The antient state of property in this county, and other circumstances connected therewith, before the whole lands of Scotland were valued in the reign of K. Charles II.

2d. The valued rents of the different landed estates, according to that valuation.

3d. The proportion between that valuation, and the present rental of the lands of this county, along with other circumstances connected with the value of landed property.

I.—With respect to the first of these, the antient state of property, before the lands were valued, in the reign of K. Charles II. It is obvious, that in a rude state of society, in which agriculture was little known, there would be no very correct method of valuing land. Accordingly we find, that no other criterion was fixed of the value of arable land, but merely its measure or extent, without the least regard to its *quality*. The antient denominations of this extent, it also deserves to be remarked, were not either decimal or octaval, but founded in tredecimal ratio, or were half, double, quadruple, &c. of thirteen. They were as follows:—

1 oxgate—13 acres, or *bolls sowing*.

4 ox-

4 oxgates, or 52 acres—1 pound land.

8 oxgates, or 104 acres—a forty shilling land.

32 oxgates, or 416 acres—1 dauch, or davoch, *davata terræ.*

There was a still smaller denomination, viz. *an husband land.* This was only half an oxgate, or 6 acres; the half acre being allowed for the house, garden, and loaning, which are commonly called the Toft.

The principle upon which our ancestors adopted these denominations, appears to have been the following:—Every proprietor of an hundred and four acres, i. e. every man who could plough an acre, or a boll's sowing, for every one of the *52 weeks in the year,* and had as much lying in natural grass, for his cattle in summer, was entitled, if the land was his own, to vote for a member of Parliament, and his land was valued at 40 shillings, which anciently weighed two pounds of silver. Every man who had only half this quantity, was supposed, whatever was the quality of his land, to have half as much as a plough of 8 oxen required for summer and winter food, and his property was called a *pound land, librata terræ,* of old extent. At that period, it must be remembered, the Scots pound was a pound of silver, or equal to three pounds sterling.

The total valuation of the county of Aberdeen was, of the old extent, as stated by that eminent antiquarian, Mr. GEO. CHALMERS, in his Caledonia, L.4448 6s. If the measure of *the pound land* had been exactly 52 Scots acres, the whole county would have contained nearly 231,300 acres of arable land. That measure, however, was very indeterminate.— The oxgate, which was the unit of these ancient properties, though reputed to be 13 *acres,* or *bolls sowing,* would be sometimes (as we find it is at present) considerably more, and sometimes less than that quantity: Consequently the pound land,

land, or 4 oxgates, would not in many cases correspond with the extent of 52 acres; although as a general average, it might not be far from the truth.

This sum of L.4448 6s. being the amount of the total extent of the lands of Aberdeenshire, included the actual rents or feu-duties of those estates in this county, which were the *personal property* of the king, as well as what belonged to those proprietors *who held of the crown.** By giving away several of these estates, the above sum was reduced nearly one-half, or to L.2588 5s. 2d. of *new extent*. At the same time, the Scots Judges, by different decisions, that were at variance with one another, found *one* pound of *old* to be equal to *four* pounds of *new* extent, (11th March, 1585) estimated an *oxengate* of land at 20 shillings in all public duties (18th July, 1541); and as land in the vicinity of royal burghs was more valuable than in the remote parts of the kingdom, they valued a husband land, which was only half an oxengate (including the toft that belonged to it,) in one case at L.3 Scots, in another case at 5 merks Scots, (1st December, 1545) and in the case of a field lying beside the burgh of Dunbar, (and probably of excellent quality) they valued it at 14 bolls of bear,

---

* Although the above account of the *old* and *new* extent has been correctly given by Mr. CHALMERS, yet by the oldest tax-roll to be found in the records of the county of Aberdeen, dated May 10th, 1554, the whole lands belonging to the laity are rated at only L.1372; those belonging to the clergy at L.321 6s. 8d.; and the Queen's (Mary) own lands at L.833 16s. The total sum was only L.2527 2s. 8d. And after the Reformation, as the lands belonging to the church no longer paid the old taxes to government, were dilapidated, or given away, the whole taxable property fell, in the time of K. James VI. to L.1821 9s. While the old extent was the rule of taxation, the county of Aberdeen paid one-eighth part; and immediately before the Reformation, it paid one-ninth part of the land taxes occasionally imposed on Scotland.

bear, and bolls or wheat. Nothing could be more unequal than the valuation of land only in money, when the corn was rapidly decreasing to one thirty-sixth part of the original weight of silver in a pound of Scots money. *The extent of land*, instead of *denoting its value*, became at last only the *rate of taxation;* and in fact, the very unequal or unfair proportion, according to which the land tax was imposed.—In short, from the decrease in the value of a pound Scots,—from the different qualities of soil,—from the lands being cultivated or neglected,—as well as from the misfortunes of the nation, and from the improvident gifts of the Sovereign,—a new valuation became a measure absolutely necessary.

Accordingly in 1674, all the lands in Scotland (with the exception of those that were devoted to charitable purposes) were valued by commissioners, who acted under the authority of the Scots Parliament. The whole valued rent of the kingdom, according to that valuation, amounted to L.3,802,563 3s. 10d. Scots money, or L.316,880 5s. 3d. 16-12ths sterling.

It may be satisfactory to the reader, to have a general view of the different *taxable* values of all the lands of the kingdom, according to what was termed the *Old Extent* and the *New Extent;* and also according to the valuation of 1674, which is the rule for proportioning the yearly land-tax that is paid by the different counties. It is contained in the following page, in 4 columns—the first includes the name of the county; the 2d, the sum of *old extent*; the 3d, that of the *new extent;* and the 4th, the amount of the valuation in 1674.

NOTE.—The ancient divisions of counties were different from the present. Hence we have only from Mr. CHALMERS the Old Extent of 20, and the New Extent of 21 counties.

ABSTRACT

ABSTRACT OF THE OLD AND NEW EXTENTS, AND ALSO OF THE VALUED RENT OF THE DIFFERENT COUNTIES OF SCOTLAND.

COUNTY.	OLD EXTENT. Scotch Money.			NEW EXPENT. Scotch Money.			VALUATION, 1674 Scotch Money.		
	L.	s.	D.	L.	s.	D.	L.	s.	D.
Aberdeen	4448	6	0	2588	5	2	235,665	8	11
Ayr	3358	19	13	1396	16	2	121,605	0	7
Argyle	.	.	.	.	.	.	12,446	5	10
Berwick	622	2	4	372	17	3	178,366	8	6¼
Banff	1510	6	0	128	16	8	79,200	0	0
Bute	.	.	.	.	.	.	15,042	13	10
Caithness	.	.	.	.	.	.	37,256	2	10
Sutherland	.	.	.	.	.	.	26,093	9	9
Dumfries	2666	13	4	882	15	4	158,687	4	6
Dumbarton	1442	9	6	96	9	6	33,326	19	0
Edinburgh	4029	16	10	3030	12	9	291,054	2	9
Elgin	.	.	.	.	.	.	65,603	0	8
Nairn	.	.	.	.	.	.	15,162	10	11½
Fife	3465	13	4	2555	0	0	362,584	7	5
Kinross	65	0	0	38	14	8	30,250	3	4
Forfar	3370	6	8	2240	6	8	171,239	16	8
Haddington	.	.	.	.	.	.	168,873	10	8
Inverness	3104	11	8	1030	11	9	73,188	9	0
Kircudbr.	.	.	.	.	.	.	114,597	2	3
Kincardine	1088	10	8	722	0	0	74,921	1	4
Lanark	4052	9	0	1755	19	8	162,131	14	6
Linlithgow	.	.	.	.	.	.	75,018	10	6
Orkney	.	.	.	.	.	.	56,551	9	1
Peebles	1274	18	6	863	13	4	51,937	13	10
Perth	6192	2	6	3067	1	7	339,892	6	1
Renfrew	.	.	.	585	9	8	69,172	1	0
Roxburgh	1133	15	0	523	17	0	314,663	6	4
Ross	.	.	.	.	.	.	75,043	10	3
Cromarty	.	.	.	.	.	.	12,887	2	7
Selkirk	99	9	10	80	18	6	80,307	15	6
Stirling	1749	19	4	687	3	10	708,509	3	3¼
Clackmannan	331	9	8	243	14	8	26,482	10	10
Wigton	1235	3	4	195	0	2	67,641	17	0
TOTAL,	23635	17	6	12624	11	2	3802,563	3	10
							Sterling.		
							316,880	5	3

By the above valuation, it appears that Aberdeenshire is

es-

estimated at one sixteenth part of the valued rent of Scotland, or pays the land-tax, and other public burdens, in that proportion. And it may just be noted, that three small properties, of which the rents are applied to charitable purposes, are exempted from *paying land-tax.* Their whole valuation, according to which they pay the *county rates,* is only L.416 Scots, or L.34 13s. 4d. sterling.

The valuation of 1674 was not a *correct* account of the rents of the lands even at that period. It was, in general, *below the truth.* Those who possessed great estates, and who knew that this valuation was to be the ratio, according to which the land-tax was to be imposed, gave in statements of their rental, as low as they decently could; and the small proprietors, who either were not in the secret, as to the danger of this valuation, or who, from motives of vanity, wished to be thought men of property, valued their lands pretty high in 1674. But the vicinity of many landed estates to the sea-coast, and the unexampled exertions which, within these 40 years, have been made in improving the lands near Aberdeen, have also occasioned a much greater rise in their present rents, compared with that valuation, than what has taken place in the more inland districts; so that there is a much greater disproportion between the valued and the real rents in the maritime, than in the interior parts of the county. The Writer of this Report knows a small landed estate, of which the present rack-rent is not double of the valuation of 1674: and he is acquainted with two other small estates near Aberdeen, the rent of which are, the first a hundred, and the second, sixty times as much as the old valuation. It may here be incidentally remarked, that the valuation of lands, like the valuation of tythes, having subsisted for more than a century, ought not to be rashly altered, as it would be extremely hard to tax a man according to the present rent of lands, on which he has expended ten times the sum which they cost him within these last 30 or 40 years, while another man is by this means to be relieved from a proportion of the land-tax, on an estate which he and his ancestors possessed for *three* or *four* centuries, and on the improvement of which neither he nor they ever expended a shilling. *Property is the creature of law*, and ought to be *protected*, not *injured* by the laws of the country; which, a new valuation would do.— There are many small properties near Aberdeen, which the spirited owners have improved with *borrowed money*; and where these enterprising men would be ruined by a new valuation, if ample deduction was not made for the trouble and expence of making not *two*, but *two hundred* blades of grass grow where one grew before. Besides, it ought to be considered, that the valued rents of estates at present, both convey privileges, and denote obligations, well understood both by the sellers and the purchasers; and which, in the actual sale of lands, mutually balance each other. The Writer of this Report knows a case, in which the proprietor of an estate, the rents of which were originally destined to a charitable purpose, and which therefore was not valued by the Commissioners in 1674, would be well pleased to have his lands valued, and subject to pay the land tax, in order that by having *the superiority of them*, he might be entitled to be a Commissioner of Supply, or have a whole, or part of a freehold qualification, in voting for a member of Parliament. He does not say in what county this estate is situated: but he states it as a fact consistent with his knowledge; and for all the above reasons, he is doubtful whether a new valuation of all the lands in the kingdom should take place. He mentions this, however, as only matter of opinion.

In stating the different proportions in which property is distributed among the different landholders of this county,

F the

the valued rent of 1674 is the only standard, to which, in all cases, he could find access: and he has obtained, by permission of the Commissioners of Supply, from HARY and HENRY LUMSDENS, Esqs. (Collectors of Cess in Aberdeenshire) the following authentic document of the valuation of the different estates belonging to the landed proprietors, according to which they pay the land-tax, and other public burdens.

"The total valued rent of the county of Aberdeen, as the
"same is instructed by the Cess Books, is
Scots money, L.235,665 8 11
"Besides the above, there are three estates
"in Aberdeenshire, not valued at all, as
"they were originally destined for pious
"uses, and therefore they are exempted
"from paying any land-tax. But it has
"been the practice ever since the nation-
"al valuation in 1674, to charge them
"for certain county rates, viz.—Rogue
"Money, or the expence incurred for ap-
"prehending and punishing public de-
"linquents; and more lately for the Mi-
"litia and Bridewell Assessments; and
"the sums at which these three estates
"have been valued for this purpose, is   416 0 0
"There are also several lands, locally situ-
"ated in this county, (which pay the
"Land-tax in Banffshire, and which are
"also liable in payment of the assessment
"for the expences of Bridewell, and of
"the Militia,) which are valued at   5850 0 0
"Total valuation of the lands which are si-
"tuated in the county of Aberdeen, and
"which pay any rates to the Collectors
"of Cess for said county,   L.241,931 8 11
"This

"This valuation is divided among the proprietors of 228 landed estates, in the following proportions:—

"*Three* are valued above 10,000 pounds Scots each, and amount to - - - - - - - - L.39,510 8 1⁷⁄₁₆
"*Six,* from L.4000 to L.10,000 each,   31,969 16 9
"*Six,* from L.3000 to L.4000 each, - 20,995 12 4
"*Sixteen,* from L.2000 to L.3000 each, - 39,136 18 9¼
"*Nine,* from L.1600 to L.2000 each, - 16,520 3 7
"*Sixteen,* from L.1200 to L.1600 each, - 22,240 1 10¼
"*Eleven,* from L.1000 to L.1200 each, - 12,249 18 8
"*Nineteen,* from L.800 to L.1000 each, - 17,058 15 0
"*Fifty-one,* from L.400 to L.800 each, - 27,278 11 1
"*Nineteen,* from L.300 to L.400 each, - 6,202 15 3
"*Thirteen,* from L.200 to L.300 each, - 3,267 9 7
"*Thirty,* from L.100 to L.200 each, - 4,141 5 9⁴⁄₁₆
"*Twenty-nine,* below L.100 each, - - 1,559 11 5³⁄₁₆

    "Sum of the above valued rents, L.241,931 8 3¹⁄₁₆
    "Lost by splitting valuations, - 0 0 7¹²⁄₁₆

    "Total valuation, as above stated, L.241,931 8 11

"Of this sum, the valuation of certain lands which are legally in Banffshire, though locally situated in different parts of the county of Aberdeen, is L.5850.—These pay the land-tax in Banff, and all the other county rates in Aberdeen, as above mentioned."

From the above authentic document, which, with other valuable information, was given in the most liberal manner by the Messrs. LUMSDENS, it appears,

  1st. That nearly *one-sixth part* of the valuation of the county belongs to *three* proprietors, who possess lands extending to above L.10,000 Scots money each.

  2d. That *two-seventh parts* belong to the *nine* land-owners whose estates exceed L.4000 each of valued rent.

                              3d. That

3d. That nearly *two-fifths* of this valuation belongs to *fifteen* proprietors, whose estates are valued above L.3000 each.

4th. That above *one-half* of this valuation belongs to *thirty-one* landed proprietors, each of whose estates exceeds L.2000 of valued rent.

5th. That nearly *three-fourths* of this valuation belongs to *sixty seven* proprietors, each of whose estates exceeds L.1000 of valued rent.

6th. That nearly *fifteen-sixteenths* of this valuation belongs to *one hundred and thirty-six* landed proprietors, each of whose estates exceed L.400 Scots, which is the amount of valued rent that entitles the proprietor, whose lands hold of the crown, to be a *freeholder of the county*, that is to elect, or be elected a member of Parliament.

7th. That the whole valuation belonging to *sixty-two* persons, who are entitled to be *Commissioners of Supply*, though not to be Freeholders, or whose valued rents are below L.400, but exceed L.100 Scots, is only about *one-eighteenth* part of the whole county.

And lastly, That the valuation of the lands belonging to the 29 small proprietors, or portioners, whose valued rent is below L.100 Scots, amounts only to the *hundred and forty-seventh part* of the whole lands of Aberdeenshire.

There is, however, another species of small proprietors, viz. those who hold their lands by burgage tenure. Their small properties are not valued in the county-books, but in the different royal burghs. There are about two hundred of these in the vicinity of Aberdeen, (besides the proprietors of houses and gardens in the city) whose rents for their burgh roods, as they are termed, or other burgage lands, are valued at L.7000, but are supposed to be worth L.14,000 sterling. And the present rents of both the lands belonging to the burghs, and of the houses in Aberdeen, may be estimated at nearly

nearly L.50,000.* Of that sum, the lands belonging to the royal burghs of Kintore and Inverury, will not exceed L.2000. There are above 200 persons who have property in the burgh lands of the city of Aberdeen; and nearly 100 proprietors of burgage lands in Kintore and Inverury. But none of these have any political authority in these burghs, except they be chosen members of the council.

On the other hand, besides the proprietors who pay the land-tax, and other public burdens in the county, there are 25 persons on the roll of Freeholders, from their having the *superiority* of lands, each of which is at least L.400 of valued rent, and which entitle them to vote in the election of a member of Parliament for the county. Some years ago there were too many electors, whose titles were nominal and fictitious; and it is to the honour of the gentlemen of Aberdeenshire, that the question of nominal and fictitious votes was brought by them before the Supreme Civil Court, who pronounced a decision, which tended both to expose and to check some gross abuses in the practice of county elections, without infringing on the rights of those who are possessed of real superiorities.

A matter of opinion must here again be introduced by the Reporter. Though the qualification of an elector is in general the possession or the superiority of an estate, valued at L.400 Scots, yet an estate which *stands retoured as a forty shilling land,* according to the *old extent,* whatever be its valued rent, affords a *freehold* qualification, in consequence of the *ancient laws and practice of Scotland.* Now, as the property of 104 Scots acres of arable land was anciently accounted a forty shilling land, and gave the owner a title of voting for a member of Parliament, it would save much trouble in splitting of valuations, and making up of titles as freeholders, that this old rule, which still holds in a few particular cases, were

were again made general. It would prevent equivocations, and to say nothing harsh, all supposed approximations to perjury, in questions about nominal and fictitious votes, if every man, who was a proprietor, holding of the crown, and who paid all public taxes, as proprietor of 104 *Scotch acres of arable land*, were entitled to vote for a member of Parliament, in the county elections. This would induce many to cultivate their *barren acres*, and would at once promote the interests of improved agriculture, of real property, and of rational liberty.

Of the different ranks of proprietors of land in this extensive county, the estates belonging to several of the noblemen, and many of the gentlemen, have been possessed by their families for several centuries, some of them nearly 600 years, or since the days of K. ROBERT BRUCE.* Perhaps one-fourth only has been purchased since the Union of the two kingdoms, by persons who had made fortunes by commerce, or nearly as much by those who were formerly proprietors in the county. Probably the same proportion is possessed under a strict entail, which, though useful to particular families, is certainly hurtful to the nation at large, as the law stands at present. The particular and very great injury arising to agriculture from entail is, that the possessor (for he can only in a very limited sense be called the proprietor) cannot grant leases, nor give allowances for building, enclosures, and draining, or plantations, which can be given by an heritor, who holds his estate in fee simple. This shall be afterwards noticed.

The

---

* The Duke of Gordon, the Earls of Erroll and Kintore, Sir Robert Burnett of Leys, Geo. Skene, Esq. of Skene, and Alexr. Irvine, Esq. of Drum, hold lands from King Robert Bruce. And Sir Alexander Bannerman's estate of Bishops Clinterty, was given by King Robert III.— Donaldo Bannerman, *medico nostra*, in 1367. The most ancient charter of any physician in North Britain.

The total real rent of the county of Aberdeen is, according to the best data, from which the Reporter could make any calculations, nine times the valued rent. And reckoning the grain at the average price of the last seven years; and including the burgage lands, L.200,000. It has increased one-third since 1793. The real rent belonging to individual proprietors, is, in some few cases, but a little more than the valuation of 1674; in a few other cases, it is about twenty times that sum. It would give just grounds of offence to be particular here. The Reporter can, therefore, with propriety, do no more than simply mention, that after many particular inquiries into the sums of money, and quantities of grain payable to the landholders of this county, and various calculations of the prices of the different kinds of grain, and after charging a reasonable rent for grounds in the natural possession of the proprietors, the whole real rents of lands in this county is very nearly nine times the valued rent, as above stated, or L.180,000 sterling, and including the rents of the burgh lands, belonging to the city of Aberdeen, and other two royal burghs, is, in round numbers, L.200,000 sterling. This rent, however, is rapidly increasing, and it is difficult to say to what sum it may increase, when the existing leases have expired. For this county contains a great extent of surface, of which only a small part has been thoroughly improved.

With regard to the management of estates in this county, most of the proprietors have personal farms, called their *Mains*, which they retain in their own possession, and many of them have their farms managed very judiciously. Some of them have a great extent of ground in their natural possession. But in general, the land is let to farmers, either in large or small allotments. It may be doubted, whether a gentleman should manage a very large farm, except it be chiefly

chiefly in grass, unless his active powers be very great, or unless he is singularly fortunate in having a good bailiff, or farm overseer. Yet no landed proprietor can live comfortably in this country, without having a small or a moderately-sized farm, which promotes his health, and employs his time in overseeing it, and supplies him with butcher meat, and other articles of provision. Besides, the example of the landlord, or of his farm-overseer, is frequently useful; and he generally endeavours to introduce new farm-utensils, and varieties of seed-corn, which are often used to advantage by his tenants.

The smaller proprietors are their own land-stewards, and transact all business with their tenants. Yet to every estate of L.200 yearly, or upwards, there is attached a ground-officer, whose business it is to attend to disputes about marches or boundaries, and to call in the tenants to perform certain services. His salary is generally small, and paid partly by the proprietor, and partly by the farmers.

In great estates, there is almost always a land-steward, or factor, who transacts with the tenants, and frees the landholders of much trouble in the detail of business. And in some of the most considerable landed properties, there is a commissioner who settles with the land-steward. Where the proprietor can attend to it in person, he generally finds it his interest to settle the terms of the leases he enters into with the tenants. If he is a man of rank and fortune, he chooses to commit all matters of detail to his land-steward, or his commissioner; and indeed this person is in general much better qualified for such matters. But it has been also remarked, that those estates flourish most when the landlord takes a concern in the choice of tenants, or persons who are to farm his property,

SECT. II.—TENURES.

This includes both the different species of property, belonging to the landholders, and the different leases or rights of occupancy, possessed by the farmers or tenants.

The property consists either in superiority of lands, which are possessed by others, or in actual possession of lands which hold of the crown; in vassalage, which holds of a subject superior, or in burgage-tenure, which holds of the magistrates of a royal burgh.

The superiority of the whole lands in Aberdeenshire, if split into exact proportions of L.400 Scots each of valued rent, would give 589 qualifications for electing the county member. It is to the distinguished moderation of the greater number of the landed proprietors, that we must ascribe the comparatively small number of freeholders, whose number now amounts to 147, or one-fourth of the number which might have been on the roll, if political litigation had been carried to its utmost limits. And it is deserving of attention, that the number of mere superiorities, to which the property is not annexed, is only about 30, and varies from 25 to 40; and even of these, a considerable number are really proprietors of land, though they have purchased the superiority of others to make up a qualification. But by far the greater number of proprietors are of the most respectable class of real freeholders. They generally reside on their own estates, and endeavour to promote the happiness of their tenants, and the improvements of their lands. Most of the small proprietors, or vassals, either hold of a subject superior, or possess the burgage lands in the neighbourhood of Aberdeen, or in the other burghs of the county. They are, in general, very industrious and enterprising in the improvement of their small pro-

properties. There are a few vassals of another description, viz. some of the great proprietors, who, from political motives, denude themselves of the superiority of a considerable part of their estates, which they parcel out among their friends or dependents, that by their assistance they may command a number of votes in the election of the member for the county. There are but a few of this description; and this practice, though beneficial to the lawyers, does neither good nor ill to the agriculture of their estates. The expence of splitting valuations, and the dangers attending it, will prevent the increase of the number of these freehold qualifications. Indeed the expence of transferring the rights or title deeds of small copy-hold property, owing to the late addition to the duty on stamp paper, is now a serious evil. For a small feu, or copy-hold property, though sold for twenty pounds, costs about four pounds for the expence of transferring it. We stand much in need of a law, by which copy-hold rights can be transferred as easily as in England.—Small properties, cottages for example, cannot now bear the expence of a transference; therefore long leases are given on these, instead of conveyances in perpetuity.

Leases for a term of years are the Tenures by which property is most generally held by the farmer, and other occupiers of land, who hold of the proprietors, as lessees or tacksmen. Trading companies in the neighbourhood of Aberdeen, hold the site of their manufactories on very long leases, most of them 99 years. In the country, some of the farmers have long leases of 57 years, which now are seldom granted. But leases of 30 and 33 years, with rises of rent at the end of every ten or eleven years, have been given by several heritors; and perhaps there is no lease which is so beneficial to the agriculture of the county. (The length of the lease affords security to the farmer during all the time that his

his active powers can be supposed to remain in their vigour; and the two rises of rent operate as stimulants on his industry.) The most usual term of a lease is 19 years. As this is the period of Metonic Cycle, at the end of which the moon returns to her former place in the heavens, and nearly to her hours of setting and rising, it has been supposed, that this number of years included all the variety of seasons.—There formerly were a great number, and there still are several farmers, who possess life-rent leases, or leases for a certain period, and after that is expired, during the life of the lessee. There is something benevolent in the idea, that a farmer is not to be removed during his life. But though such leases were formerly much relished, it is found, by a late decision of the Court of Session, that the day after a tenant dies, the landlord may turn his wife and children out of doors, and take possession of the farm, on paying the expence of labour, for the ground that has been ploughed, but not actually sown. By the former law, and invariable practice, the heirs of a life-renter, if a tenant, were allowed to remain till regularly summoned out, in the same way as if they had possessed without any lease, or by tacit relocation. This is the lowest species of tenure, by which an occupier holds any land. The tenant has a title to continue until he get forty days warning previous to the first term of Whitsunday or Martinmas, (whichever of them happens to be the term of his removal) by a regular summons before the Judge Ordinary, i. e. the Sheriff of the county. This method of removing is often troublesome. For it costs about fifteen shillings to remove a man out of the lowest cottage of which the yearly rent may be but five shillings. The English forms of warning to tenants, whom the landlord wishes to remove, are much wanted in Scotland. And the late decision of the Supreme Civil Court, with respect to life-rent leases, if it be

founded

founded on law, ought to give rise to a new Act of Parliament, by which the heirs of a tenant, who had a life-rent lease, should, as formerly, have the same advantages with those of a tenant, who possessed only by tacit relocation, or had no lease at all. The single check which might be proper to introduce would be a prohibition of breaking up grass-land, or over-cropping what is in tillage.

In giving an account of the different tenures, either by proprietors or tenants, the Reporter felt it his duty to mention every thing candidly, and to make the above remarks on these facts. No offence is intended to be given to any one, by either the statements or observations; and certainly no particular case was in view, while he wrote this section.—But he could not, consistently with a regard to truth, avoid mentioning the facts, with regard to the property of the different estates, and both their management, and their different tenures. It is with pleasure he adds, that the landed proprietors of Aberdeenshire are a respectable body of landholders. Many of them manage their estates themselves, others employ factors or land-stewards, and they find their account in doing it. For the division of office is often as useful in human society as the division of labour. There is no doubt a considerable difference, not only in the value of their estates, but also in their tenures, or nature of their rights. The same holds with respect to the farmers, or cultivators of the soil.—But in this free and happy country, whatever a man's tenure be, he can be ejected from his possessions, or deprived of his rights only by the laws, which are binding on all ranks of men.

CHAPTER III.

BUILDINGS.

SECT. I.—HOUSES OF PROPRIETORS.

IN so extensive a county, the houses of the landed proprietors have been built at very different periods, and in various styles of architecture. But it does not fall within the province of an agricultural survey, to examine whether the noblemen and gentlemen's seats be of the Gothic, of the Tuscan, or the Grecian orders. It will be sufficient to take notice in this Report as concisely as possible,

1st. Of the venerable Remains of the most distinguished Edifices, which ought not to be forgotten.

2dly. Of the principal antient Castles, which are still inhabited;—and

3dly. Of the most remarkable Modern Buildings belonging to the principal proprietors, merely stating the names of the others.

In the first class, and far beyond all the rest in point of antiquity, the Castle of Dun-o-Deer, in the division of Garioch, deserves to be particularly mentioned. It was a vitrified fort, built on the top of a beautiful conical hill, which springs about 300 feet from its base, and it is not less than 1000 years old, being the residence of King Gregory the Great (as he is called by the old Scottish historians, or of
Grig,

Grig, Maormar of Aberdeen and Banffshires, (as Mr. CHALMERS, in his Caledonia has designated him) who died in 892. The most ancient part of this Castle, of which the walls were vitrified, may have been much older than his time; and even the comparatively more modern part of the building appears to be of great antiquity. The walls of the antient part of the castle appear to have been vitrified by setting fire to irregular piles of the wood which was used by the masons, while building them, (in an age when regular scaffolding was unknown) and also perhaps to other combustible matter, collected on purpose, and laid along the building, before the roof was put on. The stones used in the work are in general small, and of irregular shapes, not seldom round; and lime mixed with water, appears to have been poured among them. while wooden frames surrounding them, prevented the lime from escaping or running out. The vitrification, though every where discernible in the old work, is not equally complete in all places, and a great part of the walls have fallen down. But the ruins of this venerable edifice, which is seen from almost every part of the beautiful and fertile valley, are still in many places hard, impenetrable, and entire. The north gable makes a beautiful vista, and may stand for several centuries. Mr. WILLIAMS, in his account of the vitrified forts, has noticed this antient castle.

Another vitrified fort on the top of the hill of Noath, has already been incidentally mentioned. From the conical shape of that mountain, and its being elevated more than 1200 feet above its base, it must have been a place of great strength before the invention of cannon; and from its commanding situation, it could still be defended against these formidable engines of war.

Next to these, with respect to antiquity, are the ruins of two edifices, that belonged to King Malcolm Kenmore, who died

died in the year 1004. One of them was his hunting-seat, and is situated at Castletown of Braemar. The other stands in a small island in Loch Kanders, and being surrounded by the lake, must have been inaccessible to any enemy, except when frozen in winter. A wooden bridge, which connected it with the land, has been fished out of the lake.

Inferior to these in point of antiquity, but far superior in point of extent and grandeur, stood the old Castle of Kildrummy.* It belonged, in 1150, to David, Earl of Huntingdon and Garioch, from whom King Robert Bruce, and all the princes of the race of Stuart, as well as the present Royal Family, have derived their descent. It was distinguished by its seven towers, (one of them called the Snow Tower, from the whiteness of its free-stone walls,) appears anciently to have been a royal messuage, and was the residence of two of the younger sons of the kings of Scotland. It was given along with the Lordship of Garioch, by K. Robert Bruce, to Gratney, Earl of Marr, who was married to his sister Christian. On the extinction of their heirs male, it was possessed nearly 50 years by the Lords Elphinston, who made considerable

* It would be improper to pass over in silence the still *more ancient*, but subterraneous dwellings, which are situated in the district of Kildrummy. These are old houses, supposed to be Pictish, scooped under ground, and lined up with stone-walls, which are from twenty to thirty feet in length, from eight to ten in breadth, and from four and a half to nearly six feet in height. They are drawn to a kind of apex in the top, the higher stones over-lapping those immediately lower, from two to three inches, and thus present to the eye a kind of arch, bound not by a key-stone, but by a larger stone in the middle of the roof, and must have been built at a remote period, when both architecture and agriculture were as yet in their infancy in Scotland. The whole of these houses are below the level of the surrounding fields, and they were probably raised as places of refuge, not only for the inhabitants, but also for their cattle, in times of great danger from an invading enemy.

rable additions to the building; and in 1555, was adjudged to belong to the Lords Erskine, who enjoyed it, with the title, and the greater part of the Earldom of Marr, till their forfeiture in 1715. It was a princely edifice, covering nearly an acre of ground; and its venerable remains still shew the power and grandeur of the opulent chieftains by whom it was inhabited.

Huntly Castle, which, upon the forfeiture of David de Strathbolgie, Earl of Athol, was given by King Robert Bruce to Adam de Gordoun, (ancestor of the Duke of Gordon, the Earls of Aboyne, Aberdeen, and a number of gentlemen of that sirname, who possess a great portion of the landed property of this county) is the third in rank as a venerable ruin. It was, along with Slains Castle, destroyed in 1594, in consequence of the Earls of Huntly and Erroll having fought the battle of Glenlivat, in opposition to the royal authority, which in that turbulent period, was too little respected. But the Duke has erected a magnificent and very extensive pile of building, at what was formely called the Bog of Gight, now Gordon Castle, near Fochabers, and on the confines of Banff and Morayshires. And the late Duchess of Gordon built in this county, and not far from the old castle, a very good modern house, which is now possessed by her grandson, the gallant Marquis of Huntly.

Slains Castle, above mentioned, was also a very ancient building, and was given, with a considerable proportion of the Earldom of Buchan, to the ancestor of the present Earl of Erroll, along with the title of High Constable of Scotland. When it was dismantled in 1594, as above-mentioned, the family of Erroll removed their seat to Bowness, in the parish of Cruden, where the edifice that now goes by the name of *Slains Castle,* a very extensive building, close by the German Ocean, has already stood for nearly two centuries.

The

The Castle of Hallforest, which was a hunting seat of King Robert Bruce, when Earl of Carrick, was given by him after he came to the crown, to Robert de Keith, Great Marischal of Scotland, for his services in the battles of Inverury and Bannockburn. It has been in ruins now nearly a century; but its owner, the present Earl of KINTORE, who is a descendant of the Great Marischal, has an elegant and spacious mansion at *Keith-Hall*. Hall-forest was many centuries ago the castle of the thanes of Kintore, a district in the lower division of Marr, in which the royal burgh of Kintore is situated: and which, in several old charters yet extant, is mentioned by the name of the *Thanagium de Kintore*.

In the division of Buchan, besides Slains Castle, above mentioned, there are the remains of several old castles of great antiquity. Of these, the powerful Earls of Buchan had Dundarg, in Aberdour parish, Rattray, in Crimond, which was built on a rock in the sea; and Kineddar, now called King Edward; at least 800 years old.

In the same division are situated the ruins of two Castles of Inverugy. The oldest stood very near to the sea; the other about 600 years old, belonged to the Cheynes of Inverugy, whose heiress married a younger son of the Marischal family; which, by a second intermarriage, obtained that valuable property. In this castle was born James Francis Edward Keith, who after his attainder in 1715, rose to be a General in the Russian, and a Field Marshal in the Prussian service. (He was killed at Hochkirchen in 1758.)—It deserves to be noticed, that Inverugy had *ice houses*, an article of luxury then very uncommon.

Ravenscraig, above 800 years old, Pitsligo, Pittully, and Cairnbulg, a seat of the ancient Earls of Ross, also belong to this class of antient castles, in the division of Buchan.

Though not belonging to the description of castles, the

ruins of the *Abbey of Deer*, ought not to be omitted in the list of ancient buildings. It was a large edifice, with considerable revenues, and excellent gardens. JAMES FERGUSON, Esq. of Pitfour, is now making out an orchard, which includes the ruins, and will be noticed, under the article, *Gardens*.

A number of Castles, of inferior note, lying scattered thro' different parts of the county, shew, by their ruins, that men in barbarous ages, are more afraid of one another, than in an advanced state of society, when they have much more national wealth to acquire or to defend. The Castle of Balquhain, in which Queen Mary spent a day, in her tour thro' the northern counties; the Castles of Harthill, Tillyfour, Knockquhsen, Coull, Fedderat, Inverallochy, and many others, might here be mentioned. But where law is established, and the plough is successfully managed, the ruins of old castles are only the monuments of the ferocity of our ancestors, which a more polished age has no pleasure in repairing

There are still, however, some antient castles, which are yet inhabited; and some of which make excellent habitations. Among these the most distinguished are,

1.—*Fyvie Castle*, the seat of the ancient Earls of Dumfermline, who were also Lords Fyvie. It is an excellent house, and belongs to the Hon. General WILLIAM GORDON of Fyvie, brother to the late Earl of Aberdeen. It has the advantages of an inviting situation, thriving wood, and a fine sheet of water.

2.—*Castle Fraser*, the principal seat of the antient Lords Fraser, now the property of Miss FRASER, a most commodious and spacious mansion, uniting dignity with elegance. And by the good taste of its owner, both the house, and the adjoining grounds, are laid out to the greatest advantage.

3.—The

ABBEY OF DEER, as it existed in 1770

3.—The *Castle* of *Cluny*, belonging to CHARLES GORDON, Esq. though, in some respects, inferior to Castle Fraser, is a very good edifice. It is also well wooded; and by the exquisite taste of another Lady, the late Mrs. BARON GORDON, to whom its former proprietor was married, it is ornamented with one of the best gardens in the county.

4.—The *Castle* of *Midmar*, a fine old building, in a romantic situation, is situated on a small rising ground, near a bending of the hill of Fare, (which rises almost 1500 feet behind it) and though not more than 300 feet above the level of the sea, it commands an extensive prospect towards the north and east. The excellent roads, and extensive court of offices, made out by its present proprietor, Mr. MANSFIELD; the highly cultivated fields of his personal farm, of nearly 500 acres; the great variety, and immense quantity of his plantations, with the irregularity of the surface of the valley, and of the hills of which it commands the prospect, are highly gratifying both to the man of taste and to the agriculturist; and shew, that while money has been expended with a liberal hand, it has been laid out to great advantage.

5.—*Delgaty Castle*, formerly the property of the Earl of ERROLL, afterwards of PETER GARDEN, Esq. and now belonging to the Trustees of the late Earl FIFE, is another very fine building, with the recommendations of wood, water, and fertile fields.

6.—*Craigston*, the seat of JOHN URQUHART, Esq.—An excellent edifice, built about two centuries ago, by the Tutor of the Earl of CROMARTY, and in a very good situation.

7.—*Drum*, the property of ALEXR. IRVINE, Esq. who is both the heir of line, and the heir of entail, of that ancient family. Its old tower, (not called a castle, but from 400 to 500 years old) is 60 feet long, 40 feet wide, and 63 feet in the height of its side wall. Adjoining to it, is a very commodious house, built

built about two centuries ago, containing two excellent public rooms, and ample accommodation for the family of its worthy proprietor; for the tower is no longer necessary as a place of refuge, (as in the time of the feuds of the clans) but though carefully preserved, remains untenanted. Some fine old trees surround this venerable pile, in the neighbourhood of which are several fertile fields, sheltered partly by natural woods, and partly by artificial plantations. A very good garden, a small sheet of water, at no great distance, and a castellated building on the top of a neighbouring hill, add ornament to this ancient building, which, from its elevated situation, commands a prospect of a considerable district, on the banks of the Dee, from 9 to 15 miles from Aberdeen.

8.—*Abergeldy*, the residence of PETER GORDON, Esq. at 30 miles nearer to the source of the Dee, is washed by that limpid river;* abounds in wood of various kinds, and commands different prospects of the valley of Strath-Dee, and of hills of different magnitudes up to mountains of Alpine height. It abounds in so many natural beauties, as are seldom to be met with in one place; and it is at least doubtful, whether the present venerable mansion would not, in this highland district, be preferred by a person of taste and sensibility, to a modern house of the most correct architecture. Where nature is seen in her *sublime* aspects, the merely *beautiful* is less attractive.

III. It would swell this article too much, to take particular notice of all the more Modern Houses, which belong to the different landed proprietors of this extensive county.— Many of them are *advantageously* situated; others are well *planned*, as places of residence for country gentlemen of mode-

* There is a very ingenious contrivance for passing the Dee, at this place, by a single rope and a pully.

HADDO HOUSE, the seat of the Earl of Aberdeen

derate fortunes; and not a few of them are *elegantly constructed*. But though the Board of Agriculture has prescribed these three subdivisions of this branch, it is really a matter of great delicacy, and in some cases might be thought invidious, to make these distinctions. One circumstance belonging to many of them, demands the attention of any reader of this Report, who is not acquainted with the county.— As the great article of the builders in Aberdeenshire, viz. granite, is much more durable than the softer kinds of stone, and the still more perishable brick of the southern parts of the island, the landed proprietors here are not so apt to pull down their houses, as if they were of less durable materials; but many of them add new buildings, where the old are not judged sufficiently ample. In fact, some of the most commodious houses in the county, are those which have been built at different periods; for example, where one or two wings, or at least two public rooms, have been added a considerable time after the first part of the house was built.— This sometimes hurts the symmetry of the building, but *fitness* and *expediency* atone for this defect.

After making this observation, a few of the most remarkable dwelling houses of the proprietors of land in this county shall be particularly mentioned; and in regard to the others, the name of the proprietor, and the place of his residence, shall be concisely stated, without any comment, or expressed in a few lines for the most part.

The houses belonging to the nobility are the following:

1.—*Haddo House*. A spacious and elegant modern edifice, belonging to the Earl of ABERDEEN, who has the most valuable estate in the county, consisting of about 50,000 English acres of arable, and half that quantity of uncultivated land. It is situated on the banks of the Ythan, is well wooded, ed, and has a deer park, 120 years old, and well stocked, in its immediate vicinity.

2.—*Aboyne Castle.* A spacious and commodious building, erected a few years ago by the Earl of ABOYNE, and retaining the name of the old castle, which along with the estate attached to it, was given to his Lordship's ancestor, by his father, the Marquis of Huntly. It stands near the left bank of the Dee, (although not in sight of that river), surrounded partly by well cultivated fields, and partly by extensive plantations; and it commands the prospect of several thousand acres of natural wood in the forest of Glentanner, on the south bank of that river; all of which are his Lordship's property.

3.—*Marr Lodge.* A very pleasant seat, belonging to Earl FIFE. It is situated near the river Dee, about twenty miles from its source, and was anciently called *Dalmore*, or the *great valley*, through which the river, here receding a considerable distance from its banks, takes its course. It is a charming summer retreat. His Lordship's principal residence is at Duff House, near Banff.

4.—*Keith-hall.* A spacious, elegant, and most commodious house, belonging to the Earl of KINTORE. It is situated on the left bank of the Ury, opposite to the royal burgh of Inverury, and near the influx of the Ury with the Don. An excellent bridge over the Don, the beautiful conical hill called the Bass, and the head of the navigable Canal, are seen from the south-west. The burgh of Inverury, and part of the course of the two rivers, with the fine fertile haughs or meadows along the banks of the Ury, in the vicinity of this mansion, reach from the south-west to the north-west; in the middle, between these two points, the mountain of Benochie, elevated about 1200 feet above the bed of the river,

lifts

KEITH HALL, the seat of the Earl of Kintore

lifts up its head in the west; and towards the north-west and north, lies the greater part of the rich and extensive valley of the Garioch. On a rising hill towards the south is an old plantation, chiefly of Scotch fir, which has begun to decay, and has been considerably thinned. But besides a number of ash, elm, horse chesnut, and other trees above 140 years old, various plantations have been made in every direction, and a beautiful bridge has lately been thrown over the Ury, about 300 yards from that over the Don. So that this elegant mansion is embellished by both wood and water, and by two bridges erected the one in 1791, and the other in 1809, which are equally useful and ornamental. The building itself consists of four houses, one built by Sir ANDREW GARIOCH of Caskieben, (the ancient name of this place) another by the JOHNSTONS, Knights Baronets, who obtained the estate, by marrying the heiress; and two others, one making an elegant front, and the other a spacious east wing, by JOHN, the 1st Earl of Kintore, in 1690 and 1700. There is an open court, and adjoining to this an excellent staircase, in the middle, which communicates both with the front, and wings of this commodious house. At a proper distance from it is an excellent court of offices; behind this a dog kennel, in a castellated form, and fine exposure, and about 400 acres of well cultivated land, being his Lordship's personal farm.

6.—*Putachy*, the seat of Lord FORBES, the chief of that ancient family, is an old, but spacious edifice, on the banks of the Don, at the termination of the fertile valley of Alford; and on the south-west corner of the mountain of Benochie.— It abounds in both natural woods and artificial plantations.

6.—*Philorth*, the seat of Lord SALTOUN, the *male* representative of the ancient house of Cowie, and the *female* heir of the still more ancient house of ABERNETHY, Lord Saltoun. This is an old but commodious house; and from the long minority of its noble proprietor, has received less embellishment than if his Lordship had been residing. It is situated near Fraserburgh, and in the neighbourhood of some fine downs, and of the German Ocean. It is also surrounded with wood and the fields near it are in general well cultivated.

So much for the seats of the nobility.

The Houses, belonging to private Gentlemen, will be best known by mentioning the parishes in which they are situated; beginning at the head of the county.

CRATHIE and BRAEMAR.—The most remarkable edifice of this description is *Invercauld*. An excellent house, with every accommodation, suited to the residence of a highland chieftain, who possesses above *an hundred thousand* acres of land, in the counties of Aberdeen and Perth. It is the property of Mrs. FARQUHARSON of Invercauld, of whose family most of the gentlemen of that name are descended. This mansion, attractive, like its fair owner, is delightfully situated on the north or left bank of the Dee, in a beautiful valley, abounding with varied scenery, and besides the view of several hills, of different altitudes, enjoying two Alpine prospects. One of these is the mountain of Lochnagar, whose *peak* is at least 3000 feet above the bed of the river, and probably 4000 feet above the level of the sea. It is about 6 miles distant on the south-east. The other is Ben-a-board, or Ben-a-vourd, the highest of the hills of Benavon, presenting its broad shoulder on the north-west, at nearly double that distance. The picturesque scenery along the banks of the Dee is here peculiarly pleasant: but the prospect, instead of being *merely beautiful*, becomes *sublime*, when we look up from the river to the distant mountains. The eye is relieved, by beholding the intermediate objects, viz. the natural woods on the skirts of the mountains, as high up as the

woody

woody region extends, which, in this latitude, is about 2000 feet above the level of the sea. Perhaps there is no place in the island, in which the sublime and beautiful are more happily blended, than they are at or near the house of Invercauld.

The Rev. THOMAS GORDON, minister of Aboyne, has a small house in this parish, on his estate of Crathienaird; but does not reside in it.

GLENMUICK, GLENGAIRDEN, and TULLICH. The most remarkable residence of any landed proprietor in these parishes is *Ballater House*. This is a very commodious and well situated dwelling, the property of WILLIAM FARQUHARSON, Esq. of Monaltery. It is in the immediate vicinity of a fine bridge over the Dee, and in front of the mineral well, and Lodge at Pananich, which are the property of its owner. It is embellished by wood, water, a fine bridge, the mineral well, and many natural beauties; and its vicinity contains not only hills, mountains, and the valley of Strath Dee, fringed with woods, but also presents such indications of lead ore, as may one day be found of considerable value to its worthy proprietor.

Garden Shiel—A retired spot for shooting quarters, beautifully laid out, and belonging to FRANCIS GARDEN CAMPBELL, Esq. of Troup, is situated in this district;—as also,

Birkhall, the property of PETER GORDON, Esq. of Abergeldy, an old house, in a very picturesque situation. But its proprietor, as already mentioned, resides in the fine old castle of Abergeldy.

ABOYNE.—*Balnacraig*, the paternal estate of LEWIS INNES, Esq. commands an extensive and pleasant prospect.— Mr. INNES is also proprietor of a beautiful estate in the neighbouring parish of

BIRSE,

BIRSE, viz. *Ballogie*, anciently Tilliesnaught. Both places are situated on the south side of the Dee, with mountains at a distance, and meadows, wood and water, in their immediate vicinity. In the parish of Birse is also situated *Finzean*, a commodious house, (belonging to ARCHIBALD FARQUHARSON, Esq. who has a very valuable and extensive property in this district) between the rivulets Feuch and Chattie, with abundance of wood in its vicinity, and a prospect of lofty mountains at a distance.

KINCARDINE.—On the north of the Dee, and nearly opposite, stands *Kincardine Lodge*, now the property of JOHN DOUGLASS, Esq. of Tilwhilly, in a most picturesque situation, and let to a gentleman for shooting quarters.—At a greater distance from the river are situated *Craigmile*, a commodious house, belonging to JOHN GORDON, Esq. surrounded by wood, (of which the proprietor has lately sold to the amount of L.1600), and *Lairney*, the property of ALEXR. BREBNER, Esq. a striking example of what human industry can accomplish. Besides improving the arable land, the judicious proprietor, after finding that Scotch firs would not grow on his estate, has added very considerably, both to its beauty and its value, by planting larches, on a large scale, and with great success; and though a merchant, who is extensively engaged in commerce, and usually resides in Aberdeen, has taught the landed gentlemen of the county (many of whom were too partial to the Scotch pine) that larches will thrive in a poor soil, and in very unpromising circumstances.—In this parish is also situated *Campfield*, the plain, but comfortable residence of JOHN STRACHAN, Esq. This gentleman, at the advanced age of 90, is both able, and is usually accustomed to walk to the top of his hills, in the morning before breakfast; while the sons of effeminacy and dissipation are generally

rally enervated at 45. If an *Alpine mountain* be accounted deserving of notice; we surely ought not to overlook an *Alpine character*—a sound mind, and a sound body, the effect of temperance and of moderate exercise, after seeing nearly a century revolve. In the *commerce of life*, luxury deals *largely in short annuities;* while *temperance* makes *fewer purchases* of stock, but they are more valuable, because they are made in *long annuities.*

TARLAND.—In the upper part of this parish, which intersects that of Strathdon, are situated *Skellater*, belonging to WILLIAM FORBES, Esq (who is proprietor of above 20,000 acres, in this, and two neighbouring parishes), *Inverernan*, belonging to ALEXANDER FORBES, Esq; *Edinglassie*, the property of CHARLES FORBES, Esq. a merchant in Bombay; and *Candacraig*, the property of ALEXR. ANDERSON, Esq. and distinguished by an extensive and fertile meadow.— They are all excellent summer habitations, and in general both well situated and finely wooded. But Mr. FORBES of Skellater resides on his estate of Balbithan, in the parish of Keith-hall and Kinkell, and Mr. CHARLES FORBES of Edinglassie is still in India. Mr. FORBES of Inverernan has quitted his house at the confluence of the Don and Ernan (to accommodate his son, Major FORBES, who was senior Captain of the 71st regiment, at the battle of Vimiera) and has built a snug retreat for himself, named *Forbes Lodge*, on the banks of the Don, in a very pleasant situation. From the great quantity of wood planted within the last 20 years, these highland mansions have derived much embellishment, while the value of the estates has been greatly augmented.

STRATHDON.—On the north bank from the source of the Don, near to Curgarff, is situated the estate of Allerg, belonging to the Rev. ROBERT FARQUHARSON, minister of Coldstone, who has let his house in this north-west extremity ty of Marr, for shooting quarters. Nearly opposite to this is the Castle of Curgarff, which was occupied by a detachment of the military for several years, as a check on the highlanders, who were supposed to be generally attached to the House of Stuart. But that family being now extinct, it has been given by government to the proprietor of the adjacent estate, WILLIAM FORBES, Esq. of Skellater.

In the lower division of this parish are situated the houses of *New* and *Bellabeg*, both the property of JOHN FORBES Esq. late merchant in Bombay. They are excellent summer habitations, and the proprietor has expended the rents of both estates in planting, enclosing, and under-draining, on a plan given by Mr. JOHNSTON; so that in a short time they will be highly embellished, and greatly increased in value.

Glenkindy.—A commodious house, belonging to ALEXR. LEITH, Esq. of Freefield, with wood, water, and well cultivated fields to embellish it.

LOGIE COLDSTONE.—*Blelack.*—A large and commodious house in the district of Cromar, which has changed several proprietors within the last thirty years. It now belongs to JOHN FORBES, Esq. of New. It is well wooded, and situated in a fertile part of the country. But the residence of a proprietor would probably tend to add both to its value and to its ornament.

LEOCHELL.—*Craigievar.*—A very spacious and commodious house, (belonging to Sir WILLIAM FORBES, Baronet), attached to an extensive property in this and the two adjoining parishes. The worthy Baronet resides at Fintray House, and has let this, which is commonly termed the Castle of Craigievar, to a most industrious and enterprising tenant, Mr. SCOTT. It is also well wooded; and by an authentic document, which the Writer of this Report has seen,

its

its excellent stone fences, or inclosures, were built 100 years ago.

MIDMAR and KINARNY.—*Corsindae.*—WILLIAM DUFF, Esq. A commodious house, well wooded. The personal farm is well cultivated by its proprietor.—*Kebbaty*—The property of JOHN DAVIDSON, Esq. Advocate in Aberdeen, is, like Lairney, a proof of what attention and industry can effect. The house plain, but commodious, with thriving wood, and well cultivated fields. The face of nature is here greatly improved in the course of 20 years. Money judiciously, but liberally expended.—*Shiels*—ADAM WILSON, Esq. of Glascoego, though in a high situation, is well wooded; and the proprietor's personal farm has been well cultivated.

CLUNY.—*Linton* and *Sauchan*, belonging to Mrs. CRAIGIE. A commodious snug mansion, with a few patches, and a ring fence of wood. The fields are well cultivated; but the house has received less ornament, as the proprietor has not resided for 20 years.

ECHT.—*Housedale.*—The residence of WILLIAM FORBES, Esq. A very commodious house, attached to a valuable and extensive landed estate, with abundance of wood, excellent roads, and the proprietor's farm in high order.

SKENE.—A spacious old mansion, built at various periods, the property of GEORGE SKENE, Esq. who was member for the county in one Parliament, and for the boroughs of Kintore, Inverury, &c. in another. Extensive plantations of wood, and complete under-draining, on a plan of Mr. JOHNSTON's, with fine old inclosures of the arable land, and a prospect of the Loch of Skene, (which, with the estate, has belonged to this family since the days of King Robert Bruce,) distinguish this venerable mansion. The proprietor has on his personal farm, and on the estate of Fornet adjoining, about 600 acres of land in his own possession, part of which

he

he has already improved, and the rest is under a course of improvement. Besides this estate, and that of Wester Fintray, in this county, he is proprietor of that of Carriestown, in Forfarshire, where he has an excellent house, in which he generally resides.—*Concraig,* the property of JOHN SMITH, Esq. a very good modern house, with some patches of wood, and a farm, well cultivated, is also situated in this parish.

PETER CULTER.—The house of *Culter,* the property of ROBERT WILLIAM DUFF, Esq. of Fetteresso, (in the county of Kincardine,) a spacious mansion, with a good garden, abundance of wood, and fertile fields, commanding an extensive prospect along the banks of the Dee.—*Binghill,* formerly the property of JAMES WATSON, Esq. (and highly improved by him, both in planting the barren, and cultivating its arable land) now belonging to Mrs. KYLE; and *Oldfold,* belonging to ARTHUR ANDERSON, Esq. are two small properties, with suitable houses, and much industry has been employed in the cultivation of both.—*Murtle.*—JOHN GORDON, Esq. a neat small house on the banks of the Dee, with a good garden, and in a fine situation, with thin sharp soil, well cultivated. In this parish also is *Counteswells,* an excellent house, well wooded, with inclosures and roads well laid out by the late proprietor's Father, JAMES BURNETT, Esq. It now belongs to JAMES GAMMEL, Esq. Banker in Glasgow.

North part of BANCHORY DEVENICK, or NETHER BANCHORY.*—*Dee Bank,* formerly Newton of Murtle, the property of ARTHUR ANDERSON, Esq. a neat commodious house, in a most attractive situation, fronting the Dee, which washes this property on the south, finely wooded, and lands well cultivated. *Cults,* GEORGE SYMMERS, Esq. a small but commodious house, where, though in a high situation, wood thrives well. The late WILLIAM DURWARD, Esq. while he was proprietor, improved this estate very much; and both

by

---

* The greatest part of this parish is in Kincardineshire.

by cultivating his personal farm, and by giving lime in great quantity to the tenants, trebled its rental in a few years.—Mr. SYMMERS has the principal lot of this estate, which has been partitioned among four proprietors.

NEWHILLS.—*Cloghill.*—JOHN GRANT, Esq. a plain but commodious house, in a high situation, with a farm well inclosed, and in high order.—*Crabstone,*—Mrs. THOM, a commodious house, with a good garden, surrounded by thriving and extensive plantations.

DYCE.—*Pitmedden.*—A very good house on a small property, originally bequeathed for the support of four bursaries at Marischal College, by Dr. DUNCAN LIDDEL, but feued out by the town of Aberdeen to one of the then Magistrates for L.16 : 13 : 4 of feu duty. Its present proprietor, Mr. INNES, has greatly increased its value. It is not only well wooded, but well watered. For the Canal from Aberdeen to Inverury passes through the whole length of this property, and the Don washes its northern boundary.—*Caskieben*—JOHN HENDERSON, Esq. (formerly Craig).—An elevated but well situated house, with some thriving wood, and fields well cultivated.

KINELLAR.—*Glasgoego.*—ADAM WILSON, Esq. a spacious and commodious house, well wooded, and in a commanding situation, with a small but valuable estate, through which the turnpike road from Aberdeen to Inverury is conducted. It is deserving of notice, that this property, about 80 years ago, was sold for L500; about 65 years ago for L.800; and that the present proprietor received that sum *of damages for injury to his property,* by conducting the turnpike through it, by a line of which he probably could not, certainly did not, approve. A more striking proof of the rise in the value of land is not to be found in any part of the kingdom. Adjoining to this estate is the more valu-

able

able property of *Glasgoforest*, on which a small house was occasionally the residence of the late Thomas Leys, Esq. at different times Provost of the city, and Convener of the county, of Aberdeen. He was a most respectable character, and his death was a public loss. His sister, Mrs. Brebner of Lairney, inherits his property.

KINTORE.—*Thainston.*—Duncan Forbes Mitchell, Esq. One of the most commodious houses of its size, to be found in the county, well wooded, having an excellent court of offices, a good garden, and fields well cultivated, with a fine prospect of the Don, which bounds this property on the north. It was built by the proprietor's father, D. F. Mitchell, Esq. who was a most respectable country gentleman, and died young and much regretted.

KEMNAY.—An old, but commodious and spacious mansion, the property of John Burnett, Esq. It is well wooded, and the grounds adjacent were laid out with great taste, by Geo. Burnett, Esq. his grandfather. He planted on a large scale, and with great success, on a peat-moss; and both by his plantations, and by cultivating what could be rendered arable, might be said to have created a valuable property, and to have introduced great ornament to a place, which it was supposed could not be made capable of receiving embellishments from art. It is situated near the Don, which bounds it on the north-west.

MONYMUSK.—A spacious and very commodious house, the property of Sir Archibald Grant, Bart. in a most inviting situation, in the midst of fertile fields, on the banks of the Don, surrounded by several thousand acres of Wood on the hills and uncultivated grounds. The proprietor's grandfather, Sir Archibald Grant, is supposed to have planted more extensively than any man in Great Britain, (above 50 *millions* of trees, in the course of 50 years,) and lived to see some

some of them two feet in diameter. His son and grandson have already sold wood to the amount of much more than the estate cost a century ago (when it was purchased from the ancestor of Sir WILLIAM FORBES of Pitsligo.) The arable lands are also extensive and well cultivated; yet it is believed that if they were to be sold separately, they would not fetch such a price, as could be obtained from the wood that remains on the uncultivated parts of the estate, altho' L.20,000 of timber is supposed to have been sold by the two last proprietors. A more remarkable evidence of the advantages of planting, is not to be met with in any part of the island.

TOUGH.—*Tonley*—A very commodious house, the property of JAMES BYRES, Esq. at the south-east extremity of the district of Alford. Its worthy owner, after spending many years at Rome, has returned to his native country, and both ornamented and improved his paternal estate in a very high degree. Large plantations of wood, land well cultivated, and every thing laid out with taste, distinguish a place, which, though it lies in a good situation, owes much to its present proprietor.

Whitehouse—PETER FARQUHARSON, Esq. A good modern house, in an elevated situation, and commanding a view of the whole valley of Alford, with its meandering river, and its bounding hills. It is finely wooded; and on the whole, very attractive of the attention of a traveller.

ALFORD.—*Haughton*—JOHN FARQUHARSON, Esq. A spacious and elegant modern house, situated in the middle of the valley of Alford, in a most inviting situation, with an excellent garden, within a few hundred yards of the Don, finely wooded, in the midst of fertile fields, and attached to a valuable landed estate. Its late worthy proprietor, FRANCIS FARQUHARSON, Esq. who died in the prime of life, was one of the most respectable characters, and best informed men in

H   Great

Great Britain. His brother, though bred in the Navy, has embellished this property by additional plantations, conducted on a great scale, and by excellent roads, which have been made out since he succeeded to the estate.

Breda, antiently *Broad-haugh.*—ANDREW FARQUHARSON, Esq. A neat and commodious house, with a small estate, adjoining to the lands of Haughton, (about a mile further up the river). The inclosures well laid out, and the land is both fertile, and of sharp soil, and well cultivated by Mr. FARQUHARSON. The property is well wooded, and the roads and avenues laid out with taste.

Opposite to these two properties, on the north bank of the Don, and in the parish of

TULLYNESSLE, is *Whitehaugh*, now the property of THEODORE FORBES LEITH, Esq. It is an excellent modern house, with similar advantages of situation, wood, water, and fertile soil. And its late worthy proprietor, WILLIAM FORBES LEITH, Esq. who had about 700 English acres of land in his natural possession, will long be remembered as a great friend to agriculture, and a man of the most unexceptionable morals, and of extensive information.

KILDRUMMY.—*Clova*—The residence of HARRY NIVEN LUMSDEN, Esq. (who has a considerable property in different parishes of this county): A commodious house, with a good garden. This place is finely wooded, and in the course of being thoroughly improved.

CUSHNY.—The estate of *Cushny*, which has been about 400 years in the same family, belongs to JOHN LUMSDEN, Esq. who has been for some time in high rank in India. The house is an old one, and situated in a glen, or narrow valley, capable of receiving great improvement.—*Hallhead*—The paternal estate of GEORGE GORDON, Esq. who resides at Esslemont, is a fine highland retreat, and well wooded.

These

These are all the principal habitations of the landed proprietors in the division of Marr. In that of Strathboggie, there are only two which deserve notice, viz.

Huntly Lodge.—Which has been already mentioned incidentally (when speaking of the old Castle.) It is a very good modern house, with elegant and spacious public rooms, situated on the banks of the Doveran, finely ornamented with ring fences, and commanding a beautiful view of Huntly Castle and the neighbouring country. The town of Huntly is about a mile distant, to which the communication is by a handsome bridge over the river. The gallant Marquis of HUNTLY, who is Lord Lieutenant of the county, now resides here during a considerable part of the year.

Avochy, the property of JOHN GORDON, Esq. a good summer retreat. The proprietor chiefly resides at Edinburgh. All the rest of this division is the property of the Duke of Gordon.

In the division of Garioch, are many gentlemens seats in the different parishes.

CLATT.—*Knockespock*—HARRY GORDON, Esq. Well wooded, and in the midst of well cultivated fields.

KINETHMONT.—*Leith Hall*, the property and residence of Lieutenant-General ALEXANDER HAY, of Rannes. An excellent house, in a commanding situation, surrounded by wood and fertile fields.—*Gordon Hall*, belonging to CHARLES GORDON, Esq. of Wardhouse. A good modern house, and with many recommendations in point of situation. But its proprietor has been long in the army, and does not reside in it.—*Craig Hall*—WILLIAM WEYMSS, Esq. A plain but comfortable house, attached to a small estate, on which the proprietor's father brought up, and liberally educated a numerous family.

KEARN.—*Drumminnor*—A good commodious house, near the ruins of Castle Forbes, now belonging to JOHN GRANT, Esq.

INSCH.—*Rothny.*—WILLIAM GORDON, Esq. A small but commodious house, well wooded, and fields in good order.

PREMNAY.—*Likelyhead*, the property of HARY N. LUMSDEN, Esq. well wooded, in a commanding situation, on the north side of Benochie, and having some fine farms attached to it.—*Overhall*—GEO. LEITH, Esq. An old house; and the proprietor does not reside; but he has built a cottage for occasional residence, and has improved part of his estate very well.

CULSAMOND.—*Newton*—An excellent and spacious house, built about 40 years ago by Captain DAVIDSON, its then proprietor, who laid out his fields and plantations with great taste, and both improved the former, and embellished the latter. On his death, this property was sold to Colonel McINTOSH, and it now belongs to ALEXANDER GORDON, Esq.

RAYNE.—*Freefield*, an excellent house, the property of ALEX. LEITH, Esq. in the middle of the Garioch, surrounded by wood, and situated in the most fertile division of the county.—*Warthill*.—WILLIAM LESLIE, Esq. A plain but commodious house, surrounded by some patches of wood, and fields naturally fertile, in a high state of cultivation.

CHAPEL of GARIOCH, (anciently *Logie Durno*.)—*Logie Elphinstone*—An elegant and commodious mansion on the left or north bank of the Ury, a little below its confluence with the Gady.* It is finely wooded, has an excellent garden, and its avenues and fields are laid out with great taste. Its proprietor, R. D. HORN ELPHINSTONE, Esq. has both embellished and improved this place, on which his elder brother, JAMES HORN, Esq. (who was much and deservedly esteemed) had expended no inconsiderable sum of money before his death. This delightful place, owing to its low situation, was but little known, till the opening of the turnpike road from Aber-

---

* It is not unworthy of notice, that JOHNSTON, next to Buchanan, the best latin poet of modern times, was born at Craigmill, about 500 yards from this place, at the confluence of the Gady and Ury; or as he says himself—" Ubi Gadus in Suit Urum."

Aberdeen to Huntly presented the traveller with the varied scenery of a spot, which at first he beholds with surprize, and will not be apt to forget.

Pittodry—A spacious old house, in an elevated situation, on the east side of Benochie, well wooded, and commanding an extensive prospect of the rich valley of the Garioch.—*Pitcaple*—Belonging to Hary Lumsden, Esq. of Belhelvie; a venerable old building, in a romantic situation, on a bending of the river Ury One of the largest thorn trees in Great Britain, distinguishes this mansion: in the vicinity there are some fine old trees, and several plantations lately made—the soil is very fertile.—*Inveramsay*—Belonging to Patrick Irvine, Esq. A small neat house, with a few old trees, and a clump of Scotch firs, with a farm adjoining, that was well cultivated by its former proprietor.—*Fetternear*—A spacious mansion, belonging to John Leslie, Esq. of Balquhain, on the north bank of the Don, with fertile fields, and abundance of wood—*Braco*—The property of Robert Harvey, Esq. who built a commodious house, and inclosed and improved his personal farm, but now resides in Aberdeen.

OYNE.—*Westhall*—An old house attached to a considerable property, belonging to R. D. H. Elphinstone.

DAVIOT.—*Glack*—A commodious and spacious house belonging to Roderick Mackenzie, Esq. surrounded by fertile fields, well wooded, and hedges.—*Mounie*—Presently belonging to Alex. Anderson Seaton, Esq. eldest son of the late Dr. James Anderson, well known for his various writings on the subject of Agriculture. The house here is old, tho' commodious; but several fertile fields, and some fine old trees, with a neat garden, surround it.—*Phingask*—The residence of Thomas Elmslie, Esq. who has expended money very liberally on buildings, plantations, and in otherwise ornamenting his property.

MELDRUM.—The residence of James Urquhart, Esq. who

who, by building a considerable addition to that part of the old house which was allowed to remain, has made out one of the most elegant and commodious mansions that is now to be found in the county. It is surrounded by abundance of wood, and has been made completely dry, by under-draining.

BOURTY.—*Barra*—Belonging to JOHN RAMSAY, Esq.— An old and spacious house, with a good deal of wood.— *Bourty*—Belonging to ALEX. ANDERSON, Esq. A snug modern house, surrounded by wood.—*Thornton*—The property of JOHN NIVEN, Esq. A small commodious house, with a ring fence, and several patches of wood.— And *Leithfield*—Another small house belonging to JOHN SHEPHERD LEITH, Esq. It is remarkable, that none of the landed proprietors reside in this parish, and that the ground in the neighbourhood of all their mansions requires under-draining.

KEITH-HALL and KINKELL.—*Balbithan*—The property and residence of WILLIAM FORBES, Esq. of Skellater.— A very good old house, well wooded, and much improved by its worthy proprietor.

FINTRAY.—Perhaps the residence of Sir WILLIAM FORBES, Baronet, Craigievar, (the principal heritor of this parish) formerly called Lamington, now *Fintray House*, is, in many respects superior to the residence of any private gentleman in the county. It is a spacious and elegant house, with excellent public rooms, and ample accommodation to the family and their guests. It is also ornamented with excellent paintings—of various kinds. Its situation is upon the north bank of the Don, above which it is elevated about 50 feet, with a prospect of that river, of the navigable Canal from Aberdeen to Inverury, and of different plantations of wood on both sides of the river, which flows through a spacious and fertile meadow. The woods belonging to this estate extend above two miles in length, not in square plantations, but happily distributed in different figures adapted to the nature

of

of the grounds, and interspersed with the fields of the personal farm of the worthy proprietor, which are naturally fertile, and in a high state of cultivation. An excellent garden, hanging at an angle of about 20 degrees above the river, is distinguished by the earliness of its produce, and the flavour of its fruits, on which its natural warmth has very considerable influence. As the opening of a turnpike road along the banks of the Ury has presented the traveller with the delightful view of Logie Elphinstone, so the opening of the Canal along the banks of the Don, has rendered the beauties of Fintray-house more conspicuous to strangers. Fintray-house is Logie on a larger scale; much taste is shewn in both.

In the division of Formartin are many good houses, viz.—
NEWMACHAR.—*Straloch*—The residence of JOHN RAMSAY, Esq. of Barra. A commodious and spacious modern house, with abundance of wood, a good garden, and several hundred acres of land, well cultivated, and in the natural possession of the proprietor.—*Elrick*—The property of JOHN DUMBRECK, Esq. a commodious modern house, well wooded, and attached to an estate, which in general is well cultivated. The proprietor occasionally resides.

Parkhill—A spacious and elegant modern house, situated on the north bank of the Don, within six miles of Aberdeen, surrounded by very extensive plantations of wood. The worthy proprietor, ANDREW SKENE, Esq. of Dyce, is now *father* of the county, that is the oldest on the roll of freeholders; among whom he has voted at the elections of a county member, for the space of 63 years, or since the death of his elder brother. With an ample fortune, a liberal education, and uniting moderate exercise to singular temperance, he bids fair to hold the rank of *father of the county* for several years longer.—*Disblair*—The property and residence of Miss DYCE, in an elevated situation, but finely wooded.

UDNY.—The principal mansion in this parish belongs to John Robert Fullerton Udny, Esq. a venerable and lofty pile, well wooded, and surrounded by some fertile fields, which his grand uncle, the late Commissioner Udny, spared no expence in improving.

BELHELVIE.—*Menie*—George Turner, Esq. a commodious house, with a few trees, not far from the German Ocean, surrounded by fields, which in general have been well cultivated.—*Belhelvie Lodge*—The residence of Hary Lumsden, Esq. of Belhelvie; an excellent commodious modern house, with a very good garden, and surrounded by fields in a high state of cultivation. On both, money has been laid out unsparingly, and judiciously. The German Ocean is distant two miles from this place—*Orrock*, formerly Colpna—The property of John Orrock, Esq. a commodious modern house, with excellent offices, in the midst of a farm of about 120 English acres, at present possessed by Mr. James Gordon, one of the best farmers in the county.—*Drumside*—A neat commodious house, the property of Charles Davidson, Esq. whose estate borders on the south with the German Ocean.—*Eggie*—The property of Dr. Fordyce, of Culsh, in the same circumstances with Drumside, and with the grounds attached to it, rented by another excellent farmer, Mr. John Lumsden.—*Blairton* and *Hopshill*—The property of John Walker, Esq. who has not as yet built a house upon the former, but resides upon the latter. At a greater distance from the sea, and in the hilly parts of the parish, are situated *Ardo*, the property of John Dingwall, Esq. which has been much improved by him; and *Smiddyburn*, the property of Pat. Irvine, Esq. of Inveramsay.

FOVERAN.—The principal resident proprietor, Andrew Robertson, Esq. has an excellent and commodious house at Foveran, where, by steady perseverance, he has been pretty ty successful in raising trees, which is generally done with great difficulty on the coast of Belhelvie.—*Tilleray*—The residence of JOHN CHALMERS HUNTER, Esq. A plain but commodious house, at a greater distance from the sea, and where trees have been raised with less difficulty.

LOGIE BUCHAN.—*Auchmacoy*—The property of THOS. BUCHAN, Esq. in whose family this property has continued for many centuries; an excellent and spacious house on the north bank of the Ythan, well wooded, and with a commanding prospect.—*Ranniestown*, the property of ARTHUR DINGWALL, Esq. and *Watridgemuir*, the property of THOMAS BLACK, Esq. at a greater distance from the sea, are modern houses, in situations somewhat exposed, but surrounded by fields, which have been well cultivated.

ELLON.—The Honble. WILLIAM GORDON, inherits from his father, the late Earl of ABERDEEN, the spacious mansion of Ellon House, sometimes called *Ellon Castle*, with a considerable estate attached to it. But he does not reside at this place, which, from elegance of architecture, pleasantness of situation, and the fertility of the adjoining lands, has many attractions.—*Esslemont*—The excellent house of GEORGE GORDON, Esq. is well situated, has abundance of wood, and a valuable property attached to its manor.—*Arnage*—Belonging to JOHN LEITH ROSS, Esq. An old house, on which, till lately, no proprietor has resided for many years; and *Dudwick*, the property of JOHN ROBERT FULLERTON UDNY, Esq. of Udny and Dudwick, shew that the residence of a proprietor is intimately connected with the state of his manor. —*Turnerhall*—The property of JOHN TURNER, Esq. is chiefly distinguished from the age and strength of the dry stone walls by which the personal farm has been for nearly a century inclosed.

TARVES.—The only resident Heritor in this parish is
ALEX-

ALEXANDER IRVINE FORBES, Esq. of Schivas, eldest son of ALEX. IRVINE, Esq. of Drum. *Schivas* is a plain but commodious house, well wooded, and situated on the north bank of the Ythan. A valuable property is attached to this mansion.

FYVIE.—JAMES LESLIE, Esq. of Rothie, has a commodious house suited to the extent of his property, and well wooded.

MONTQUHITTER.—*Auchry,* belonging to JOHN CUMINE, Esq. An excellent house, well wooded, with a fine sheet of water, and fertile fields, both in no small degree occasioned by the active and public spirited exertions of the late Mr. CUMINE of Auchry, one of the earliest and best improvers in the county of Aberdeen.

TURRIFF.—*Balquholly*—A fine old house, commonly denominated a Castle, well wooded, and advantageously situated, about a mile from Turriff. It is the property of GARDEN DUFF, Esq. who resides at Hatton, in the parish of

AUCHTERLESS—where he has also an excellent house, and uncommonly good offices, with abundance of wood, and fertile soil.—In the same parish lies *Blackford,* the residence of JOHN FORBES, Esq. a commodious house, and well wooded.

FORGUE.—*Frendraught*—The property of THEODORE MORISON, Esq. of Bognie. An excellent house, surrounded by abundance of wood, fertile fields, and having an excellent estate attached to it.—*Boyndsmill*—A small property belonging to JAMES ALLARDYCE, Esq. one of the most spirited and extensive farmers in the county, who rents nearly 800 acres of land from different proprietors.—*Auchaber*—Belonging to JAMES WILSON, Esq. A plain house, suited to the extent of the property, which is in the natural possession of Mr. WILSON. —*Corse of Monelly*—The property of Captain HENRY. Good farming in general in this parish.

DRUM-

DRUMBLADE.—*Lessendrum*—An old house, but well wooded, the property of Mrs. ABEL BISSET.

KING EDWARD.—*Byth*—A commodious house belonging to Miss ELIZABETH URQUHART, on whose property great improvements have been made by under-draining.—*Eden*—Belonging to WILLIAM GORDON DUFF, Esq.

In the division of Buchan, there are also many good houses.

NEWDEER.—*Nethermuir*, belonging to JOHN GORDON, Esq. An excellent house.—*Brucklaw*, belonging to JOHN DINGWALL, Esq. A large old house, formerly called a Castle, with a commanding view of the adjacent country.—*Artamford*—A very convenient house, surrounded by ring fences of planting, and fields well cultivated.

OLD DEER.—*Pitfour*, belonging to JAMES FERGUSON, Esq. Representative of the county, well wooded, with good office houses; but the house itself, though commodious, is not proportioned to the extensive estate of the owner. He has got a plan of a new one on a much larger scale.—*Aden*—An excellent modern house, belonging to ALEXANDER RUSSEL, Esq. with a semi-circular court of office-houses, and abundance of wood.—*Kinmundy*—JAMES FERGUSON, Esq. A large and commodious house, well wooded.

LONGSIDE.—*Cairngall*, the property of JOHN HUTCHINSON, Esq. A small but commodious house, attached to a property which is well cultivated by its owner.—*Faichfield*—An old house, the property of Lord NEWTON, with abundance of wood.—*Kinmundy*—The property of JOHN ARBUTHNOT, Esq. used by the Messrs. KILGOURS as a cloth manufactory.

PETERHEAD.—*Dents*, the property of JAMES ARBUTHNOT, Esq.—*Invernetty*—The property of ALEX. GORDON. Esq. and *Invernetty Lodge*, belonging to JAMES SKELTON, Esq. all situated near the German Ocean, on lands which in general are well cultivated.

CRUDEN.—*Auchleuchries*, a commodious house, the property of Geo. Gordon, Esq. with a small patch of wood; and *Auquharney*, Colin Gillies, Esq. which he has greatly improved.

SLAINS.—*Gordon Lodge*, the property of General Gordon, of Pitlurg, the male representative of the ancient family of Gordon, a commodious house, attached to a property which has been much improved by its owner and his tenants.

STRICHEN.—A very good house on the principal estate of that parish, belonging to Alex. Fraser, Esq. well wooded, with fertile fields.

CRIMOND.—*Logie*, the property of Lieut.-Colonel Alex. Tower, Member for Berwick in the last Parliament. A commodious house, in a low situation, attached to a farm which was lately well cultivated by Mr. James Gordon.—*Broadland*, the residence of Alex. Harvey, Esq. [A very good house, attached to a farm, which the proprietor has improved, and part of which he has let to great advantage. An excellent garden, and fine thriving wood, notwithstanding many obstructions, distinguish this place. In its neighbourhood are the finest downs in the island.—*Haddo*—James Laing, Esq. A commodious house, with a good farm, well cultivated by a brother of the proprietor's.

LONMAY.—*Cairness*—An elegant and spacious modern building, the property of Thomas Gordon, Esq. of Buthlaw, with abundance of wood, and surrounded by fertile fields. Though its situation, within a mile of the German Sea, is not equal to that of some houses in the interior parts of Aberdeenshire, yet Cairness is, no doubt, both the largest and best house belonging to any private gentleman in the county. —*Crimonmogate*—A commodious house, belonging to Patrick Milne, Esq. attached to a personal farm in very high order, and inclosed with excellent stone walls. It is surrounded

rounded by abundance of wood; and it may be noticed, that in improving its soil, besides the dung of the farm, and shelly sand from the sea, above 70,000 sea dogs have been used as manure.—*Craigellie*—WILLIAM SHAND, Esq. An old house, well wooded.

RATHEN.—*Mormond House*, lately Cortes—A new edifice, the property of JOHN GORDON, Esq. of Cairnbulg.—*Auchiries*, the property of his brother CHARLES GORDON. These houses are attached to two valuable properties, that have been lately purchased from the former proprietors. On Auchiries, though near the sea, is a den of thriving wood, equal to any in the most inland parts of the county.

TYRIE—*Boyndlie*, belonging to GEORGE PETER IRVINE, Esq. A neat small house; and *Upper Boyndlie*, belonging to JOHN FORBES, Esq. in a more inland situation.

PITSLIGO.—The principal estate in this parish is the property of Sir WILLIAM FORBES, Baronet. His late father, one of the best characters in the island, and whose memory will long be revered in this county, expended very considerable sums of money in adorning his house, and improving his property. The old castle is not now inhabited.

ABERDOUR.—*Penman Lodge*—Belonging to the Hon. WILLIAM GORDON, is finely situated; and *Aberdour House*, belonging to Mr. GORDON, of Aberdour, is commodious, and the farm well cultivated.

The Houses of the landed proprietors in the parish of Old Machar, or Old Aberdeen, have been purposely delayed till the last, because this extensive parish, which surrounds the city of Aberdeen, reaches nearly six miles on both sides of the Don, and nearly three miles on the north bank of the Dee, which is here the boundary with Kincardineshire. On the north or left bank of the Don are

Fraserfield, the property of Miss FRASER. A spacious and com-

commodious house, in a commanding situation, with a good garden, abundance of wood, attached to a valuable estate; on a part of which are cottages, with excellent gardens, belonging to three gentlemen in Aberdeen, in most inviting situations. One of them, belonging to the late Provost Leys, is perhaps the finest cottage in the island.—*Scotstown*—The property of Alexr. Moir, Esq. Sheriff-Depute of the county. An excellent modern house, finely wooded, with a good garden, and both avenues and fields very well laid out, where not many years ago a large peat moss was situated. No where in the county has more industry been exerted, more taste displayed, nor money more liberally or judiciously laid out, than in reclaiming this moss, and in building and ornamenting this house, and improving the fields in its neighbourhood.—*Grandholm*—A spacious and commodious mansion, belonging to John Paton, Esq. well wooded, and in a fine situation, close by the Don, and surrounded by a valuable landed estate.—On the south bank of the Don is *Seaton*, the property of James Forbes, Esq. (who has also several other estates in this parish) an excellent edifice, situated in a beautiful meadow, finely wooded, with an extensive garden, and a valuable estate, let at a high rent to a number of small, but industrious farmers.—*Woodside*—Patrick Kilgour, Esq. the proprietor of other two contiguous properties. A commodious house, with offices, in a castellated form, on a good situation, well wooded, and surrounded by 200 acres of land, in a high state of cultivation.—*Grove*—James Lumsden, Esq. A neat house, with good garden, and thriving wood, in an inviting situation.—*Tanfield*, Alexr. Shand, Esq. A neat modern house; and another, adjoining to it, belongs to Robert Garden, Esq. with a very good garden, both in a commanding situation.—*Hilton*—Sir William Johnston, Baronet. A commodious old house, with abundance of
wood,

wood, and lands of a thin soil, but in general well cultivated. *Ashgrove*—Major Daniel Mitchell. A neat and commodious house, with a good garden, in the midst of a few well cultivated fields.—*Bushybank*—Mr. Annand. A spacious house, with a fine prospect, an excellent garden, and thriving trees.—*Berrybank*—Right Rev. Bishop Skinner. A commodious house, surrounded with wood, and a fine garden.—*Sunnybank*—Heirs of Maj. Mercer. A good house, in an inviting situation, and land in high order.—*Berryden*—Mr. Leslie. A very good house, with thriving wood, laid out with great taste, and well cultivated.—*Powis*—Hugh Leslie, Esq. An elegant modern house, attached to a valuable property, with ground laid out with taste, and highly improved.—*Fountainhall*—Mr. Professor Copland. A commodious house, with some fine trees, a neat garden, lands well cultivated, and valuable stone quarry.—*Westfield*—An elegant modern house, with thriving shrubbery, and an excellent garden.—*Newlands*—William Duguid, Esq. A good modern house, with a ring fence, and land well cultivated.—*Friendville*—The Rev. Dr. Shirrefs. A commodious house with a good garden, and some young plantations.—*Broomhill*—William Donald, Esq. A convenient house, in a fine situation, with a few trees.—*Arthur Seat* and *Pulmore*.—Dr. Dingwall Fordyce. Two convenient houses, in a fine situation, well wooded, near the banks of the Dee.—*Ferryhill*—Lieutenant-Colonel Alexr. Tower. A commodious house, to which great additions are making by Colonel Tower, who is also proprietor of Logie. Adjoining to this is another good house, belonging to his brother, George Tower, Esq.—*Rosebank*—Mrs. Dyce. A neat house, with a good garden, and shrubbery in high order.—*Barkmill*—Mr. Raft. A small house, with neat and well cultivated garden.— *Clerkseat*—Mr. Black. A convenient old house, with some good land annexed to it, and a neat garden.—*Lochead*—William Burnett,

NETT, Esq. and Mr. CHEYNE. Commodious houses, on lands reclaimed from being a complete morass about 50 years ago.—*Cornhill*—ALEX. YOUNG, Esq. An excellent house, with an uncommonly fine garden, and hot houses, surrounded by a valuable property.—*Forresterhill*—Mr. BLACK. A commodious old house, in an elevated situation, and well wooded.—In its neighbourhood, a neat house, belonging to Dr. HAMILTON. A fine retreat for a philosopher.—*Burnside*—ALEXR. DUTHIE, Esq. A commodious house. The estate of Ruthrieston belongs to its proprietor.—*Raeden*—GEORGE MORE, Esq. An excellent house, well wooded, with a fine garden, and valuable property.—*Maestricht*—JOHN LEITH ROSS, Esq. of Arnage. A plain but comfortable house, attached to the property of some adjacent fields.—*Springfield*—W. GIBBON, Esq. A good modern house, and an excellent garden.—*Shedocksley*—A plain, but commodious house, on a valuable farm, that has been highly improved.—*Whitemyres*—A commodious house, with a good garden, and lands formerly very marshy, now well cultivated.—*Hazlehead*—Dr. A. ROBERTSON. A commodious house, well wooded, with thin sharp soil, well cultivated.—*Rubislaw*—JAMES SKENE, Esq. Advocate in Edinburgh. A very old house, with some good trees, and attached to a considerable landed estate.—*Den of Rubislaw*—Dr. DAUNEY. An excellent house, with fine hanging gardens, and an observatory; much taste displayed in ornamenting this villa, and the adjacent fields.—*Summerhill*—ALEXR. DAVIDSON, Esq. A very commodious old house, in a commanding situation.—*Villa Franca*—Miss PEACOCK. A neat rural retreat, in a pleasant situation.—*Belvidere*—ALEXR. DUGUID, Esq. A spacious modern house, with an excellent garden, and field in high order.—*Lonehead*—JOHN THOMSON, Esq. A neat villa, to which additions are now making.—*Cherryvale*—Mr. JAMIESON. A neat house, in a pleasant low situation, with a good garden.

SECT. II.—FARM HOUSES, OFFICES, AND REPAIRS.

It is with particular pleasure, that the Reporter can state, that there is in this respect a very great and general improvement in the county of Aberdeen, within the last thirty years. Yet much still remains to be done by individuals; and even an alteration must be made in the law, before the Farmers' Houses, and Offices, can, in many estates, be kept in proper repair.

Till within the period just mentioned, their condition was very bad. None of the common farmers' houses were built with stone and lime; and comparatively, a small proportion was built with stone and clay; they were generally built about four or five feet high, either with stone and clay, or with stones filled with earth, instead of wrought clay, and one or two feet of turf, (provincially *feal*) placed above the stones of the side walls, and the gabels built of the same perishing materials. The couples, or supporters of the roof, were built in the wall, the feet of them about a yard above the surface of the ground. Strong spars, called *run-joists*, were laid along side of the roof, and a number of small spars for top-pieces, of fir wood, called *watling*, across these, from the top of the roof to the turf walls. Above all these thinly pared turf, made by the breast plough, (provincially, *divots*, cast by the flaughter spade) were laid on like tiles, but covering each other very closely; and a thin coat of straw, or thatch, tied on by straw ropes, and pinned to the top of the turf wall by wooden pins, was put on every two years. Two or three small panes of glass in the top, and two wooden leaves in the lower part of a small window frame, were in each apartment of the farmer's house, which usually consisted of three divisions, all on the ground floor, and without either

any

any ceiling or upper-storey for the most part. The farmer and his servants ate in the kitchen, or place in which the fire was kindled. The master commonly sat on a kind of wooden sofa, called a *long seat*; from the back of which a deal or board of wood, three feet long and one foot broad, fixed by a hinge, was let down at the time of meals, to supply the place of a table. At other times it was fixed to the back of the sofa by a wooden sneck, to keep it from falling down. On particular occasions, a fire was kindled in the other end of the house, and a small table, made of Scotch fir, and a few chairs of the same kind of wood, or sometimes of ash, with always one, and sometimes two bedsteads, constituted the furniture of this room. A few stones, laid on the ground with clay and sand, and a large thin stone set up before these, also constituted the hearth or fire place, and a hole in the roof, containing a wooden box of four sides, fixed to the divots and roofing, but open at both ends, constituted the chimney, and was called the *lumb*. The middle division was used as a cellar, both for milk and ale; and also as a bed-room for the children. The barns, stables, and byres were generally built of dry stone, with loose earth poured into the heart of the wall, for the first three feet above ground; and above these stones two or three feet of feal, or turf, were placed, with the same kind of roof with that of the farm-houses, but commonly of weaker timbers, and without any light except what came from the door. The reason why these houses were all so bad, was, that the farmer, on leaving the premises, at the expiration of his lease, or on being warned to remove, where he had no lease, was allowed nothing for the *walls* of either the *farm-house*, or *office-houses*. He had a *certain value* of wood afforded by the proprietor, on what was called a *dead inventory*; and on the houses, of which the proprietor thus had a part, he was entitled to get a *melioration*, if

this

this was expressed in his lease, and if their value exceeded the *dead inventory* of the proprietor. But he got no allowance whatever for his houses, unless it was *expressed in writing* in his lease, or otherwise, that he should have the value of *his wood* on his removal; and the landholder's *dead inventory* was always very trifling. It very seldom amounted to *one-fourth*, and often to less than *one-sixth part* of a year's rent of the farm. In this situation it was impossible that there should be good, or even decent farm houses; and still more, that the offices should be either good at first, or kept in repair, especially toward the end of the lease.

About 50 years ago, some of the more enlightened proprietors began to give their tenants some encouragement to build *decent farm houses*; but it was not till they were compelled by the misfortunes, which befel the native farmers, that any indulgence of this kind became general. After the calamitous season of 1782, which was so fatal to many of the old tenants, it became a matter of consequence to the landholders to get a few farmers from the southern counties, to teach, by their example, a more improved system of husbandry; by which, owing to a mixture of turnips, and sown grasses, with the white, or corn crops, the farmer might not be so much dependant on the latter. Before these farmers would agree to leave their own country, it was necessary to offer them not only reasonable, and even favourable bargains of their farms, but also good farm-houses. A few of the more enlightened and liberal-minded proprietors, accordingly gave encouragement in this way, some allowing a year's rent, and some a stipulated sum of money, which exceeded a year's rent, in addition to the old trifling *dead inventory*. The native farmers of the county looked with dissatisfaction, and demanded the same kind of encouragement, or at least, desired, that on their removal, they should be paid for all their walls (if either stone and lime, or stone and clay, snecked or pinned with lime) and also for wood, slates, tiles, and stob-thatching, i. e. for a coat of straw from 9 to 12 inches thick put on the roof. These reasonable demands were for the most part, complied with; and, in some cases, the landlords generously came forward and offered these encouragements as far as was in their power. For such of them as possess *entailed estates,* are not, by the existing laws, able to give in this way, what encouragement or accommodation they would otherwise chearfully give to their tenants.

At present many of the landed proprietors allow a year's rent for building farm-houses or offices; others allow a larger sum for these purposes; but many deduct nothing out of their rental, and only *permit* the farmer to lay out his *own money* to a *certain stipulated extent* in building and repairing his houses, obliging themselves to *allow that sum* on his removal, if the houses are worth as much at that period, and the farmer must lose whatever sum more than this amount he may choose to expend. It seldom happens that this allowance is sufficient; but the farmers are now in a situation so much better than their former condition, and are so sensible of the advantage and comfort which arises from possessing good houses, that they are willing to lose a proportion of their value, on account of their great additional accommodation.

On the subject of Farmers' Houses, the following observations, made by Dr. ANDERSON, in the original Report of 1794, are highly deserving of attention.

" One of the greatest inconveniences to which the tenants
" in Scotland are at present subjected, is what regards farm-
" houses. The houses, for the most part, are very poor,
" and many proprietors throw too great a burden on the te-
" nants. This is highly impolitic. Nothing contributes
" more

" more to the content and conveniency of a farmer, than good
" and well disposed buildings. It elevates his mind, gives
" him spirit to pursue his operations with alacrity, and con-
" tributes in many instances to augment his profits. On
" these accounts he ought to have them (and indeed I never
" yet saw a thriving tenant who had not good houses.) But
" on no account should he be induced to expend that stock
" upon building houses, which should be employed in extend-
" ing his own proper business. It ought always to be done
" by the landlord; and in general a good set of houses upon
" a farm, will bring him much greater additional rent than
" the interest of the money expended upon them. If, then,
" a farmer, possessing a farm, or bargaining for one, wishes
" to have houses, these should be built by the landlord in a
" good substantial manner, on a plan suggested, or at least
" approved of by the tenant, the tenant furnishing carriages,
" and paying at the rate of 5 per cent. for the money so ex-
" pended; becoming bound to uphold the houses in good re-
" pair during the currency of the lease, and to deliver houses
" to the landlord to the value at least of the money expend-
" ed; taking the tenant bound likewise to pay along with
" his rent annually the premium of insurance for the va-
" lue of the houses against losses by fire. Any other restri
" tions are unnecessary even in this case, than those abo
" specified, respecting the last years of the lease." " No-
" thing will prove such an allurement as good buildings; and
" long leases on equitable terms. And as many gentlemen
" in East Lothian, and other southern districts, intoxicated
" with the prospects that open, in consequence of the rapid
' improvements that have there taken place, refuse to grant
" leases but on very hard terms, they have disgusted many
" of the best farmers, who, by proper encouragement, might
" be induced to go to Aberdeenshire, where they would soon

" find

"find their situation altered much for the better. Of all the plans that could be suggested for the improvement of the county, this would be found the most efficacious."

It is here proper to observe, that since 1794, when Dr. ANDERSON wrote the above paragraph, we have got farmers from Berwickshire, Angus, Mearns, and other southern districts, who have taken farms in Aberdeenshire, and many of whom have shewn excellent examples in agriculture, as well as improved their own capital; and that the native farmers of the county, in consequence of their example, both in asking good houses, and in raising good crops, are now in a much more flourishing situation, than they were previous to 1782, when one very bad season compelled them to abandon a wretched system of husbandry. But we still want very much that a law should pass, by which the possessors of entailed estates may be empowered to bind their successors, in regard to the expence of farm-houses, and other buildings necessary for the farmer, as fully as if they held their lands in fee simple. It is but candid to add, that we, in some cases, want farmers, who have so much good sense, as to set down their offices, so as to form three sides of a square, with a straw, or dung-yard in the middle. For we too often see the barns in one place, the byres in another, and the stables in a third place, quite detached from each other, while the dung is washed away by the rains, and not attended to as it ought by many of the native farmers. In short, we *often want a proper distribution* of farm-offices, in consequence of their being *no plan agreed* upon between the landlord and the farmer previous to their being built; a thing that *should never be omitted.*

The following description of a farm house and offices at Wester Fintray, (in the parish of Fintray) belonging to GEO. SKENE, Esq. of Skene, and in the possession of Mr. ROBERT WALKER,

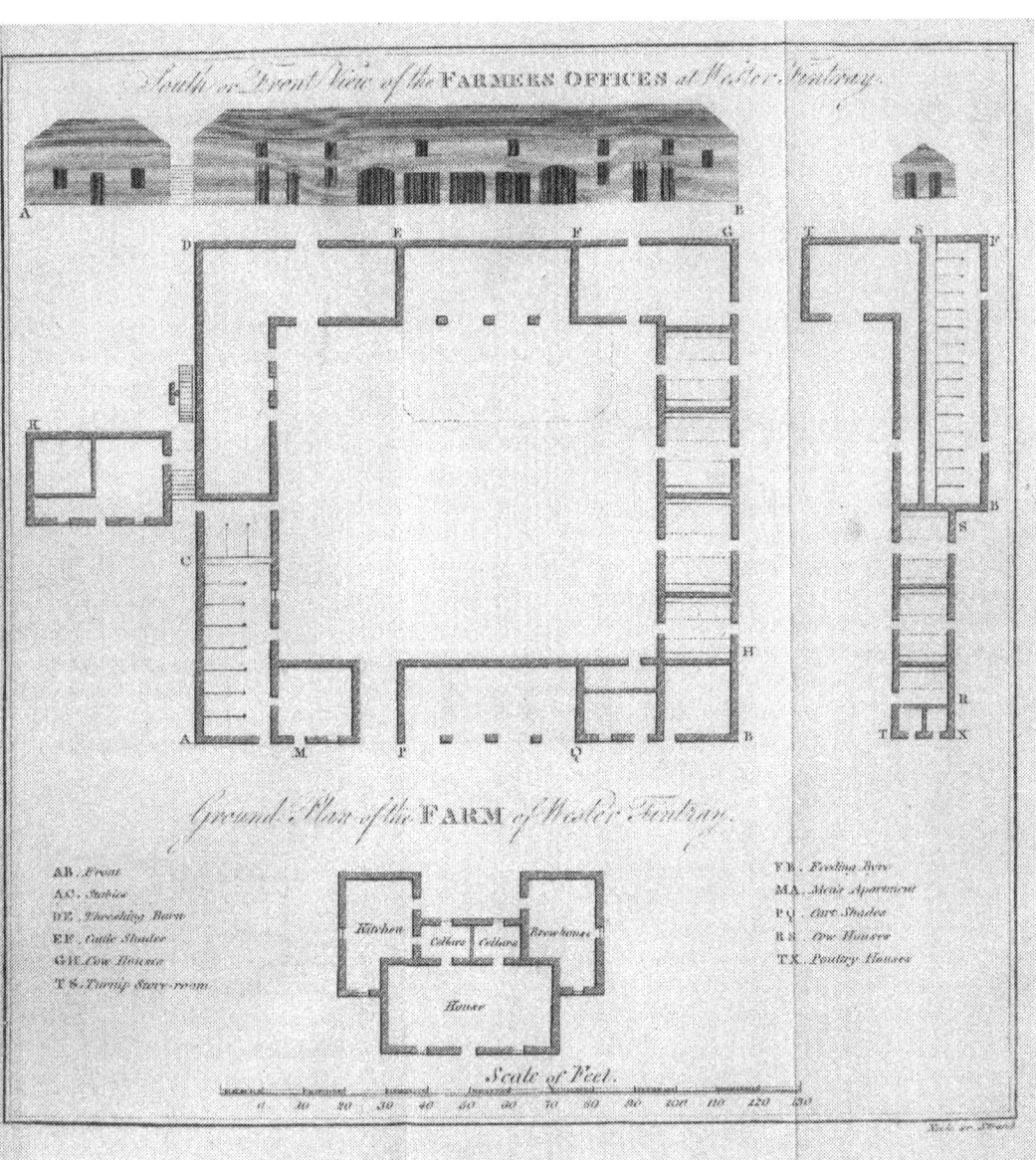

Wester Fintray

WALKER, a native of the county of Angus (who was for nine years bailiff, or farm-overseer to the late Earl of KINTORE,) will shew, that the farm house and offices are both well set down, and properly distributed. It needs only to be mentioned, that the proprietor, beside *the old dead inventory*, allowed L.150 in money, with wood from his own forest for roofing, and the price of foreign wood for doors and windows of the farm house. Also that the tenant was originally bred a house carpenter, by which means the whole work was both planned and executed in a better manner than it would probably have otherwise been. The total advance made by the proprietor was about one year and a half of the present rent, or nearly two years of the old rent of the farm. The lease of this farm was for 33 years, with two rises of rent. For the first 11 years the rent was L.130; for the second 11 years is L.175; and for the last 11 years will be L.215, including the value of some meal and bear, converted at the fiar prices. The contents of it are nearly 276 acres, besides 89 acres of moor reserved by the proprietor for planting.

The farmer's house consists of two stories and garrets. It is 43 feet in length by 21 in breadth, has a milk cellar, and an ale cellar, like the sculleries of the English farm houses, adjoining to the backwall, and communicating with the lower storey; a kitchen and brew-house, 30 feet by 20 each, making two wings, and also placed at the back of the farmhouse; 60 feet farther back, the front of the court of offices is situated. This is a square, or rather oblong court, 130 feet in front, and 120 feet deep. The different offices are 16 feet within, or 20 over walls in width, and from 12 to 13 feet high. The front contains an apartment for the *farm servants* on the one side, and a house for holding *farm utensils* on the other, each 20 feet by 16. Next the *mens house* is *an arched entry* to the dung yard in the middle of the court;—adjoining to it, and in the middle of the front are *three cart-sheds*, and between the farthest of them, and the house for utensils, is a fourth cart-shed, with an arched entry, the others being flat, and having long stones of excellent granite, nine feet long, used as lintels. These fill up the space that is in the front, and within the line of the two wings. The whole is floored above the arches; and grain lofts are in the second storey.

The west wing contains the *farmer's stable*, 46 by 16, and a common stable, 12 feet by 16. On the end of the last is a *threshing barn*, which includes the machinery of a *threshing mill, that is driven by water*, and threshes *eight bolls of grain in an hour*. This barn is 68 *feet long on the west wing*, and extends 56 *feet along the back of the court*. Adjoining to that end of the barn, are sheds for young cattle, 40 feet in length by 18 in breadth, there being no wall in front except pillars in the divisions. A byre on the end of the sheds extends along the remainder of the back part of the court, and the whole east wing, 120 feet in length, is filled with cow houses for oxen, cows, and calves, except an apartment for a wright's stable and shop, 18 feet by 20. These are the contents of the four sides of the court. But in the inside, instead of making *the whole a dung court*, the farmer has judiciously cut off about *one-fourth part for the cattle*, and has erected two rows of small sheds opposite to the large ones; and when he finds the dung is sufficiently treaded, it is removed into the open space of the inner court, which is by this means untouched by the cattle.

Besides the accommodation by the houses of this court, another row of offices stands parallel to the east wing, and at 30 feet distance from it, so as to allow a loaded cart to pass to the stack-yard, which adjoins the back-court; also a store-room for turnips is built in a line with the back wall. In this

this row a large cow-house for feeding cattle with turnips, 17¼ feet broad by 64 long, to which there is access from the store room, has a division for the servants to carry the turnips into stone troughs, into which water can be introduced at pleasure; and is one of the best models of a house for feeding cattle in the kingdom. On the end of it are other two cow-houses (provincially named *byres,*) for more black cattle; and adjoining are three poultry houses, one for geese and ducks, one for hens, and one for turkeys, of which Mrs. WALKER breeds a considerable number.

This description will shew, that the farmers of this county not only have better farm houses and offices than formerly, but also that in a county in which cattle are reared in great numbers, and turnips are now raised for feeding them, a great deal of more accommodation is requisite than is necessary in the corn-raising counties of Scotland.

In large farms, where turnip fields are at a considerable distance from the farmer's offices, it is very expensive to carry the turnips first to the cattle, and then to cart the dung of the cattle to the field. To remedy this, Mr. WALKER has lately built a cow-house, 60 feet long by 17 broad, and two wings or small byres, as they are called, 32 by 17 feet, for feeding part of his cattle on the turnips, which are raised at half a mile's distance from the farm-steading. A large avenue leads from it to these feeding houses; and eight fields are so near to them, that the turnips can be carted at a very small expence, and in all seasons. This will soon pay what is expended on building the houses.

All these offices of Mr. WALKER's are stob-thatched at present, i. e. are covered with a coat of straw, a foot thick, fixed to the roof. But he much regrets that he did not slate them at first; and he may still be induced to slate the whole,

if

if the covering of thatch, which was very thick, were once exhausted.

He has, however, lately slated a kiln-barn, which he has built near to the west wing of his court of offices. This contains a granary and a loft in one end, and an iron-plated kiln in the other, which dries four bolls, or three quarters of corn at a time. It is very complete of its kind, and cost above L.150. And the farm-house, and whole offices could not now be built for L.1000. The only thing wanting is a slate roof to the square court of offices, and this may in due time be supplied.

SECT. III.—COTTAGES.

These were formerly built with dry stones, filled with earth for three feet, and the rest of the side walls and gabels with feal or turf, like the farmer's offices. Sometimes they were built with turf entirely, and were very mean and uncomfortable dwellings. The Reporter remembers to have seen the cottager and his family in the one end, and a cow in the other end of some of the meaner cottages. They are now generally built with stone and clay; and artificers, or the better sort of cottagers, have the walls of their houses snecked with lime. They are generally from 11 to 13 feet wide within walls (12 feet is most usual) and from 24 to 36 long, having always two divisions, or what they call *a butt and a ben,* the one for a fire-place, with a bed for some of the children, if there be several of them; the other for a press and a bedstead, sometimes two bedsteads, one for the father and mother, and the other for such of the children as are young. These cottages have generally no ceiling above, and on-

only an earthen floor below; but a fire is sometimes kindled in both ends, when any of the family is sickly. They are thatched with straw once in two years, like the farmer's offices formerly; have generally two windows, about 2 feet by 18 inches, containing four panes of glass. But some of them have three or four windows, and are ceiled above in the one end, either with the thin backs of deals, or with a coarse coat of plaister. They cost from L.8 to L.10 for the inferior kind, and from L.10 to L.20 for tradesmen's houses. And since the late great rise in the price of wood, and reward of labour, they now sometimes expend from L.25 to L.40 on the houses of artificers. Though they do not look so well as the cottages in the Lothians, yet having the advantage of two apartments, no cottager in Aberdeenshire would choose to be confined to one; and indeed, when the children are grown up, or any of the family sickly, one apartment is by no means convenient. The cottages are seldom rented, but belong to the cottager, who pays a ground-rent for his cottage and kail-yard,* if he have no croft (from 5 to 10 shillings for the most part.) This holds good of cottages in the country; but in the burghs of Kintore and Inverury, as well as in the towns of Peterhead, Huntly, and Oldmeldrum, and in several other villages, they are rented, but their rent is extremely various. A number of villages have been built within these 30 years, and two of them belong to JAMES FERGUSON, Esq. of Pitfour, Member for the county, who has paid particular attention to the best plans of cottages. It needs only to be added, that the great rise on the price of wood, joined to the cheapness and excellent quality of granite in this county, has occasioned those plans of cottages and farm-offices which had pavilion

roofs

---

* A kail yard is a small inclosure of a few falls of ground for planting coleworts, or raising garden roots.

roofs, to be less eligible now than they were formerly; for wood is above six times the price for which it was sold thirty years ago, from the forest of the landed proprietor; and mason-work, though dear, has not risen in the same proportion, and yet is more durable.

After thus stating the fact, the following remarks are submitted to the judgment of the reader, on a subject which affects a very numerous, and generally industrious class of labourers.

Excepting cottages for those married servants who are attached to a farm, and for a few day labourers, who get constant employment from the farmer, neat villages, which contain from 100 to 500 inhabitants, are, in many respects, preferable to cottages, which are scattered over the country in all directions. Where great division of labour is necessary in certain manufactures, cities, and populous towns have peculiar advantages, both from the number of labourers which they contain and employ, and from the emulation and competition which take place where there are many persons of the same rank and occupation. Such persons ought not to reside in the country, at any rate ought not to hold their houses of a farmer. But for regular day labourers, or married farm-servants, cottages, built on the farm, cannot be dispensed with. And where the division of labour is not necessary, or where a little *emulation* or *competition* for public favour is useful, at least *two* blacksmiths, *two* plough-wrights, or house-carpenters, and two sets of artificer's shops in a village, are most beneficial to the country. Pure air, simple manners, and cheap education, fit their children for being useful to the farmer, who finds that their parents also are generally more to be relied on, than the stranger from another district, or hireling from a town, whom he employs to cut down his corn in harvest, or to assist him occasionally in hoeing turnips, or in hay-making.

It is, however of the utmost importance, that cottagers and day-labourers should be lodged in houses which are dry and comfortable. Their health, and consequently their usefulness, depend, in no small degree, on this; and from motives of policy and expediency, as well as of humanity, the cottages of the peasants in Aberdeenshire deserve more attention than is always paid to them by either the proprietor or the farmer. And the best and simplest remedy which the Reporter knows for this defect, is to allow every cottager melioration for the stone and clay, as well as for the wood of his cottage. The farmers have got this allowance to themselves, but it is too seldom given to their sub-tenants or cottagers. Indeed the same allowance ought to be made for well disposed and substantially executed cottages, as for other buildings, by both the proprietor and farmer.

Perhaps the time is not distant, when the *walls* of the houses of cottagers will be built with *lime;* and when, from the durable nature of our granite, they will not require to be rebuilt for a century. At present, stone and clay walls are all that can be generally expected. But both humanity and good policy require that the cottager's habitation should be at once comfortable and desirable.

CHAPTER IV.

MODE OF OCCUPATION.

SECT. I.—SIZE OF FARMS, AND CHARACTER OF THE FARMERS.

Size of Farms.

THERE is the greatest diversity in regard to the size of farms in this county. Not to mention the large crofts, from 6 to 12 acres, which an indolent farmer subsets at as high a rent as possible, in order to lessen both his own rental and his labour (a very prevalent error, which was noticed in the preliminary observations) some of the landholders let their estates in large farms, others have broken them down into small ones, and not a few retain them in their former dimensions, without regard to their being large or small; consequently they are even on the same estate extremely unequal. Yet the practice of letting larger farms than formerly, is, in general, gaining ground. In the Original Report, drawn up by Dr. ANDERSON, this diversity in the size of farms is treated at great length; and the Writer of this corrected Survey feels it his duty to quote, as far as it bears on the question, all that Dr. ANDERSON has said on this important, and keenly agitated subject.

" There is not perhaps an extensive corn-county on the " globe, in which the farms are in general so small as in " Aberdeenshire, these running, in general, from two pounds
" of

"of rent, to one hundred; but, as farms of this last size are
" rare, they cannot, at an average, exceed fifteen or twenty
" pounds. As this district is by no means in that state of
" cultivation which admits of small farms being managed to
" advantage, it becomes a matter of some importance to en-
" deavour to trace the cause of this peculiarity.

" About the middle of the last century, the farms in Aber-
" deenshire were of much greater extent than they are at
" present, and from many incidental circumstances that oc-
" curred to me during my residence in that county, it seems
" evident to me, that farmers were then in general a more
" wealthy and respectable body of men than they are at
" present; and it is very obvious, that many extensive tracts
" of land, which were then under the plough, are now aban-
" doned as wastes, and covered with heath.

" The circumstance that seems to have occasioned such a
" remarkable change in the internal state and domestic eco-
" nomy of this county, was a scarcity of grain for a succes-
" sion of about seven years, which prevailed about the close
" of the last century. This series of scanty seasons are spoke
" of till this day, by the name of *the ill years*. The crops
" during that time were so defective, and the true principles
" of commerce were so little understood here, that a fa-
" mine prevailed to such a degree, as to impoverish all, and
" greatly to diminish the numbers of the people. Many sub-
" stantial tenants were at that time reduced to poverty; their
" farms were thrown into the hands of the owners; and
" other tenants, who were capable of stocking them, could
" not be found. To induce poor men, the only persons
" then to be found, to take farms, they were made small, to
" adapt them more nearly to the circumstances of the te-
" nants; and even then it frequently happened that the land-
" lord was fain to give a team of oxen, or some milk cows in
" a pre-

"a present to induce the tenant to accept of these small
" farms. At that period, sheep were entirely banished from
" the lower districts of the county; for although when the
" farms were large, such flocks were kept as could afford
" the expence of a shepherd to tend them; when they be-
" came small this could not be done; and as money to pur-
" chase a flock could not be easily commanded, these two
" considerations concurring, banished the sheep entirely,
" and with them the corn that was to be produced from their
" fold. Since that time, the stocking manufacture affording
" a ready resource to those who had a small bit of land to
" produce a little meal, and a cow to give the family milk,
" many persons chose rather to betake themselves to that
" mode of life, than to continue labouring servants. Ser-
" vants, of course, became scarce, and their wages high.—
" The business of a large farmer became then a very disa-
" greeable one. From the difficulty of obtaining extensive
" markets, farming became a retail business instead of a
" wholesale. Servants soon acquired as much money as to
" enable them to take one of these small *possessions*, from
" two to five pounds rent, for farms they are not called; and
" many proprietors tempted by an offer of rent, seemingly
" high, for these small patches, broke down their farms into
" minute allotments, which has gradually reduced the occu-
" piers of land to that degraded state they hold at present;
" the consequences of which, as to this county, have been
" extremely deplorable.

" Many treatises have been written by speculative men,
" to prove the numerous advantages to be derived to the com-
" munity from small farms, rather than large ones, without
" their seeming once to have taken a glance at those cir-
" cumstances which are necessary to fit land to be farmed
" with advantage in small lots, and those which render it
" im-

" impossible to cultivate the soil with any degree of proprie-
" ty, unless when it is done upon a large scale; for there
" are natural bounds in this case which cannot be disregarded
" with impunity.

" When ground, naturally good, has been enriched by
" frequent manuring and good culture, so as to have obtain-
" ed a high degree of fertility, it can then be cultivated with-
" out the aid of beasts of labour of any sort: And if it be near
" a market for all the articles of produce, it may then be let in
" allotments that are no bigger than is just sufficient to main-
" tain the single family that cultivates them. In these cir-
" cumstances, and under this management, the greatest pos-
" sible population may be maintained by a given extent of
" soil. But where the lands are poor, and never have been
" fertilized by good culture, the case is quite otherwise. The
" scanty herbage naturally produced on these bare fields
" must be consumed by domestic animals, which must have
" a wide range to pick up their poor pittance. These ani-
" mals must be tended by some person, and their dung col-
" lected into a narrow compass to fertilize a small portion of
" the soil. A considerable number of these creatures can be
" guarded nearly at the same expence as a few; so that if one
" flock were divided into ten, the loss that must be incurred
" by this want of economy is obvious. Beasts also must be
" kept for the purpose of labouring; and the teams must be
" of such strength, as to perform the rugged operations that
" are necessary. The same expence nearly must here be in-
" curred for plowing a few acres, as for doing as many as the
" team could manage properly during the season. Nor can a
" single team only be employed with economy in many cases,
" as a variety of the necessary operations in these cases,
" require occasionally a much greater power than that
" could afford. If stones are to be taken out of the soil,

K " where

"where these are of great magnitude, they require many
" hands, expensive tools, much time, and great dexterity to
" effect it. The assistance of all the force that can be given
" by three teams, is often required to assist operators skilled
" in that work, hired on purpose. When these, therefore,
" cannot be commanded, the work cannot be done. In le-
" velling high ridges by means of the plough, at least a do-
" zen of men are necessary to admit of this operation being
" performed, in the most economical and perfect manner.—
" In the same manner, draining, inclosing, leading manures
" from a distance, cannot be done at all, in many cases, but
" upon a farm of considerable extent. To talk then of small
" farms, in these circumstances, is absurd. Wherever they
" take place, the improvement of the country must be at a
" stand, and it must be doomed to perpetual sterility.

"Nor are these the only operations in which a command
" of considerable strength upon a farm is required; nor is a
" state of great sterility the only circumstance that makes it
" necessary to have a certain extent of farm to carry on the
" ordinary operation with economy. To give a single exam-
" ple. On every farm, the dung produced upon it must be
" led out to the fields. To perform this operation with due
" economy, things must be so arranged, as neither to allow
" the beasts, nor the persons employed, to waste their time in
" idleness during the working hours; but if few persons be
" sent to fill the carts at the dunghill, the beasts will stand
" idle a long time at filling; and on the other hand, if too
" few carts be employed, the fillers must stand idle during
" the intervals that they are absent. To carry on this ope-
" ration with spirit and economy, four persons at least, toge-
" ther with the driver, must be employed in filling; these
" five, in a very short time, fill the cart, so that the horses
" are not allowed to stand idle; and if the fold be at hand,
" and

"and a person employed there, to draw out the dung into
"small heaps, three carts will be sufficient to keep a con-
"stant succession at the danghill, and no more.* Thus eight
"persons, and three carts, are the fewest that can be em-
"ployed to perform this common operation with the strictest
"economy; but if the field be at a distance, one, two, or
"more carts, in proportion to the distance, must be added,
"in order to obtain the same economical dispatch. An at-
"tention to economical arrangements of this nature is neces-
"sary in all the operations of farming, where the greatest
"profit from the farm is looked for. Thus in plowing ground
"for barley, where it is necessary to sow it as quickly after
"the plough as possible, in order to insure its germinating,
"one person must be kept for sowing, another for harrowing,
"and so many for ploughing. The proportion between the
"ploughs and harrows, &c. must be so preserved, as that
"the whole shall have constant employment, and no loss of
"labour be incurred. It is plain, that if the farm be of such
"size only, as that a single plough can be kept, this cannot
"be done. Before enough could be ploughed to admit a
"harrow to go, in many seasons, and on various soils, it
"would be already too much dried on the surface for sow-
"ing: Even were this not the case, the plough must be stop-
"ped till the ground be sown; the same horses must then
"be put into the harrows, and again put into the plough af-
"ter that is finished. Every idea of proper culture, and due
"economy, in such a case, must be abandoned: yet often
"have

---

* There is a small mistake here;—only three persons can be profitably employed in filling a dung cart; but other three are necessary to
assist in taking the dung out of it, and spreading it on the ground. So
that the Doctor is right as to eight persons in all; and the ploughs
ought, on large farms, to work at the same time.

"have I seen this wretched shift adopted in Aberdeenshire,
"with the single alteration, that a starved cow was sometimes
"obliged to be substituted, instead of a better beast, to
"scratch the surface of the soil.

"From these considerations, it is obvious that no determi-
"nate size of farms can be considered as invariably the best,
"but that this must be regulated by circumstances; and it is
"evident that Aberdeenshire in general is far from being in
"that state which can admit of their being reduced with eco-
"nomy to a small size."

The importance of the subject will justify the length of this quotation from the Original Report, which may be considered in three lights:—1st. As recording the matter of fact.— 2d. As assigning causes for this; and, 3d. As adducing arguments against the small size of farms in this county.

1st. As to the fact. Since 1794, a number of small farms have been thrown together in larger ones; and the rent of land has risen considerably since 1794. Several farms, no doubt, have been broken down into villages, and this alteration has been of equal service to the proprietors, to the neighbouring farmers, and to the villagers. But setting aside all villages, and also those broken and detached pieces of ground, which have been generally let to small farmers, on account of their local situation, or distance from the larger farms on the same property, the average size of farms is not now under 100 acres of both arable and pasture lands; and the rent in the lower parts of the county from L.30 to L.50, according to the value of the arable, and the proportion of barren land. The knitting of stockings, to which Dr. ANDERSON ascribes the breaking down of farms, was an employment *only of women* and *children*, not of *men*, except of those who were unable to endure servile labour. While meal and other provisions were cheap, the knitting of stockings employed a *cottager's tager's wife and daughters,* and *he himself* acted occasionally as a *day-labourer* to the landholder, working at his *Mains* or *personal farm,* or sometimes assisted the neighbouring farmers, when called upon by them, and received in return *no wages in money,* but *the use of their cattle* in ploughing and harrowing his land, or in carting the dung to his small possession. When this consisted only of two or three acres of arable land, exclusive of the pasture which maintained one or two cows, he was a useful member of society, and by his own and his family's industry, he generally improved his situation, and brought up his family suitably to his rank in life. But as his sons grew up, he wished to extend the size of his farm, and feeling that he was much more independent, when he held his few acres directly of the landed proprietor, than while he rented it from the farmer, he offered a high rent, as things then were valued, for a *large croft,* or *small farm,* of 10, 12, or 16 acres. Too many landed proprietors, in order to increase their rents, broke down their farms from the year 1760 to 1782; and these small tenants getting one or two weak horses, united them with their cows in a wretched plough. Their carts also, like their cattle, were diminutive, and their farming utensils scanty or imperfect. They were no longer day-labourers; and they conducted their farming operations in a superficial or slovenly manner; while their sons, who assisted them, were worse fed and worse clothed, though they had less hard labour to undergo, than if they had acted as servants to the large farmers. The very unfavourable season of 1782 wasted their little capital; and such as survived that shock were unable to get the better of the two calamitous years of 1799 and 1800. The landholders in general saw the bad consequences of letting their lands to men, who had neither knowledge nor capital; and a number of these small possessions, have been re-united into their ancient farms. In some places, where there is abundance of peat moss, and where labourers are scarce, the landholder is still induced to prefer small tenants to large farmers. But in general the breaking down of large farms into small possessions is on the decline at present. This is the fact in 1810, considerably different from what it was in 1794, according to the most accurate observation which this Reporter could make, and the best information which he could procure, with respect to the small tenants, or farmers who hold of the proprietor. There are still too many large crofts, which hold of the farmer, and where the sub-tenants plough their own grounds, as has been already mentioned in the Preliminary Observations. But a cottager, who has only an acre or two of land, and pays only L.2 or L.3 of rent, and acts as a day-labourer, a farm servant, or an artificer, engaged in an occupation necessary in the country, is a useful member of society, whether he hold of the landed proprietor or farmer.

With respect to the causes of the small size of farms, the Reporter does not entirely coincide in opinion with Dr. ANDERSON. He has seen no reason to believe, that the *farmers* in Aberdeenshire were in general, in the middle of the 17th century, in a more wealthy and respectable state than they are in at present. It is a fact, not generally known, that (owing to the civil wars, and to the effects both of famine and pestilence, which prevented the tillage of the land) the price of corn *was higher from the union of the two crowns,* in 1603, to the accession of Queen Anne, in 1702, than from *that period to the year* 1768, when the knitting of stockings, and the breaking down of farms became more general. The *civil* wars, in the reign of Charles I. and before the restoration of Charles II. *injured the agriculture* of Scotland very deeply. In *one case only,* the *tenants,* at that period (or rather the persons who *held the lands from many of the heritors*) were more *repectable* than

in

in 1794, because a number of these were either *gentlemen*, the *younger sons* of landed proprietors, or *monied men*, the creditors of the landholder, who obtained *wadsetts*, or the *temporary property* of farms for their patrimony, or as a security for money due to them, and for the interest of their capital. But many of these let their farms in *sub-leases*; and agriculture was not at all studied in the 17th century. Farther—with respect to the unfruitful seasons in the end of the 17th century, i. e. from 1693 to 1700, as far as the Writer of this Report has been able to trace the fact, these had not the effect of breaking down, but rather of enlarging farms, in order to induce the few persons who had any capital remaining to take them from the proprietors. In the Reporter's parish, a year's rent was given to one farmer, that he might be prevailed on to take another farm adjoining to his own;—a pair of oxen was next presented to him, to induce him to take a lease of a third; and a chalder, or 16 bolls of oats, to bribe him to take a fourth farm into his possession. By this means he became the tenant of 4 farms, or of 9 ploughs of land, which in his old age he divided among his sons. It was because he *had a considerable capital*, and could be *depended on for paying the rent when due*, that these *douceurs were offered to him*. From this fact, of which he has the clearest evidence, and from others of which he has been credibly informed, he presumes that the landed proprietors in 1700, after the *seven ill years*, as they are called, *bribed* those who had a little stock to take as *large farms as possible*; instead of *giving presents to men*, who had *no capital*, to take *small farms*.

Though it does not bear directly on this question of the size of farms, yet it deserves to be noticed, that the many traces of the plough, which appear on the ridges of high hills, where the ground is now covered with heath, are not to be ascribed to the desertion of farms from 1693 to 1700;

but

but to a different cause, and a much more remote period.—In the antient and uncultivated state of the lands of this country, the *lower part* of Aberdeenshire, and almost all the *glens* or valleys, were one *vast forest*. The *little arable* land was on the sides or *ridges of the hills*. When the country became populous, and the inhabitants more attentive to agriculture, after the valleys were cleared of wood, the plough was *brought down to the plains*; and as the last were more fertile and more easily cultivated, the *hills were then deserted by the farmer*. The Reporter has not been able to trace any considerable extent of arable land in the hills, which was deserted only a century ago, except a few hundred acres in the glens of Foudland, and division of the Garioch, the inhabitants of which were said to have been destroyed by famine, and which was never afterwards re-peopled. But he knows, that in King Robert Bruce's time, the forest of Kintore, the forest of Drum, and the forest of Stocket, above Aberdeen, covered a great part of the lower division of Marr, in this county. The first of these gave name to the royal burgh of Kintore, which signifies the *head of the wood*, or *head of the oak*; and all the three forests, though spread over extensive valleys, are called Forests in the Charters of K. Robert Bruce.

To return to the important question of the size of farms. The conclusion drawn by Dr. ANDERSON, though he clearly favours large farms, is certainly well founded, that no determinate size can be considered as invariable, but that this must be regulated by circumstances; though in Aberdeenshire in general, the farms are too small in the internal parts of the county. The case is very different with respect to the cultivated lands *near Aberdeen*. There it is entirely owing to their *small size*, to the great quantity of *street dung*, and of other manure, and to the *mixture of the spade with the plough*

plough husbandry, that the poor thin soil is so very productive. No large farmer, cultivating the soil with the plough alone, could raise nearly so great a crop on the *three thousand acres* which ly in the neighbourhood of that city, as is raised by at least *three hundred small* farmers by the *spade alone*, or by the *spade and plough* united, or used alternately. And no power of horses, with a number of men attending them, could remove those masses, and level those hills of granite which the spade, the mattock, the lever, the wedge, and sometimes the boring-iron, followed by gunpowder, can easily reduce. In short, when the plough alone is used, large farms are preferable in most cases. Where the ground is broken, and lies detached, or when the spade alone, or a mixture of the plough and spade is used, small farms are most eligible.

Character of the Farmers.

THE cultivators of the soil (independently of those in the vicinity of Aberdeen) belong to three distinct classes, who have each their discriminating qualities, or characters. The antient farmers of the county, i. e. such of the native farmers as are far advanced in life, or dwell at a distance from the towns, are generally honest and sensible men. Some of them are very simple in their manners, and plain in their dress. Others of them are naturally shrewd and acute.— But all are distinguished by civility to strangers; though many of them are rather indolent in the management of their farms, especially in some of the higher districts. They are generally good judges of black cattle; and many of them can also judge pretty accurately the qualities of horses, though this faculty is not so common—probably from raising fewer of them. They are too little attentive to a proper rotation of cropping, because very few of these old farmers have studied the principles of agriculture, but retained

the

the wretched husbandry of their fathers, till dire necessity made them better practical farmers than they were formerly. They are in general fond of going to markets, and public places of resort; and they mix together frequently in society, especially in the winter months. When employed at shooting at a mark, which was much practised 60 or 70 years ago, their fathers were excellent marksmen. And they decided in this manner all raffles or lotteries; which about the Christmas season are now most commonly determined by the less manly amusement of cards and dice. Also playing at foot-ball was generally practised at Fastens-even. In some places, the first practice has been revived; and from the necessity of the times it deserves to be encouraged; but, in too many cases, cards and dice are substituted in the room of the more manly exercise of the fowling-piece and foot-ball. This was owing to the absurd and unjust law in 1746, which disarmed the highlanders; and is still occasioned by the multifarious game laws, which do not promote the strength of the nation. These antient farmers formerly used a great deal of malt liquor, and more lately, a considerable quantity of ardent spirits. But though they occasionally indulge in a social glass, they are now in general temperate; though they are sometimes accused of being irascible, and fond of law suits.

Of a character in many respects opposite to these, is the class of farmers, who have been induced to leave the southern counties, and to take leases of farms in Aberdeenshire. Most of these are very intelligent, industrious, and deservedly esteemed for their private characters, and for the good example which they give to their neighbours, the native farmers, of an improved system of husbandry.—Superior to these in many respects; though generally inferior in point of natural acuteness, and in the intuitive and accurate judgment of the value and qualities of cattle; they dwell in good houses,

houses, live hospitably and comfortably, and generally take the lead in agricultural societies, or ploughing matches, which they are active in forming, and impartial in judging. Their conversation turns chiefly upon their own profession, in which they are desirous to excel; and they now live without any jealousy, and for the most part in high estimation among the native farmers, (who were somewhat jealous of them on their first arrival:) while they are made welcome at the tables of many of the landed proprietors. On the whole, they have no reason to regret their having left their native counties and settled in Aberdeenshire. For they are now generally worth ten times as much of capital, as they had when they first came to this county. At the same time it must be acknowledged, that the landed proprietors gave them advantages, with regard to their houses, their bargains, and even as to the choice of their best farms, which were not commonly granted to the native farmers; and that the former, on rugged lands, or on unfruitful farms, have not made so great exertions as the latter.

Between the extremes of these two classes, is the more numerous one of younger farmers, or of men not beyond the middle of life, who have contracted the greatest zeal for agriculture, attend the meetings of agricultural societies, and both endeavour to understand the principles, and pay great attention to the practice of agriculture. They readily acknowledge their errors, when they commit a blunder in this respect; feel pleasure in associating with their neighbours, who have come from the southern counties, and pay them that respect which is due to their greater experience. At the same time, with that acuteness which generally characterises the natives, they quickly perceive any defects in the farming practice of these strangers, which they are happy to discover, and desirous of concealing, but which they take

care

care to avoid. Many of them have also a spirit of enterprize, which is not possessed by their neighbours, who have come from the southern counties. Hence the greatest and most expensive improvements have been effected by the better sort of the native farmers, who have spared neither money nor labour. Many of these, who are desirous to excel in agricultural knowledge, read what books they can get on Husbandry; and The Farmer's Magazine, (a most useful periodical performance) is held by them in very general esteem. They also attend the markets in their neighbourhood pretty regularly; are very skilful in rearing, and also both in buying and selling black cattle; sometimes take a social glass, but are very rarely drunk; and though frequently merry, are seldom riotous. In the higher parts of the county, the young farmers, like their fathers, are more irascible than those on the sea coast; and they are very expert in dancing and managing a cudgel, without being taught by any master of these arts. In the lower division, the dancing master has been introduced; and in the winter evenings has found in many places a good many scholars. But where the young farmers are as zealous as many of them are at present, this amusement is not allowed to interfere with the seasons of labour, or with the hours which nature intended for rest. In the wearing apparel of both sexes, there is a very great change; and the young farmers have, *at least, one suit of good English cloth*; yet they dress in plain and strong cloaths in the *hours of labour*: and it is often found, that those who *dress most genteelly* at church or market, are most *indefatigable* and *active* in conducting the plough or cart, or wielding the hoe or scythe. Their religion, and indeed that of the other classes of farmers, is generally sincere, without any bigotry. Not only Protestants, of all sects, meet together in amity, but in this moderate county, where religious disputes

were

were never carried so high as in the southern and western districts, the Roman Catholics, of whom there are a good many in the higher parts of Marr, cannot, by their conversation, be distinguished from other professions or sects. Such is the general character of the farmers in the country, to which there are no doubt a few exceptions. May they all long possess these good qualities, and may those who are defective learn to acquire them.

The characters of the small farmers, or cultivators of the land, in the immediate neighbourhood of Aberdeen, differ, in many respects, from those of the farmers in the country. Many of them are opulent and intelligent merchants, who have erected elegant villas, in the midst of a few acres of highly cultivated land, formerly covered with heath, or with masses of granite. Their gardens are laid out with taste, or loaded with produce, and the grapes in the hot-house, with the melon, pine-apple, and native productions of a warmer climate (here raised artificially) indicate both the accumulation of capital, and the enjoyment of riches. Intermixed with the elegant villas of the opulent, are the neat, but commodious houses, or cottages of those manufacturers or tradesmen, who, having made a competence, choose to employ part of it in building a small cottage, and in dressing their little gardens. Some of the cultivators of these lands are poor gardeners, who maintain themselves, and their families by raising vegetables of different kinds for the citizens.— Their rents are extremely high, yet by their industry and their sobriety, they are enabled to pay them. Others of them are workmen in the shops of different artificers, who rent a little ground to afford them both exercise and articles of provision. The Reporter has seen, among this number, half a dozen of woolcombers at work by four o'clock in the morning, hoeing potatoes for family use; and was informed by these indusdustrious than, that in order to counteract the effects of their confinement, and unwholesome employment, they spent the summer mornings, before six o'clock, in thus promoting their health, and providing for their families. It would be as absurd as it would be *false*, to draw all those *characters* in the *same colours.* In all there is *more acuteness and intelligence* than is generally found among *plain country farmers.* In none is the *same simplicity of manners* which are usually found among the inhabitants of the more inland parts. But *manners are different from morals;* and private virtue, public spirit, and every estimable quality, are found in all ranks of society, whether in the *light of the city,* or in the *shade of the country.* The *cultivation of the soil has a natural tendency* to promote *health, industry,* and *happiness.* The *merchant* retiring from business, to enjoy his rural retreat in the evening, *sleeps much sounder* in the pure air of his villa, than if pent up in a city in the summer months. The *manufacturer* and *tradesman,* find *retirement, health, and competence* in their cottage and small gardens. The *gardener,* by the labours of the day, *enjoys pure air,* and *wholesome exercise,* and *returns* to the city to his family in the *evening.* While the artificer, and even the woolcomber seek health in the *morning,* by *cultivating the soil* in summer, and raising vegetables to *support their family.*—When one considers these elegant villas, commodious houses, and neat cottages, with all the various scenery of these highly cultivated, but naturally barren fields, and reflects on what they were 40 years ago, (covered with masses or quarries of granite, interspersed with heath, and altogether unproductive,) one cannot avoid applying to Agriculture in particular, what our amiable Poet has said of Industry in general—

   All is the gift of Agriculture which
   Exalts, embellishes, and renders life
   Delightful.—

People

People employed in Agriculture.

Although the population of the county of Aberdeen exceeds 123,000, yet there is nearly one-third part of that number of persons who inhabit the county-town, or the different boroughs and villages, and who chiefly depend for their support on their various trades and occupation.

The population of the city of Aberdeen, and of the parish of Old Machar, which includes Old Aberdeen, Gilcomston, and a considerable district of country, is, in round numbers, - - - - 27,500
That of Peterhead, Frasersburgh, and the villages of inferior note in Buchan, is - - 9,000
That of Kintore, Inverury, Oldmeldrum, and other villages in the Garioch, - - - 2,000
That of Newburgh, and other villages in Formartin, 500
That of Huntly, and other villages in the division of Strathboggie, - - - - 2,000

Persons residing in towns, nearly one-third of the whole population, - - - 41,000

Yet a number of these, though resident in towns, and also a number of the inhabitants of Old Machar, who really live in the country, though in the vicinity of Aberdeen, depend almost entirely on agriculture. On the other hand, a number of women and children, in various districts of the county, derive their subsistence from spinning lint or cotton, or from knitting stockings, though this once valuable manufacture is much on the decline. Hence in the country parishes a very small proportion of the inhabitants derive their support from either of these sources. In the Reporter's parish above eleven-twelfths were found to depend on agriculture alone. (It was otherwise 30 years ago.) And yet from the improved mode

mode of farming, though their rents be nearly double, the people live more comfortably.

Estimating the number of those who derive their subsistence from commerce and manufactures in the country, to be equal to that of the inhabitants of Aberdeen, and other towns, who derive their subsistence from agriculture, which the Reporter believes to be very near the truth, two-thirds of the whole inhabitants of Aberdeenshire, depend almost entirely on raising corn, and rearing of cattle. At any rate, the number of those who at present depend upon the produce of the land is not below 80,000.

When we look at the number of persons resident in individual families, we find that on large farms from 250 to 400 acres, a farmer's family varies from 15 to 20, according as he has many children, and few married servants, who indeed are not so common as in the southern counties. On smaller farms, this number varies from 8 to 15; and in large crofts, or cottages of different descriptions, from 8 or 9, where there are many young children, to 2 where there are none. Where the number of children exceeds six, and the family consists of the father, mother, and also sometimes of either grandfather and grandmother, the older children go to service. In the Reporter's parish the greatest number of persons residing in a farmer's house is 20, and in a cottager's 9. The average in a farmer's house is 8, and in a cottager's 4. In these respects there is probably little difference between the farms in Mid-Lothian and Aberdeenshire. For in the latter, though the farms are less in point both of value and extent, fewer married servants are kept, than are kept in the Lothians.

Dis-

Distribution of Crop.

Here he difference is very great between the corn-raising counties of the south, and the cattle-rearing counties in the north of Scotland.

Mr. George Robertson, in his very able account of the Agriculture of Mid-Lothian, has candidly stated the manner in which the different crops were distributed on his own farm, which he then rented at Granton, in the vicinity of Edinburgh. It may be proper to state in what way the land is cropped in the county of Aberdeen, on a farm of nearly equal dimensions. The buildings both of the farmer's house, and of the offices at Wester-Fintray have already been described. The distribution of crop on the same farm, with the number of servants, horses, and black cattle, shall now be stated; and that the reader may see at once the different state of agriculture in the south and north of Scotland, the corresponding articles on Mr. Robertson's late farm of Granton, shall be marked in columns opposite to those of Mr. Walker's farm of Wester Fintray.

It needs only be remarked, that though wheat makes a part of Mr. Walker's crop, and though he raises a proportion of barley, and also of pease and beans, yet by far the greater number of farmers in Aberdeenshire raise only bear or bigg, and oats, without either wheat, pease, or beans, except in the division of Buchan, where drilled beans are commonly raised in greater or less proportions.

CROP.

	GRANTON. Scotch Acres.	WESTER FINTRAY. Scotch Acres.
Wheat,	50	7
Barley	35	2
Bear or Bigg	0	23
Pease & Beans	35	4
Oats	50	80
Hay from sown grass	50	25
Past. from do.	20	90
Natural Pasture	5	3
Potatoes	15	4
Turnips	10	27
Summer Fallow	15	11
Total Acres,	285	276

HORSES AND BLACK CATTLE.

	GRANTON.	WESTER FINTRAY.
Horses in plough and occasionally in cart	10	4
Constantly in cart	2	2
Threshing Mill	2	0
Horse for riding	1	1
Colts	4	4
Total Horses	19	11
Bull	1	1
Cows	4	16
Oxen in the plough	0	6
Black Cattle of different ages	0	62
Total	5	85

Families on the Farm of GRANTON 16 WESTER FINTRAY 5
Of which wholly maintained - 12 - - - - - 5
Souls in do. - - - - - - 59 - - - - - 33
Families partly maintained - 4 - - - - - 0
Souls do. - - - - - - 11 - - - - - 0

                              Total, 70   33

By

By comparing the different quantities of corn and green crops, raised on these two farms, it will be found that the produce of Granton, in wheat, barley, pease, and beans, at a moderate conversion of the prices of grain, amounts to L.635 more than that of Wester Fintray, but is L.150 inferior in respect to the price of oats; consequently is superior on the whole corn crop by L.485; but that on the green crops, viz. hay, pasture from sown and natural grass, potatoes and turnips, Wester Fintray is superior by the sum of L.211; therefore the produce of Granton is only L.274 superior on the whole to that of Wester Fintray. On the other hand it deserves to be noticed, that the rent of Wester Fintray is at present only L.175, while that of Granton, in 1793, was L.636, or L.461 more than the present rent of Wester Fintray, and L.505 more than its former rent (viz. L.130) previous to 1802.—That the farm of Granton is now rented much higher than L.636, (no less it is said than L.1200) that Mr. ROBERTSON has left it, and come to the parish of Arbuthnot, in Kincardineshire, where he has got the management of a valuable estate; while Mr. WALKER, who in 1791, took the farm of Wester Fintray, began with less than L.200 of capital; but from the indulgence of his landlord for some years after the commencement of the lease, from the good opinion of his industry and character, which procured him credit, from living with economy till he had acquired some capital, and from the particular attention paid to the rearing and feeding of black cattle, by the sale of which he has drawn L.500 in one year (while he paid only L.50 for what part of these was not reared upon the farm) has raised his whole stock of horses, black cattle, and farm produce to L.3700; has inclosed the greater part of his farm, has embanked a thousand yards of the river Don, which frequently overflowed, and destroyed the crops of the old farmers; has improved his land by lime,

drain-

draining, dung, and green crops, and laid out so much on building houses, that the remainder of his lease, of which 15 years are to run, (and of which for the last 11 years the rent is L.215, or L.1000 less than Granton is at present) would now sell for at least other L.3700; while the landed proprietor at the expiry of the 33 years, will, at least, *quadruple* his *rent*, and the *real value* of the farm will be *more than double* of what it would have been, if it had continued in the possession of *the five old farmers who* formerly rented it. These facts, respecting the farms of two very intelligent farmers, convey lessons of instruction both to the landholders and to the farmers, in both the southern and northern districts of Scotland, in respect to rent, produce, distribution of crops, and the advantage of large over small farms, superior to what are to be deduced by any arguments or speculations. It is only necessary to add, that Granton, 60 years ago, was occupied by three farmers, each of whom kept two cottagers, (making only nine families in all) and 11 horses, or 33 altogether; and that Wester Fintray, at the same period, contained 7 farmers; that Mr. WALKER rents four of these farms, and nearly the half of the 5th; but that the population was then more than double of what it is at present. Corn-farms require more labourers and servants, than where a large quantity of grass is used, and cattle are reared in great numbers.

This comparison will shew the great difference between the stock raised and maintained on a farm in Aberdeenshire, in general, compared with that in the Lothians. But the distribution of crop is considerably different on the coast of Buchan, from what it is in the other divisions of the county of Aberdeen, because there is a great proportion of clay in the soil of Buchan, and pease, and drilled beans, which seldom succeed in the other districts, are here raised in considerable quantities. Therefore in addition to the distribution of of crop used by a farmer, who is a native of the county of Angus, and who has farmed so successfully in the parish of Fintray, near the boundary of Garioch and Formartin—it may be proper to state the distribution of crop which is practised by the best farmers in the other divisions of the county.

The following may be considered as a specimen of the mode of cropping in the division of Buchan. They are given by four farmers, who rent their farms from JAMES FERGUSON, Esq. of Pitfour. One of these, Mr. THOMAS LOGAN, is a native of Berwickshire, who, by obtaining a good bargain of two farms, and reasonable encouragement for building houses, was induced to settle in this county. The other three, Messrs. ALEXANDER WATSON, GEORGE FALCONER, and ROBERT SCOTT, are all natives of Aberdeenshire.—The measure is in Scotch acres.

Crop 1809.	Mr Logan's Farms.	Mr Falconer's	Mr Watson's	Mr Scott's
Wheat, acres,	8	3	0	0
Barley,	8	0	2	0
Bear or Bigg,	13	16	11	6
Oats with Seeds,	58	35	24	28
Oats from lea,	67	40	44	65
Pease or Beans,	5	3	5	0
Hay,	52	34	15	24
Grass for soiling,	10	4	0	0
1 year's pasture,	0	10	6	17
2 year's do.	58	27	18	52
3d, or old Grass,	74	45	21	70
Turnips,	13	10	10	17
Rutabaga	3	0	0	0
Potatoes,	5	3	2	0
Tares,	1	0	0	0
Fallow,	45	34	10	26
Total Scotch acres,	420	264	168	305

These

These farms lie in the north-east corner of Buchan, and in the north-east extremity of the island. On that account the distribution of their crops has been selected.

Towards the north-west lies the parish of Forgue, in the division of Formartin; and adjoining to that of Strathboggie, and immediately below it is situated the parish of Auchterless, both of which border on the county of Banff. In these parishes reside some of the earliest and best improvers of land, all native farmers. The following is the distribution of crop on the farms rented by Mr. JAMES ALLARDES of Boyndsmill, in Forgue, and of Mr. CHARLES BARCLAY, in Mill of Knockleith, in Auchterless; both of whom, besides small properties of their own, hold in lease pretty extensive farms from different landed proprietors.

Crop.	Mr. Jas. Allardes.	Mr. Chas Barclay.
Bear or Bigg,	20	10
Oats, with seeds,	50	29
Oats after lea,	60	32
Oats, second crop,	60	25
Pease and Beans,	0	4
Turnips,	47	21
Potatoes,	6	3
Hay,	50	14
Cut for soiling,	10	6
1 year's pasture,	80	20
2d year's do.	42	48
3d, or old Grass,	10	130
Fallow,	5	16
Total,	440	358

In the interior parts of the division of Marr, where a proportion of *wheat* has been lately raised by a few industrious farmers, two brothers, Messrs. JAMES and WILLIAM SCOTT, de-

deserve to be selected, on account of their successful cultivation of the soil. The first rents the farms of Achath, in the parish of Cluny, at 16 miles, and of Black Chambers, in the parish of Kinellar, at 11 miles distance from Aberdeen.— The second rents the farm of Mains of Craigievar, belonging to Sir WILLIAM FORBES, at 27 miles distance from that city. They are native farmers; but their father, Mr. DAVID SCOTT, came from Angus or Forfarshire, and was factor, or land-steward over three large estates, and also was one of the earliest and best improvers in this county. Their distribution of crop is as follows:—

Crops.	Mr. James Scott.	Mr. William Scott.
Wheat,	19½	9½
Oats, with seeds,	22½	0
Oats from lea,	57	32
Oats, second crop,	38	19
Bear or Bigg,	13	11¾
Turnips,	16	8
Potatoes,	4	3
Hay,	37½	15
1 year's pasture,	60	7½
2d year's pasture,	41¼	21
3d, or old Grass,	41	22
Fallow,	22½	12
Total,	372	160

It may be here remarked, that of the wheat on Mr JAMES SCOTT's farm, 7½ acres belong to the late farmer, Mr JOHN ROBERTSON, who has let the farm of Black Chambers in a sublease to Mr. SCOTT, at four times the rent which he pays for it:—a proof both of the rise of the rent of land, and of successful cultivation of the soil.

In the higher districts of Marr, or towards its south-west

extremity, as far as the plough has reached, turnips and sown grasses are little used. This is partly occasioned by the difficulty of restraining the sheep, of which there are considerable numbers, from destroying both these crops in winter; and partly by the impossibility of preventing, in the neighbourhood of the high mountains, the flocks of deer, (whose depredations are so destructive,) from eating off a great proportion of a field of turnips in the stormy nights, when hunger renders these animals both vigilant and enterprising. Potatoes are raised with success, but turnips cannot be preserved; and we are not to expect a regular distribution of crop in such circumstances. It must be at the same time acknowledged, that agriculture is not well understood, and that the farming practices of this district as they are conducted by *the common farmers, are not to be commended.* The personal farms of the landed proprietors are however kept in good order. On that of PETER GORDON, Esq. of Abergeldie, the Writer of this Report saw 17 acres in green crop; and was informed, that of 140 Scots acres of land, now arable, 41 acres had been trenched out of rocks and moor, at an expence of L.16 per acre.

In the lower extremity of Marr, viz, in the immediate vicinity of Aberdeen, where trenching and improving poor barren soil has sometimes amounted to L.100 per acre, various kinds of green crops, namely turnips, potatoes, coleworts, and garden roots, are cultivated with the greatest success; and very weighty crops of sown grasses, and of bear or bigg, and oats have been raised. Two rowed barley, though repeatedly tried, has seldom succeeded; but attempts are now made to raise wheat in this district. It is doubtful, on this light soil, whether bigg and grass seeds be not more profitable. But wherever the soil has any tenacity, as street dung is a good manure for wheat, it is to be hoped that more of this will be sown in the vicinity of Aberdeen.

In this district, a regular rotation or distribution of crop is not to be expected, where the land is let in very small parcels; and where the farmer, impelled by the stimulus of high rent, often takes two crops in one year, or three crops in two years, according as he can afford to purchase street dung (night soil) or has occasion to raise any particular crop of garden stuffs, or other vegetables. He generally contrives to make the most of his land, while he pays from L.4 or L.5 to L.15, and even L.20 per Scotch acre.

SECT. III.—RENT.

The rent of land in the county of Aberdeen is extremely unequal, and very different in different places. Indeed it depends more on the local situation of the ground than on the fertility of the soil. The average rate of every acre, taking into the account the moors, and unimproveable, as well as the arable land, does not exceed three shillings and sixpence, or at the utmost of three shillings and sevenpence per *English*, or four shillings and threepence per *Scotch acre*. But in fact, the mountains are not let by measure, a right of pasturing upon them being enjoyed in common by all those who possess any arable land in the neighbourhood. Even those farmers, or rather graziers, who have an exclusive right to the pasturage of certain districts, rent them as bounded by hills, or mountains, and not according to any known extent or measure. The most considerable portion of land, that is rented by one person, is the farm of Dallavorar, (or the Earl's haugh) about 64 miles from Aberdeen, situated on the banks of of the Dee, near the influx of the Geoully, and rented by Mr. CHARLES McHARDY, from Earl FIFE. This farm includes the Doubrach, and all the mountains of the Scarsoch betwixt the Dee, and the boundaries of Perth and Inverness shires. The whole arable ground is about 18 Scotch, or 22½ English acres. But attached to this is the pasturage of nearly 32,000 Scotch, or 40,000 English acres. The rent is L.260; about *twopence* per Scotch, or a little more than *three halfpence* per English acre. The highest rent of any acre of land in the county, is a small patch of land belonging to Mr. CHALMERS of Westfield, in the vicinity of Aberdeen, and rented by Mr. GEORGE WALKER, nurseryman and seedsman, at L.20 per acre. At two miles distance from that city, WM. STEVENSON, another nurseryman pays L.17 per acre to JAMES FORBES, Esq. of Seaton. Hence the highest rent near the city of Aberdeen, is about 2500 times as much as in the mountainous regions in the S. W. extremity of the county.—And yet from the small number of black cattle in that district, Mr. McHARDY finds it difficult to pay his rent, however low it may appear.

The arable lands near Aberdeen are of two descriptions.—The first is that in the immediate vicinity of that city, and in the parish of St. Nicholas, lying near to the village of Footdee. These rent above L.10 per acre, *on an average*;—some of them as high as L.14. The other description of land is that which has been trenched at a great expence, mostly within these 40 years, and rents from L.2 to L.12. The average may be about L.7. (A few patches that have been thoroughly cultivated, rent as high as the old cultivated lands.) These lands are situated in the parish of Old Machar, or Old Aberdeen. Within the last 40 years, by trenching at a very great expence, no less than L.10,000 a-year has been added to the rent of lands in these two parishes.

The

The arable lands in the other parts of the county may be classed under three different denominations. 1. The old croft land, termed *infield*, which was formerly dunged every third year, and constantly kept in tillage. 2. The secondary kind of land, called *outfield*, which was *tathed*, or manured with the dung of the cattle who were folded upon it *every tenth year*. 3. The faughs, or inferior quality of arable land, which received no manure at all, but were ploughed every four or five years, after lying as long in natural grass.

These different kinds of land are variously rented in different places; and, where the tenant has an old lease, are let at a much cheaper rate than where they have been set in a lease within these few years, and especially where the lease is expired, and where the farmer enters upon a treaty for a new lease of 19 or 21 years. The rent of land is now rising rapidly, except where this is paid in corn or meal; in which case, the additional price of the articles paid in kind, has virtually become a rise of rent. After making these distinctions, the rent of land in this county, may now be stated as follow:—

In the lower parts of Aberdeenshire, the old croft land, called Infield, is rented in new leases at from L.1 : 10 to L.2 : 10 per acre. The burgh roods, or ground in the vicinity of the burghs of Kintore and Inverury, and that in the neighbourhood of Peterhead and Oldmeldrum, is about L.2 : 10, and in some cases, lets at L.4 per acre: but only a small proportion of the arable land of the county, is rented at this high rate. In the head, or upper part of *the Garioch*, where the soil is rich, and the ground is sheltered by the hill of Foudland, and other bounding hills, the best infield land is let from two guineas to fifty shillings per acre in several new leases; but the old croft-land, at an average of the county, is not above thirty shillings per acre. The outfield, in new

new leases, varies from ten to thirty shillings; at a medium of the county, is probably fifteen shillings. But where the soil is good, and where the ground has been improved, and the distinction between outfield and infield abolished, whole farms let at 30 shillings. The faughs (here including low wet lands, called *laighs,* and *burnt-lands,*) vary from four to ten shillings, in new leases, and are perhaps eight shillings at a medium. The hills, *riesk* or *benty moor,* and other coarse lands, not worth the expence of cultivation, are generally thrown in the lump along with the arable lands. Even in the higher parts of Marr, and in Strathboggie, at a great distance from the sea-coast, the old croft, or infield lands, are let in several instances at L.2 per acre; the outfields at fifteen shillings, and the inferior arable land at seven shillings and sixpence per acre. But in general great tracts of hilly land, and a liberty of pasturage in common with others, on the neighbouring hills and mountains, are added as privileges belonging to the arable lands of the highland farmers. These afford nourishment to their sheep and black cattle for several months of the summer; and when the *heath flower,* (provincially the *heather-bell,*) is in season, the mountains produce very fattening nourishment to sheep and black cattle. Hence in the autumn, the mutton of the highland sheep is very delicious, and finely flavoured. The goats also feed on various herbs and grasses on the mountains; and both their milk and their carcases enable the highlanders to pay their rents, which, in their remote situation, are comparatively high.

One general remark, concerning the rent of land, presents itself to the mind of the Reporter, and may deserve the attention of the reader. It is this—

The land-measurers, or appraisers, almost uniformly value the bad or inferior land too high, and the good or fertile lands

lands at too low a rate. As land is worth no more than what its produce will pay, after clearing all expences of cultivation and management,—from the late great rise in the money price of labour, and the comparatively inferior produce, as well as the greater expences of cultivating and improving unfertile and rough soils, the farmers who rent poor lands are never opulent. Even in fruitful seasons they can save but little—they can pay their rents only when the season is not unfavourable;—and one calamitous year, both by the defective crop, and the bad quality of the seed, generally overwhelms them. On the other hand, where the soil is fertile, and the farmer has once acquired a capital, he can always pay a much higher rent, and in unfruitful seasons he is able to endure an occasional loss (even although he feel this for a few years) without being overset by one or two unprofitable crops.

The rent of land in Aberdeenshire has been doubled within these thirty years, and is still rapidly increasing: But it may be a long time before it can rise so high as that of the southern counties of Scotland, which are now so much employed in raising of wheat and barley; and where the demand is so great, and the access to market easy. Here oats and bigg only can be generally raised; and the culture of wheat and barley is not so congenial either to the soil or climate of Aberdeenshire, while the demand is less, and the markets are more distant. Yet as our soil in general is well adapted to the raising of turnips, and of the sown grasses, if, from the increase of luxury, the price of butcher meat rise in proportion to that of corn; and if the landed proprietors encourage their tenants during the first years of their leases to improve their lands, the farmers of this county will be able to pay a much higher rent, than they pay at present, under their new leases. But if from any cause the price of black cattle,

cattle, and of butcher meat shall be permanently reduced, the present rents will not long be paid by many of the farmers.

It deserves here to be noticed, that owing to the introduction of the turnip husbandry, black cattle are now fattened at a much less expence than they were 200 years ago; when they were fed on corn chiefly; and that the price of butcher meat bought for the household of Henry, Prince of Wales, son of K. James I. of Great Britain, was 5d. per pound.— From various causes, the money price of corn has not risen in proportion to the money price of labour since 1609; and since the introduction of the turnip husbandry, the price of butcher meat has not increased in the same proportion as that of corn. Wheat also pays much better than any other crop; and hence land adapted to the raising of wheat, will always be rented much higher than land that is fit only for raising barley, oats, turnips, and sown grasses. Luxury in Britain has a different effect from what it had in ancient Rome.— There, from the largesses of corn given to the common people, the price of wheat was rendered too low, the *most valuable* crop was *good pasture*, and *the least valuable* was *good tillage*. Here, from the high prices of wheat, lands fit for raising this species of corn are most valuable; and except in the neighbourhood of cities, or large towns, turnip and grass are of comparatively inferior value.

SECT. IV.—TITHES.

No tithes are now paid in this county. Formerly there were two kinds of tithes paid by the landholders and farmers of Scotland. The first was called *the great tiends*, and consisted of the tenth part of the produce of corn, which was taken in kind; and secondly, the small tiends, which comprehended certain payments for cows or calves, the tenth fleece of the sheep, and one-tenth of the lambs. The dignified clergy, and rectors of parishes, drew the former, and the inferior clergy, who alone generally resided, drew the latter, which were called the vicarage tiends. At the Reformation, these taxes on agriculture, which were both extremely high in their amount, and vexatious in regard to the mode of collecting them, were generally abolished. A provision for the clergy is now made out of the tithes in the following manner: The landholders are allowed to value their tiends, one fifth part of the rent at the time of valuation, is taken as the whole tithe; which can never be increased: and the Court of Tiends is vested with a power of granting what proportion of the tithes or tiends, they judge proper for the maintenance of the minister of the parish. The tithe laws of Scotland are very beneficial to the interests of agriculture.

The small tithes were paid in many parishes, till within these 50 years; and in some parishes only 30 years ago.—Before the union of the kingdoms in 1707, they were a great support to the clergy, as the Scotch wool, particularly in Aberdeenshire, was of excellent quality, and bore a high price on the continent. But the exportation of Scotch wool being prohibited by the union of the two kingdoms, the value of the small tithes fell, with the decrease of both the price and quality of wool; and the clergy were obliged to get an

ad-

addition to their livings out the great tithes. The vexatious office of collecting the *small tiends* is now happily abolished; and the only exceptionable part of the tithe-laws of Scotland, namely, the apportioning the minister's stipend among the different landed proprietors in his parish, where the burden does not fall equally, may be inconvenient to the minister, who is often obliged to wait several years before he can get payment of the money due to him, and oppressive to those proprietors who have no right to their tithes, but it cannot, in any case, affect the interests of agriculture.

One remarkable fact, founded on the agriculture of a portion of the lands of this county, demands the attention of the legislature. It has been already stated, that the lands in the immediate vicinity of Aberdeen, have been improved at an expence unequalled in any part of the kingdom, from soil naturally very unfertile. At least 2000 acres have been trenched, manured, drained, inclosed, and highly cultivated, and not less than L.10,000 of rent, and L.25,000 of produce, are drawn from these lands. But if the rector of the parish, or the lay impropriater of the tithes, had been entitled to draw the tenth part of that produce, there is not the least reason to suppose that 200 acres would have been improved.—Many of the spirited cultivators would have been well pleased to have had *one-tenth* of the produce *for their profit*. Most of them were enabled to improve their fields of granite, heath, or sand, merely by having nothing at all, or at any rate a *few pence*, to pay for their tithes. Here, while the English farmer may derive an important lesson of mixing the spade with the plough husbandry, where the land is barren, and situated in the neighbourhood of great towns,—the British Legislature may see, that the value of the tithes should be ascertained, and that those for unarable lands should be fixed

and

and valued at a moderate rate, before waste lands can be taken into culture.

The clergy of the church of Scotland are placed in that happy mediocrity between opulence and poverty, which calls on them to look for the respect of their people, only from the purity of their morals, and the cultivation of their intellectual powers. And in those parishes in which the tithes do not amount to L.150 yearly, a late act of Parliament has, by a donation of L.10,000 annually, raised the smaller stipends to that sum.

In return for this liberality of the English and Irish Members of Parliament, the Writer of this Report hopes to be excused for expressing a wish, that all the tithes of England and Ireland were valued as they are in Scotland; that the English clergy may enjoy them at this value, without any diminution; and that the landed proprietors and farmers may derive all the benefit of their farther exertions in cultivating waste lands, or improving more highly the lands which are now under cultivation. If the tithes were valued *in quarters of wheat, barley,* and *oats,* instead of *money,* the clergy could sustain no injury, and agriculture might derive great advantage.

SECT. V.—POOR'S RATES.

Poor's Rates are unknown in this county. We have Acts of Parliament, and Proclamations of the Privy Council of Scotland, (which in ecclesiastical matters had the force of law,) appointing assessments to be made, wherever this was found to be necessary. But no assessments are ever actually made

made in any parish of this county for the maintenance of the poor. Even in the calamitous seasons of 1782, 1783, 1799, and 1800, the poor of Aberdeenshire were supplied without any assessment. The exemptions from tithes and poor's rates are the principal causes of the late rapid improvements in the agriculture of Scotland. While England pays about 4 millions annually for poor's rates, the industry of Scotland produces at least four hundred thousand pounds worth of additional labour (or labour which would not have otherwise been exerted) owing to there being no poor's rates established in North Britain. And while the certain resource of being maintained by the parish lessens both the industry and the economy of the lower classes in the sister-kingdom, that honest pride or independency of spirit, which renders the peasants of this county so averse to receive public charity, as to exert themselves to the utmost, both by labour and parsimony, to supply their own wants, and those of their nearer relatives, produces the happiest effects both on the character of individuals, and on the interest of the community at large. It excites industry, it cherishes filial piety, it rewards paternal affection; it is accompanied by private competence and contentment, and tends not a little to promote public prosperity and opulence. In populous towns, to which the widow and fatherless retire, there is often a greater degree of penury and distress, than what can be relieved by the united labours of a family;' but there the hand of affluence is open, and the poor are supplied, in various ways, by the rich and benevolent: and even in country parishes, in seasons of peculiar distress, subscriptions are made by all ranks, and donations are generally given by the landed proprietors; so that those who are really objects of charity, are supplied in this county without making any assessments. In the general management of the poor's money in Scotland, the

only

only expences incurred, are the *very moderate salaries* to the Session Clerk and officer, which are, for both persons, from L.3 to L.4 annually. And it deserves to be noticed, that in country parishes the Session Clerk is also precentor, without any additional reward for his trouble. So that the annual allowance for the Session Clerks and Church Officers in North Britain is below L.4000. The Elders, or Church-wardens, receive no recompence—not even a *dinner* from the *funds of the Church Session*, which are applied solely to the relief of the poor. The parochial clergymen, in country parishes, generally give them their dinner twice or thrice a year; and the only reward of these worthy men, who manage the poor's funds in Scotland, arises from the general esteem of their neighbours, and the approbation of their own minds.

The Writer of this Report should have been very happy, if a regard to truth, and to the interests of the poor, did not oblige him to state some facts with regard to the Poor's Money of Scotland, which are not of so pleasant a nature.

While the agriculture of England *is loaded with poor's rates*, to the amount of several millions annually, the property tax of 10 per cent has *been levied from the funds which belong to the poor in Scotland*; and the Poor's Funds have been taxed in various ways, certainly without any intention on the part of our Legislature.

In Scotland, the property belonging to the Church Sessions is usually lent out on bills, with personal, or on bonds, with landed security. By an omission in the Property Act, (owing to there being no money in England belonging to charitable institutions lent out on bonds or bills) 10 per cent. of the interests of money belonging to the poor of Scotland, has been paid to Government.

Owing to a similar cause, viz. the special Commissioners for the Property Tax being Englishmen, and residing in

London, and consequently being unacquainted with the institutions of N. Britain, they have refused to exempt the rent of lands which belonged to Church Sessions, from the tax on property, though all rents of lands, which are applied to charitable institutions, are expressly exempted by Act of Parliament.

Owing to an oversight in regard to the mode of collecting the auction duty, though the *goods and household furniture* belonging to the poor are exempted from this duty, it has been rigorously exacted when *lands*, belonging to them, were sold in times of *great distress*; when Church Sessions were obliged to sell them, and when they could sell them regularly only by public auction.

The Writer of this Report has been at a good deal of trouble to get these evils redressed. They have been already acknowledged to be errors; and if not remedied before this work is all printed off, they will be stated very fully in the Appendix.

SECT. VI.—LEASES.

When treating of the different *tenures*, it was already mentioned, that the most usual period of lease is for 19 years;—though longer leases, with one or two rises of rent, are frequently granted,—but that life-rent leases, which formerly were used, are now seldom asked, owing to a late decision of the Supreme Court respecting life-rents.

The usual term of entry to farms in this county is at Whitsunday, i. e. May 26th, as the terms for removal of tenants are regulated by the old style. This is really a very inconvenient term; as the tenant, in general, has no right to any thing at entry.

entry, excepting the small garden for planting cabbages or coleworts, and the grass of the outfields, where they are not in crop. And a number of tenants, by ploughing and liming their outfields, contrive to get them treated as infield, or old croft lands, and plough up the greatest part of them.— Under the old system of husbandry, the half of all the arable land, except the infields, was left to the entering tenant, and belonged of right to him. Till the landed proprietors change the term of entry from Whitsunday to Martinmas, and fix certain rules respecting the quantity of land, (i. e. the proportion on each farm) to be left in grass, and also in turnips and fallow, the tenants who enter into new leases for farms which they did not formerly occupy, will be in a very disagreeable situation for the first sixteen or seventeen months after their entry. The only remedy is, for the landholder to bargain with the removing tenant, and get a right to the farm manure, to the ground allotted for turnips, and a certain proportion of land for grass and fallow.

Some of the landed proprietors, though with a temporary sacrifice of their interest, or the loss of part of a year's rent, have already changed the term of entry and removal, from Whitsunday to Martinmas—a practice deserving of general imitation.* And a few of them have already bound their tenants, on their removal, to deliver up the manure of the farm for the turnip crop, to their successor, and a certain proportion of sown grass, both on appretiation, by arbiters mutually chosen. The other usual conditions (or *prestations*) commonly

---

* Some intelligent farmers prefer an entry at Whitsunday. But except in a case in which the entering tenant has a right to the grass, straw, and dung of the farm, an entrance at Martinmas is, on the whole, the most eligible. In that case, the tenant may very properly enter at Whitsunday.

commonly contained in a lease, besides the payment of the money rent, are to deliver that part of the rent which is paid in oatmeal, or other grain, generally farm-bear, at the port of Aberdeen, or any place at the like distance from the farm, to pay a certain number of poultry, and a certain measure of peats or of coals, carried to the heritor's mains or dwelling house—to drive a certain stipulated number of carriages to the same place, and also the tenant's proportion of slates, lime, clay, and sand, either to the landlord's house and offices, or to the church, minister's manse, and offices, and to the school-house of the parish. Some insert restrictions against breaking up meadow grass; and others lay down certain regulations in regard to cropping. But it must be acknowledged that these are either too general and obscure, which gives rise to law-suits, or that they are not suited to the soil and climate of this county. When they are expressed in *general terms*, they bind the tenant not to *out-labour* or *mis-labour*—but to manage the tillage in a husband-like manner. When they *enter into minute articles*, they commonly allow two white crops to be taken; even on the thin gravelly soil along the river Dee, where two successive white crops ought not, on any account, to be allowed. They also frequently insist on three crops of grass being taken, two of which must be pasture, before the ground is broke up for sowing corn. This is establishing a seven shift course, which is neither so profitable to the farmer, nor so easy for the ground as a *five* shift one. Others, in their partiality for a grass crop, admit of only one crop of oats—before turnip—then bear or oats, laid down with grass seeds, followed by one crop of hay, and two year's pasture. This is certainly an excellent method for improving the ground; though a tenant on this *six shift course cannot* pay above *two thirds* of the rent

rent which *he could pay on a four shift one,* nor above three-fourths of what he could pay on a five shift course.

Two landed proprietors, viz. the EARL OF ABERDEEN, who inherits in the divisions of Marr, Formartin, and Buchan, the most valuable estates in the county, and JAMES FERGUSON, Esq. of Pitfour, Member for the county in this and four former Parliaments, have printed regulations, which their tenants or farmers are obliged to adopt; and as the former has above 40,000, and the latter near 30,000 acres of arable land, it may be useful to insert these regulations at full length, and to subjoin a few remarks on each.

Those of the EARL OF ABERDEEN are as follows:—

Art. 1.—" All assignees, whether legal or voluntary, and all subtenants are excluded; heirs-portioners are also excluded, the eldest daughter being to succeed without division; but power is given to tenants having children, to appoint any one of them, whether son or daughter, to succeed them in their leases; or in case the children should be minors, or decline the business of farming, or in case the tenant shall have no children, then power is given to assign the lease, on condition that such assignee shall be, or have, an actual farmer resident constantly on the farm, shall keep a sufficient stock thereon, and shall become bound to perform the whole conditions and obligations prestable on the original tacksman, who has only power given to assign, in case of his death, under the circumstances above-mentioned. This clause is not to extend to prevent tenants from accommodating their servants with dwelling houses and yards on their farms; the tenants must reside with their families on their farms, and always have a sufficient stock thereon.

Art. 2.—" The lands shall be managed and cropped by the

tenants during their tack, in the following manner, viz.—After the expiry of the first five years of the lease, one-fourth part of the land on the farm, which shall be ploughed in any year, or as near that proportion as the size of the fields will admit of, shall be yearly in fallow, turnips, cabbages, or potatoes, and must get, at least, four ploughings, and a harrowing after each ploughing, in the proper season, and be manured with dung, as far as the whole collected or made on the farm, will go, at the rate of at least thirty single, or fifteen double cart loads to the acre; and such part of the fallow not receiving dung, shall be limed, at the rate of at least twenty bolls of shell lime, or sixty bolls of slacked lime, of one hundred and twenty-eight Scotch pints each boll, to the acre. The crop following shall be oats, bear, or wheat, sown with at least twelve pounds of good clover seed, and one bushel of perennial rye grass, to the acre; and shall only be cut for hay once; three-sevenths at least shall be always in grass; and the grass land, when broke up, from one or two year old grass, shall only carry one white or grain crop; or if from three or more year-old grass, no more than two grain crops, and again fallowed, manured, limed, and cropped, in the manner above-mentioned; the tenant being at liberty, when he takes two grain crops from three or more year-old grass, to have pease or beans intervening between such two crops, on the land receiving at least two ploughings before sowing, or the crop being sown in drills, for the purpose of being horse and hand-hoed.

In terms of the mode of management before specified, the tenants shall be obliged to leave, at the expiry of their leases, three-seventh parts of the whole arable land on their farms in sown grass—one of these seventh parts being infield; and also to leave one-fourth of the ploughed land for fallow or green crop—one fourth of which fallow must also be infield;

and

and the remainder of the grass and fallow shall be in whole fields, or as contiguous as can be done conveniently. And the tenant shall be obliged to allow the proprietor or incoming tenant, to sow grass seeds along with the last crop; and not to allow his bestial to pasture thereupon, after the separation of the crop of grain from the ground.

"Lint is only o be sown as a first crop after grass, in place of oats, or after fallow, with grass seeds: Declaring, that if the tenant shall at any time contraveen any of the articles above mentioned, he shall be obliged to pay the sum of three pounds sterling of additional rent, for each acre managed contrary to the regulations above specified; and that along with the first rent falling due, after the contravention has taken place; and shall also pay six shillings for each boll of lime shells, and six shillings sterling for each cart load of dung, that is not laid upon the fallow or field for green crop, as above specified.

Art. 3.—" The tenants shall be bound to consume with their cattle, upon their respective farms, the whole straw and fodder that shall grow thereupon, excepting clover, hay, and the straw of the last crop of the lease; and to lay on their ground the whole dung that shall be made upon the same; and upon no account to sell or give away any of their fodder or dung. And they shall be obliged to leave the whole dung made upon their farms the last year of their leases, carefully gathered together, for the use and behoof of the proprietor, or his incoming tenant, to whom the same shall belong, on payment of the value thereof, to be ascertained by two persons, to be mutually chosen by the outgoing and entering tenant: Declaring, that if notwithstanding hereof, they take upon them to sell or give away, any of the said fodder or dung, they shall be bound to pay to the proprietor the sum

of

of five shillings sterling, for every threave of straw or fodder, or for every cart load of dung so sold or given away.

Art. 4.—" Whatever the term of entry may be, the term of removal in the last year of the lease, shall be—from the kailyards, at Candlemas; from a fourth part of the lands under tillage for fallow or green crop, on the first of March; from the houses and pasture grass, at Whitsunday; from the hay fields or new sown grass, at Lammas; and from the lands in corn, at the separation of the crop from the ground, of each respective field: so that whenever one field is cleared, the incoming tenant may enter to it, for the purpose of ploughing only, although the crop may not be separated from the other fields; and the incoming tenant shall have full liberty to plough the fallow, as often as he pleases, after the first of March, with a view to his having a green crop thereupon.

Art. 5.—" If any tenant shall inclose and subdivide any part of his Farm, the same shall be done with dykes built of stone, three feet thick at bottom, eighteen inches at top, and four feet high, and covered with feal of six inches height; or inclosed with ditches, six feet wide at top, one at bottom, and three feet deep, with a quickset thorn hedge planted on the thrown up bank, within nine inches of the surface of the ground, or lip of the ditch; and that on a plan to be previously approved of by the proprietor or his factor; and for which inclosing, the tenant shall be allowed the value at the end of the lease, as the same shall be ascertained by neutral men of judgment and skill, to be mutually chosen by the landlord and tenant: Declaring, that in no case shall the allowance exceed sixpence per ell lineal measure, for such dykes and fences; nor shall the allowance on the whole exceed half a year's rent of the farm; nor shall any allowance be made for said inclosing, unless the fences shall be in complete fencible repair, forming entire fields or inclosures at the time.

Art. 6.—" The tenants, at their entry to the houses, shall be obliged to accept of them in the condition they may happen to be in; and to receive the timber of them, and such of the houses as are contained in the proprietor's inventory, at a fair valuation, and to pay to the outgoing tenants any melioration which they may be entitled to: and at their removal, they shall leave the timber and the houses of at least the same value they received them, and be entitled to receive from the proprietor or his incoming tenant, the sum they paid to the outgoing tenant, with such additional sums as may be equal to the value of additional buildings constructed of stone and lime, or stone and mortar, sneck-pinned with lime, on a plan to be previously approved of by the proprietor or his factor for the time: Declaring, that such additional buildings shall in no case exceed half a year's rent of the farm; and in estimating the value, no carriages of any kind are to be reckoned upon, as the tenants are to furnish carriages without any allowance. In cases where the proprietor shall allow additional timber, or the prime-cost of slates, for any new houses, the value of such timber and slates is to be added to his inventory of the value of the timber and others entered to at the commencement of the lease: Declaring, that the allowances before-mentioned, for inclosing and buildings, are to be made from the rent of the farm the last year of the lease; which shall be considered as the rule for ascertaining the amount.

Art. 7.—" The tenants shall be bound to maintain, in good condition, the whole houses, office-houses, dykes, ditches, and drains, or gates, that are upon their respective farms, at their entry, or that may be built thereupon during the currency of their tacks, and to leave them, at their removal, in the like good repair—all upon their own charges. And where any ditches, drains, or water-runners, form, or are necessary as parts of, a general drainage for other farms, or for a district of the estate, the whole tenants, having such drains passing through or along the boundaries of their farms, shall be obliged to clear them at the same time, if the factor on the estate shall deem such clearing to be necessary.

"In case the proprietor, (or his factor or doer for the time) shall at any time find the houses, dykes, drains, hedges, ditches, and gates, neglected, and in disrepair, power and liberty are expressly reserved to him, to order the same to be put into proper and sufficient repair and condition; and the expence thereof, being ascertained by the workmen's receipts, or by accounts certified by the factor, the tenants shall be obliged to pay the same, with interest from the date of the receipts or certified accounts.

Art. 8.—" In case it shall be judged proper to make any alteration in the farms, either by streighting marches, or excambing lands with neighbouring proprietors, or tenants, the proprietor or his factor shall have power to do so, and the tenants shall be obliged to concur and acquiesce therein; and the variation thereby occasioned in their rent, whether increase or decrease, shall be determined by two neutral persons, to be named by the proprietor and tenant.

Art. 9.—" The tenants are to be thirled to the mills to which their respective lands have been in use to be astricted, and shall pay multures, mill dues, and mill services, conform to use and wont, and that until the end of the current leases of the respective mills, but not longer, unless provided for by a clause in their leases. And the tenants are to be bound to furnish carriages, for leading materials for building or repairing kirks, kirk-yard dykes, manses, and school-houses.

Art. 10.—" The tenants are to have liberty of casting, winning

ning, and leading home peats, for the use of their families, and servants residing on their farms, from such moss or mosses, or parts of the same, as the proprietor, his factor, or moss-grieve, shall set apart to them yearly for that purpose; and with which allotment they shall be obliged to rest satisfied, whatever the former practices or usages may have been.

Art. 11.—" The tenants are restricted from cutting up any of the surface of the ground, for muck feal, or compost middens, or for wall feal, turf, or divot, except from such parts of their farms, as shall be set apart for that purpose, by the proprietor, or his factor or doer for the time. And if any of them contravene this article, they shall forfeit and pay five pounds sterling, for every such trespass, along with the first term's rent falling due after the discovery of the contravention, or a charge being given therefore, by the factor, or the baron officer.

Art. 12.—" The proprietor reserves to himself the whole game, of every kind, and all the fish in the rivers and burns within his property, with power and liberty to himself, and others having his leave and permission, to hunt, shoot, fish, and sport thereon, at all times and seasons. And he further reserves all mines, metals, and coal quarries, of stone and lime, marle, and other fossils whatever, of the like kind, within the bounds of the lands set, with liberty to work, win, and carry away the same; and for that purpose to sink pits, build houses, make roads, and erect any necessary works thereupon; the tenants being allowed such surface damages, and such abatement of rent, as shall be determined by two neutral persons, to be mutually appointed by the parties:— Declaring, that this is not to prevent the tenants from digging for limestone and marle, in their own farms, for the improvement thereof only. The proprietor also reserves power

to

to shut up, alter, or make new roads, through any part of the lands; to plant trees around the yards or gardens on the estate, without any allowance, and which the tenants shall be bound to preserve; also power to take off pieces of ground for the purpose of planting, on allowing damages, which are to be ascertained by two neutral persons to be named by the proprietor and tenant: But declaring, that in ascertaining damages to be allowed for making roads, the advantages that may accrue therefrom to the tenants, are to be considered; and the same are to be modified accordingly."

It is to be observed, that in this lease his Lordship allows two successive crops of corn to be taken, after breaking up the land from three years old grass. It may be doubted whether two white crops should be permitted as a general regulation on so extensive an estate, and in which there is such diversity in the nature of the soil. It may be also doubted, whether a third years grass should be uniformly insisted on. For there are many soils in which the grass of the third year is very poor, as the natural grasses seldom appear in any great quantity till the fourth year. It appears to the Writer of this Report, that this five-shift course would be preferable to the *seven shift* one, which is fixed by the above regulations. —The allowance for the expence of *additional* houses, and inclosures, appears to him to be by far too little; and where the rent is paid in *corn* or *meal*, to be also too *undeterminate*. A good farmer should have liberal encouragement in laying out *his own money*, on his *landholder's property*.

Mr. FERGUSON of Pitfour has shewn a greater partiality for grass crops, in the following regulations:—

Marches.

"The marches of the several farms and possessions let, and to be let, having been previously laid off, and properly fixed

on

on the ground, as well as particularly delineated on a plan of the lands, the tenants are to observe said marches, as the boundaries of their respective possessions, and keep to the same, until the proprietor shall think proper to make any alterations thereon, which he has power to do, in manner after-mentioned.

Mines, Quarries, Roads, Canals, &c.

" The proprietor reserves to himself all mines, whether of metal or of coal, or other minerals or fossils; all quarries, whether of lime-stone, or other stone, or slate; all pits or banks of marle, shells, clay, sand, or gravel, within the lands let; and all kelp, and other seaware, and shell-sand, or limestone gravel, which may be found thereon; with power, by himself or others authorised by him, to search for, dig, and work such mines and quarries, pits, and banks; erect the proper engines or machinery, lay out, make, and use the proper roads; to dig, make, and use such canals, watercourses, and reservoirs, as he shall think fit, and to collect and carry away all the articles aforesaid; and to take off, wherever he shall think proper, the ground proper for all the above purposes, and for roads and towing paths along the canals, and for storehouses and repositories for fuel, and other articles, near the quarries and pits, and for houses and gardens there, or for the persons thereat employed, or others, along the canals: And further, to take off such part or parts of the lands let, as he shall judge proper for irrigation, or being converted into watered meadows, not exceeding in whole one-tenth part of the possession. The tenant is not to be entitled to any damages, or deduction of rent, for what ground may be taken off for roads, nor for materials taken for making and repairing them, nor for ground taken off for canals, or for the banks and paths, or horse and carriage-ways along the sides of them: But, for what ground may be taken off for

for any other of the purposes above-mentioned, the tenant shall be paid damages, or be allowed a deduction of rent, according to the estimation of two persons of skill, to be mutually named by the parties; and failing such nomination, within eight days after requisition, by persons to be named by the judge ordinary of the bounds; but in making these estimations, the possible or contingent profits which may ensue from the improvements, by irrigation, or watering, are not to be considered: But in case the tenant shall have laid down any of the land, fit for irrigation, in a proper manner, before the end of the seventh year of the tack, and shall use it only for pasture and hay, it shall not be in the power of the proprietor to take off such land. The tenant shall have no right to take water out of the reservoirs or canals, for any other purpose than for the use of his family or cattle, but not for irrigation or driving machinery, without a written licence from the proprietor. When the line of any turnpike road or canal, to be made through the lands let, shall be properly ascertained, the proprietor shall be at liberty to take off such pieces of land near the marches, not exceeding one-tenth of the possession, as may be separated from the farm by such turnpike road or canal, and cannot be inclosed or converted into a field or fields, proper for the farm, and to dispose thereof as he shall choose; he always allowing to the tenant such a deduction of rent therefore, as shall be ascertained in manner aforesaid. The said JAMES FERGUSON farther reserves to himself, and his successors, the whole game and fish which may be found on the lands let, or in the waters thereof; with power to them, by themselves, or others having authority from them, to hunt, fish, and fowl thereon, at all lawful seasons; but that the tenant shall be entitled for damages he may sustain by

his

vate, and crop the same properly, and in a fair and regular manner, and agreeably to the rules of good husbandry; and particularly, that the old meadow grass, or links of the farm, shall never be ploughed; and that the proprietor has it in his power to fix the quantity of land to be used as meadow, so as it shall not exceed one-seventh part of the whole farm; that at no time, shall there be above one half of the other ground of the farm under tillage at once; that wheat, oats, barley, bear, or big, rye, and lint, shall be reckoned white or exhausting crops; that the fallow crops are, clover, fed or mown, but not allowed to seed, turnips, beans, carrots, potatoes, cabbage, kail, sown in drills, and thoroughly horse and hand-hoed, or a real naked fallow. The tenant is bound to have one of either of these fallow crops, between every two of the above-mentioned white crops; and when potatoes are made the fallow crop, there must only be one-half of the break of that kind of crop; neither are potatoes to be repeated upon land which has carried a potatoe crop before, oftener than once in seven years. All land to be managed with a naked fallow must be five times ploughed, and properly harrowed between each ploughing, and three of these ploughings must be performed between the first of May and the first of September; and that such fallow shall be manured with not less than fifteen loads, of a cubic yard each, of good reduced dung, or ninety-six bushels of lime or shell-sand, or one hundred and forty cubic yards of marle, to each English acre. That in case the fallow crops shall be foul or weedy, the tenant, at the desire of the proprietor, must plough them up, and turn them to a real fallow; and such part of the land as should be, by the above-mentioned rules, under a fallow crop the last year of the tack, must be left at the Whitsunday of expiry, to be occupied by the proprietor, or incoming tenant, and they shall likewise be and to preserve the trees planted in the hedgerows, and yards, or any where else on the farm. The tenant shall have no allowance, or deduction, for the ground occupied by the inclosures or drains, or the damages which may arise in making or repairing the same: but if the hedges shall be cleaned and cut properly, the tenant shall be paid by the proprietor half the expence thereof, at the ordinary rates at which such work is done in the country. If the proprietor chooses rather to have any of the above work, relative to fences or drains, done by people of skill to be employed by himself, he may cause it to be done by them; and the tenant, upon having the accounts of the expence attested by the persons who have done the work and presented to him, shall be bound to pay such persons, or to the proprietor, the whole expence of repairing the drains, and half the expence of putting and keeping the fences in order, without any other proof whatever. In case the proprietor shall inclose or drain the farm or possession let, or any part thereof, the tenant is bound to pay to the proprietor, along with his rent, six per cent. yearly, on all the money disbursed by him in so doing, commencing from the time of the expenditure: but if any of the inclosures made shall be on the outward marches of the farm, or a ring fence, then the tenant is only to pay three per cent. yearly, on the expence disbursed in making such ring fences, and one-fourth only of the expence of keeping it in repair; and if any part of the inclosures run along the side of a road or canal, the tenant is bound to pay the six per cent. on the expence of making that part of the inclosure, and bear one-half of the expence of keeping it in order.

Residence, and Mode of Culture.

"The tenant is bound to reside personally, with his family and cattle, on the lands let, and to labour, manure, culti-

vate,

vate, and crop the same properly, and in a fair and regular manner, and agreeably to the rules of good husbandry; and particularly, that the old meadow grass, or links of the farm, shall never be ploughed; and that the proprietor has it in his power to fix the quantity of land to be used as meadow, so as it shall not exceed one-seventh part of the whole farm; that at no time, shall there be above one half of the other ground of the farm under tillage at once; that wheat, oats, barley, bear, or big, rye, and lint, shall be reckoned white or exhausting crops; that the fallow crops are, clover, fed or mown, but not allowed to seed, turnips, beans, carrots, potatoes, cabbage, kail, sown in drills, and thoroughly horse and hand-hoed, or a real naked fallow. The tenant is bound to have one of either of these fallow crops, between every two of the above-mentioned white crops; and when potatoes are made the fallow crop, there must only be one-half of the break of that kind of crop; neither are potatoes to be repeated upon land which has carried a potatoe crop before, oftener than once in seven years. All land to be managed with a naked fallow must be five times ploughed, and properly harrowed between each ploughing, and three of these ploughings must be performed between the first of May and the first of September; and that such fallow shall be manured with not less than fifteen loads, of a cubic yard each, of good reduced dung, or ninety-six bushels of lime or shell-sand, or one hundred and forty cubic yards of marle, to each English acre. That in case the fallow crops shall be foul or weedy, the tenant, at the desire of the proprietor, must plough them up, and turn them to a real fallow; and such part of the land as should be, by the above-mentioned rules, under a fallow crop the last year of the tack, must be left at the Whitsunday of expiry, to be occupied by the proprietor, or incoming tenant, and they shall likewise be entitled to sow grass seeds with the last or way-going crop; and the out-going tenant shall leave to them the whole straw and chaff of that crop, together with one whole year's dung, not to be previously laid on the land, without any allowance: that grass seeds shall always be sown down with the crop first succeeding a fallow, and that the quantity shall not be less than seven and one-half pounds of red clover, seven and one-half pounds of white clover, and one bushel of rye grass seeds, to each English acre; that from the ground laid down in grass, hay shall not be cut oftener than once in three years; that there shall not, in any one year, be above one-fiftieth part of the arable part of the farm in lint and hemp together; that no hay, straw, chaff, turnips, or other green crop, or peats, shall be sold or carried off from the farm, but that they shall be consumed thereon; and the dung produced from them; or otherwise, shall be laid on some part of the farm, under the pain of forfeiture of the value of the articles carried off. That all whins, or broom, already growing on the ground, shall be rooted out, and none shall be sown therein for the future. That no rabbits shall be introduced into the farm, nor allowed to burrow or remain there: that none of the surface of the ground shall be burnt, nor any turf, feal, or divot, cast from it, under the penalty of one shilling for each feal or divot, and five shillings sterling for each rood of thirty-six square ells of burning. The tenant is bound, full three years before the issue of his tack, to have one-half, at least, of the arable ground of his farm laid down in good heart, and sown with a sufficient quantity of white clover, and other grass seeds, fit for pasture; and shall not, at any time thereafter, during the remainder of the lease, cut hay from, but pasture on it only. The proprietor shall have it in his power to choose the one-half of the ground, which is to be laid down in grass as aforesaid,

and

and the tenant the other half; the part to be chosen by each must lie contiguous in itself, but may be in a different part of the farm from that chosen by the other. For enabling the proprietor more easily to observe whether the lands be properly managed, the tenant is bound to divide the farm into regular fields, as he shall be directed by the proprietor, or any person having authority from him, and to have these fields in one kind of crop only in one year; and no white crop shall be sown in any part of the same field, or break, where there is or ought to be fallow, green crop, or grass.— In case the tenant shall fail to lay down, and leave in pasture, the half of the arable land of his possession for the last three years of the tack, or shall crop or manage any part of his possession, at any time, in a manner different from what is before prescribed, or shall plough up any part of the land allotted for meadow, perpetual or pasture grass, previous to, or in the last year of the tack, or otherwise neglect or counteract any of the regulations herein before prescribed, he is bound, from the time of contravention, till the end of the tack, to pay five pounds sterling of additional rent yearly to the proprietor, for each English acre which shall not be laid down for pasture, or shall be otherwise managed or cropped than is before stipulated, and ten pounds sterling for breaking up meadow, or perpetual grass, in the said last year, along with his other rent, and over and above the same, and so proportionally for any part of an acre; that certificates from any surveyor of land, of good reputation in his profession, shall be sufficient vouchers for instructing the quantities of ground in white crop, in meadow, or perpetual and pasture grass, green crop, and fallow, and which shall be managed in a manner contrary to the rules before prescribed, and shall be considered as parts of the tack, and when registered along with it, shall be sufficient ground for summary diligence for the additional rents before stipulated, alike as if they were inserted in the tack itself. The tenant is bound to defend and protect the possession let to him, and all its privileges, from the encroachments of all others, during the lease; that the tenant shall not be at liberty to retail any ale, or spiritous liquors, on the possession let to him, nor allow others to retail the same thereon, without a licence in writing from the proprietor. The tenant shall have no right to any privileges beyond the bounds of the farm or possession let, either of manure, fuel, or the like, unless specially granted in his tack; nor shall he have title to any allowance, or consideration, at the issue of his tack, for managing, laying down, and leaving his farm, in the order before stipulated; nor for any other articles, except what are herein stipulated, or shall be specially and specifically stipulated in his tack, whether he shall have paid for such articles at his entry, or been at the expence of the same, during the subsistence of his tack or not. The tenant is bound to attend any mill belonging to the proprietor in the neighbourhood of his farm, to which he shall be directed, with all the victual, the produce of his farm, which shall be ground for the use of the persons residing on it, and to pay the usual bannocks and perquisites for the same, but no multures; but where the lands let are in the sucken of, and thirled to, any mill not belonging to the said JAMES FERGUSON, or his foresaids, the tenant is bound to attend such mill with all grains, the produce of his farm, as the rules of thirlage to the said mill oblige, and to pay the ordinary multures, sequels, and knaveships therefore, and to perform the accustomary services to the said mill; or if the proprietor shall choose to value such multures, sequels, &c. as is now allowed by law, the tenant shall be bound to pay such sums, or articles, in proportion to his farm, as the said multures, se-

sequels, &c. shall be valued at by the jury, or agreed to by the proprietor.

Houses.

"That all houses shall be built in a situation, and upon a plan, approved of by the proprietor, or persons having authority from him, in writing : that the walls of these houses shall be built of stone and lime, or stone and mortar, outer course laid and sneck-pined with lime, and be two feet at least in thickness, and shall be covered with slate, tile, or stob-thatch only, and not with divot; that the windows of the dwelling house shall all be of glass, and shall not be less than eight square feet of window. For the walls of the houses so to be built by the tenant, he shall have allowance from the proprietor of two pounds sterling per rood of thirty-six square ells, for building and lime, and a proportional allowance for quarrying the stones, if it shall have been necessary to quarry them; but no allowance is to be made for carriage: and the allowance for quarrying, building, and lime, is not to exceed in whole, one year's rent of the farm or possession let; and the one-half of it shall be payable at the term the first payment of the second year's rent is due, provided the building is completed, and done on the terms before expressed, and the other half within three months after the issue of the tack; but only, in case he leaves the houses in good repair. The tenant is to have no further allowance for walls of houses, although he shall have built to a greater value, and shall not, notwithstanding, be at liberty, either at his removal, or for three years preceding, to throw down or demolish any additional walls built: it is to be understood, that if the tenant has furnished or paid for the timber, and glass windows, and slates, (if good Esdale) they shall remain his property, and he shall, at the issue of the tack, either be paid for them, according to the appreciation of persons of skill, to be mutually named by the proprietor and him, or to be allowed to take them away, in the proprietor's option. No delay of the proprietor, in executing any of the articles he has undertaken, or in making any of the payments he stands bound for by these regulations, or by the tack, shall afford ground to the tenant for refusing to pay his rents punctually at the proper terms, or for not removing punctually at the issue of the lease, the tenant always being entitled to the same execution to follow hereon, and on the tack, for implement of the proprietor's part, as the proprietor has for implement of the tenant's.

" LASTLY, the said JAMES FERGUSON, Esq. and the persons hereunto subscribing, whose tacks to these regulations refer, CONSENT to the registration hereof in the books of Council and Session, Sheriff Court Books of Aberdeenshire, or any other competent register, that letters of horning, on six days charge, and all other execution necessary, may pass hereon."

The great object of these regulations is to admit only one white or corn crop in succession, and to have grass always three years old, besides a great proportion of old meadow grass. But it may be doubted, whether a third year's grass should be always insisted on. For there are many soils in which the grass of the third year is very poor, as the natural grasses seldom appear till the fourth year. It appears to the Writer of this Report, that if the farmer is obliged to keep *one-third of his arable land in pasture*, he should be left to judge where he would break up some fields after two years grass, and on the other hand, where he would allow, in balance of this, other fields to remain four years in grass. The nature of the soil and state of the fields must be considered in such cases. To obviate this objection, Mr. FERGUSON, in fact, has never refused to allow his farmers to substitute a crop of pulse in place of the third crop of grass, which may always be done with

with propriety where land has been laid down and in good heart. On the same accounts it may be doubted whether in some farms one-seventh part is not too great a proportion of meadow. For the finer grasses in many soils disappear after eight or ten years; an objection to meadows of the utmost consequence, which is too little attended to by the landed proprietors. He does not oblige his tenants always to have one-seventh of their farms in meadow, but the proprietor has the choice of the quantity; only he cannot exceed that proportion. On the other hand, nothing can be more judicious than the prohibiting two white or corn crops to follow each other in succession. For it is to this that the great improvement of West Norfolk is chiefly to be attributed.

GEORGE SKENE, Esq. of Skene, Member for the county in one Parliament, and in another for the burghs of Kintore, Inverury, Elgin, &c. in his agreement with Mr. WALKER for the lease of Wester Fintray, stipulated, that within a certain number of years from the commencement of the lease, the tenant should specify what fields he chose to keep in a four-shift course, what in a five-shift, and what in a six or seven-shift course, but that he should have no liberty of deviating from the course thus fixed by himself, without paying a considerable additional rent, which would operate as a penalty on thus changing the rotation of any particular course of cropping. For a long lease of 33 years, this was the most proper rule, as the farmer, in the course of a few years, would become acquainted with the quality of the soil in all parts of his farm, and would be able to specify what would be best for himself. And the probability is, that what he chose to abide by for 28 or 30 years, would be no bad permanent regulation. In fact, different regulations should be introduced in an improved farm, and a corn-raising county, from what are proper in a farm that needs improvement in a cattle rearing one.

SECT.

SECT. VII.—EXPENCE AND PROFIT.

This is a most important article; but from the very unequal rents of farms, according as they are under new or old leases, or more or less fertile—and also from the accidental circumstances of not only good or bad seasons, but of few or many deaths among the black cattle, (who are more numerous than in the corn counties of Scotland) it must be extremely difficult to fix any thing like a general scale—still more a particular account of expence, receipts, and of the difference between these two, profit and loss. The Writer of this Report has been long convinced, that supposing the soil of two farms to be equally productive, and the bargain with the landholder equally fair, or equally favourable, there are three articles that cannot be charged as *items* in the account, which have a wonderful effect upon augmenting the profit of the farmer. They are *skill, credit,* and *capital*. The first is essential, for without skill in his profession, a farmer may soon ruin his credit and waste his capital. The second is also necessary in point of character, for punctuality in making payments, even if a man had more capital than the farmers in this county generally have; and if the farmer have but little capital, he must have the good opinion of his neighbours for his industry and honesty, as well as punctuality in payments. The last is extremely advantageous, and a certain proportion of capital is in general requisite to a farmer. But capital alone would soon be dissipated by an unskilful farmer—who would raise too little produce, and be deceived in the manifold transactions of buying and selling cattle, which occur so frequently in this county.

Without therefore pretending to accuracy, or giving a distinct idea of every man's expence, receipts, and profit—the fol-

following is given as the expence and profit of such a farm as that of Wester Fintray, which has already been selected in this Survey.

Wheat	L.125	Horses on the farm	L.280
Beans	24	Cows and Bull	296
Pease	11	Working Oxen	180
Barley	28	Other Black Cattle	508
Bear or Bigg	300		
Oats	900	Total live stock,	L.1174
Clover Hay	300	Add farm produce,	2340
Pasture	315		
Potatoes and Turnips	337	Total stock and prod. as valued by Mr Walker	L.3514
Total farm produce	L.2340	Implements of Husbandry, including the machinery of a Threshing Mill, which goes by water, at least	L.200

Hence the total capital on this farm, supposing that the farmer were to remove from it (besides the money due him for melioration of houses, for inclosures of various kinds, and for 1000 yards of an embankment of the river Don) is, in round numbers, L.3700.

He has therefore a claim for the interest of this capital, at 5 per cent. - - - - L.185

His rent at present, converting his victual at the fiars of the year, is - - - - 175

Four married servants cost him of board and wages yearly - - - - - 100

Seven unmarried men-servants cost him of wages alone yearly - - - - - 105

Four women servants, of wages annually - 25

Clothes for his own family, five persons - 50

Carry over, L.640

Brought over,	L.640
Expence of education of two sons and a daughter	60
Articles of provision for his family and unmarried-servants, not raised on the farm	80
Malt-tax, road-money, tolls, and travelling expences for himself and servants	60
Ironmonger, blacksmith, plough-wright, stone-mason, house-carpenter, yearly	75
Labourers in hoeing green crops, and for cutting down his hay and harvest crops	75
Tear and wear of threshing machine, and implements of husbandry	20
Annual loss on the wear of his horses	30
Incidents	20
Total charge against the farm, besides money sunk on building and inclosures	L.1060

To pay this, he sells of farm-produce and live-stock yearly,

Wheat, at an average of years, to the amount of			L.100
Barley	do.	do.	20
Bear or Bigg	do.	do.	250
Oats	do.	do.	400
Dairy and Poultry		do.	40
Hay	do.	do.	150
Cattle, chiefly when fed on turnips			400
Total sold by him yearly,			L.1360

Leaving a profit, or yearly accumulation of capital to a skilful industrious farmer, of L.300 annually.

But this cannot be regarded as a *general* estimate of profit. It is, in fact, only the probable rate of the accumulation of capital of one of the best farmers in the county of Aberdeen, who has an old lease, a large farm, and a soil of an excellent quality.

The following contains the first expence upon entering to a farm in the Garioch, at Whitsunday, 1806, upon a new lease for 19 years, with the subsequent charges incurred till Martinmas, 1807, when the tenant, who is one of the native farmers of the county, but from modesty does not choose his name to be mentioned, had harvested the first crop eighteen months after his entry.—Its extent is 190 Scotch, above 240 English acres. His rent is L.170 in money, 20 bolls of oatmeal, (equal to 10 sacks of 280 averdupois) and 4 bolls of bear or bigg, above three Winchester quarters.

At his entry he brought with him
Household furniture, and implements of husbandry, worth
L.234 5

He paid to the heirs of the former tenant,
For melioration of his farm-house and offices - 380 0
For the machinery of a threshing mill - 45 0
For hay and pasture grass, to prevent breaking up the land - - - - - - 270 0
For old corn, and straw of crop 1805 - 33 0
For the rent of 2 acres of bear, sown after turnips 14 0
For seed to do. - - - - - 2 0
For grass seeds sown among his corns, and putting them in - - - - - - 13 4

His expences at entry were in all - - L.991 9
His 6 horses were worth - - - 150 0
His 6 cows and 2 calves - - - 50 0
His 12 oxen - - - - 180 0
22 other young cattle - - - - 184 0
He paid for road assessments, and other taxes 4 0

His expences and stocking united, were L.1559 9
For the first six months after his entry
His servants' wages, exclusive of their board wages 35 0

Carry over L.1594 9

Brought over	L1594	9
His family expences, and board-wages of servants	78	0
His fuel cost	9	0
At Martinmas, 1806, he paid		
For corn and fodder, i. e. corn with the straw	154	0
For lime got in the last 6 months	44	0
His expenditure at Martinmas, 1806	L.1879	9
In the course of next year before he had a crop of his own, he paid		
For fuel, meal, and malt to his family	59	18
For wheat seed to a mall patch of land,	1	1
For turnip seed in 18 months	1	4
For an additional horse bought	27	0
Wages for one year to 6 men and 3 women servants	94	0
To harvest shearers and day-labourers	21	0
Farming implements, and blacksmiths and carpenter's accounts since his entry	71	10
Lime in 1807	121	0
Incidents, and travelling expences to his servants	15	15
Total expences, and value of stock	L.2291	17

When the interest of capital, and the farmer's personal expences are added to the above, the total sum will amount to L.2500.

By the above statement it appears that a much greater capital is now necessary to stock a farm, than was necessary under the old system of husbandry in this county; and even than is requisite in a farm of equal extent in the corn-raising districts in the southern counties of Scotland. Yet the Reporter has not the least doubt, that the farmer, who is a very good one, will find this lease a lucrative bargain.

It is not however from the statement of profits of the large farmers, nor of the capital necessary to stock a farm of considerable

siderable extent, that a correct general idea can be given of the capital and profits of the farmers of Aberdeenshire. The Writer of this Survey believes that the best way of making a general estimate of these things, is to take a more comprehensive view of the subject, by looking at the stocking and produce of a pretty large district, which contains large, middle-sized, and small farms. He shall therefore state the stocking and produce of the whole farms in his own parish, (Keith-hall and Kinkell) which he took last summer (1809), both by conversing with the farmers themselves, and also with their nearest neighbours, in cases where this last was judged expedient. The total extent of this parish is nearly 10,000 English, or 7869 Scotch acres. The farms on the banks of the Don and Ury, are from 160 to 240 feet above the level of the sea, and contain a considerable proportion of old croft or infield land, and also of outfields, of which the soil is of a good quality. The back parts of the parish rise from 240 to 500 feet above that level, and the soil in general is wet, and comparatively unproductive. In fact, both the best and worst soils in the county, are to be met with in this parish. The rent, including what is in the actual possession of the two residing heritors, is, at a fair conversion of the meal and bear, which are paid in kind, L.2000 sterl.

The value of horses, including carriage, saddle, farm-horses,
  brood-mares, and colts, is -   -   L.3578
That of black cattle, of all ages   -   -   6549
Sheep and hogs   -   -   -   -   135

Total live-stock, a little more than 5 year's rent, L.10262
Wheat raised in the parish, 12 acres, at L.10 per
  acre,   -   -   -   L.120
Barley, and bear or bigg, 208 acres, at L.7
  per acre   -   -   -   1456

Carry over L.1576 L.10262

	Brought over	L.1576	L.10262
Oats, 1201 acres, at various prices	-	6304	
Potatoes, 36 acres, at L.10	-	360	

Total food for man above 4 year's rent L.8240

Hay, 236 acres at different prices L.1331

Sown grasses pastured, 596 acres,
　at L.2　-　-　-　1192

Natural grass of all kinds worth　200

Turnips and yams, 227 acres, at L.6 1362

Green crops above two year's rent　　4085

Farm produce above six and one-fifth year's rent　　12325

Total stock and produce about eleven and one-fourth
　year's rent　-　-　-　-　L.22587

This parish is occupied by two residing heritors, by the minister, who rents a small farm, besides his parsonage lands, by 29 farmers, who have in all 35 sub-tenants, and by 69 small farmers, or day-labourers, who rent their lands from the proprietor, besides 48 cottagers who have no land, but only a house, and for the most part a small garden, in all 184 families, or 874 persons.

The live-stock and produce belonging to the heritors and minister, amounts to　-　-　-　L.3707

That belonging to the 29 farmers and their sub-tenants　-　-　-　-　14325

That belonging to 48 small farmers, or day labourers, who hold their ground immediately from the heritors, one-fifth of the whole very nearly　4525

Total live stock and produce, as above mentioned　L.22587

It is impossible to calculate with tolerable accuracy the annual charges on the agriculture of this district on an average of years. But the Writer of this Report can, from his intimate

mate knowledge of a parish, in which he has resided above 32 years, state that of the 29 farms, mentioned as in possession of large farmers, nine persons, owing to the calamitous season of 1782, were either entirely ruined, or induced to take small possessions; and that 20 of the small farmers were reduced by the same unfavourable year, and bad harvest—but that owing to the introduction of the turnip husbandry after 1782, not a single farmer of any description, was reduced by the unproductive crops of 1799 and 1800, though several of these got a pretty severe check at that period; that the high price of cattle soon enabled them, by their turnip and other green crops to repair their losses;—that the farmers houses and offices, and also the inclosures made by several of them, at the expiration of their leases, will be worth L.4000, or two year's rent, of melioration, to be paid by the proprietors or the succeeding tenants;—that the amount of live-stock and crop is L.4500 more than when the Reporter took an account of it in 1793, when the turnip husbandry was by no means general; and that the whole farmers have at least L.2000 of lent money, besides their stock; and consequently that the whole parish is at least ten thousand guineas richer than in 1793—and is at present accumulating capital in the best possible manner, by increase of livestock, improving the agriculture, and increasing the quantity of green crops; while that of the white or corn crops, with little more than half the seed sown, is not diminished, but on the whole, rather increased. The Reporter believes that this is the case of the farmers in the county of Aberdeen in general; but he would rather state the question of profit and loss, from the evident increase of capital that is visible, independently of the expence of draining, trenching, liming, and such improvements as are not visible, than to

cal-

calculate from uncertain data, which could only lead to a doubtful conclusion.

As long as there is encouragement for raising green crops, and rearing black cattle, the agriculture of this county will probably continue in a progressive state of improvement; and more land of inferior quality will be brought into cultivation, because the quantity of putrescent matter obtained from turnips and sown grasses, especially where sown grass is allowed in general to continue for three years, is much greater than can be obtained from a greater proportion of white crops, whatever attention be paid to the rotting of the straw, and making of dung. But whenever the rent is raised higher than what turnips and black cattle can pay, the farmers will have recourse to the raising greater quantities of corn. The rearing of cattle will then be more generally practised in the highlands; unless the progress of luxury shall render hay and pasture as valuable as a crop of wheat or barley. In the mean time, the Writer of this Report is of opinion, that the expence of stocking, and the profit on cropping a farm can be more correctly stated from the above documents given by two respectable farmers, and by exhibiting a general account of the live-stock and produce of a whole parish, and of the situation of the various classes of farmers, than by hypothetical calculations, which can hold good only in a few particular cases.

On a general view, the *total capital employed in agriculture* in the county of Aberdeen, calculated on the data of the proportion of the valued to the real rent, and of the live-stock and produce of the parish in which the Reporter has resided, and which on that account he selected, is *very nearly two millions sterling*; and shews the *importance both of the rearing of cattle*, and of an *improved system of agriculture*.

CHAP.

CHAPTER V.

IMPLEMENTS.

There has been a most rapid improvement in the construction of the implements of husbandry, which are used in Aberdeenshire. In order to do justice to the present race of farmers, it is necessary to quote a few passages from Dr. ANDERSON's Original Report.

"The mode of ploughing is as bad and slovenly as the
"team is aukward with which it is performed. The plough
"itself is beyond description bad; and it is of so little con-
"sequence to perpetuate the memory of what can never be
"imitated elsewhere, that I shall omit the description of it.
"I shall only observe, that it makes rather a triangular rut
"in the ground than a furrow, leaving the soil for the most
"part, equally fast on both sides of it."—" Harrowing is al-
"ways performed by horses; and it is customary for a far-
"mer, who keeps a plough of 10 or 12 oxen, to keep also
"6 horses for harrowing and carriages. This is an immense
"expence of labouring beasts, which operates as a burden
"of inconceivable weight upon the shoulders of the farmer.
"With a view to diminish this expence, the horses were
"poorly kept, and of little value; which by rendering them
"unfit for even the work that was required of them, aug-
"mented the expence. The harrows formerly were often
"made with wooden teeth; now they are generally made of
"iron. All kinds of corn are sown broad cast."—" In the
"small possessions, which I have had so often occasion to
"mention, the ploughing is universally performed without
"oxen.—Two, three, or four neighbours join to make up a
"plough: each furnishing one or more beasts. These are

"ge-

"generally four or six horses, yoked two abreast. Sometimes a cow or two are substituted from necessity in the stead of horses. The miserable scratching performed by the weak creatures in these cases scarcely deserves the name of ploughing."

"There is no such thing as a wain, or wheel machine of any kind drawn by cattle, among the old-fashioned farmers in Aberdeenshire; though some few of these have been introduced of late among modern improvers; but all loads are either carried to market on horses backs, or on carts drawn by horses."

Such was the state of Aberdeenshire with respect to ploughing, harrowing, and implements of husbandry in 1793, which Dr. ANDERSON has correctly stated as they were at that period. The Writer of this Report remembers to have heard a plough-wright, who had great employment in the division of Marr, about 40 years ago, boast, that "he could from June to January, (i. e. at all seasons of the year) make three ploughs every day, and get his two groats (eightpence) for each of them, or two shillings a day." The only tools he used were a saw, an axe, and an adze.

The whole plough seldom cost above 3s. 4d. for the wood, or 4s. for wood and workmanship—besides the coulter and sock, which, as iron was then cheap, cost about 3s. 6d. Harrows cost about 16d. or 18d. each, and 3s. when the teeth were made of iron. The old tumbling cart wheels cost from 2s. to 2s. 6d. and very little iron was used about them. Even after wheels were made to go round upon the axle, and were ringed with iron, the box cost 12s. the wheels and axle 12s. and the iron rings seldom amounted to 18s. or L.2 in all, for the carts which were used from 1760 to 1775. Now things are greatly altered for the better. Ploughs of the best construction, generally swing ploughs, on the model of Small's, and adapted either to two horses or two oxen, are in ge-

general use. Sometimes in stiff or rugged land, or in tearing up barren ground, four or six cattle are used. Cast-iron mould-boards, and sheaths of the same metal, are almost universally used by the farmers. Even where the tenants are poor, and where two of them unite their cattle, to form a plough team, the old Scotch plough is disused.—They either have a plough of a regular construction with cast-iron furniture; or where they plough only a few acres each, they use a carved mould board of wood, or have their small ploughs in a kind of intermediate construction between the Rotheran and old Scotch plough. The harrows also are greatly improved, as well as the ploughs; in consequence of several of the landed proprietors getting both ploughs and harrows from other counties—some from Angus, some from Leith, and a few from England. Almost every farmer has a roller and a break, or as it is sometimes called, a drag-harrow. The roller is usually made of granite, and mounted in a plain but substantial manner. Some use wooden rollers, very neatly fitted out, and containing a box on the top, which is filled with stones occasionally, when it is wished that the land be severely pressed. A plough costs from two to three guineas, a break-harrow from thirty shillings to three guineas, and a roller from ten to twenty-five shillings.

Various machines have been used for drilling; but the unevenness of the surface, in a county, in which there are many rising grounds, will for ever prevent the complex and expensive hoeing and drilling machines, which either hoe or drill six or seven furrows at a time, from being used in Aberdeenshire, except in very level fields. There is a very simple instrument for sowing turnips, which is used among the small farmers of Aberdeenshire. Dr. ANDERSON was very partial to it, on account of its simplicity, and his account of it deserves to be quoted.—" It is a small box made of plated " tin, called white iron in Scotland. This box is about nine " inches

"inches long, or one inch or more in diameter. It has a slip
"cover at one end, which, when slipped in, is kept from
"coming off by means of a little catch fixed to the body of
"the box, which is received into a slit made in a kind of
"appendage to the lid, by having a hinge upon it. This
"end of the box is perforated with three or four holes just
"large enough to allow one seed at a time to pass through.—
"It is made flat on one side to admit of being fixed to a
"handle. This box complete costs between eightpence and
"a shilling. When it is to be used, it is filled about half-
"full of seeds; the slip cover is then put on, and fastened
"by the catch, and being fixed to a small rod, having a
"knob or natch at the end, to keep it from slipping through
"the fingers, it is then gently shaken, as the person follows
"the plough. The seeds are scattered in the drill about
"three or four times thicker than they are allowed to stand;
"but as no two of them are ever deposited close to each
"other, the thinning of the plants is afterwards performed
"with much more ease than by any other mode of sowing I
"have ever seen practised. To those who are proud of fine
"and expensive implements this will appear a very trifling
"matter, but many hundreds of persons have I known, who
"neither could have purchased nor used a fine apparatus, but
"have been induced to enter keenly into the cultivation of
"turnips, which they would have found it a very difficult
"matter to accomplish without it."

This implement is not now used by the better sort of farmers. They have various drilling machines, some more and some less expensive; but the most common is called the turnip seed barrow. It has two sides about six feet long, with three cross-bars, and an axle with two small wheels about 14 inches diameter, is fitted to the extremity of each of the sides, or stools of the barrow. Round this axle is fixed a tin-plate

of

of about four inches diameter, the circumference of which is distant about an inch from the axle, and forms a cylindrical box inclosing the axle, and able to hold nearly a pound of turnips in the hollow between the axle and the tin-plate.— Several small holes are perforated through the tin-plate, at equal distances from the wheels, and the turnips run through these holes into the drill, or open space that is made for receiving the seed. A larger hole in one of the wheels, sufficient to admit of a common cork, or nearly an inch in diameter, is made for pouring the turnips into the empty space of the cylinder. The Writer of this Report got one of these barrows made, having both the wheels and cylinder larger than usual, and making the holes as large as would allow beans to run out on the drills. Above the tin-plate cylinder he put a slip cover of the same metal, with holes of different diameters, but at the same distance from each other as those in the inner case of the cylinder. If he wished to sow turnips he moved the slip-cover over the other, till the small holes covered the larger ones, so as to allow only the turnip seed to run out. If he wished to sow wheat, he moved the slip-cover over the other, till the holes intended for allowing the wheat to run out, were placed above those of the tinplate. By this means, any kind of seed may be drilled with a machine which does not cost above ten shillings, and a woman or a boy can drill an acre in a few hours. An ingenious wheel-wright, by adding a board fixed by a pin to the cross-bar, and fixing three pieces of iron, about three inches long in this board, has made it answer the purpose of covering as well as drilling the seed, by putting on about a pound of lead, fixed to the back of the board. This machine could easily be made to sow two drills, by making the axle and box three feet long, and perforating holes at the distances wanted—covering with the slip of tin-plate the holes not meant to be open But the Reporter, who never had above nine acres of turnips in one year, was satisfied with sowing one drill at a time, or two acres in a day.

Two other kinds of drilling machines for sowing turnips are becoming pretty common, viz. 1. One sowing one drill at a time, drawn by a horse, commonly called the *checking* barrow, has two wheels and a seed box, which does not turn round, but receives a jirk from the spokes of one of the wheels, which makes the seed fall out. The objection to this machine is, that it makes a noise which frightens some horses.—2. The other kind of sowing barrow for turnips sows two drills at a time, has two shafts for a horse, a wooden roller which goes before, and levels the tops of the drills, two seed coulters following, which deposit the seed to any depth that is judged proper. Two seed boxes are turned by means of cranks from the roller, and introduce the seed down the middle of the coulters. Attached to this machine are two small rollers following the coulters, which cover up the small drill or scratch made by them. This machine is one of the most perfect and expeditious sowers of turnips that could be desired for this county; where complicated machines, that sow a number of drills, are not suited to the irregularity of the surface; but where two drills at a time are in general all that can be done with any degree of accuracy. Mr. WHYTE, at Bridgend of Longside, in Buchan, has one; Mr. JAMES GORDON, at Orrok, in Belhelvie, another; and Mr. PIRIE, at Watertown, near Aberdeen, has a third; which differ but little in their construction, and are excellent models for the farmers in these different districts of the county.

Another cheap implement has also been mentioned by Dr. ANDERSON, and is deserving of the commendations which he gives it.—" It consists of a piece of broken scythe, about five

"five or six inches in length, fixed by two clink nails to a
"piece of iron, in shape like a very small hoe, having a
"socket joined to it, for receiving a handle, like an ordinary
"hoe. This piece of scythe is cut over at both ends, at
"right angles, the corners being left quite sharp: The
"square ends too are grinded down to an edge, and the edge
"made as sharp as a knife, by a touch on a grindstone eve-
"ry day before going to the field. This is a neat, clean,
"light tool, which is so sharp as to cut all kinds of weeds
"without requiring a stroke, and almost without being felt
"by the operator; and the edges being so sharp, and the
"corners so clean, and the whole implement so light, that
"the operators, with a little practice, acquire an inconceiv-
"able dexterity in separating the turnips from each other,
"and from the smallest weed that comes close upon them,
"so as not to leave almost any weed whatever. Of all the
"implements in agriculture, I have ever seen, this I think
"the most complete for the purpose for which it was in-
"tended. "Let those, however, who intend to adopt this
"tool advert that it is very slight in its construction, and
"cannot bear improper treatment, so that it will often hap-
"pen, that one may give way in the field. To obviate the
"inconvenience that might result from this circumstance,
"by laying people idle at a busy season, it will be proper to
"have in readiness on the field a considerable number of
"spare hoes; and as these can be made and repaired at the
"most trifling expence, this can be attended with inconve-
"nience to those only who neglect to adopt such a necessary
"precaution."—It is only necessary to remark, that these
scythes are less used for this purpose than formerly; and
stronger hoes are now in general use. This implement was
pretty generally used in 1793, but the better sort of farmers
now use either hoes made in England, and called patent hoes,

or

or other kinds of this implement, which are made by the blacksmiths residing in their neighbourhood.

The expence of hoeing turnips entirely by hand, and the difficulty of procuring labourers being now very great, several of the larger farmers have got hoeing or paring ploughs, which clean two sides of the furrow; and an ingenious man, PETER DUNCAN, bailiff, or farm-overseer to PETER GORDON, Esq. of Abergeldie, has added a triangular plate of iron for cleaning the bottom of the furrow. This tends very much to abridge the work of the manual operators, who need only set off the plants, and hoe an interstice of six or eight inches.

Although many of the implements of husbandry are incomparably better than they were 30 years ago in this county, yet it is but candid to acknowledge, that we are here in several respects inferior to the more improved counties of Scotland, and that there are several implements which we have need to borrow from England. The Kentish shim could be used in our turnip land, where the ground is too valuable to allow horse hoeing. The trench plough of Ducket, except where granite interferes, would be much cheaper than using the spade and mattock; and the Miner of Mr. ECCLESTONE would be of essential use in many fields, where the subsoil is hard and stiff, and does not allow the surface water to sink. The landed proprietors have done much good in introducing useful implements among their tenants; yet a general subscription to purchase a set of several of the newest and most approved implements of husbandry, would be productive of many advantages.

Carts are in general use, and one very rarely sees a horse carry a load of corn or meal on his back. Some of them are slight, and owing to their small size and light mounting, cost only L.3 or L.4; but a good cart costs from L.8 to L.10; and

and where frames for carrying hay and corn, with fodder or straw, are made at the same time, a very good cart, with an iron axle, costs about L.12. Since the time that turnpike roads were made in this county, single horse carts are chiefly used.

Fanners, for dressing corn at home, and larger ones for cleaning the hulled oats, and other grain at the mill, have been pretty generally used for these 15 or 20 years; and every good farmer now has them.

Threshing Mills are not as yet general, but considerable additions are made to their number every year. And in consequence, mill-wrights are in much request. The greater part of these machines go by water, and on many accounts are preferable to those which are driven by horses.—Many of the proprietors, and a great proportion of the better sort of farmers, have erected threshing machines, which thresh from 4 to 12 bolls, or from 3 to 10 quarters of corn in an hour; and have frequently a pair of fanners and shakers attached to them. Several farmers, whose rents do not amount to L.40 a year, have got threshing machines, which thresh from 2 to 3 bolls in an hour. And both from their threshing the corn more completely, and from having it quickly ready for market, even such farmers find, that the expence of the machine is soon paid—by the quantity of its produce, and by its expeditious dispatch.

Agriculture is at last calling in the aid of mechanics; and it is difficult to say, what improvements will not be effected by the inventive genius of our mechanists. The Writer of this Report was not long ago amused by seeing the churn driven by water: and observing that this more equable motion is favourable to the making of butter. The dairy maids would rejoice if the practice were general.

Per-

Perhaps it was carrying the mill-wrights attention to a place where it really was unnecessary, when he was employed in the *nursery.* Within a few miles of the Reporter's house, he has seen the *cradle rocked by water.* This is not quite so agreeable to the feelings of nature. The songs and caresses of the mother, and even the lullaby of the nurse, when not much removed from a monotony, owing to the want of a musical ear, have not the cold and lifeless uniformity of the sound of the small streamlet, and of the cog-wheel, which acts as the *moving* power, in swinging the cradle without any sensibility.

CHAPTER VI.

OF INCLOSING.

In a cold and northerly climate the benefit of inclosures is very great; and their effect, when well laid out, and properly executed, in sheltering land, and rendering it more productive, is hardly credible by those who have not experienced it.

That kind of inclosures which is most favourable to warmth, viz. raising of hedges, is not so generally practised in Aberdeenshire as could be wished. Where granite is near at hand, it is not to be expected that thorn hedges should be often raised. But there are too many cases in which thorn hedges are wanting, where stones are not to be had, and where thorns could be raised to great advantage.

Ditches, with a low earthen fence are raised in many places, because they are the cheapest—sometimes two feet of stones are placed above the front of the ditch, and the loose earth thrown behind them. Sometimes a wall of three or four feet at the bottom—the same in height, and eighteen inches on the top, of turf (provincially feal) is used as an inclosure, but is seldom durable. At other times, after a *backing*, as it is called, of turf and earth has stood 6 or 8 months, a stone breast or front, of about eighteen inches at the base, fourteen on the top, and four feet high, is built close to the turf, but made to rest upon its own base. This is commonly called a half-dyke, and when well executed, is a warm inclosure, and very durable. It costs from fivepence to tenpence per ell, of thirty-seven and one-fifth inches.— Where there is no facing but turf and earth only, it costs from threepence to sevenpence, according to its quality and other circumstances. When the fence is raised very little above the ground, and the ditch is faced with stones, it generally costs about sixpence per ell, or three shillings per rood of six ells. This species of inclosure is called a *ha ha;* and answers very well for keeping *in* cattle, but is good for that purpose only. When on the south of a gentleman's house, and in a warm exposure, it answers very well, but nothing can be more improper than to sink these *ha has* on the north, west, and north-west of cold lying fields on the east-coast of Scotland. Thorns three or four feet high above the surface of a ditch and wall, add much to the warmth and fertility of the inclosure. It is because the Writer of this Report has observed sometimes wide low ditches in this county, where dykes of a moderate height would have been more proper, that he makes this remark, while he mentions the fact.

The most durable, though the dearest fence, is a stonewall, or double-dyke, as it is called, three feet six inches wide, from four to four feet six inches high, and two feet at the top. It is no doubt true, that this wall is not so warm a fence as a hedge and ditch above ground; but it is ready at once, and is very durable. Many of the landed proprietors have got excellent stone walls built on their *mains*, or personal farms. The late Earl of Aboyne did more in the way of inclosing with good walls of dry stone, than any other person in the county. His inclosures extended many miles in length; and though many of them are 40 years old, they are still good fences. One advantage of stone walls is,

that

that when they require to be re-built, no new material is wanted, if they had sufficient thickness of stones formerly.—Dry stone walls cost from eightpence to two shillings per ell, according to the distance from which the stones are carried, the height of the walls, and goodness of the work. An uncommonly fine wall lately built by PATRICK KILGOUR, Esq. of Woodside, near Aberdeen, cost half a crown per ell, for building only; and will probably stand for ages.

In several places in Buchan, and in the Garioch, thorn hedges are raised; but very few are found in Marr or Formartin. The Reporter knows of none in Strathboggie, except in a few gentlemen's gardens. Earthen fences are more common than they were. But stone dykes are raised as generally as it is possible to get undertakers to execute them. And in a country, where granite can so generally and easily be had, the inclosure on the outside ought to be of stones, though it were but what they call an half-dyke.

The great defect of this county is the want of belts of wood on the north, north-west, and west, and on the north-east and east, where this can be done; leaving open the south, south-west, and south-east. It is seldom that wheat or two-rowed barley can be raised to advantage at present; but were these belts of wood generally raised in proper places, the local heat of a great part of the county in the seasons of vegetation, would be increased from 5 to 8 degrees more than the general heat of these districts is at present, from April to October. And the advantages of this to our agriculture, independently of beauty and ornament, would be very great indeed. When one sees all the scenery of England in the valley of Strathern, and observes the Dee, by the natural woods near its bank, calling on men to plant, one cannot help regretting that we have too few ring fences

or

or belts of wood, as well as of larger plantations, compared to what we need in this county.

GATES.—These were formerly very mean—now they are better executed. Near the houses of the landed proprietors, stone-pillars, with excellent gates, bound by diagonal bars, are very common. And even the farmers are improving the gates of what inclosures they have. Where they cannot afford expensive gates, they by various means attempt to keep cattle from breaking into, or getting out of their inclosures.

CHAP. VII.

ARABLE LANDS.

SECT. I.—TILLAGE.

THERE is very great variety with respect to the manner in which the ploughing of ground is performed. In general among the younger farmers and ploughmen there is a great zeal excited, which will have the happiest effects on our tillage. Already the ploughing matches have produced a very great reformation in this respect: And the Reporter feels a pride in remarking, that a far greater proportion of straight and well ploughed ridges is to be found in Aberdeenshire, than what he could see in Yorkshire, or indeed on any of the great roads from Scotland to London. He was indeed very much surprised to find the ridges of all the improved lands in Scotland straight, and much of the ploughing executed in a most correct manner, while those in that country (of which he expected so much till he saw it) were generally crooked, and the work very ill executed. The old farmers of this county however could not hold the ploughs which are now generally used; and the old serpentine ridge made by the oxen a century ago, is still to be seen, but only in a few places.

Ridges were formerly made of great breadth, and so constantly gathered up in ploughing, that the sides of them, which occupied one-half of the surface, had very little soil left upon them, and produced a small proportion of either grass or corn. At present the average breadth of ridges is 15 feet, in a few 12, and in others 18 to 20. When the ground is wet, and the subsoil retentive, the ridges should never exceed 12 feet; and in many cases where surface-draining is the only security against the clover being thrown out by the frost and moisture, 10 feet ridges are preferable to 12. Indeed there are other advantages attending a ridge of this breadth: though it would be improper to lay out very dry land, either in grass or corn in so narrow ridges. The old ridges were raised so high on the crowns, that two men *sitting* in the furrows, with only a ridge between them, could not have seen each other.* And a considerable loss was incurred by the Writer of this Report in too hastily levelling these high ridges, and thus burying the soil which was in a state of vegetation, and bringing up dead earth, which required a year's exposure to the sun and frost, as well as a considerable quantity of manure, before it became productive; besides the risk arising from the harvest being later.

In

---

* The practice of *gathering* the furrow, or raising it to the crown of the ridge, was no doubt carried too far. But a gentleman, who was the proprietor of an estate in the Garioch, carried his desire of *scaling* ridges, or of cleaving them in the crown to the opposite extreme; and one day observing a farmer, who was ploughing a field to one of his cottagers, and gathering the crown of the ridge, in a great rage called to him to *scale*, or *cleave the land he was ploughing*. The farmer calmly replied, that he must *obey the orders* of his employer, which were *to gather the ridge*. The gentleman, in great agitation, cried out—*You blockhead— some ignorant ass like you has gathered up the hill of Dun-o-Deer*. This is a beautiful conical hill, which rises about 300 feet from its base, in the neighbourhood of the field which the farmer was ploughing: and this anecdote is mentioned to shew the impropriety of carrying either our passions or our prejudices to an immoderate height.

In the old system of husbandry, the motion of the 10 or 12 oxen was very slow; but their furrow was very broad on the surface, and so narrow in the bottom, that a section of it would have made an equilateral triangle. Hence, though they took five successive crops of oats from their outfields, these 5 ploughings did not exhaust or expose the soil so much as three ploughings with 2 oxen or horses made by a plough of the present construction. At an average, they tilled half an acre in about 5 hours. The Reporter once knew a whole acre ploughed on the banks of the Dee, by 12 oxen, in about 6½ hours. But this was accounted a great exertion. At present, a plough with two horses tills about one-third of a Scotch acre in four hours, which is equivalent to an English acre in 10 hours. When the horses are good, and the season is pressing, some farmers work their horses ten hours a day, except in the winter months. But few horses will endure this without being injured. The ordinary horses, except in a hurried season, plough only one-fourth of an acre in about 3¾ hours, and work only 7 hours a day to the small farmers. It may be laid down as a general rule, that in the lower part of the county, five ploughs out of six have no driver to their horses. Nor are any drivers used to the two oxen ploughs; but several of the farmers use four oxen in the plough, at least for a part of the season. Two of these are either young ones, whom they are training, or old oxen whom they are to feed after their stiffest lands are ploughed. The Reporter knows only one farmer in the Garioch, who still ploughs with 12 oxen. This singular circumstance, which thirty years ago was a very common case, occasioned a facetious gentleman, at a public meeting, to propose as a toast, the *Garioch farmers, with four times three.*

The depth of the furrow is from 4 to 6 inches, according as circumstances render proper, and the breadth varies from

7 to 12 inches. In a late ploughing match, a 15 feet ridge, by the worst ploughman, had only 18 furrows, while the best had 22, or 11 bouts of his plough. The ground was broke up out of grass. On the whole, the tillage, though in some places defective, is much improved, and in the lower parts of the county is well performed. Dr. ANDERSON, in the Original Report, complained of the ploughmen near Aberdeen. But they are very much improved since 1793, though it cannot be supposed that an Aberdeen carter, who ploughs only occasionally, should be as correct in his practice as a ploughman in the country, who has so much experience in ploughing, and makes it his study.

Oxen are used in ploughing, even when the farmers intend to fatten them. But they are never over-wrought; and their work is not inferior to that of horses. For one purpose they are generally superior—viz. for carrying of large masses of granite in a paddock, or kind of sledge made for carrying these off. The Reporter has seen six horses break their tackle, and from their impetuosity unable to carry off a very large stone, which eight oxen, without being fretted, and seemingly without being overloaded, carried off in a slow and steady motion, to the distance of 200 yards. In breaking stiff and coarse land, full of stones, interspersed with clay, six oxen with a strong plough are more to be depended on than as many horses, because they are more steady, and do not fret when unable to move the plough. In short, the ox is useful in tillage, and therefore is used in this county wherever this is expedient, though the horse is now more generally employed, and both now execute their work much better than formerly. Though the works of ploughing, harrowing, rolling, drilling, horse-hoeing, and striking of the furrows, are in general performed in a workmanlike manner, the scarificator has not been used in this county; and owing

to

to the number of small stones in the greater proportion of the arable lands, it could not be used to advantage. But for the most part the hand-hoeing and weeding of turnips, and other roots, is correctly performed. The hoeing of white crops has not as yet been introduced into Aberdeenshire.

SECT. II.—FALLOWING.

FALLOWING is not much practised in this county; and indeed it is not proper for the greatest part of our arable lands. Perhaps there are not as yet above 400 Scotch, or 500 English acres sown usually with wheat; and at least one-fourth part of that is after potatoes, or drilled beans; at any rate, not sown after fallow. We formerly had a kind of summer fallow on a part of the faughs, or inferior land, which got no manure. After lying five years in natural grass, it was torn up at midsummer, and from that circumstance was called *riven ley*; next year, and the three following years it was sown with oats. That wretched system is now laid aside. But the thin loose soil of our outfields, and lands of inferior quality, has too little tenacity to bear the frequent ploughings of a summer fallow; and even the greatest part of our old croft land is adapted for the raising of turnips rather than lying in fallow. The Reporter saw a field, which he knew from his infancy, summer-fallowed by a farmer from the county of Angus, who took a 57 years lease of the farm to which that field belonged. The effect of the frequent ploughing on land, which had no tenacity, was not merely a complete pulverization of the soil, but the following year

year one of the greatest crops of yarr (or spergula) ever seen in the county, was produced, and choaked the bear that was growing on the field. The farmer, who is a very intelligent one, and had but newly come to Aberdeenshire, soon saw his error, raised turnips, and left off the summer fallowing of this light and loose land. By suiting his mode of tillage to the nature of his soil, he has, in 24 years, accumulated a very considerable capital, improving an extensive but rugged farm, which he has subset at above four times his rent (for L.200 a-year of profit on a farm of L.60 of rent) for the 33 years of his lease which are yet to run. This circumstance is mentioned merely to shew the absurdity of laying down an invariable system of agriculture, whether flatly condemning, or indiscriminately extolling, summer fallow for example. Great Britain is not all Norfolk, nor is it all the soil of East Lothian, any more than of Aberdeenshire. A judicious farmer must study the nature of his soil, as a master studies the character of his scholars. In Buchan, where there is much stiff land, particularly near the sea-coast, summer fallow is used by many of the farmers, and may properly be used, where drilled beans cannot be raised. In several other places of this county fallowing may be used occasionally; but the fact is, that very little fallow is used, and comparatively a small proportion of it is necessary in Aberdeenshire. In breaking up waste lands, a summer fallow is often very beneficial, and even there, when the texture of them is loose, a crop of turnips has been raised to advantage among the half rotted heath, and been succeeded by an excellent crop of oats or bigg, without more fallowing than from midsummer of the preceding year, and three ploughings next spring.

SECT. III.—ROTATION OF CROPS.

Within the last 40 years, very different rotations of crop have been adopted, not only in different places, but even in the same districts, and on the very same farms. For several ages previous to the year 1750, there was a dull uniformity in all the rotations of cropping. The old croft land being dunged every third year, when it bore a crop of bear or bigg, followed by two only of oats; and the other lands of inferior quality, whether they got any manure or not, bearing no other crop except oats, for four or five years, at the end of which period they were allowed to go to natural grass. For although we had inclosures about a century ago, we had no sown grasses till about the year 1750. It is proper, here following the order of time, to quote Dr. Anderson's account of the mode of management previous to 1793; then to subjoin a very few remarks on that account, and afterwards to state the various rotations of cropping which have taken place since that period. Dr. Anderson, in the Original Report, describes the general management of the farms in this county, in the following words:—

"Perhaps no county in Britain can exhibit such perfect "specimens of extremes of good and bad management as "Aberdeenshire. Of the lands in the immediate neighbour- "hood of Aberdeen, some hints have been already given.— "At present this particular patch is out of view; it is the "practice of the country at large beyond the barren zone of "Aberdeen, that I am first to describe.

"Throughout the whole district, the general practice that "has prevailed for time immemorial, is to divide the arable "lands of each farm into two parts at least, infield and out- "field.

" field. The infield, as the name implies, is that portion of
" ground which is nearest to the farm-stead, and usually con-
" sists of about one-fifth part of the whole arable ground on
" the farm. This is kept in perpetual tillage; and the inva-
" riable system of management was, and still is, with few
" exceptions, to have it divided into three equal parts, to be
" cropped thus: First, *bear*, with all the dung made by the
" beasts housed on the farm, laid upon it. Second and third,
" *oats*; then *bear* again, and so on in the same unvarying ro-
" tation. The mode of tillage under this management, is, for
" bear, to scratch the ground with a slight furrow, any time
" during winter, when convenience permits, and to turn the
" loosened earth upon the surface of the stubble, so as to co-
" ver it. This is called *ribbing*. It lies in this state till the
" end of April or beginning of May, when it is harrowed
" down, and the dung spread upon it. It is then ploughed,
" or rather hashed over, for ploughing it can scarce be call-
" ed, and sown with bear,

" For oats, the ground is ploughed as soon after the grain is
" cut down as possible. Often some part of the ridges are
" ploughed the day the corn is cut down, while the stooks
" are standing on the field. In this state it remains till the
" first dry weather in the spring, usually the month of March,
" when the oats are sown, and immediately harrowed in.

" That part of the farm called outfield, is divided into
" two unequal proportions. The smallest, usually about one-
" third, is called *folds*, provincially faulds; The other large
" portion is denominated *faughs*. The fold usually consists of
" ten divisions, one of which each year is brought into tillage
" from grass. With this intent it is surrounded with a wall
" of sod, the last year it is to remain in grass, which forms a
" temporary inclosure, that is employed as a pen for confin-
" ing the cattle during the night time, and for two or three

" hours

" hours each day at noon. It thus gets a tolerably full dung-
" ing, after which it is ploughed up for oats during the win-
" ter. In the same manner it is ploughed successively for
" oats for four or five years, or as long as it will carry a crop
" worth reaping. It is then abandoned for five or six years,
" during which time it gets by degrees a sward of grass,
" when it is again subjected to the same rotation.

"The faughs never receive manure of any sort; and they
" are cropped exactly in the same manner as the folds, with
" this difference, that instead of being folded upon, they are
" broke up from grass, by what is called a *rib-ploughing*,
" about midsummer, one part of the sward being turned by
" the plough upon the surface of an equal portion of ground
" that is not raised, so as to be covered by the furrow.—
" This operation on grass land is called faughing, from
" whence this division of the farm takes its name. It is al-
" lowed to ly in this state till autumn, when it is all ploughed
" over, as it can be done, and is sown with oats in spring.—
" It produces a poor crop, and three or four succeeding crops
" still poorer and poorer, till at last they are forced to aban-
" don it by the plough, after it will scarcely return the
" seed.

" Every farm in Aberdeenshire contains fields of the three
" kinds above enumerated. There are still two other deno-
" minations of fields that frequently occur, some part of
" which is indeed found upon most farms, though not always
" so. These are called laigh lands, and burnt, vulgarly
" brunt lands. The laigh lands are a kind of low lying moist
" meadow ground, sometimes with a mixture of moss. They
" are invariably ploughed three years for oats on one furrow,
" and are allowed to be in grass for three years, and so on al-
" ternately, without ever receiving any dung. Brunt lands are
" now very generally managed after the same manner.—
" They

' They are always of a mossy nature. / The turf, when
" broke up from grass, used to be gathered into heaps, and
" burnt, and immediately sown with bear, after which two
" crops of oats were taken in succession, and then it was suf-
" fered to run to grass. But this practice of burning having
" been found prejudicial, is now very generally prohibited,
" so that the practice is now rare, though the patches, that
" had been formerly subjected to it, still retain that name."

Concerning the facts stated in this long quotation, which it would have been unfair to abridge, it is necessary to observe,

1st. That the common rotation of cropping the old croft or infield land, was, before 1782, pretty uniformly as Dr. ANDERSON has described it. 1st, Bear or bigg, with dung; 2d and 3d years oats; then bear again, with dung as before. But in several parts of the county, pease were introduced, and generally without dung, then the bear, with or without dung, followed. Where the bear was dunged after pease, it was laid down with grass seeds, pretty generally after 1750, previous to which time they were not kept for sale by any of the merchants in Aberdeen. In the division of Buchan, either pease or beans were sown, and followed by bear and grass seeds. This was the first approximation to improvement in husbandry.

2dly. Although very few farmers sowed turnips before 1782, yet before 1793, a considerable number of turnips was raised; and the quantity of sown grass, after turnips, potatoes, pease, or beans, was very considerable. Perhaps one-half, certainly one-third, of the infield, or old croft land, was taken out of the old rotation, and subjected to a new one, of which grass seeds, after bear thus prepared, made an essential part.

3dly. After the year 1782, but before 1793, a considerable

ROTATION OF CROPS.   235

...ble quantity of lime was driven from Aberdeen, and laid on the outfields, and other inferior lands, which also changed the course of the rotation of cropping in these.

The old rule was five crops of grass, on the last of which the cattle were kept all night, and a few hours in the middle of the day for about five months in summer, and then five crops of oats, *(never more than five)* but sometimes only three or four were taken, if the farm had only eight or nine folds attached to it; if the ground seemed unable to carry another crop, or if the farmer, from the inclemency of the season, could not overtake the ploughing of his poor land.

4thly. The inferior land, besides the outfields, was denominated faughs, if only *ribbed* at midsummer; was called *one fur ley*, if the whole surface was ploughed, or *burrel ley*, where there was only a narrow ridge ploughed, and a large strip or baulk of barren land between every ridge. All these denominations, as well as the laighs and burnt lands, had no manure laid upon them; but they seldom bore more than three crops, never more than four, as far as the Writer of this Report observed, or could learn. Even this was bad husbandry; but better was not then known.

On a general view of the husbandry of Aberdeenshire before 1750,

The infield bore, 1st, one crop of bear after dung.

  2d, oats,   3d, oats, and bear as before.

Or, 1st, Pease. 2d, bear.

  3d, oats; and 4th, oats; then pease as before.

On the introduction of sown grasses after 1750 till 1760.

1. Pease.   4. Grass.
2. Bear, with grass seeds. 5. Grass.
3. Grass.   6. Oats.
  7. Oats, followed by pease.

On the introduction of turnips, (which did not take place on
            the

the personal farms of the gentlemen, till after 1760; and very few farmers had so much as an acre on their farms till 1782) the rotation was,

1. Turnips dunged.
2. Bear and grass seeds.
3. Clover hay.
4. Do. or pasture.
5. Pasture.
6. Oats.
7. Oats or bear without dung.
8. Oats, and then turnip as before.

[This was the most general rotation of all the improved infields in 1793, when Dr. Anderson wrote the Original Report, and continued to be so till 1800. The depriving the farmers of Aberdeenshire of the power of disposing of their bear to distillers, by abolishing the privileges of the intermediate district, occasioned so low a price of that grain, and also so dull a sale, that the farmers gave up sowing bear on the second crop after breaking up out of grass; and the excessively high malt-tax, joined to the imposing an unfair proportion of that tax on malt made from bigg, compared that from barley, has now induced the farmers to lay down their grass seeds very frequently with early oats. And the following rotations are now most commonly adopted.

1. Turnips.
2. Early oats with grass.
3. Clover hay.
4. Do. or pasture.
5. Pasture from grass.
6. Oats, generally potatoe oats.
7. Oats.
Turnips as before.

This rotation leaves out bear or bigg altogether; and this mode of cropping, which was introduced by Lord Sidmouth's altering the *ratio* on malt from Scotch, compared to that of English grain, must soon hurt our agriculture very deeply.— It is but justice to many farmers to add, that they continue to lay out their grass seeds with bigg, (though they know that oats are more profitable) because bear is more favourable to their grass. The bear is not sown so early, nor lies so

long

long on the land It is generally removed before the equinoctial storms set in, and the grass seeds usually rise better among the bear, which has fewer leaves on its stalks, than among the oats, which has generally a much closer crop, and sometimes chokes the grass by the closeness of its leaves, and by its being lodged in rainy seasons.

Although the grass in general lies for three years—1. for clover hay, 2d, for soil, or cutting to cattle, and a 3d for pasture, yet some farmers allow their grass fields to remain in that state for five years, in which case they take two crops of hay, and three of pasture, and when they break up this old pasture, they always take three white crops, before they make turnips. So that their rotation stands thus—

1. Turnips.
2. Bear and grass seeds.
3. Hay.
4. Hay.
5. Grass cut for soil.
6. Pasture.
7. Pasture.
8. Oats.
9. Bear or oats.
10. Oats—then turnips as before.

This is a bad rotation; but to a dealer in cattle, who wishes a great deal of pasture, and who has an extensive farm, and finds he cannot get dung for more than one-tenth part of it, for his turnip fields, it is not without its recommendations. The faulty part of it is the succession of three successive crops of corn, after being five years in grass. A black crop, i.e. pease or beans, kept perfectly clean, the second year after the land is broke up, would be a very great improvement, and would be followed by an excellent crop of oats, after which the turnips would come in very properly.

Opposed to this rotation of ten years, some of the best farmers have adopted the Flemish husbandry of alternate white and green crops, sometimes called the Norfolk, and some-

sometimes with more correctness the *four shift* course of cropping. This is,

1. Turnips.
2. Bear and grass seeds.
3. Clover.
4. Oats or wheat.

or 1. Potatoes.
2. Early oats & grass seeds.
3. Clover or hay.
4. Oats.

This is by far the most profitable course to a farmer.— But few of the landed proprietors will permit it, where they give new leases. Attached to the raising of grass and rearing of cattle, and knowing from dear bought experience, the danger of over-cropping, they are determined to put it out of the power of their tenants to injure their lands. And it must be acknowledged, that there is too little putrescent matter for the purpose of manure, by adopting this course in a county of which so large a proportion is still uncultivated, and still more in a poor state of cultivation. A five shift course of

1. Turnips.
2. Bear and grass seeds.
3. Clover hay.
4. Pasture from sown grass.
5. Oats—and then Turnips as before,

is perhaps more proper for a considerable proportion of the inferior croft, or old infield, and for the greater part of the outfields of the county of Aberdeen. But on the very inferior soil, where the land, till very lately, got no manure, the faughs, the laighs, the burnt lands, and burrel leys, which have been cleared of stones, drained, levelled, and both subjected to the operation of the plough, and admitted to a share of lime, manure, and of the farmer's indulgent attention, a six shift course, consisting of

1. Turnips.
2. Oats and grass.
3. Clover.
4. Grass.
5. Grass.
6. Oats, then turnips as before,

would be a more adviseable rotation, at any rate for some time

time after they are first improved.* Unfortunately the landed gentlemen in their leases, which are commonly given, allow the tenant to take two crops of oats after ley, without any respect to the poorness of certain parts of his farm; and they oblige him to allow it to continue three years in grass after it is laid down, although the ground be of the most fertile quality. In the new leases which are now given to tenants, this seven shift course of turnips, bear, grass three years, and two white crops is very generally adopted. It would be a great improvement to it, to take away the second white crop after ley, although one of the grass crops were also subtracted from the rotation. For a five-shift course, including two crops of grass, two of bear or oats, with one of turnips between them, is preferable to a seven shift one, which admits of three grass crops, but allows two white crops to be taken in succession.

If the Writer of this Report might take the liberty to suggest any rule for the course of cropping to the landed proprietors and farmers of Aberdeenshire, it would be this—
" In the richest lands, which will bear this rotation, let a
" four shift course be fixed along with a corresponding high
" rent, say for example—three pounds an acre. Let a 5 shift
" course be established on land of a secondary quality, from
" one pound ten to two pounds five shillings an acre; and
" where the land is poor, and requires a 6 shift course, let it
" be fixed to that rotation of cropping, and pay from five to
" twenty shillings an acre." By adopting this rule for 19 years, the value of the farm will be greatly increased. But at all events it should be laid down as a rule, that a farmer should never be allowed to take more than one white crop in succession, with

one

* Wheat certainly could be introduced in place of oats, on all the best lands.

one year's clover on the best lands, two years grass on the next, and three year's grass on the poorest lands.

It may be necessary to mention, that though turnips are here uniformly mentioned as the green crop, yet potatoes or drilled beans are equally proper, and are frequently used, especially drilled beans, in the division of Buchan. And that when the season is early, spring wheat may be sown to as much advantage as barley, though it is not so soon ripe in harvest as the bear or bigg is for the most part.

A farmer in East Lothian will be surprised that a more decided preference is not given to the four shift course, to which he is so partial. But the soil of Aberdeenshire is very different from that of the Lothians. Of pure clay we have very little, and the light soil of this county is very different from what is called light soil at Upper Keith, where Mr. BRODIE has successfully followed a four shift course. The Writer of this Report was extremely partial to this rotation; and ten years ago trenched and improved four Scotch acres, on which he had excellent crops for the first four years, and tolerable crops on the second rotations; but last year its want of tenacity was evident, and it is now in turnips, will be sown with bear and grass seeds, and will ly at least two years in grass. It will then be kept in a five shift course, which is best adapted to the greatest part of the arable lands of this county.

Next to prosecuting any plan with success, and proving its advantages to the conviction of every impartial man, is the frankly confessing an error, or renouncing an opinion which is found to be erroneous.

Let thus much suffice for the different rotations of crops which are practised in the districts of Aberdeenshire, and for suggesting what may tend to establish different courses of cropping suited to the different soils. By allowing the

te-

tenants to take what crops they pleased, some of them, by first liming, and then over-cropping, have greatly injured their farms, and the landed proprietors are now awakened to a sense of their danger. But adopting the same indiscriminate rule of cropping for all qualities of soil on a large estate, must either prevent the farmers from paying such rent as they could afford to pay, without injury to their land, or must establish a rotation, which is not generally adapted to their different farms, and even to the different soils which are found on the same field.

"In the immediate vicinity of Aberdeen," as Dr. ANDERSON justly observes, " the crops raised on the high-rented
" fields, that have been long in culture, are bear, grass, con-
" sisting of a mixture of clover and rye-grass, cabbages, tur-
" nips, greens, carrots, parsnips, and potatoes, and scarcely
" any thing else. The ground is for the most part open and
" uninclosed. Rent from six to eight pounds per acre : And
" as the whole of this ground is occupied by men, who farm
" it for profit, it is very obvious, that great attention and
" skill are wanted to make it afford a reasonable return for
" rent and labour.

" There is no general system, nor plan of rotation of
" crops there adopted, every person varying his crops in the
" manner he thinks will turn out to best advantage for him-
" self; but the culture is invariably good, the crops clean,
" and the produce highly luxuriant. The soil is a deep mel-
" low loam. In some places near the sea it has been original-
" ly sand; but is now a light warm loam that is highly pro-
" ductive.

" Grass is cultivated either for the purpose of being cut
" and consumed green during summer, or for hay. When
" it is cut for hay, it is in general ploughed over the moment
" it is cut, after being slightly dunged on the stubble, and
" either

"either planted with greens, or sown with turnips. If the
"latter, it is allowed to be a few days to wither in the sun,
"then ploughed a second time, across or diagonally, if the
"field will admit of it, and harrowed, so as to destroy the
"roots of the grass; then gets the seed furrow. As the
"ground is very tender, the whole of these operations can
"usually be finished in the course of one week. As soon as
"the turnips are off the ground early in the spring, it often
"happens that cabbages, carrots, or parsnips, succeed them,
"as these crops can be put into the ground, as early as the
"season will permit. If it is in greens, which may be al-
"lowed to stand later in the spring, it is usually cropped
"either with potatoes, early turnips, or bear, as circum-
"stances will admit.

"After early turnip, which are off the ground in June,
"the ground is immediately turned over, and sown with win-
"ter turnip, or planted with greens, though more frequently
"the first, as greens can admit of being planted after some
"later crops.

"Potatoes are taken up by the spade as they become rea-
"dy for lifting. Whenever they are dug up, greens are
"immediately planted in their place."

Such was the state of the old cultivated lands near Aberdeen in 1793. Since that time many villas have been built by the opulent merchants, and a great deal of barren ground has been improved. The Writer of this Report has been favoured with an account of the crops since 1804, which will be found in the Appendix.

SECT. IV.—CROPS COMMONLY CULTIVATED.

OATS.

These are first mentioned, because cultivated in greater quantities than any other kind of corn. They grow upon most soils; and from the comparative rise in their price, partly owing to the demand for them as food to horses, partly owing to there being no importation of oats from the north of Europe, and partly owing to oatmeal being almost the only kind of food consumed by the great body of the people, (who, formerly used a considerable proportion of bear or bigg made into oatmeal,) they are raised in far greater quantities than all the other kinds of grain. Nay more ground is sown in oats than is ploughed annually in Aberdeenshire for all other purposes. By computations taken from the quantity of oats sown in his own parishes, and also from allowing an acre of ground to be sown with oats for each individual in the county, the number of Scotch acres, in round numbers, is, 120,000; and English acres 152,500.

Although a number of persons use wheaten bread, yet the great quantity now used for horses, dogs, poultry, and for the seed of the succeeding crop, render this a moderate computation of the quantity of oats sown, especially when we consider that a great proportion of oats is sown on land of a very indifferent quality.

1. *Preparations, or situations in which they are sown.*— They are sown on the old croft lands universally as the first crop after ley, after being once ploughed; also by many farmers on the second crop after ley; thirdly, by those who use bear as a second crop, oats are too often sown as a third crop after ley; and, lastly, since the tax on malt became so high, and that tax also comparatively higher on malt made

from Scotch bigg than from barley, early oats are very frequently sown after turnips or potatoes, along with grass seeds. 2. On the outfields, they are sown as the only crops previous to cultivation by turnips, and very generally after turnips along with red and white clover, rib grass, and rye-grass. 3. On the inferior soils, when ploughed up sometimes after an imperfect fallow, but more generally after a single furrow, they are the only crops of corn that are raised. For it is found, that even after turnips, bear does not answer so well as oats. In 20 or 30 years hence, if the ratio of malt-tax be altered, and this poor land properly manured, and gently cropped, the grass seeds may be sown out along with bear; but this is very rarely attempted at present.

2. *Sort.*—We had a number of different varieties of oats imported into Aberdeenshire after 1782, when our crop was so defective, and when the oats on our best farms, which were then in flower, were destroyed for the purpose of seed, by a severe frost in the beginning of August.

1. The antient native oat, commonly called small oats, were very commonly raised on the poorer lands in the Garioch and Formartin, previous to 1782. It was not raised in the lower parts of Marr, and the Reporter never saw it till 1778, when he settled in the Garioch. It has a long beard like wild oats, from which it is easily distinguished; and in the braes of Angus is call Shiacks. It has been on the decline since 1782; and since our agriculture was improved is now entirely discontinued. The Sheriff could find no proof of its being bought or sold for the last three years. It yielded from 6 to 8 pecks of meal from the boll, which was measured by hand waving, and was equal to the bulk of an English quarter. It is not to be wondered that a species of oats was discontinued of which a quarter yielded only 56 pounds of meal. Yet before 1782, the *farm meal* was commonly paid

of this inferior oats: i. e. The landlord in many places of the county, got part of his rent paid in kind from meal made from this grain.

2. Of the large white oat, which is commonly raised in the county by the old farmers, we have several varieties. That which is in greatest esteem, comes from Kildrummy; but it is doubtful whether the excellence of this oat be not occasioned by the fine soil, and favourable situation of the place from whence they come: For both bear and oats are of excellent quality in that parish. A boll of this oats yields generally a boll, or 140 English pounds of meal.

3. The Halkerton oat, brought from the How or Hollow of the Mearns—a very large corn, but thick hulled, was used a few years, but was found inferior to the Kildrummy.

4. The Dumbennan oat, from the district of Strathboggie, is a short and plump corn, but was generally ten days later than the Kildrummy, and is now less used, except in Strathboggie. It was not inferior to the other in point of weight, though it did not ripen so early.

Besides these native kinds, a great variety of oats was brought from other places.

5. The Blainslie oat came highly recommended; but was soon disused. It was longer but less plump than our native oats, and was found to be later than any of them.

6, 7, 8. The Dutch oat, the Polish, and the early Essex, were in succession introduced, but were so apt to shake, that they are now seldom to be found.

9. The Chief Baron oats, or Peebles, or red oats, which Mr. MONTGOMERY of Magbiehill, first brought into notice, and the late Lord Chief Baron cultivated on his personal farm of Wheam, in Peebles-shire, continue to preserve their character. They are early, and not apt to shake, and grass seeds are frequently sown out along with them. Instead of degenerating

nerating they have improved since they were first imported. The Writer of this Report got six bolls of them about 24 years ago from the Lord Chief Baron's farm, when he found they weighed only 14 stones and 4 pounds per Aberdeen boll, which is five-sixths of the English quarter. Next year their produce weighed 15 stones and three quarters, and the year following 16 stones, which is nearly 42lbs. avoirdupois per Winchester bushel, above 4lbs. per *bushel* heavier than when he imported them. But they become less early, if not exchanged, or sown on different soils.

10. The Potatoe oats. This is the most valuable oat which we have, only it requires land in good order. The Writer of this Report got from Mr. Brown of Markle, by favour from Mr. Burns, architect at Haddington, potatoe oats, which were 16½ stones per Aberdeen boll, and which he raised in one favourable season to 17 stones, but they did not continue at that weight, though he has had 21 pecks of meal from the boll of them.

11, 12. A species of oat, a native of the county, and also a variety of it which was imported, called the Barley (or Birley) oat; first used in sowing burnt lands and laighs; and afterwards on account of its earliness sown along with grass seeds, was for a few years acceptable on account of its plumpness, and the quantity of meal which a boll of it yielded. But the meal had a taste resembling that of bear or barley, whence it derived its name; and the country people not relishing it on that account, it is now very little used.

13. The common oat raised in the Garioch, when leased, i. e. when the black or inferior corn is picked out, and then sown in good land and multiplied, is preferred to all others, except the early and the potatoe oats; and in the opinion of some, is not inferior to them.

Seed

Seed.—From three firlots to a boll, of county measure, to an acre.

Harvest.—From the 1st of August to the end of October. Oats sometimes cut down with the scythe, commonly with the sickle.

Produce.—Extremely variable, from 2 to 16 bolls; the first in poor lands in the upland districts; the last in the vicinity of Aberdeen. In good years, a boll of oats from good land gives a boll of meal at an average. In 1782, three bolls at an average, yielded only a boll of meal.

BEAR OR BIGG.

This species of barley was very generally raised in the county of Aberdeen, till the alteration of the ratio of the malt tax from bigg, compared to that from barley, occasioned early oats to be in many cases used for sowing along with clover and rye-grass seeds. It is distinguished from what, by way of eminence, is called barley, by having four rows of corn on its stalk, (and a particular species of it, called *packman rich*, has six rows) while the species which is most generally raised in England, and in the south of Scotland, has only two rows. From its being a more hardy grain than barley, from its lying in the most favourable seasons three weeks less time on the ground, and in the late seasons at least five weeks less time than the two-rowed kind, and from its being generally cut down before the equinoctial storms in September, it is peculiarly adapted to the nursing of grass seeds, and ought to be cultivated more generally than it is in Aberdeenshire. During the time that the excise laws permitted legal distillers to take out licenses, on terms which were suited to their capital, and to the state of this county, our agriculture advanced rapidly, no illegal distillation was countenanced by either the landholders or farmers, and a considerable

rable revenue, capable of a still greater increase, was received by Government. But by the alteration of the ratio of malt-tax imposed upon Scotch grain, malt made from bigg pays three shillings and one-eighth of a penny per bushel, while that from Scotch barley pays three shillings and eightpence, or eightpence more; and the mode of levying the tax, viz. imposing the greatest part of it on the contents of the still, occasioned all the distillers in Aberdeenshire either to give up the business, or after struggling with taxes which they could not pay, to stop payments entirely. In consequence of this much less bear is sown than was sown ten years ago; and at least one-half of our grass seeds is sown with early oats. Indeed many farmers sow above three-fourths of their land that is laid out with clover and rye grass, with oats and not with bear. This is particularly the practice in the division of Buchan.

The quantity of bear now sown, does not probably exceed twenty thousand bolls, but is rather below that number.— The produce, estimating the value of the grain and fodder at an average of the last seven years, is worth about seven pounds an acre. An acre of oats in such preparation, as to be fit for laying down grass seeds, is generally worth more money for the price of the corn, while the fodder is of much more value.

Time of sowing from 20th of April to 26th of May; or as soon as the oats are put in the ground.

Seed.—Three firlots or five bushels to a Scots acre.

Harvest from 1st of August to the middle of September.

The quantity of bear raised on an acre belonging to the small farmers, or to the large farmers where the soil is of inferior quality, is below five bolls. On good land belonging to the best farmers, it exceeds seven bolls. But on an average

of

of produce, and of the price, it may be stated at five and one-half bolls, or L.7 an acre.

One fact respecting this species of grain deserves to be known. This is, that though a much greater number of bolls is now raised on the acre than was raised under the old husbandry; and though the bear is much *cleaner*, or freer of wild oats and weeds than it was formerly, it is not now so *weighty* a grain. The boll of it seldom exceeds 17 stones, when correctly measured by an Aberdeen standard firlot; which is about 15 stones and a quarter to the Linlithgow boll. It was seldom so light as 18 stones 30 years ago. In the year 1782, the stipend bear paid to the Reporter was above 19 stones, and yielded above 30 pecks of meal, though ground only on a common corn mill. In 1779, this stipend bear weighed 19¾ stones; but on account of the number of black oats it contained, was at first rejected by brewery companies, till its weight was seen. But what is yet more remarkable, about the year 1770, a wager of ten guineas was laid by the then EARL OF ERROL, against Mr. GORDON of Wardhouse, that his bear in Kildrummy did not weigh 22 stones per Aberdeen boll, which is above 57 pounds per Winchester bushel—after allowing Mr. GORDON's tenant to pick out all the wild oats or weeds, as the bear, though very weighty, was very foul or full of weeds. It was properly cleaned, and a boll of it weighed 23 stones Dutch, or 60 pounds per bushel. No bear, nor even barley, of such quality is now raised in Kildrummy, although very fine corn, both bear and oats, are raised in that district. But the quantity produced on an acre 40 years ago, seldom exceeded 3 bolls or 20 bushels. There was room for the action of the sun's rays, and for a free current of air to blow around the heads of the grain, which was never lodged, and the hot dung applied directly to the bear crop, hastened its vegetation,

tion, and the harvest was earlier by ten days or a fortnight. The ears of the corn in that coarsely ploughed land, which is now completely pulverized, if one might borrow an image from a Roman poet, were at a considerable distance from each other

   Apparent raræ stantes in cespite vasto.

From all these causes, the weight or quality of grain, when thoroughly cleaned, was really better than it is now, though little more than half the quantity was raised on an acre.— The Writer of this Report, who has paid particular attention to weights and measures, has weighed his stipend bear regularly for above 30 years. From 1778 to 1788, it weighed above 19 stones per boll; from 1788 to 1798, it was only about 18 stones; and from 1798 to 1808 inclusive, it weighed only 17 stones. No doubt it is not so well dressed within these last ten years, since the *fanners* were generally used in place of the weght and riddle. But a man must accept of marketable grain; although the taxes are now so high, that every means should be used to dress bear so fine, as that the brewer or distiller may pay them more easily.

BARLEY.

This two-rowed species of barley is very little cultivated in Aberdeenshire. Even in the vicinity of Aberdeen, where agriculture is carried to as high a degree of perfection as any where in Great Britain, as Dr. ANDERSON justly observed in 1793 (before there was any dispute between the southern and northern counties about these two kinds of grain) " the " bear they cultivate is the four-rowed kind called bigg, " and seldom barley, which they find never yields such a " profitable crop; so that though it has been frequently " tried, it has been always abandoned." It is also deserving of notice, that those farmers from the southern counties,

who

who took farms in Aberdeenshire, and who were partial to barley, where both the corn and the straw are much more valuable than bigg, have almost always given over raising any quantity of the two-rowed barley. Mr. Watson of Wester Fintray, has only 2 acres in barley to 23 of bear; and several south country farmers in this county raise no barley at all. Indeed there is generally a fourth part more increase from bigg than from barley, and not seldom a third part more of the former than of the latter, which is not nearly so proper a nurse for the sown grasses. Yet whenever barley can be raised to advantage, it should, and indeed it will be cultivated, both as the grain is much fitter for pot-barley, and as the straw is incomparably better for horses.

At present there are not above 400 acres sown with barley in the county, and the greatest part of what is sown is in Buchan. The barley is generally 6 pounds per bushel weightier than the bigg, i. e. when bigg is 42, barley is 48; and when bigg is 44, barley is 50 pounds, in a dry and favourable season.

The produce of barley is very unequal, being in late seasons below four bolls, and on early soils nearly six bolls an acre. And when the equinoctial storms come on before it is harvested, it is so liable to be injured, that less is sown of it than was formerly. As our agriculture improves, and as soon as the ratio of taxation on malt made from bigg is fairly proportioned to that from barley, it is probable that more bear or bigg will be raised, and even wheat will be cultivated on dry and strong lands; but barley will never be sown in great quantities in this county.

What is raised is chiefly manufactured into pot barley, for which it is incomparably superior to bear or bigg. Its corns are larger, and more equal in point of size; and the barley, when manufactured, is both of a much larger body, a fairer colour, and swells more in the pot. When bear or bigg is made into pot barley, it is in all these respects defective, and its produce is generally a third part less than that of barley, especially of south country barley, which we import for this purpose. In every view it is desirable that the farmers in Aberdeenshire should cultivate as much of this two-rowed grain as will serve us for pot barley.

Both barley and bear are manufactured into bread, though not in great quantity, for the use of private families. Formerly one-third part of the meal used by the farmer's servants was bear-meal, and was ground on the corn mills, which are used for the grinding of oats. Now that our servants will not eat this kind of meal, except when the oat crop is very deficient, it is generally ground on a flour mill, after being previously made into pot barley, and boulted as wheaten flour. In this way it makes very excellent bread; and several gentlemen both in Aberdeen and in the inland parts of the county use it in their families. Where it is wanted for porridge to children, it should be ground on a corn mill, or the second flour should be kept for this purpose. For what is ground too small, and has no portion of the hull, does not make good porridge. In consequence of the very high prices of flour in 1795 and 1796, and the still higher prices of it in 1800 and 1801, the Writer of this Report made the greater part of the bear paid him as his stipend, and also of what he raised on his glebe or parsonage lands, into meal or flour of different qualities; the particulars of which will be found in the Appendix. He shall only state in this place, that when the bear was of excellent quality, and having been six months in his garret, 20 stones of it yielded $18\frac{1}{2}$ stones of meal; when it weighed 18 stone per Aberdeen boll, 3 bolls of it weighed only 50 stones, after being kiln-dried, and yielded only 42 stones and 2 pounds

pounds of meal, when ground on a flour mill, and boulted through a half-guinea cloth, or bran-cloth. But when carried to a common meal mill, it produced from 10 to 14 per cent. less than when ground at a flour mill. Therefore it is evident that oats only should be ground in that manner; and that barley and bear, as well as wheat, ought to be ground on a flour mill, and then boulted or dressed to the degree of fineness that is wanted. It may be proper to add, that the bran of bear or barley, ground in this manner, makes excellent food for horses, when mixed with a few boiled potatoes; and that work horses thrive much better on this food than on raw potatoes, or on oats alone.

The price of barley was generally one-fourth part more than bigg, and sometimes a third, or 33 per cent. dearer. But by the fiars of the year for the last eight years, the price of bear or bigg, reduced to Linlithgow or standard measure, is only eighteen shillings and fivepence farthing, while that of oatmeal is eighteen shillings and tenpence eight-twelfths, or fivepence five-twelfths more. For the eight years preceding 1802, when the ratio of malt-tax on bigg was altered, the price of a Linlithgow boll of bear was nineteen shillings and twopence, and that of a boll of meal was only sixteen shillings and ninepence halfpenny. Thus while the money price of labour, and the price of every other kind of grain has increased, that of bear or bigg has decreased; and in particular, when we compare it to oatmeal, it is now $2\frac{1}{2}$ per cent. cheaper, while the boll of bear was formerly $11\frac{1}{2}$ per cent. dearer, than the boll of oatmeal. It is not therefore to be wondered that less of it is sown than formerly, or that grass seeds, though bear is the best crop for nursing them, should now be sown with early oats.

WHEAT.

WHEAT.

Although wheat is the most valuable kind of corn, yet in Aberdeenshire it is so little cultivated, that it is here placed after oats, and the two different species of barley.

Neither the soil nor the climate of this county in general is adapted to the raising of wheat. It is often thrown out of the ground by the storms, or hard frosts, succeeded by sudden thaws in the winter or early in spring; and in late seasons it is often exposed to the equinoctial rains in September. At present it is supposed that there are not above 400 Scotch, or 508 acres English of wheat sown in this extensive county. Wheat after a summer fallow does not succeed so well on our light lands as after potatoes, when they are taken off the ground in September. Besides a crop of turnips, and another of either bear or oats, pay better than a crop of wheat only; and the farmers do not always find a ready sale for wheat, especially when it is of inferior quality. On all these accounts they prefer the certain profits of a crop of turnips, followed by oats or bigg, to the chance of a favourable season, and a high price for wheat. The relative price of wheat is now however become so very high, that the sowing of that kind of grain is this year upon the increase; and it appears far more probable that wheat will be more generally cultivated than the two-rowed barley. Independently of the high price of wheat, two other circumstances will occasion it to be more frequently sown. The first is the increase of luxury among the farmers. Though they live with far greater economy than those in the southern counties, yet wheaten flour is occasionally purchased by them.—Some of them sow half a bushel or a bushel for family use; and the chance is, that, as turnips were first sown for this purpose in the garden, and afterwards raised in the field in

quan-

quantities, the farmers will be tempted to sow a greater proportion of wheat, where they have their ground in good order, and where the autumn is mild. The other circumstance that will occasion wheat to be more generally sown, is, that it has been found that grass seeds, sown on wheat land, and harrowed in with the wheat plants in the spring season, not only do no injury to the wheat, but that even a slight hoeing to cover the grass seeds is favourable to the tillering of the wheat plants; and that these make as good nurses to the young clover, as early oats do, though not equal to bear or bigg in this respect.

But while these circumstances indicate an increase of the quantity of winter wheat, it is extremely probable that spring wheat will spread more rapidly. A species of this wheat, of the bearded kind, and very hardy, has been sent in small quantities by the President of the Board of Agriculture; and on this respectable recommendation several patches of it have been sown by some of the landed proprietors of this county. JAMES FERGUSON, Esq. got a quantity of it from Sir JOSEPH BANKS. Mr. FORBES of Echt had a few bushels of it last season, (1809) which looked exceedingly well, and is supposed to have been very productive. In the Appendix, the particulars of this crop, and of any other crop of wheat, which deserves to be mentioned in this Report, will be added in a separate article.

In the meantime it is proper to state, that spring wheat should be fairly tried in various places, where it is expected to answer well with the soil of this county; and that the seed of all wheat should be changed every third year by importation from England. This was the practice of the late Robert BARCLAY, Esq. of Ury, who informed the Writer of this Report, that he bought generally as much seed from England yearly, as sowed about a tenth-part of his wheat lands;

thus

thus the crop of this wheat was usually ripe ten days earlier than what had been a year in Scotland; and that he sowed all his wheat lands, except the above tenth part, with seed from the produce of what he had thus imported from England. By this means he raised wheat successfully in the northern district of Kincardineshire, and within 14 miles of the city of Aberdeen. It therefore can admit of no doubt, that by adopting the same mode, a very considerable quantity of land in this county could be occasionally sown with wheat, when perfectly clean and in good order.

To prevent any of our farmers from engaging too rashly in the culture of wheat, it is necessary to warn them, that even in the Lothians, wheat is not raised to advantage when more than 400 feet above the level of the sea; and that therefore in this county, which is nearly two degrees farther north, and projects a considerable way into the German Ocean, it will not be advisable to sow wheat on land that is more than 300 feet above that level. In a few exposures, such as Kildrummy, where a soil and subsoil, naturally good, are almost surrounded by an amphitheatre of hills, and owing to that circumstance, have several degrees of local heat more than is usually found in the same parallel of latitude, wheat may be raised to advantage at 400 feet above that level. On the other hand, in many exposed situations, at half of that elevation, it will not succeed. The reason for inserting these minute articles in this Report is, that the farmers have, in some cases, been deterred from making the attempt, because a few persons had not succeeded in cultivating this grain, where, from various causes, there was no probability of success. Yet there is every reason to believe, that by sowing wheat early, i. e. in the end of July, or beginning of August, in ground that has been properly manured and cleaned by a summer fallow, a considerable propor-

tion

tion of our land might raise wheat in most seasons, to great advantage.

The sorts of wheat, which seem most proper for this county, are the red winter, and the bearded summer wheats. The white kinds have scarcely ever answered in Aberdeenshire.

The few farmers who sow wheat in this county are very careful in steeping or pickling it. Some prefer steeping in a pickle made as strong as to swim an egg, and then drying it with quick lime. Some mix soot and peat-ashes along with the quick lime, which makes the seed of a grey colour, but they think more hardy; and others sprinkle it with urine, and, after turning it frequently, dry it in either of the above ways. Complaints of the wheat being smutted, are by no means general; the principal danger in this cold climate is that the seed does not vegetate, or is thrown out in spring.— The best preventatives here are surface-draining, and early sowing.

As seed is dear, it is not uncommon to drill wheat. And when peat-ashes are thrown in the drills, they assist much in promoting the growth of the plant. But when it is sown in broad-cast and pretty thick, it is earlier ripe in harvest.— This circumstance in a northerly climate, is in favour of thick sowing and broad cast, although the wheat does not tiller so much when sown in this manner.

It is seldom sown till the beginning of September; and when put after potatoes, not till October, or as soon as the land is dry, and cleared of the potatoes. When the season admits of it, wheat is sown after drilled beans; this makes the seed-time in the end of October.

Seed, when sown in drills, half a bushel; when in broad cast, from 2¼ to 3 bushels per acre.

When sown in drills, and thrice hoed, and when the earth is

is gathered to the plants at the last hoeing, if the season be early, a very weighty crop is raised.

The little wheat raised in this county is seldom ripe before the beginning of September, and, as repeatedly mentioned, is often exposed to the equinoctial storms.

The produce is very variable, and the price fluctuating, owing to the very unequal quality of the grain. But when a good season, a good crop, and high prices, all meet, a crop of wheat even in this county, pays well. A farmer in the parish of Bellelvie drew L.200 for the produce of nine acres in 1800. But six bolls, or L.12 per acre, is the general average. Sometimes more, sometimes much less is raised.

Reaping, even of wheat, is now frequently performed with a scythe, which is adjusted for the purpose. Of twelve acres in the parish in which the Reporter resides, eleven were last year (1809) reaped by the scythe. The harvest at first was very precarious, which occasioned the scythe to be applied to wheat, as well as to every other kind of corn, where the bottom was soft, and either level or uniformly sloping, and where the grain lay nearly in one direction. But in general, wheat is cut down by the sickle; as indeed most kinds of corn are in this county. The scythe, however, was introduced a few years ago, and both from the high wages, and the difficulty of procuring reapers in a precarious season, was employed last harvest (1809) on many farms of great extent: And from the uncommon care taken by the persons who were employed in gathering, and binding the corn, it is very probable that oats will be generally, and wheat only partially, reaped in that manner. Bear or bigg is still cut with the sickle; and from the shortness of its stalk, and brittleness of its straw, it is not probable that much of it will be reaped by the scythe.

Threshing

Threshing.—Wheat is generally threshed by the flail, as is also the greater part of all kinds of corn that are raised in this county. In addition to what has been said on this subject, it may be observed that threshing mills have been erected and are building in various places; but as yet not above one-sixth in most parishes, and in many not one-tenth of the produce is threshed in this manner. It also deserves to be attended to, that barn-men or lot-men, as they are called in the southern counties, who thresh corn for a certain ratio of the produce, (from one-twentieth to one-thirtieth part, according to the quality of the grain) are unknown in Aberdeenshire; that threshing is not paid for by the boll; but either by day-labourers, hired on purpose, or most commonly by the farmer's servants, who thresh in the mornings, or after dinner, in the winter months, what is sufficient for a day's provender. Some of the landed proprietors have lately employed day-labourers to thresh out their corns, either at a stipulated hire per day, or at a certain sum per boll; but the practice is by no means general. And as the servants have always been accustomed to thresh, they are very ready to assist, even at spare hours, in working the threshing machines, which are every year becoming more common.— Those masters of corn mills, who have erected threshing mills, frequently allow the small farmers to carry their corns to these; and the grain, after being threshed, is dried on the kiln, and made into meal. If threshing mills were contrived of a simpler form, and on less expence, they would be within the reach of small farmers, which at present they are not. The late rise in the price of wood has prevented many from building them, and the introduction of cast-iron machinery has added a little to that expence; but still more to the value and durability of the threshing mills which have been lately erected.

R 2   Bearded

Bearded wheat, like bear or bigg, is either put twice through the mill, or afterwards beat by the flail, to take off the *awns*, as they are called, i. e. the beard or long spire that adheres to the grain. Some of the threshing mills, which cost but L.30 or L.40, dress bear much better than others which cost thrice the sum. And from the ingenuity of the mill-wrights, we have reason to hope both for better work and a simpler construction. But a considerable moving power will always be necessary, and we cannot expect that threshing mills, which are driven by the hand, can be so contrived as to be of general utility, even to the smaller farmers.

BEANS AND PEASE.

Very little of either of these kinds of pulse is sown in Aberdeenshire. Of both probably not above 2000 bolls, of which 500 of beans, and 1500 of pease. The greatest part of both, but especially of the former, is raised near the sea-coast. The latter crop has been so very precarious, that it is less used than formerly. In the division of Buchan, drilled beans are, on the contrary, on the increase. In Marr, the Garioch, and Strathboggie, they are little used except in gardens.

They are sown chiefly as a meliorating crop. But drilled beans often pay very well, except when the season is unusually late. And on the other hand, when the harvest is early, they make a very good preparation for wheat, though they should always be dunged, when this is intended, and topped when it is wanted to check their growth.

Of pease two sorts are used, the white and the grey; and of the latter an early kind, called *Hasterns* by the country people, is imported generally from Angus, sometimes from Peebles.

When

When pease are sown for a *crop*, a boll only is used for seed to an acre; when the sole purpose is to *meliorate the soil*, six and even seven firlots of seed are allowed.

Beans are either hand-hoed at 14 or 16 inches distance, or horse-hoed at the distance of two feet and an half; and the interstices are hand-hoed, by picking out or cutting off the weeds. Horse-hoeing is now most general, and certainly is preferable to narrow drills.

Time of sowing.—As soon in January or February as the season will admit. Pease are ploughed down, where it is meant to plough them in, or they are sown on a rough surface in the beginning of March. The ground is harrowed afterwards, and generally gets a second harrowing a fortnight after they appear above ground. A small quantity of beans is usually sown among the pease to support them when growing.

Harvest.—Pease are generally partly cut, partly torn, and gathered in parcels, and repeatedly turned over till ready for the stack-yard, and are very easily damaged by rain.— Beans are first cut, sometime afterwards bound up, and are a very hardy grain, if the farmer has only patience to wait till they are fit for being taken in. The following ingenious contrivance, by the Rev. Dr. GEORGE MOIR of Peterhead, for preserving a crop of pease in a very inclement season, extracted from the Statistical Account of the parish of Peterhead, deserves to be here recorded, as also *his patience* in allowing his beans to stand till he could take them in with propriety.

" The greatest part of the pease of this crop, 1784, was
" entirely lost, and never taken off the ground for want of
" good weather to dry the straw. I had myself that year a
" field of between six and seven acres, of which four acres
" were beans, and two acres and a half were pease. The
" pease

"pease, after they were cut down, were frequently covered
"with snow; and in turning them, a great part of the grain
"was lost. I waited till near the end of November; and
"seeing no prospect of preserving them in the ordinary
"manner, I took three Norway trees, of 10 or 12 feet in
"length, tied them at the top with a rope, and extended
"them at the base. The pease were built round the trees at
"the outside; four small arches were left at the bottom: all
"was hollow within the trees, and open at the top before
"the rick was thatched. The air rushed in and went to the
"top; and in a few days the straw which was formerly wet,
"was as dry on the inside as on the outside, the whole crop
"of pease was contained in three ricks, the grain that re-
"mained and straw was entirely preserved; and if the expe-
"riment had been tried sooner, the whole of the grain would
"have been saved.

"Many, by hurrying in their beans to the corn-yard,
"lost both them and the fodder. I delayed touching mine
"till the last day of November; and in the night time, with
"moonlight, and a brisk breeze of wind from the west, got
"them all out of danger. Beans are a hardy grain, and will
"remain long on the field, without receiving injury from
"the weather. I know no part of husbandry in which far-
"mers are more apt to err, than in the management of their
"beans, by taking them too quickly off the field. If not
"sufficiently dry, and even blackened, the fodder and grain
"are both spoiled."

Produce.—The crop of pease is extremely precarious and unequal in its produce. Sometimes scarcely the seed; sometimes four or five bolls an acre, being raised. That of beans, though variable, is, at an average, seven or eight bolls, while pease, at an average, is below three bolls to an acre. But the straw of both is extremely valuable, being little inferior to hay.

Pro

Produce how applied.—Formerly the meal of pease was commonly used by the farmer's servants for *bread* or for *brose*. Even beans were made into meal. And a gentleman, who had an extensive estate in the inland part of Buchan, and another on the sea-coast, was asked by a friend, who paid him a visit, "How it happened that his people in the in-"land parts were not nearly so stout men as those on the sea-"coast."—He replied, "that those on the sea-coast were fed "with bean meal, the others only with oatmeal."—Both pease and beans, however, are now used only as food for horses.

POTATOES.

Though this valuable root was brought from America in 1565, it was not used in this county, even in the garden, till about 1750, and not in any considerable quantity till after 1782, when one calamitous season taught us many lessons in agriculture. Still they are not cultivated nearly in the same proportion as on the west coast of Scotland. Turnips are the favourite green crop of our farmers.

In the whole county of Aberdeen, there may be about 4000 acres planted with potatoes. Of this 1000 may be in the vicinity of Aberdeen, and of the other towns on the sea-coast. These last are generally dibbled into the ground at the distance of 13 or 14 inches. But the greater part of those used in the country are put in after the plough, with a little horses dung, then covered, harrowed, and generally thrice horse-hoed, between the furrows, and hand-hoed in the interstices between the plants.

The gardeners about Aberdeen raise very great crops of potatoes, as they do of all other roots. The particulars will be stated in a separate article. In the neighbourhood of Peterhead, they were raised for many years in a peculiar

way,

way, which deserves to be generally known. Dr. Moir, whose Statistical Account was so lately quoted, frequently raised 400 bolls, or 100 tons annually, and set an excellent example of cultivating that valuable root. The plan he adopted was to trench the ley ground, that had borne three crops of sown grass, to dibble his seed without any dung into the trenched land, at 12 inches distance, to hoe them carefully with a hand-hoe, and dig them with a spade.—" The ex-" pence of the management of this acre," he says, " was " not less than five guineas ;" but adds, that " it is not un-" common to have 50 bolls, of 32 stones Dutch, or 35 En-" glish, (i. e. 12½ averdupois tuns) from the acre. An ordinary " crop is estimated at 40 bolls." Bear and grass seeds succeeded this crop; and the ground, after lying other three years was again trenched as before, and planted again with potatoes. In this rotation the land might have continued for any number of years without dung or any manure whatever. And as a single ploughing only was necessary after the potatoes were taken up with the spade, a crop of bear, worth at least six guineas an acre, followed by three crops of grass, (one mowed for clover hay, and other two of pasture, worth in all at least fifteen guineas) would have preceded the trenching out of ley, and a crop of potatoes worth fifteen guineas, in 1794, when Dr. Moir wrote the very excellent account of his parish; and the whole five crops would now be worth at least L.60. The only crop of which Dr. Moir complained, was the third grass crop; and by the advice of a neighbouring farmer, he afterwards took a crop of oats before trenching. Perhaps he would have done better, if he had trenched his ground after the second crop. For red clover is a biennial plant, and in land that is rendered completely clean from weeds, can only endure two years. In strong, but comparatively foul land, where it is seen in considerable

quan-

quantity the third year, the plants that then appear do not spring up the first year, but are checked by weeds, or various causes ; and as they spring up in the second year, they, of course, last until the end of the third. The only objections to this course of cropping by the spade only, or with very little assistance from the plough, are, that the ground tires of a frequent repetition of the same crop; and that this mode of cultivation was too expensive, and required too many labourers. But in answer to the first objection, it is to be observed, that as the ground was trenched from 12 to 14 inches deep, that portion of the soil which carried the weighty crop of potatoes, was laid down by the first trenching after the third grass crop; and that it was therefore only once in ten years that the same portion of soil carried a crop of potatoes. As to that part of the second objection, which regards the employing so many labourers, it is a recommendation to the rotation ; for it fed the labourers and their families in the winter months, when they had little other employment. No doubt the expence has at last become so great, from the alteration of the money-price of labour, that land which was trenched for 3d. the rood, or L.2 the acre, now costs 1s. the rood, or L.8 per acre ; and therefore with the rise of rent, and of the additional charges on hoeing and digging the potatoes, the plan cannot now be generally adopted. But though trenching once in five years for a potatoe crop is now too expensive, it might be attended with the best effects to use the trench plough once in ten years, and the spade and shovel to bring up new soil, alternately with the trench plough at the end of other ten years. It has already been shewn, that the gardeners near Aberdeen, by the mixture of the plough and spade husbandry, have taught important lessons to the farmers in the neighbourhood of great cities, and populous towns; and it may now be added, that

the

the Minister of Peterhead, without *much aid from the plough*, and with *no manure*, has given a most useful lesson to those who *plough and dung too much*, but neither *trench nor clean their land* properly.

Various kinds of potatoes are used by the gardeners near Aberdeen. 1. A small round kind of early potatoes is generally ready about the middle of June; though sometimes owing to the coldness of our climate, near the beginning of July. 2. The kidney, which still maintains its character as the driest or mealiest, though it is not so productive as some other kinds. 3. The large white, which grows to a great size, but is not so dry as the former. 4. The pink-eye, purple, or calimancoe potatoe, which is very productive, and is generally eaten in the spring and beginning of summer. 5. The Dutch cluster, a watery potatoe, but very prolific. 6. The common red, which is also prolific, but whose flavour is not much relished. 7. The Irish red, a very mealy potatoe, lately imported from Ireland, which is much esteemed.

In the country, the kidney for eating in autumn and winter, and the pink-eye for spring food, are the kinds chiefly used. Near Aberdeen they are generally dibbled, as above mentioned, and in the country they are horse-hoed; but the question whether dibbling or horse-hoeing be preferable, and also the particular distance at which they should be planted, has been much disputed.

As this valuable root is now so much cultivated, and is so much used as a very considerable part of the food of the lower classes, as well as a favourite vegetable at the tables of the opulent, the Writer of this Report shall state as concisely as he can, a number of experiments, which, since the year 1772, he has made on potatoes.

In the year last mentioned, he planted a boll of them in

a quarter of an acre of ground which had borne a crop of bear the preceding year. They were dibbled into land that had been previously dug with the spade, and on a thin light soil, with a bottom of pure sand. They were carefully hoed, and when ripe were taken up with the spade. The produce was 12 bolls and a firlot, or 49 bolls (12¼ tons) to the acre. The kinds were the kidney, the common white, and one peck of red. Some of them were planted at the distance of only a foot, others at 14, 16, and 18 inches.— That which was planted at the distance of a foot gave the greatest produce on an acre, as the soil was but thin, and the stem neither choaked by being too close nor too luxuriant, from the distance of the plants. He continued for other four years to plant from half a boll to a boll, according as he could get proper land to rent from the neighbouring farmers, but as he did so, only to amuse himself and to serve a friend, he gave over planting any kind, except the kidney, as this kind was mealiest, or of the most farinacious quality.— These experiments were made in the county of Kincardine, on the banks of the river Dee, within four miles of the city of Aberdeen.

In 1778 he removed to his present settlement; but the only ground in which he could plant potatoes that season, was about two-fifths of an acre, very thin out-field, that had been inclosed as a garden. On the 19th of May, as soon as he got the ground ploughed after he was settled, he planted a boll of potatoes, three-fourths of which were of the kidney kind, and a firlot of Dutch cluster. The produce was only eight bolls, for the ground had got no manure, was very thin, and had borne a crop of oats after turnips. His object was merely to get a quantity of potatoes, not only for family use, but for making into spirits. It was necessary for him to lay in a stock of these; and the state of his finances required

quired that this should be done at as moderate an expence as possible.

He mixed, in various proportions, 7 firlots of bear that had been made into malt, with 4 bolls of potatoes. The malt would weigh 14¼ lbs. Dutch to the peck, consequently the firlot weighed 58 lbs. Dutch, and the 7 firlots 406 lbs. very nearly one-fifth of an English averdupois tun. The potatoes were all weighed at 32 lbs. Dutch, or 35 lbs. English, which is the standard of the county, when sold by weight; and the 4 bolls were therefore exactly a tun weight. The produce in spirits, reduced to 10 per cent. above proof, was four ankers very nearly, and probably a little more than four ankers, reckoning the drams given to the operators. The anker is 20 *Aberdeen* pints, of 108.875 inches to the pint, or nearly 38 gallons of spirits, 1 to 10 above proof. They were distilled very slowly in three small stills, about ten gallons each, which private families were permitted to use without paying duty till 1779.

The quantity of spirits however yielded by the *cluster potatoe* was *not above*, but a *little below* two-thirds of what he extracted from an equal weight of *the kidney*. Of this last he fermented six pecks, without any malt, except the weak worts of a few pecks, which were brewed into ale for making malt spirits; and he had above six gallons of potatoe whisky, which was not so pleasant as malt spirits, but superior in point of flavour to the corn spirits which were then distilled in the south country. (The spirits of the corn distillers have since been greatly improved.)

The law having prohibited the distillation of spirits in private families, he could only plant potatoes in different modes and at different distances, marking their produce carefully, without attempting to distil them.

Being

Being informed by Dr. Moir of the method which he practised of trenching out of ley, he trenched in 1796, a quarter of an acre out of rich clover pasture, which had lain three years in grass; and he formed it into plats of 30 feet by 23 each, so that he should have two falls, or one-eightieth of a Scotch acre in each plat. He planted each of these at different distances, from one foot square to two feet two inches, by two and one-half feet. They were dibbled in, and every hole was marked by his own hand, for the following reason :—The farmers who drill and horse-hoe them have generally nine or ten inches between the potatoes in the drills, and the distance between the drills being two and one-half feet, every potatoe set contains two square feet, and is in the corner of a parallelogram of thirty inches by nine and six-tenths. The gardeners generally plant them by guess, but as near to squares as they can, moving their feet along the ridge, and dibbling a set before each toe. But the most correct and mathematical way of planting any thing, is what the gardeners call *quincunx*, i. e. putting all the plant holes in the corners of equilateral triangles, or one plant in the middle between the line of the other two. When potatoes are planted fifteen inches distant in the row, and at thirteen inches between them, they form almost exactly an equilateral triangle, if one plant is placed opposite to the middle of the other two. For although it be only thirteen inches between the rows, yet owing to the *bay* as it is called, or oblique direction, the plants are all fifteen inches from each other. The reason of this is, that a perpendicular let fall upon the base of an equilateral triangle is almost exactly thirteen-fifteenths of the length of that base. And in planting cabbages or coleworts, that proportion should always be observed, as the plants both occupy less space, and cabbage better. To return to the result of these experiments.

The

The produce of the whole quarter of an acre was 11 bolls 9 pecks. But as three of the plats were dug up for food to the family before they were ripe, at least 5 pecks were lost by premature digging, the first row taken up being only 12 lbs.; and three weeks after, another row produced 19 lbs. of the same kind of potatoe, planted at an equal distance. The average therefore was at the rate of 48 bolls, or 12 tuns English to an acre. But the greatest crop of the white cluster on the 2 falls, was 1 boll 5 pecks and 21 lbs. or 3-5ths of a peck, Aberdeen measure. This is at the rate of 108 bolls to an acre. The greatest crop of the kidney kind was 13 pecks and a quarter of a peck, which is 68 bolls 3 firlots to an acre; and the distance at which the sets were planted was thirteen inches between, and fifteen inches in the rows.— The least weighty crop on the acre, though the separate stems appeared to be rich, was where there was two feet two inches between the rows; where the produce of the kidney was a little more than 7 pecks, or 35 bolls to the acre; and that of the cluster 10 pecks, or exactly 50 bolls to an acre. The plants that were set in squares did not turn out so well as those which were in equilateral triangles, where every plant is in the center of a hexagon. They could not be so evenly drawn up into the shape of a sugar loaf, which every potatoe plant should be. The plants that were set like parallelograms, or nine and one-half inches in the row, by twenty-six between the rows, and dressed up like horse-hoeing, turned out worst, because a number of their potatoes were small sized, not having had time to ripen after the last hoeing had covered up the stem as far as it could reach.

While the Reporter feels it his duty to state these experiments, he would by no means recommend quincunx planting to farmers, but horse-hoeing only. But he would recommend the other method to gardeners, day-labourers, and others,

others, who have but small patches of ground;—and to farmers he would suggest, that they should either plant and horse-hoe their potatoes like a crop of turnips, or plough their land in the following manner, after being well pulverized. 1st. Let there be two furrows of ribbed land, only fourteen or fifteen inches from each other, and let the potatoe first, and then the dung (which should always be laid above the potatoe) be put into these narrow ribs, or furrows. Afterwards let a third furrow be made from two feet to two feet six inches, as the farmer chooses, distant from the former; that there may be room for horse-hoeing.— By this means, the whole breadth taken up by the two rows is about three feet and an half. (It will simplify the work, if the dung be spread on the surface before the ploughman begins his ribbing, and be raked or laid on the potatoes in the two narrow furrows, while the third is ploughed without being planted.) Then let the field be harrowed when the potatoes are beginning to get above ground; let the earth be taken away where the horses can go, when the plants are a few inches up; and let the interstices be handhoed, as well as the small row between the plants; let the earth then be laid back gently, so as to cover the plants a little, but let the top of the plant always be left uncovered. In three weeks after, let the plough raise a little more earth to the neck of the plants, and the interstices again handhoed; and lastly, (as late in the season as the plough can be introduced) let a third covering of earth be raised out of the bottom of the furrows, and laid as high on the stem as possible without injuring it; and let the interstices be hoed, and the stems drawn up as much as possible into the shape of a sugar-loaf. The consequence will be, that the potatoes will be drier, and will apple better; and what is often of great consequence, the crows will not be able to do them much injury, which they generally do where only one hoe-

ing,

ing, or even two hoeings, laying the earth to the plant, are given. This is also approaching as nearly as possible to the Cheshire method of raising potatoes, which appears to be the best of any that is noticed in any of the printed reports from the different counties which this Writer has seen.— In the present advanced state of the Corrected Reports, where much information has been communicated, he feels it his duty to insert in the proper place any correct practice of another district, that it may be imitated in his own county; and he also thinks the following remarkable fact worth recording.

It appears from the Statistical Account of Kilsyth, that on the 21st of April, 1762, ROBERT GRAHAM, Esq. of Tamraer, planted a peck of potatoes, and that the produce of that peck on the 26th of October following, measured on the ground in the presence of Messrs. JOHN MARSHAL of Townshend, HENRY MARSHAL of Ruckhill, and ALEXR. MAXWELL, baillie of Kilsyth, was 264 pecks, or 16¼ bolls. Each set was distant from eighteen to twenty-one inches lengthways, and between twelve and fifteen inches across the ridge.— The measure of the land ought to have been stated, but is not mentioned. This was a most material omission; and to supply if possible this defect, the Writer of this Report wrote to the Rev. Dr. RENNY, minister of Kilsyth, who informed him that the soil was gathered up very high to the top of the plants, and that the ground, if planted in the common way, would have served for three pecks of seed potatoes.

It may be of service to the cultivators of potatoes to know the method of covering them, which is adopted by Mr. ECCLESTON, contained in Mr Holt's valuable Survey of Lancashire.

" After the potatoes are gathered, and sufficiently dried,
" they are put together in heaps in the shape of the roof of a
                                                     " building,

" building, covered closely with straw, which should be
" drawn straight, and to meet from each side, in a point at
" the top, about six inches in thickness; and then covered
" with mould closely compacted together, by frequent ap-
" plications of the spade; after which Mr. ECCLESTONE
" makes holes in the mould, at the sides and tops of these
" repositories, as deep as the straw, and about three yards
" distant, to permit the air, which, he says, visibly arises
" from the fermentation, to escape; after the fermentation
" has ceased, the holes are closed, to prevent the effects of
" frost or rain."

It is owing to the neglect of this last precaution, that po-tatoes often acquire a bitter taste after they are covered up, and before April are often very unpalatable. By keeping holes in the sides of his potatoe pits, for several days after they were covered, and plugging these holes at night with round pieces of wood, with which they were made (old spade shafts) the Writer of this Report preserved 30 bolls last season, (1808) and the last of them was as sweet and fresh as when dug up in autumn.

To conclude this long account of a most valuable root.—When the season is early, and the apples are beginning to form (or while the flower fades) on the top of the stem, as ma-ny children as can be got, should be employed in topping, that is picking off the flower and about half an inch of the top of the stem. The soil is much injured by allowing the apples to form on the tops of potatoes, which do not injure the land more than any other green crop, when the flower is thus topped. This does not stop the growth, and rather adds to the dryness of the potatoe; but the cutting of the stem for food to cattle, before the flower has completely fad-ed, always checks the growth of the crop.

YAMS.

This root is commonly called Horse Potatoes by the country people, from being usually given to horses. It is also given to black cattle; but great care must be taken to wash them clean, and not to give too many of them at first, till the animal is accustomed to them; otherwise the effects may be often very dangerous.

Yams are cultivated in the same manner as horse-hoed potatoes; only they are generally allowed three feet between the drills, and a foot between the plants, as they grow more luxuriant, if properly dunged. Fifteen tons, or sixty bolls of them, have been raised on an acre; and twelve tons is a moderate crop, when they are properly treated. Even when they get little dung, there is a tolerable crop; yet it is not in this county judged prudent to plant them without some manure; but roots of cabbage, and rotten straw are used when dung cannot be spared. They are longer in ripening than the potatoe, and not so easily hurt by the frost. But except in very early seasons, they are too late in ripening in Aberdeenshire to admit of a crop of winter wheat to succeed them.

They are given either after oats, or instead of boiled bear, which was the common food of horses that were fed very high; and half the former quantity of oats, with a proportion of yams, is found to answer as well as oats alone.

Of late, several experiments have been made as to the quantity of starch that may be produced both from potatoes and from yams. In the white mealy potatoe, from an hundred pounds of the root, sixty-eight pounds of water, and thirty-two pounds of meal, were found, of which from fifteen to seventeen were starch. This makes about one-sixth part

of

of starch nearly from the best kind of potatoe. But one-seventh of starch is the highest that can be stated as the average proportion in that root. Some experiments seemed to indicate that there was one-sixth part of starch in yams.— That, however, is now known to be over-rated. And if one-ninth part of the yams be taken as the medium for the weight of starch, it will not be far from the truth. Even at this rate, it would be an object to make our starch rather from the green than from the white crop. The laundress, however, does complain, that the starch obtained from the potatoe, and still more that from the yams, does not stand or keep so well as that from flour. When applied to clothes that are to be worn in a few days, the complaint is ill founded. If they are intended to ly for some time, the objection may deserve some weight.

It may not be improper to mention, that *potatoe starch* has lately been given as a medicine by some physicians of character, to persons of a *delicate constitution*, when threatening a *consumption*. It has the recommendation of being, at least, *harmless*, and in no small degree nutritive without loading the stomach. If our advertising quacks were satisfied with vending *placebos*, or *harmless medicines*, potatoe starch might, under *Divine Providence*, be more beneficial than either the *Balm of Gilead*, or the *Balm of Quito*.

TURNIPS.

In a county, in which the raising of cattle is so much attended to, the cultivation of turnips merits the first attention of the farmer. Accordingly, although an improved system of husbandry was only introduced by the calamities of 1782, and although there were not 200 acres in the county, except what was raised by the landed proprietors, they are now cultivated with the greatest success. There are not less than 20,000 acres of turnips raised annually; that number is increasing rapidly, by the application of almost all the dung, and by importing great quantities of lime, and other manure, to nurse the turnips, which are considered as an essential crop in the farmer's rotation.

Turnips, on the west coast of Scotland, are much less cultivated than potatoes are. On the east coast it is the fault of the farmers if they are not cultivated more generally than the other root, however valuable. In this county, the farmers raise, at least, ten acres of turnips for one of potatoes. But from the great quantities of the latter raised by the gardeners near Aberdeen, and by the inhabitants of all the towns and villages, and the cottagers scattered over the county, the whole crop of turnips is only about five times as much as that of potatoes,

Preparation.—The best farmers sow their turnips on the second crop after ley, called the Olland in Norfolk, and the Awald, or Yawald, in Aberdeenshire. Others take two crops; and not a few in the upper parts of the county take three white crops, before they sow turnips. The taking two crops is not good, but the taking three is wretched husbandry. When one considers that a boll of lime shells, measuring 128 Scotch pints, (or $6\frac{1}{4}$ bushels) costs 4s. 8d. at Aberdeen, or 9d. every bushel; and that from 16 to 24 bolls, or from 100 to 150 bushels of lime are requisite to an acre, which, with the expence of carriage, costs from five guineas to eight pounds, and that a subsequent liming does not answer so well as the first; also, that one-half of the quantity of dung will answer for turnips sown on the second crop after ley, compared with what will do for the fourth crop, after the land is exhausted by the previous white crops, one should think

think that those who reside at a distance from Aberdeen, and from lime quarries, would be careful to take but a single crop before turnips; although those in the neighbourhood of lime were so injudicious as to take two, or even three crops before cleaning their ground, and laying it down to grass.—Yet the fact is just the reverse. The gardeners in the vicinity of Aberdeen, do never—and the farmers, who live within a few miles of it, very seldom take more than one crop, while those who are fifteen or twenty miles distant from lime, take frequently two, and sometimes three crops of corn before turnips.

When ground is intended to be in turnips, it is ploughed immediately after harvest. Sometimes it gets only a ribbing. In the spring it gets a deep ploughing, either diagonally, or across the former. If it has borne only one crop after ley, it is (after being severely harrowed, and cleared of its weeds, which are gathered into heaps, and either burnt, or carted off the field) shaped into narrow drills of two feet and a half or three feet wide, by two furrows, or one bout of the plough. The dung is then put into the open drill, after which the plough reverses the furrows, and lays the dung in the middle of the drill. The turnips are then sowed and covered, and if the lime has not been laid on before, it is spread over the field, and both it and the turnips are rolled in.

When the land has borne two or three white crops after ley, or after it has been fallowed, or otherwise cleaned, it requires three or four ploughings before it is formed into drills, or five or six ploughings in all. A rainy season also obliges the farmer to give an additional ploughing to land that had borne only one or two white crops. But in general, he is careful to have his ground as free of weeds as possible before the turnip seed is sown; and also to sow immediately after the dung is ploughed in. Sometimes these are above ground in five or six days; at other times they ly for some weeks. In this case the fly and the caterpillar are blamed, when it is entirely owing to the season. Those who are impatient sow a second time, which is seldom found to succeed. Those who wait coolly, generally find that when the rain comes the turnips have received no injury. Sometimes, however, they do suffer, either by the grub, a short worm, or by the slugg, a snail that is very destructive. It is, however, after they have sprung up; and even after they have got on the four blades, or rough leaves, that the slug is most pernicious. Rolling at midnight is a never-failing remedy, and is generally known to be so. (This was first discovered in Ireland, and published by a Mr. PETERS, about 1768. Afterwards Mr. HENRY VAGG, of Chilcompton, who perhaps had not heard of Mr. PETERS' book, published it, and obtained a reward of L.2000 for the discovery from the farmers in Norfolk.)

Great care is paid to the ground and weather being moderately dry. In rain, both turnip and clover are in danger of bursting by swelling too rapidly. In dry weather, it is the practice of every good farmer to sow his turnip seed deep, and to roll them well.

When they are fit for the first hoeing, the plough is introduced, the earth is *taken from* the plants, and the interstices, or small strips of eight or nine inches wide, are hand-hoed, and the plants set out at eight or ten inches of distance. When the weeds have got up again, a second hand-hoeing is executed, the young turnips that may have sprung up since last hoeing, and any of the former turnips that are too thick in the drill, are hoed out. But if there be any blanks, or empty spaces in the row, two turnips are left pretty near each other, and both generally succeed very well. In rainy seasons, a third hand-weeding is sometimes requisite; but in general the turnips meet in the drills, and choke the weeds

in

in a few weeks after the second hoeing. At first the farmer's servants were afraid of wounding the young turnips, or of hurting them, by loosening the soil about the plants; but now they are much more hardy, as they find, that if not pulled out of the ground, a plant thrives the better for being freed of weeds and clods, and fixed only by its root to the soil. They were also at first afraid to thin the young plants much, or to set them out at nine or ten inches from each other; and they were desirous, as indeed some of the cottagers still are, to obtain a great quantity of weedings for food to the cows; but they found, that if not set out early at a proper distance, the turnips shot up like lettuces, and never formed a large bulb. Now when they find that the leaves of a turnip, at the second hoeing, are not sufficiently flat, owing to being confined by other plants, they hoe out the worst one, and by a gentle pressure of the foot, flatten the top of the turnip, after which it apples very rapidly.—In short, there is a great zeal among the farm servants and day labourers of all ages, to hoe out the turnip properly.—The only defect is scarcity of hands, where there are a great number of turnips, and only a few dry days at the time when hoeing is wanted. And the best remedy, as already mentioned, is to induce the women to assist in this useful, but not too severe exercise.

In the end of November it is usual to pull up a proportion of the turnips, and after cutting off the tops and roots, to pit them, till a storm renders it difficult to get access to the turnip fields, or the frost to pull up the plants; and some persons take up a quantity in autumn, to keep in pits, or in store-rooms, till the spring. But common field turnips cannot be so well preserved in this way, as ruta baga, or Swedish turnip can. On large farms, a proportion of this species of turnip should always be raised for spring food. And in

stormy

stormy weather, a quantity of common turnips should be kept in a store-room. For when turnips are frosted, or have icy particles adhering to them, they are hurtful to the cattle, especially to milch cows. In this county, when a judicious farmer is suddenly overtaken by a storm, he causes his turnips to be thrown into a stream, or small quantities of them to be put into a vessel filled with *cold water*. This extracts the frost, and they can then be given to his cows with safety. Where *warm water* is applied for this purpose, it injures the flavour of the turnip.

Sort.—Various kinds of turnips are used. The *early white* are sown in April by the gardeners in Aberdeen, or by those in the employ of the landed proprietors, for the table. Of field turnips we have the white globe, and Norfolk tankard, the green, the red topped, and the yellow. The white globe is now in most general esteem among those farmers in the country who prefer the *greatest weight* per acre. The yellow turnips are in greatest request among the *cow-feeders in Aberdeen*, and those who wish to raise turnips of *the finest quality*, or to *remain long on the ground;* for the white, the red, and green topped, are all more easily injured by the frost, than the yellow turnips are.

Time of sowing.—Field turnips are sown at all periods, from the last week of May till the beginning of July. The farmers wish to sow one-half of their turnips before the 10th of June, and the whole of them before the end of that month. Those who have extensive farms, begin to sow their turnips by the 1st of June, and continue till the whole is finished, in the course of three, or at most, four weeks. They choose to sow them at *different* periods, that the *hoeing* may not come upon them all at once, and that part of them *may catch a favourable season*, whether that be earlier or later. They also have a proportion of their turnips pretty late for spring food.

For

For a turnip field sown in the beginning of July, though it carries not so weighty a crop, is abler to resist the winter frost than what is sown earlier. Yellow turnips are commonly sown for this purpose, where the farmer does not raise ruta baga. In general every farmer has *three sowings* of turnips, *at eight or ten days from each other.*

Most of the farmers preserve and raise their own seed; or they purchase it from persons in the country, who are known to pay particular attention to the raising good kinds of seed, and to the keeping of the different kinds unmixed.— They are afraid of being imposed upon by the seedsmen in Aberdeen; and they have reason to be cautious in this respect. For when a seedsman has several kinds of turnips running to seed in the same field, in the flowering season, they are very apt to mix together, and to produce varieties which are inferior to the parent plants. They also find it necessary every fourth or fifth year to sow their turnips in a spot where they are to be allowed to run to seed, instead of being transplanted. The seed thus raised is hardier and more prolific, only a little thicker in the neck. To prevent the fly from injuring their crop, some of the most judicious farmers mix the seed of two separate years before sowing.— The oldest seed lies a day or two longer in the ground, and the fly, at most, destroys only one of the kinds.

Although turnips are generally drilled by all the farmers, they are not so generally horse-hoed by the small tenants.— They sow them in drills of eighteen or twenty inches, by gathering two narrow furrows together: And while the one furrow is ploughing, a person with a *seed-box,* described by Dr. ANDERSON, or sometimes with an old pen-case of plated tin, having three or four small holes in the end, and a small rod, or walking cane in the mouth of it, sows the turnips on the back of the drill, and they are covered by the return of

the

the plough. In a precarious season, some of the larger farmers sow them in this way, when they are afraid of great drought. The great recommendation attending this method is, that the turnips sown in this manner seldom fail. For the back of the drill in which they are sown forms a kind of inclined plane, and when this is covered by the return of the furrow, some of the seed is always inserted at such a depth that it must succeed. For if the drought continue, they spring from the side of the inclined plane nearly an inch below the surface, owing to the moisture at that depth. (Where the sowing machine, with double rollers, is used, there is less danger of injury from the drought; but this is not as yet generally introduced.) If rain fall soon after the turnips are sown, they spring from various places, and form a narrow strip of three or four inches in wideness, which the farmer hoes in a zig-zag direction, always sparing the best turnip. This method is most suited to small farmers; and it is to be noticed, that the dung and lime are spread on the surface immediately before ploughing, when this plan is adopted.

Other farmers sow their turnips in broad cast, the dung being previously ploughed in; and then hoe them out at eighteen or twenty inches distance, in the following manner—One man goes before, with a large hoe of ten or twelve inches, and cuts a space of about fifteen inches wide, in a straight line across the ridge; a second cuts out another line of the same breadth, leaving an interstice of three or four inches between the two; and a third follows at a small distance, and hoes out the superfluous plants, leaving always the best ones in this narrow stitch, though not always in the middle of it. This was the method adopted by the late Mr. BARCLAY of Ury, one of the best practical farmers in Scotland, who travelled five times on foot through the great farming counties

ties of England, that he might see the agriculture of the sister kingdom more thoroughly. The recommendation of this plan is, that in clean land it produces a greater crop than is produced by horse-hoeing. This quality, however has been denied, and considerable betts were laid by several gentlemen in Forfarshire, in favour of a field of very fine turnips in the county of Angus, which had been horse-hoed, and were said to be superior to Mr. BARCLAY's. A square chain of each was weighed, and Mr. BARLCAY's *hand-hoed* turnips were *one-fourth part weightier* than the other which were *horse-hoed*. It cannot admit of a doubt, that in extremely clean land, or in a dry season, when land can be thoroughly cleaned by hand-hoeing, a greater crop may be raised, than could have been obtained by horse-hoeing. But it is also indisputable that on an average of soils and seasons, and in raising turnips on a great scale, horse-hoeing cleans the land best, and is in strong lands, otherwise unfit for raising turnips, the only judicious method of raising them. A farmer in the Lothians would act very unwisely, if he did not horse-hoe his turnips; as this is the only proper way in which this valuable root can be cultivated to advantage in a tenacious soil: And even in Aberdeenshire, where the soil is light (yet the infield or old croft land generally foul) a crop of horse-hoed turnip, though *somewhat less weighty* than hand-hoed turnips sometimes are, is always more to be depended on for *cleaning* the land, and preparing it for being laid down in grass seeds.

While some farmers in the country continue to adopt Mr. BARCLAY's mode of dressing their turnip fields (which is something intermediate between drilling and broad-cast,) the gardeners in the vicinity of Aberdeen, both sow their turnips in broad-cast, and hoe them in the same indiscriminate manner, setting them quickly out to a proper distance, only
leaving

leaving always a good plant that seems to be near the dung, and very thriving. They have no ploughs of their own; they pay a very high rent; and as the operation of hand-hoeing gives employment to themselves and their children, they raise in general the greatest crop that can be raised on their land. They also have access to night soil, or street dung, though this is a costly manure. Their example, therefore, though highly deserving of imitation by all who are similarly situated, cannot, as far as regards sowing turnips in broadcast, be copied by the large farmers in the inland parts of the county.

Dr Anderson supposes, that the size of the turnips generally raised in this county is nearly double of those which are raised in Norfolk. Without pretending to a superiority in this respect, the farmers in Aberdeenshire claim to be placed on a foot of equality with those of South Britain, in their management of turnips. But it may be observed, that when turnips are of too large a size, they are seldom able to endure the frost in winter; and not only the specific gravity of the bulb is less, but the crop is not so weighty on an acre. In the end of last year, (1809) and the beginning of the present, the Writer of this Report made various excursions through the county, and both numbered, measured, and weighed the turnips in different places, and in several instances ascertained their specific gravity. He uniformly found *the greatest crop per acre* was where the *drills were narrow*, or from two feet three, to two feet six inches, and where the turnips weighed, each at an average, *about four pounds averdupois*.

The produce of an acre of turnips, raised in a favourable season by the gardeners in the vicinity of Aberdeen, is not overcharged when stated at 50 tuns averdupois per Scotch, or 39 tuns 7 cwt. on the English acre. The land here is thoroughly

roughly manured with from 20 to 25 tuns of *night soil*, provincially called *street dung*; and the turnips, as above-mentioned, are sown in broad-cast. The greatest crop, in some cases, is not below 60 tuns. In 1808, ALEXANDER DUGUID, Esq. sold 300 beds, as they are called, (i. e, three eighths of a Scotch acre) for L.9 sterling. This was at the rate of L.24 per Scotch, nearly L.19 per English acre. Yet the cow-feeder and his wife, who bought them, both informed the Writer of this Report, that "they were a good bargain'; for that "they maintained three milch cows about twenty weeks."—This crop could not be under 60 tuns per Scotch acre. But the crop of turnips near Aberdeen, in 1808, was uncommonly good. In 1809 it was much injured by the fly. The best turnips, of the red topped kind, which the Writer of this Report saw near Aberdeen, belonged to DANIEL MITCHELL, Esq. of Ashgrove, a major in the East India Company's service. They were sown in broad-cast, and weighed 42 tuns 12 cwt. per acre. Another field, belonging to PATRICK KILGOUR, Esq. of Woodside, and horse-hoed, weighed only 31 tuns; but though the turnips were good, considering the season, the drills were too wide, being nearly three feet distant from each other. Mr. KILGOUR had spared no expence in manuring and dressing the turnips; but the sowing in broad-cast, where land rents so high, and dung can be had in any quantity, is always preferable to horse-hoeing, unless the drills be very narrow, i. e. not exceeding 26 or 27 inches.

In the country, the produce of a Scotch acre of turnips, varies in different seasons from 30 to 40 tuns. The average is probably 36. Some persons who consider that our soil is generally light, and particularly adapted to the raising of turnips, and that this crop is cultivated with more care in Aberdeenshire, than in those counties in which the raising of wheat engages the principal attention of the farmer, may

think

think this too low an average. But the Reporter cannot state it higher, because he has measured and weighed the crops in different parts of the county; and because he knows that 36 tuns is a weighty crop of turnips, if they be either of the yellow, or red-topped kinds. Of the white globe and tankard, which are of a more spungy nature than the red, and especially than the yellow, above 40 tuns are raised on lands in good order. Nothing, in fact, can be more inaccurate, than to speak of a crop of turnips, without specifying the particular kind that is raised. Wherever an excessively high estimate is made, there is generally a misapprehension; although no misrepresentation was intended. An example of this is marked in the note below.*

In order to give as distinct a view as can be given of the quantity of turnips raised on the Scotch acre, in different parts of this county, the Reporter shall exhibit what he actually

---

* In the excellent Survey of Lancashire, the surveyor, Mr. HOLT, was shewn a piece of ground, 30 perches, at 8 yards to the perch, (*equal to* 63$\frac{1}{2}$ *statute perches*) the early potatoes raised on which had been sold for L.30; after which in the same year (1793) a crop of turnips had been raised, which at 6d. per bushel, was worth L.50 per acre. A most respectable and well informed gentleman, Mr. HARPER, remarks in a foot note, calculating the number of bushels very properly at 2000, (but forgetting that the Lancaster perch contains 64 square yards, while the statute one is only 30$\frac{1}{4}$, and that the provincial acre raised from this large perch, viz. 10,240 square yards, is equal to 2 acres, and nearly 18$\frac{1}{2}$ perches) that this was more, by 800 bushels, than he had ever heard of, for either large or small lot, either by hoeing, or any other way. If the measure of the acre, and contents of the bushel, had been defined, this apparent misrepresentation would have been removed.—And if either Mr. ECCLESTON, to whom agriculture owes much, or Mr. HOLT, whose Survey is a most valuable one, happen to read this note, they will see that a *misconception*, and a *misrepresentation*, are very different things.

tually weighed, measured, and even numbered, on an acre, computed from the produce of a fall or perch, distinguishing the species of turnips, as well as the weight per acre.

	tun.	cwt.	qr.
Pitfour, personal farm of JAMES FERGUSON, Esq.			
worst crop of white globe,	38	2	3
Do. best crop of do.	44	9	3
Do. yellow turnips,	44	8	3
Mains of Pitfour, red topped turnips raised by the farmer,	38	9	3
Glebe or parsonage lands of Old Deer, the Rev. JOHN CRAIGIE, white globe,	42	11	3
Mill of Old Deer, JAMES MITCHELL, yellow turnips,	45	7	0
Glebe of Cruden, the Rev. ALEXR. COCK, white globe,	50	3	0
Do. do. red topped turnips on the same field,	36	8	0
Mains of Ellon, Mr. CUMMING, white globe,	50	6	1
Do. yellow turnips, same field,	44	2	0
Mains of Orrok, Mr. JAMES GORDON, white globe, horse-hoed, *best*,	50	5	0
Do. sown broad-cast in the head-ridge of same field,	56	5	0
Mains of Broadland, ALEXR. HARVEY, Esq. yellow turnips,	36	10	0
Rattray, Mr. DAWSON, red tops, *best*,	41	2	0
Do. partly yellow, partly red,	40	1	2
Crimonmogate, PATRICK MILNE, Esq. red tops,	40	10	0
Glebe of Rathen, the Rev. WILLIAM COCK, red tops,	41	18	2
Hopeshill, JOHN WALKER, Esq. red tops,	38	5	0
Mains of New, the Rev. GEORGE FORBES, white globe,	50	4	0

These

These turnips were much superior to the average of the county in 1809, for the fly was very destructive, and the Reporter went to those places which had escaped its ravages.

One circumstance, however, demands the reader's attention, viz. that on the glebe, or parsonage lands of Cruden, Mr. Cock had *thirteen tuns, and fifteen hundred weight less* upon the acre of *red tops*, than upon the acre of *white globe* turnips, that were sown in the *same field.* There had elapsed a few days between the sowing of the two kinds, the red being sown latest. But the white globe was certainly the best kind, not only as producing the weightiest crop, but also in the estimation of a very competent judge, viz. a cow that Mr. Cock was feeding, which would not taste a red topped turnip, while she could get one of the white globe. The same thing happened in the Reporter's parish, where a farmer's feeding oxen, after being accustomed to the *white,* would not eat the *red* turnip. The rind, or skin of the white globe is more tender, which perhaps renders it more acceptable to the cattle. On that account it ought to be preferred.

There are three distinct recommendations of any species of turnips. The first is the *superior weight of crop*; the second is the *preference* given to any kind by the cattle; and the third is its *specific gravity,* which generally indicates its fattening quality, and enables it to endure the winter frost. The red topped turnips of Mr. Cock's field appearing to be of a greater specific gravity than the white, on being thrown into a vessel filled with water, the Reporter saw that it might be useful to examine more accurately the specific gravity of different kinds of turnip. On comparing this with some degree of accuracy, he found that the yellow turnip was heavier, and the Norfolk tankard lighter than any of the other kinds of field turnip; that the red topped, the green, and

the

the white globe occupied the intermediate space between the yellow and the tankard; that all these kinds were from one-sixteenth to one-eighth part of less specific gravity than water; and that carrots were weightier than water, ruta baga weightier than carrots, and potatoes the weightiest of all.—The particulars, being too minute for insertion in this place, will be found in a separate paper in the Appendix.

It is here proper to mention, that a person who has any cattle to feed, should begin as early in the autumn as he can, giving them white globe first, for a month or six weeks; turnips afterwards, then ruta baga, and last of all, feed them off with potatoes. This is the method adopted by Mr. JAMES GORDON at Orrok, one of the best feeders in the county;—and although it is an expensive mode, it generally pays better than feeding imperfectly with common turnips.

Consumption.—The turnips raised in this county are generally drawn, very rarely consumed on the field, or carried to be used in an adjoining one. "Hurdling of turnips " by sheep," as Dr. ANDERSON remarks, " could never be " generally practised for want of a market, but the turnips " are almost universally consumed by cattle in the house, " which produces a great deal of excellent dung." In fact, the great object of the best farmers is to collect dung by every possible means, and to apply all that dung (except a small quantity for raising potatoes, yams, coleworts, and a few garden roots) to the cultivation of turnips, where animal dung is always found to be the most valuable manure. As we have so great a proportion of unimproved lands, and need yearly to raise the greatest possible quantity of turnips, hurdling them off by sheep would for some time be improper as a general practice. Where a man has sea-weed, shelly sand, or night soil, or any other manure within his reach, he may feed off his turnip lands with sheep, as a preparation for spring wheat;

wheat; and these animals, by treading the light soil with their feet, will render the soil more tenacious. But still, in adopting this method, a considerable portion of their dung is lost, or greatly injured by exposure to the atmosphere, and the influence of the sun, wind, and frosts. When a greater proportion of the land has been cultivated by turnips, then feeding off by sheep may be more safely and generally practised.

Turnips, on their first introduction from the garden to the field, were applied solely to feeding cattle for the butcher. And the landed gentlemen, who were the first raisers of turnips on their personal farms, before their tenants imitated their example, made very great profits by consuming them in this manner. They bought in lean cattle in the end of summer, or beginning of autumn, and after feeding them two or three months, they generally *doubled*, and not seldom *trebled their price*. But after turnips were raised more commonly, these profits gradually decreased; and it was discovered, from the rise of the price of lean stock, that a greater profit was to be made, by rearing young cattle, than by feeding for the butchers. The competition between these two modes of consumption again raised the price of turnips; and they are now either employed in rearing young cattle belonging to the farmer, or in feeding off the old ones; or they are sold to the butchers in Aberdeen at a certain price per acre. This varies from L.8 to L.10; and when sold in this way, they are fed on the farm. The farmer generally affords the cow-houses, or feeding byres, with straw for litter; and he also maintains the butcher's servant. In return for this expence, besides the stipulated price of his turnips, he receives all the dung made by the cattle, on which he rests his hopes of the succeeding crop of turnips.

It deserves, however, to be remarked, that turnips may
be

be raised without dung, on land that is in good order. In 1781, the Rev. WILLIAM WILSON, minister at Dyce, raised a good crop of turnips, on a field of old croft land, without any animal dung. He laid 15 bolls of shells on a corner of his field, mixed this with rich old croft land in a dunghill, which was turned repeatedly, and then spread on his turnip fallow when the turnips were sown. In 1809, the Rev. GEORGE FORBES, of Strathdon, raised an excellent crop of turnips without any dung, in a rich field that had been a garden about twenty years before. This method of raising turnips, however, can be practised only where the land is very good, and where it has borne only one crop after being some years in grass. In the examples that were given of two weighty crops at Mains of Ellon, the personal farm of the late EARL OF ABERDEEN, (of one quantity of white globe, that weighed above 50 tuns per acre, and another of yellow turnip, weighing 44½ tuns,) the field had been above a dozen of years in grass, and had borne only one crop of oats.

The quantity of manure laid on an acre of turnips is extremely various. At first, when a farmer raised only one acre of turnips, he gave it from *forty to fifty loads of dung.*— Now when the farmer lays the greatest part of his dung on the turnip field, he allows only half that quantity, or from *fifteen to twenty loads* to the acre. This may vary from 12 to 15 tuns: For a load of dung seldom exceeds 16 cwt. and sometimes it is below 12 cwt. when the field and road to it are steep. But where dung is laid on the second crop after grass, a much less proportion of it is sufficient for manuring an acre, than where two or three white crops are taken after it is broke up.

In reporting facts, nothing ought to be suppressed. Too many of our farmers neglect to plough their turnip grounds in winter, immediately after the turnips are carted off. The better sort of farmers are careful of attending to this, which a farmer in Norfolk never neglects. By this means they preserve several of the rotted leaves that would otherwise be lost, as well as a kind of slime on the surface of a turnip field, which is of great service to the land. In general the farmers of this county are careful to cross-plough, or to angle their turnip drills, in order to mix the dung more equally with the soil; a practice worthy of imitation.

Taking this important article into general view:—Turnips before 1782 were raised in very small quantities. Since 1793 they have been generally raised, as far as the farmers can get dung or manure of any kind. They are now, for the most part, horse-hoed, and well cleaned. Red topped turnips were long in greatest repute, but the white globe is now found to carry the weightiest crop; and the yellow turnip, a hardy plant, (and far superior to the *yellow ox* turnip of England) is preferred on account of its quality. All turnips are raised to greatest advantage, when sown on land that has been lately broke up out of grass, and borne only one white crop.

SECT. V.—CROPS NOT COMMONLY CULTIVATED.

1. RUTA BAGA, OR SWEDISH TURNIP.

THIS is a valuable acquisition to the turnip farmer; and though lately introduced, will probably be soon raised in considerable quantities. It has many and peculiar recommendations. 1. When the turnips have been destroyed by the fly, the slug, or any other cause, when they are found defective,

fective, owing to bad seed, or an unfavourable season, ruta baga admits of being transplanted like coleworts, and is well adapted for filling up the vacancies in a turnip field, if sown in the end of April or beginning of May. 2. When other kinds of turnip, not excepting even the hardy Scotch yellow turnip, are deeply injured, or completely destroyed by the frosts in winter, the ruta baga remains untouched, except by the hares. And even when they eat off a part of the bulb, the rest remains entire in the hardest frosts. 3. When other kinds of turnip are either exhausted, or are become of little service in the spring months, the ruta baga not only supplies their defects, but is a far more nutritious kind of food, and better calculated for *finishing off fat cattle*, (i. e. completely fattening and firming their flesh) than the common turnips are, when used in their best condition. 4. When the ruta baga has perfected its seed, its bulb contains a considerable degree of nourishment, having lost only about *one-tenth part* of its specific gravity. 5. Its leaves, in the earlier part of spring, when other vegetables either are not fit for use, or have been destroyed by the inclemency of the preceding winter, make excellent greens for the table. 6. Its bulb, when kept in a pit, will keep till the beginning, and sometimes to the middle of June. This makes them valuable either for the table, or as food for horses, or for cows in calf. 7. By using ruta baga as food to horses, a great deal of oats may be saved to the farmer.

The first of these recommendations, viz. its being fit for filling up blanks, or vacancies, in a turnip field, is so great, that every farmer ought to sow a quantity of ruta baga in the month of April, in land that is thoroughly prepared, and has been well manured. It may be proper to add, that when the young ruta baga plants are taken up in order to be transplanted, for filling the vacant spaces in a field of common turnips, the

Writer of this Report has found it beneficial to bedaub their roots in a composition of peat ashes and water, which makes them take root more quickly. By this means the growth of the plants will be less checked, for by the best management it must always be retarded by transplantation.

The second valuable quality of ruta baga has been denied, because this plant has in a few cases been injured by the frost. But this happens only when the seed has been adulterated, or when the land has been left in a wet condition, and not surface-drained by the plough. No plant can endure both wet and frost, as the formation of ice must destroy its juices.

The third recommendation of this root is so powerful, that every farmer, who feeds a number of cattle, ought always to raise a considerable proportion of ruta baga. When an ox begins to feed he eats voraciously, and common turnips in this state are very proper. But when he is nearly fat, his stomach has lost its power, and he must be finished off with more nourishing food.

The peculiar quality of ruta baga, viz. its bulb remaining entire and fit for use, after perfecting its seed, has been exaggerated by some who have asserted, that it was in this case as good as ever. The Writer of this report has found the specific gravity of the best ruta baga, in Decmber, to be 1.035, and in the month of June, only 94, or equal to that of the Scotch yellow turnip, and superior to all other field turnips, when in their best or weightiest condition.— And a crop of ruta baga is *weightiest*, and *most valuable* in the *month of April*, when its leaves are shot, and its root not too much exhausted.

The fitness of ruta baga for supplying two dishes at table, or serving in place both of greens and of turnips, will increase the demand for it as a garden root, in a climate in which

other

other greens are often destroyed by the alternate frosts and thaws, which we experience in the spring months.

When ruta baga is taken up and pitted, it not only enables the farmer to sow the land on which it was raised, but will be fresh and palatable, as above-mentioned, till the beginning of June. Yet the same precaution should be used in making air holes, for some days after the ruta baga has been covered with earth, as was recommended with regard to potatoes.

Besides all these recommendations, it deserves to be noticed, that ruta baga will prosper in stronger lands than what is proper for raising any other kind of turnips. This deserves the attention of those farmers whose soil is heavy and tenacious.

The objections to the raising of this root are, that it is extremely difficult to get good kinds of its seed; that it requires the land to be good and well manured; and that its crop is much lighter than any other kind of turnips.

The first objection of the difficulty of getting good kinds of ruta baga seed, will be gradually removed, when it is more commonly raised in the country. But till the farmers in general raise their seed, or purchase it from those who live in their neighbourhood, and who preserve it from mixing with other seeds, the objection is a very serious one; for a bad species of this valuable root is commonly sold by the seedsmen in Aberdeen. The reason why ruta baga seed is so often bad, is the great difficulty of keeping it unmixed with others growing near it in the flowering season. For it can be adulterated not only by all the kinds of common turnips, but also by coleworts, and any plant of the brassica tribe that happens to be in flower. It should therefore always be planted for seed in the remote corner of a field, which is at a distance both from turnips and coleworts, that are raising for the same purpose.

The second objection, that ruta baga requires both good land, and a great deal of manure, is well founded. And no man should attempt to raise this valuable root, without putting it into the best prepared, and best manured land, if he expect a good crop.

It is also true, that after all, he cannot expect so many tuns per acre of ruta baga, as of common turnips. But a tun of ruta baga contains as much nourishment as two tuns of the common white, green, or red kinds, and, at least, a tun and a half of the best yellow turnips.

The Writer of this Report distilled ruta baga into spirits, and was satisfied with the quantity, but disliked the flavour of what he extracted. But after keeping the spirits for 12 months, that flavour ceased to be disagreeable.

Another quality of ruta baga deserves to be mentioned.— While common turnips become spungy and soft, when they grow too bulky, and are very easily injured by the frost, ruta baga, even when it grows to eight or ten pounds weight, (which it seldom exceeds) still continues so hardy as to resist the severest of our frosts in winter.

The *soil* most proper for ruta baga, is a light soil; although, as already mentioned, it will grow on strong land, and resist the hardest frost, if horse-hoed, or surface drained.

The *tillage* ought to be as complete as it can. One ploughing before winter, another early in spring, a third before the dung is laid on, when the land is formed into drills; and a fourth, when the furrows are split and reversed, before sowing the ruta baga, where horse-hoeing is intended.— Where a patch of ruta baga is sown merely for filling up the gaps, or vacancies for turnip crop, the practice is to sow it in broad cast.

The

The manure is generally given with an unsparing hand, at least 24 tuns to the acre; and indeed the value of the crop of this root depends in no small degree on the quantity of manure applied. Common turnips will often be raised on good land to advantage, but ruta baga will not thrive on the best soil without a considerable quantity of manure.

Seed.—The time of sowing proper to be observed in this county, is the end of April, or beginning of May. If the farmer delay sowing ruta baga till the month of June, when he sows common field turnips, he will have only half a crop. The Writer of this Report received from Chevalier EDDELCRANTZ, when in this county in 1803, a small quantity of the best ruta baga seed. He sowed it next year in his garden, allowed it to run to seed, and had a great crop. But in 1806, his servant, by mistake, sowed, with this seed, instead of common turnips, two acres of good land; and though the ground was very well manured, he had only nine or ten tuns on the acre, owing to the lateness of sowing. Although it cannot gratify vanity to record mistakes, or to confess errors, he thinks it his duty to state this fact, for the following reason.—Sir THOMAS BEAVOR, in one of his valuable communications to the Bath Agricultural Society, states that he sowed ruta baga like common turnips, in the month of June, and that they answered well, when sown so late in the season. Sir THOMAS is perfectly correct in preferring the *sowing* to the *transplanting* of ruta baga (for every plant must be checked in its growth, by being transplanted); and in Somersetshire he might raise a good crop of this valuable root, though sown in June. But the Reporter's experience has taught him, that in Aberdeenshire, ruta baga should be sown in April, when the land is properly prepared, and the season permits; at any rate about the middle of May.

Sort.—There are two kinds of ruta baga used in this county, ty, the white and yellow. One plant of the white, on examination, was only 1.022 of specific gravity; while another of the yellow kind was very nearly 1.035. This therefore was the best, but that difference might have been accidental between the *plants*, not between the *two kinds*. As so much depends on the quality of the sort preserved, every person who attempts to raise his own seed, should be at great pains in selecting the cleanest and best roots.

Transplanting should never be used, except for filling up empty spaces in a turnip field; for though ruta baga may be transplanted, a weightier crop is obtained without transplanting.

Horse-hoeing.—This is certainly the most proper mode for cleaning the land, and no large farmer will sow ruta baga in broad cast, except for filling up the gaps in a turnip field.

Hand-hoeing should, however, be used between the interstices, and ruta baga being sown earlier than field turnips will pay best when it gets *three* hoeings.

The fly, owing to their being sown earlier, is not so apt to injure the ruta baga, as it is injurious to common turnips. But as the land is generally well manured for ruta baga, and as the slug is most destructive where the land is in the highest order, midnight rolling should never be omitted, where a slug is seen, or is apprehended to commit his ravages.—Rolling can never do hurt, when properly and timeously applied, and may often be of use on the soils which are prevalent in this county.

Application.—The bulbs of ruta baga have been used very successfully as food to horses, by a gentleman in the neighbourhood of Aberdeen; and are sliced down for this purpose. They are also used for milch cows, and as food for black cattle, which, as Mr. BARON COCKBURN observes, will eat no other turnips while they can get ruta baga. Mr. JAMES Gor-

Gordon at Orrok, uses them with great advantage in feeding off his black cattle. The landed proprietors chiefly use them for spring food to milch cows; but they are of the greatest value when applied as food for horses, or for finishing off cattle, as above-mentioned.

Quantity.—The Reporter last year measured ruta baga at three places. The most weighty per acre, were raised at Logie Elphinstone, by R. D. Horn Elphinstone, Esq. and weighed per Scots acre, 38 tuns 10 cwt. 1 qr. Here the ground was naturally fertile, and had been dunged very thoroughly.

The next belonged to Mr. James Gordon at Orrok, and weighed, 37 16 3

The third belonged to James Ferguson, Esq. of Pitfour, and weighed 27 12 2

But though the land was in high order, the ruta baga had suffered by being sown a few days later than the others.

The average of the county cannot be estimated above 20 tuns per acre.

Remarkable Crop of Ruta Baga.—Mr. James Gordon, now in Orrok, at that time in Mains of Logie, in the parish of Crimond, in Buchan, produced in a field of good land, which had been highly manured, and kept free of all weeds, the greatest crop of ruta baga, that is known to have been raised in any part of Great Britain. The Writer of this Report did not see, far less measure it; but he knows that several persons went a considerable way to see it; and he has no doubt of the fidelity of Mr. Gordon's statement, which he gives in his own words.

"My crop of ruta baga was greater this year, (1798) "than any former crop which I ever raised. It was sown 'on the 26th of May; and on the 15th of May following, "the produce of one fall weighed 791 pounds 14 ounces averdupois,

" dupois, being at the rate of 56 tuns, 11 cwt. and 1 qr. per
" Scotch acre. Some of the single turnips, with their leaves,
" weighed 17 pounds, and many of them weighed 16 pounds
" each. None of this turnip was used till the spring was far
" advanced, and notwithstanding the uncommon severity of
" the season (1799) there was not a single plant of ruta ba-
" ga in the least injured by it. The soil on which this crop
" was raised, was pretty strong, and rather damp, which
" must shew the superior value of this vegetable."

It ought here to be noticed, that the weight mentioned, (viz. 56 tuns, 11cwt. and 1 qr. per Scotch acre) is equal to 41 tuns, 10 cwt. and 25 pounds to the English acre; and that it weighed several tuns more on the 25th of May 1799, owing to its load of leaves, than it would have done in January, before the plants began to shoot. Yet it certainly was an extraordinary crop; being nearly one-third part more than the Duke of Bedford's crop of ruta baga, at Woburn, in 1798, viz. 33 tuns, 10 cwt. and a half, per English acre.

Value.—This depends very much on the purpose to which they are applied. When used as food for horses, their value is equal to that of the oats, or other food which they save. Here, supposing a bushel of oats to weigh *thirty-eight and one-half pounds*, and to cost 33 shillings per quarter of eight bushels; also that a horse is so severely wrought as to require 14 pounds of oats daily, but can be as well kept by getting only 7 pounds of oats, and 27 pounds of ruta baga per day; the quantity of oats saved in 44 days, is exactly one quarter, or eight bushels. This is a saving therefore of thirty three shillings in 44 days, or ninepence per day. Consequently, 27 pounds of ruta baga being estimated at ninepence, every three pounds are worth one penny, and eighteen pounds are in real value, whatever be their nominal price, worth sixpence. At this rate, supposing that 18 tuns,

tuns is the average produce of an acre of ruta baga, when properly managed, the value of this acre, as far as this is ascertained by the saving of oats, is *fifty-six pounds*. If, on the other hand, this ruta baga be applied to the feeding of cattle, when nearly fat for the butcher, its value consists both in the additional number of pounds of beef and tallow, and the additional value to the quality of the meat. Farther, if it be applied to the feeding of milch cows, its value is to be estimated from both the additional quantity and richer quality of the milk, and on the greater nourishment given to the cow. These two last articles cannot be so easily calculated by a mathematician: But a butcher knows, that there is much more *profit in finishing off* a half-fed animal, than by the common imperfect feeding; and a cow-feeder also knows what addition is made both to the quantity and quality of milk by ruta baga, more than by common turnips, or any other of the brassica tribe of coleworts.

Comparison with Turnips.—It is a prostitution of ruta baga, to give it at first to a hungry, voracious ox, when only beginning to feed. While he is in this early stage of scouring, as a preparation for being fattened, common turnips are sufficient for this purpose; and these, as before-mentioned, may be succeeded by the best yellow turnips. Dissimilar things cannot be compared, as to real value: But when the yellow turnips have fattened as far as they can till they begin to shoot, every pound of ruta baga in the spring months, is worth three pounds of common turnips.— And an acre of the former ought to give, in the market, double the price of the latter. Or where common turnips cost L.10, ruta baga ought to cost L.20 per acre.

At present there are not, perhaps, above 100 acres of this valuable root raised in the county. And it is because the Reporter knows its value, that he has been so particular

in

in this account of a root, of which every farmer ought to raise an acre, for every six, or perhaps every four acres that he raises of turnips. The most valuable root as yet discovered no doubt is potatoes. The next is ruta baga. Carrots follow next in order. And then the different kinds of common turnips. The specific gravity of potatoes is nearly 1.091, ruta baga 1.032, carrots, 1.018, and turnips from .84 to to .94. These things shew the order of precedence; but the real value, or quantity of nourishment to any animal, or of saccharine, or fermentable matter contained in a tun, or any given weight of each of these roots, is much more than is indicated by the proportion of their specific gravities.— It is more nearly as the cubes, or third powers of these gravities. The Reporter does *not assert* that it is *exactly*, but he *believes* it is *very nearly* in this proportion. It would have been improper in him to have omitted stating, what *he believes* to be the *ratio of comparative value* between these roots: And if what he has hinted be the true standard of their relative values, it will not be *the less true* that he does not *assert it positively*. Let this suffice for a comparison of ruta baga with turnips.

2. RYE.

This being a species of corn should have taken the place of any roots. But it not only belongs to the class of those plants, or seeds, which are little cultivated, but which were once pretty generally cultivated, and now are rarely to be seen. It was formerly sown in end-ridges, or head-lands, which were apt to lodge, and was then made into meal, soured and baked into a thin cake, or coarse loaf, which was used at Christmas. But the increase of luxury has banished the *rye loaf*, and *sour sconns*, substituting wheaten flour, baked ed in a Christmas cake, with raisins, currants, and the rinds of oranges.

In the upper parts of Marr, i. e. both in the districts of Braemar and Curgarff, rye is still to be seen, sown along with oats; and the highland farmer's servants occasionally use it when baked into cakes. In other places it is sometimes, though rarely, sown like wheat in autumn. But there are not 100 bolls sown in the whole county; as there is no market for it, and as wheaten bread is now used by those who do not relish oat cakes. It has a real and considerable value, when used in distillation; but that value is unknown in this county. And there is little probability of its being again raised in any quantity, as long as our agriculture continues in a flourishing state. It is unnecessary to say more of this species of grain.

3. TARES.

These have been repeatedly tried in Aberdeenshire, but have never succeeded well, and are now generally given up. Owing to the lateness of our spring, we can seldom get more than one cutting of our summer tares; and winter tares very rarely stand the alternate frosts and thaws of spring.— An Aberdeenshire farmer is astonished when he hears of three or four hundred acres of tares belonging to a farmer in Norfolk, but will never think of imitating his example. It is highly probable, that the DUKE OF BEDFORD's crops of ruta baga, which were first read with pleasure, and next excited emulation in this county, will have a similar effect in all the northern counties; and that ruta baga will, in time, become a pretty general article of spring food: but the laws of climate, more formidable than an Act of Parliament, forbid the cultivation of tares in Aberdeenshire.

4. BUCK

4. BUCK WHEAT.

This has been raised in the Reporter's parish, and in some other places of the county, where it was used as food for poultry; and as it grows on poor land, may deserve to be encouraged. If, however, its application continue limited to that purpose, it cannot be raised on any large scale.— We have a great deal of inferior land, which has much need of any meliorating crop, that requires no manure. Even though such a crop were ploughed down, it would still be valuable. Grey, or hog pease, have been tried for this purpose, but have seldom succeeded, except where the ground has been limed. In fact, there is at present no substitute for the small oats, or thiacks, that were raised fifty years ago, on our worst arable lands. Blainslie oats were tried in the Garioch, but did not succeed. There is said to be a species of yellow oats in Caithness, imported originally from Denmark, which yields seven or eight pecks of meal; but they have not as yet been brought to this county. If buck wheat could be raised on poor soil, and were generally ploughed in when green, (saving only as much as would answer for next year's seed) it would be a great acquisition.

5. FLAX.

Thirty years ago a considerable quantity of flax was raised in this county. But after the breaking out, first of the first war with America, and then of the wars with France and Holland, it was found that flax degenerated, if the seed was raised more than one year in Britain; and that the lint manufactured from it became still coarser in point of quality, if the flax seed was not exchanged, or imported at the end of of every two years. The long continuation of the war with France—the shutting up of the ports of Holland—and the rise in the price of flax and lint seed, to which a circuitous and expensive sea carriage of this commodity unavoidably contributed, would have necessarily occasioned a considerable decrease of the quantity of flax raised in Great Britain. But the great increase of the importation of cotton, and the extensive manufacture of cotton cloth, by means of machinery, have supplied us with a number of articles, less durable, but both cheaper and more attractive than what were formerly made from lint: and both from the high price of the latter, and the cheapness of the former, the linen manufactory has received a great check, and the demand for flax is now much more limited at present. Yet this branch of commerce may revive; and strong, though not fine cloth, made from our own flax, and this too from seed raised at home, may again become in more general use. For it has been found, that seed raised in Great Britain and Ireland, has not degenerated so much as was at first apprehended; and it is believed, that if it were regularly exchanged between Peterhead and the north-east coast of Scotland, on the one side of this island, and Dingle Bay, in the south-west part of Ireland, the evil complained of would be palliated, if not entirely removed. When flax is sown on good land, and meets with a favourable season, it is a most valuable crop. At the same time it must be acknowledged, that it is a precarious one, though every farmer formerly raised as much as served his family, while economy was the order of the day.

The *soil* proper for flax is strong and heavy. As we have not much of this, except in Buchan, and some particular places of the other districts, flax was generally sown on ground newly broke up from clover-ley, or at any rate the second crop after breaking up. In some places it was

sown

sown after turnips with grass seeds, where the land was newly cleaned, and pretty close in its texture. But if a rainy season followed, both the flax and grass seeds were injured, and it was considered as best farming, to sow it from clover-ley, where this was only one or two year's old;—and if older, on the second year after breaking up, though two white crops, one of oats, and another of flax, certainly scourged the ground too much.

The *tillage* was, in general good, the furrows very narrow, and the grass and weeds carefully cleared off. The land was reduced to a fine tilth, or well pulverized, and partially harrowed before being sown, and completely harrowed afterwards.

No manure, except occasionally peat ashes, was applied to the ground. But the land had either been lately manured, or was in high order.

About nine pecks of seed were applied to the acre, if the flax was meant to be used for common purposes. Ten pecks were sown if fine lint was wanted.

The time of sowing was in the end of April or beginning of May, according to the season, or immediately after the *oats*, and before the *bear* was sown.

It was carefully cleaned of all small stones on the surface when sown, and afterwards weeded by women and children, where this was required.

It was pulled in the end of July, or beginning of August, or before the seed *was ripe*, indeed before it was filled, when fine lint was wanted, or allowed to remain till it was pretty ripe, or fully ripened, according as lint of a middling quality, with a moderate quantity of seed, or coarse lint, with a great proportion of seed, was wanted. It was laid very even in the roots, and bound up into small sheaves, then set up in stooks, or shoks, like corn. Sometimes the

bolls

bolls containing the seed were taken off immediately; at other times they were allowed to stand a few days, or weeks, till properly dried. If meant to be kept as seed, they were taken off by a comb, with iron teeth, provincially termed a rippling comb. If this was not intended, the whole was put into a pond of water, that had been kept standing for some days, and stones and trees were laid over the whole, to keep it covered with water. At the end of from eight to twelve days, according to the state of the weather, softness or hardness of the water, when the lint parted easily from the stem upon breaking it, it was taken out of the pond, and spread very thin upon dry land, on which there was nothing but short heath, or grass eaten very bare. Here the rains bleached the lint, and washed off the impurities constructed in watering it. After lying on the grass till its straw was quite brittle, and the lint parted easily, it was bound up, and put up in stacks till winter. It was then carried to the lint mill, and milled, or broken, heckled, and dressed, at a moderate expence. Sometimes it was *skutched*, or broken, at home, by the small farmers; and most lint was obtained by this means with great labour. But after lint mills were erected in different parts of the county, the home dressing of lint was laid aside. The expence of breaking and dressing, varies from twopence to fourpence per pound, according to the fitness of the dressing that was required.

The value of a good crop of flax, thirty years ago, was greater than one of wheat; and the Writer of this Report has known four guineas paid by a dealer for the rent of a Scotch acre, or an hundred guineas for twenty-five acres, of clover ley, for raising a crop of flax. But the farmers in this county seldom let their land for this purpose, though they sowed as much lint as their families could manufacture.

After all it must be acknowledged, that the produce of an acre

acre of flax is extremely variable. The Writer of this Report has had at the rate of 45 stones to the Scots acre; and he has had only one-fourth part of this quantity. The price of a stone of flax thirty years ago, when only roughly dressed, was eight shillings. It has now risen to above sixteen shillings, and very coarse lint sometimes sells for twenty shillings.

The spinning of flax afforded much employment to the women, in the divisions of Buchan and Strathboggie, where the linen manufactory was carried on to considerable extent. In the other divisions, lint was chiefly manufactured for family use. In the city of Aberdeen, a great deal of employment was given in the manufacturing of flax. And a manufactory for spinning it is in the neighbourhood. But that of cotton, as already mentioned, has now become the principal manufacture.

The repetition of the same crop has sometimes been carried so far, as to have flax seed sown on the same field, first after clover ley, and then the year following; and the first year's crop was intended for coarser lint, the second for finer. But this was ever considered as wretched husbandry. And it seldom happened that one of the crops was not a bad one.

It needs only to be added to this article, that the earnings of a woman, thirty years ago, if she was a good spinner, and could use both her hands at once, amounted to sixpence, and in a very few cases, to sevenpence a-day. Notwithstanding the rise in the price of labour, she can make no more at present. For the price of lint spinning is less than it was fifteen years ago; and the linen manufactory has declined rapidly within these ten years. What is made is not a profitable branch of commerce.

In the whole county of Aberdeen, there are not above an hundred acres of flax raised at present. Twenty years ago, there were above five hundred acres sown annually with flax.

But

But this county never raised so much flax as was raised in the neighbouring counties of Angus and Mearns.

6. HEMP.

This also was sown in far greater quantities than it is now. Fifty years ago, many of the cottagers had small quantities of hemp, as well as of lint, which they reared for domestic use. The progress of luxury has occasioned even this class to neglect the cultivation of both. Printfields, rope-makers, and sail-cloth-makers, supply the cottager with articles, which in a less expeditious mode, he manufactured for himself. There are not now ten acres of hemp raised in the county of Aberdeen; and few of the young people have occasion for hemp growing. It is unnecessary to treat of its culture.

It is to be regretted, however, that hemp is not raised in a county in which couch is so common a weed; as it appears from Mr. Holt's Survey of Lancashire, that hemp is an excellent crop for destroying couch, though it be ever so abundant. And in a national view it is still more to be regretted that hemp, which is so necessary a part of our naval stores, is not more generally cultivated in every county of Great Britain. In the present state of Europe, whatever contributes to the *strength*, ought to be attended to, in preference to what promotes only the *wealth* of the nation.

7. CABBAGES.

These are raised in the gardens of most of the farmers and cottagers, but are not planted very generally in the fields. The late Dr. Chalmers, Principal of King's College, raised several acres of them annually on his farm at Sclattie, within

within four miles of Aberdeen; and his cattle were always in good condition, and preferred cabbages to turnips. ALEXANDER HARVEY, Esq. of Broadland, also raised them to a considerable extent, and with great success. Several gentlemen in the county raise a small proportion of cabbages for the use of their milch cows, because that milk has none of the disagreeable flavour when fed with cabbages, which it often has when they are liberally nourished with turnips.—And it may be observed, that in a distillery at Oldmeldrum, from the years 1787 to 1796, where a number of hogs were fattened, when boiled potatoes, cabbages, and turnips, were placed before them, these animals, who must be considered as competent judges of the most palatable kind of food, first eat up all the potatoes, next the cabbages, and the turnips last of all. Cabbages no doubt afford more nourishment than is given by common field turnips, but probably not so much as is given by ruta baga, which the hares always prefer to cabbages. They produce less urine, but more dung than turnips. But they require more manure in order to produce a weighty crop; and except when raised in deep soils, they seldom are, in point of weight, per acre, equal to a crop of turnips.

There are, however, two distinct sorts of cabbages used for feeding cattle, viz. the red cabbage, which were formerly much used by the farmer's servants, and the large grey, or Alnwick cabbages, of which the seeds are imported from England. The former contains much more nourishment in a given weight, than is contained in the latter; but it does not raise so weighty a crop. It probably holds the same proportion to the other, that ruta baga holds to common turnips. And it seems to be peculiarly adapted to the soil and climate of Aberdeenshire. For it has been remarked, that when this species of cabbage is carried to the southern or northern counties, it loses its fine red colour, and becomes both more grey in its appearance, and less nourishing; at any rate, less palatable. Whether this be owing to the great proportion of decomposed granite with which our soil abounds, cannot be ascertained; but the fact is generally admitted, that the red cabbages of this county do not retain either their colour or their character, when carried either to Angus or Morayshires. But for the feeding of cattle, the Alnwick cabbages are now more generally raised in the country; while the cow-feeders about Aberdeen prefer either coleworts, or red cabbage, although they esteem yellow turnips above both, on account of their weightier crop.

White cabbages are raised for the table only, and are used during the summer and autumn. For they cannot resist the winter frosts.

Preparation.—The ground for cabbages is ploughed at least once in autumn. Some persons plough thrice, and plant their red cabbages before winter; but most farmers prefer doing this in the spring, after other two ploughings in that season. Before the last ploughing the land is thoroughly harrowed and dunged. It is also formed into ridges or drills, and the plants are dibbled in by the hand. Sometimes the ground is left flat after being planted in spring, and afterwards ridged into drills, when the plants have got about eight or ten inches above the surface. This method has this recommendation, that the stem of the plant is shorter, and it is supposed to resist the winter frost better. With those who are too sparing of their labour, the plants are sometimes laid in along with the plough, to save dibbling; but this is reckoned bad management, as the plants are often injured, or shifted from their places by the horses feet, unless the ploughman be very dextrous.

Culture when growing.—The same as is practised in horse-hoeing

hoeing turnips and potatoes, with this difference, that when the cabbages are planted in autumn, or very early in spring, they get at least one hoeing more than the other. In the last hoeing, the furrow is laid as high upon the neck of the plant as is practicable; and any leaves, that happen to be covered by the plough, are disengaged very carefully by the hand.

Consumption.—They are carted off, and given either to milch cows, or to feeding cattle in the house, never fed on the land in this county. They are used from October to April.

Mode of Preserving.—As cabbages are much exposed to the winter frosts, and are often deeply injured, if not completely destroyed, it is of consequence to adopt a method of preserving them, which seldom fails of success. It has been long practised by the gardeners and farmers of this county for preserving their cabbages in winter; and is said to have been originally borrowed from the practice of Italy. It is to dig with the spade pretty close to the root of the plant, and to place the cabbage in a sloping direction to the north. The most correct method is to take up the plant, and set it again very deep, almost to the neck of the cabbage, and to lay it in a sloping direction to the north. By this means the cabbage is not exposed to the *alternate frosts and thaws* in February and March, which never fail to destroy it, if it ly facing the south. For it is not a *hard dry frost*, but the *changes* from freezing to thawing, and from wet to dry; that is so hurtful to cabbages, and indeed to most of our vegetables; and from the practice of this county, by those who are attentive to the preservation of cabbages, there is no doubt that most plants could be preserved that are taken up, and laid in this manner.

Produce per acre.—This is extremely variable. Of the

red

red cabbages from 15 to 30 tuns. Twenty is perhaps the average. Mr. HARVEY of Broadland is the only person who is known to have raised 30 tuns on a Scotch acre. Of the Alnwick, or large grey, from 33 to 56 tuns are raised, when well cultivated, and properly planted.

It is here necessary to add, that the only proper mode of planting cabbages, which are intended to form good bulbs, (or, as the gardeners express it, to raise weighty *stocks*) is what was already described when treating of potatoes, planting in *quincunx*, i. e. one plant in the middle between the two in the opposite row, every three plants, by this means making an equilateral triangle, and six plants making a regular hexagon, with one cabbage in the center. When they are planted in squares, there are always four very broad blades, or branches, as they are called, at the diagonals or corners, which prevent the stock from forming early into a cabbage, and which, if not gathered by the hand, are frequently rotten before the cabbage is ready for service. It has already been mentioned, that the proportion is 13 inches between the rows, for every 15 inches distance between the plants in a row. Hence red cabbage in general should be horse-hoed at 2 feet 2 inches, and the plants in the drill should be at 2 feet 6 inches distance. At this rate, a Scotch acre contains 10,219 plants, and an English acre 8040. Where the ground is in high order, and has been thoroughly managed, the stitches or drills may be 39 inches, and the distance between the plants 45. Alnwick cabbage may be safely planted at this distance, at which a Scots acre will contain 6732, and an English acre 5297 cabbages. But at whatever distance they are planted, let the plants in every second and fourth row, be always in the middle between the first and third. For by this means they will produce several tons of more cabbage. This is particularly necessary to be mentioned,

tioned, because a number of gentlemen who plant cabbages in drills, and horse-hoe them, frequently order them to be planted at 4 feet between the rows, and only 2 feet between the plants. This method occasions a loss of 1-6th of the weight of produce on an acre, and even that weight contains a greater quantity of broad leaves, but a much less proportion of hard cabbage. The reason of fixing on the proportion of 13 inches between the rows, and 15 between the plants in the drill, is marked in a note below.*

8. TURNIP CABBAGE.

This valuable root has been but lately introduced, and is still too little known. It has been only raised in gardens; and as yet little can be said of its culture.

There are two recommendations of it, however, that deserve the particular attention of every intelligent farmer.—These are both its value and its weight per acre.

Its value, not its money price, which depends entirely on the demand in the market, but its intrinsic value may be estimated from this circumstance, that the hares are said to prefer it to ruta baga, which they eat in preference to every thing else.

The quantity of it that may be raised on an acre, compared with that of other roots, was ascertained by a very accurate

* If a perpendicular be let fall from the vortex to the base of an equilateral triangle, whose side is 30 inches, it divides it into two right angled triangles, where the hypothenuse is 30 inches, and one of the sides, viz. that which was divided by the perpendicular, is 15 inches. The squares of the hypothenuse, (or 900,) is equal to that of the two sides. The square of 15 is 225. This leaves 675 for the square of the other side, and the square root of 675 is very nearly 26, i. e. in proportion of 13 by 15.

curate experiment made by the late EARL FIFE; when 100 square yards were sown with carrots, 100 with mangel worzel, or root of scarcity, 100 with common turnips, and 100 with turnip-rooted cabbage. The comparative produce of each was published by his Lordship's orders, and a calculation of the quantity per Scotch and English statute acre is added by the Reporter.

	ENGLISH.				SCOTS.			
	ton.	cwt.	qr.	lb.	ton.	cwt.	qr.	lb.
Carrots,	35	0	1	9	44	10	0	0
Turnips,	34	1	3	0	43	6	1	19
Turnip cabbage,	33	5	2	0	42	5	3	0
Root of scarcity,	29	2	1	7	37	0	0	7

The above experiments were made in garden ground of equal quality, and the culture of all was the same. But as turnip cabbage produced only a tun less per acre than common turnip, this valuable plant ought to be raised wherever it can be raised advantageously. And the trouble of taking up and cleaning the roots, ought not (as it does at present) to occasion so valuable a plant to be neglected.

9. COLEWORTS.

(Provincially Kail or Greens.)

They are cultivated in very considerable quantities by the gardeners near Aberdeen; and when these industrious men raise three crops from their land in two years, coleworts are the intermediate crop. They are used in the country in small patches, generally in the garden by every family. For the poorest peasant in the inland districts, even when he keeps no cow, and has no corn, yet is allowed a small garden annexed to his cottage. They are chiefly used for

for human food; but the cow-keepers about Aberdeen use a small proportion of them for their milch cows.

When the turnips failed, or when there was a number of blanks in a turnip field, the farmers generally planted greens into the vacancies; but since the introduction of ruta baga, that more valuable root is preferred to coleworts.

They form a most delicate and wholesome dish for the table; but the great objection to their being employed as food for cattle, is, that they generally produce a very light crop; for however singular it may appear, a crop of potatoes is often as weighty as one of greens. They very rarely amount to 12 tons per acre; at an average of the county they do not exceed 9 or 10 tuns. Since yellow turnips were raised in the field, even the cow-feeders use them very rarely.

One circumstance respecting this vegetable deserves to be known. The coleworts raised near Aberdeen, where night soil is used very liberally, though more abundant in quantity, are not esteemed so palatable or so sweet as those which are raised in the country. It has been asserted, and in several instances seems to have been proved, that the land tires of the frequent repetition of the same crop, and produces less quantity of what has been frequently sown on it. But here it would seem, that nature, harrasssed by raising two crops in a year, though strengthened or stimulated by manure, does not yield her produce of so fine a quality.

Coleworts are sown in the corners of a garden, transplanted into the field, and cut down as food for man or beast, at all seasons of the year. But no rape cake is made of them in this county.

Some persons have asserted that coleworts were equal to turnips for the purpose of feeding milch cows. The practice of the cow-feeders in Aberdeen directly contradicts the assertion;

sertion. They use only a very small proportion of coleworts; and to prevent any bad flavour from using turnips to their milch cows, they do not, as Dr. ANDERSON recommended, use a solution of saltpetre and water mixed among their milk, but they pour a proportion, generally a fifth or a sixth part of boiling water among their new milk, which never fails, by disengaging the essential oils, to carry off the turnip flavour. The Reporter learned from one of them this practice, and has often experienced its good effects, in correcting the disagreeable flavour of milk, when his cows had great quantities of turnips.

10. CARROT.

This valuable root is not so generally cultivated as it deserves. The gardeners in Aberdeen raise in the vicinity of that city, as many carrots as they find demand for in the market; but this is too limited to encourage the raising of carrots on a loose soil. In the county, besides what is raised for the table, many of the landed proprietors, and several of the farmers, raise a proportion of carrots for food to their horses. The great objections to them are, that they require a considerable depth of soil, and are difficult, or at least very expensive to keep clean. By ploughing them up in stitches of three feet wide, and horse-hoeing them, both these objections may be in a good manner removed. And where a farmer can spare half an acre of the land, that has been lately cleared of turnips, he will find it much his interest there to sow carrots in narrow drills, and horse-hoe them.— For they are most useful as food for horses, who are extremely fond of them.

A principal cause of their being not sown so frequently as they would otherwise be, is, that the seed of the carrots

is

is frequently very bad. We have, as yet, none of the large whitish carrot that is raised for feeding horses in Suffolk, and the Isle of Wight.

One great recommendation of carrots is, that they can be raised in rich and deep soil, without dung, and for several years in succession. Yet great crops of them have been raised with pigeon dung; although that of black cattle is not found to succeed with carrots.

Ardent spirits, of an excellent quality, may be extracted from carrots. The Writer of this Report having mentioned to the Scotch Distillery Committee, in 1798, that very good spirits could be extracted both from potatoes and carrots, was asked to undertake a series of experiments for that purpose; and a recommendation was given to the Scotch Commissioners of Excise in his favour. He made a number of experiments in 1799, an account of which he gave in to the Scotch Distillery Committee of that year. And by other two sets of experiments, in 1802 and 1803, conducted under the authority of the Board of Excise, he found not only that a considerable quantity of spirits could be extracted from either of these roots, but that the flavour of the spirit is excellent. Indeed he knows none equal to it, in respect of the pleasant taste of its essential or flavouring oils. The carrot however does not produce, from an equal weight, above two-thirds of the quantity of spirits that is produced by potatoes. But an acre of carrots, distilled into spirits, after paying all taxes, would be equal to at least four acres of wheat; and if the crop was very good, to a much greater quantity. And when Britain attends more to its agricultural interest, than to its colonial commerce, it may merit the attention of our Legislators, whether they ought not to encourage the distillation of roots raised in Great Britain and Ireland, rather than

that

that of sugar from the West Indies, when corn is either scarce or dear in this country.

The produce of carrots is extremely variable; from seven eight, or nine, to above forty tuns. When the seed can be depended upon, and the land is in good order, it may be reckoned double to that of potatoes, in similar circumstances, or twenty-five tuns. As yet, there are not above 100 acres raised, except in gardens, in the whole county of Aberdeen.

11. PARSNIP.

This root was more cultivated before the introduction of turnips than it is at present. It is seldom raised in the field, although it has sometimes been so with success.

Cows are extremely fond of parsnip; and it might be used for this purpose to the same advantage as it is in the north of France. The deep soils on our meadows, on the banks of the Don, Ury, and Ugie, and indeed of most of our rivers except the Dee, which is too shallow, and full of small stones, would answer very well for raising parsnips.

A very good ardent spirit may be extracted from parsnips. This the Writer of this Report proved in 1802, when he distilled fifteen different kinds of spirits, from various combinations of roots. In point of flavour, the spirit of parsnip is inferior to that from carrot, when distilled separately.—But it may be of essential use if combined with other roots in the British distillery.

12. SALSAFY.

This root, (the tragopogon perrifolium of Linneus) is generally raised only in the garden. But the Writer of this Re-

Report, on learning that the cows, who are very good judges of the comparative value of roots, preferred this plant, when thrown over the garden wall, to every other vegetable, raised a few drills on the field, and afterwards fermented and distilled the produce. It yielded an excellent flavoured spirit, which was inferior only to that extracted from carrot. It may certainly deserve more attention, if the Legislature should revise the Excise Laws, and attempt, by encouraging the distillation of fine flavoured spirits, to beat the French Brandy and Dutch Geneva out of the market of Britain.

13. MANGEL WORZEL, or ROOT OF SCARCITY.

This root, which had been so extravagantly extolled by Dr. LETZOM, was, after repeated trials, found not to be deserving of its high character, although every sort of justice was done to its cultivation. By EARL FIFE's experiments, above detailed, it was inferior in point of produce per acre, to both carrots and field turnips, as well as turnip cabbage.— It is not now raised in any quantity.

14. BEET, or BEETRAVE, BETA.

These roots are chiefly raised in the garden, and used for making pickles. One species of them, viz. mangel worzel, has just been noticed. Another of them, the sugar beet, or beta sica, was attempted by Mr. ACHARD, and others in the north of Europe, as a rival to the sugar cane of the West Indies. On that account the Reporter fermented and distilled the red, white, and green beets, as he could not get any of the beta sica. They were not deficient in quantity of spirits, and the quality was the third in order of fineness of

those

those that produced any considerable proportion of alcohol, i. e. they were next to salsafy, carrots being the first.

15. SCORZONERA.

This root is much used on the continent, as a dish at table; and when the Reporter distilled it, he found it produced a most agreeable flavoured spirit, and well adapted for making a *liqueur*. But its root is small, and the quantity raised on a given measure of ground, is too little to render it an object to the agriculturist, though it might have its value in making *liqeurs for the ladies*.

16. MUSTARD.

This seed is chiefly sown by the gardeners, and sold by them to the apothecaries, who are contented with 700 per cent. of profit. But an apothecary must charge for his time in weighing out small parcels. Two kinds of it were sown formerly, viz. the white, which was used by those in opulent circumstances, and the red kind, which was ground in a wooden bowl, with an old four pound shot, and then mixed with skate or ling, at country feasts. But now patent mustard is to be found at the farmer's table; and the white kind is now chiefly raised. In the neighbourhood of bee hives it is very useful, when sown at different seasons, for the industrious bees.

17. RHUBARB.

Of this root we have two kinds, but only small quantities of either are raised. A few plants of it are seen in the garden of a great landed proprietor, and sometimes a single plant in that of a farmer or cottager. It is however very deserving

serving of attention on two accounts. Its root is well known to be a gentle purgative medicine, and the value of its leaves, towards the end of summer is also considerable, for they are found to be of considerable service in baking a pie, or tart. It often happens, and to persons who reside in the country it is vexatious when it does happen, that our gooseberries are destroyed by the caterpillars. But neither the gooseberry nor the cabbage caterpillar molest the rhubarb; and the cook considering *kitchen physic* as superior to medicine, takes precedence of the apothecary, making excellent tarts from the leaves of rhubarb, and allowing him afterwards to dig up the root. Both the stalk and the flower of this plant are beautiful, and it grows so rapidly, that its progress can be marked every day. Why should we kill the beaver, or use strong mineral cathartics? Rhubarb can be easily raised, and no doubt it will be cultivated more generally.

18. HOPS.

Of this plant we have still less than of rhubarb. Twenty years ago a few poles of hops were to be seen in many gardens; but now they are for the most part neglected, or removed altogether. If the extremely high and ill-proportioned tax on malt, that is made from the bear raised in this county, had not checked the cultivation of that species of barley, the farmer in Aberdeenshire would cheerfully purchase his hops from England, which is better adapted to the raising them of a good quality.

No other seeds, or roots, or other vegetables, are raised in such quantity in this county, as to merit to be particularly mentioned in this Report.

CHAPTER VIII.

GRASS LANDS.

SECT. I.—MEADOWS—AND NATURAL GRASS.

In the county of Aberdeen there is a much less proportion of what may be properly termed meadow land, than is found in any of the neighbouring counties. The surface is much diversified, containing many sloping fields, but an inconsiderable proportion of flat grass land or meadow. In the higher parts of the county, among the mountains of Marr, there are several narrow glens, which have never been cultivated, being remote from human habitations, where the natural pasture is very good. But in general, the mountains press on each other; and the soil toward their base is either peat moss or marshy; and situated in too remote a district to repay the expence of forming it into a regular meadow. Along the banks of the Dee, and the streamlets near its source, the meadows are for the most part narrow strips of unfertile land. From that circumstance, and from the overflowings of them being often destructive, but never fertilizing, some persons have been led to suppose, that there was something in the nature of the water of the Dee, that was
un-

unfavourable to vegetation. There is however nothing hurtful in this; but from the nature both of the soil and subsoil, which are chiefly composed of granite more or less decomposed, and from the great rapidity of its course, which is confined by mountains on each side, the waters of the Dee are extremely pure; and cannot deposit so much sediment as is left by those streams which ooze through peatmoss; and which when swelled by rains, carry off the richer particles of the soil in their turbid streams, which they deposit in another place where they move with less rapidity, or with a reflux occasioned by any obstruction to their current.— Hence the meadows of the Dee are comparatively unfertile. The Don, on the other hand, (as its course is comparatively slower, and as this river in many places recedes to a considerable distance from the hills,) contains a considerable proportion of fertile meadows near its banks. These are provincially termed haughs; and the greater part of them has been cultivated. In several cases they would have been more valuable, if they had been embanked; or if those parts of them which lay nearest to the rivers, had been formed into sloping banks, and laid out in perpetual grass, while the rest was properly inclosed and cultivated. The Ury, the Ythan, and the two Ugies, with the bounding rivers, the Bogie and the Deveron, have each a considerable portion of meadow, or flat grass land lying along their banks. The various inferior rivulets, and small streams, provincially named burns, have also small patches of land, of very unequal figure and extent, according to the size or rapidity of their currents, and the nature and situation of the adjoining grounds.

In general, only a small proportion of these vallies, or flat lands, is left untilled by the plough; but yet there are many patches of grass, along the sides of the different streams,

streams, from a narrow bank of only three or four yards broad, to a small meadow extending from thirty to forty, and sometimes an hundred yards. The grass is occasionally overflowed by the swelling of the stream; and the farmer is afraid to break up this land, lest not only the crop should be injured, but also the soil carried off by the rain in summer and autumn, or by the sudden thaws, or more destructive masses of ice in winter and spring. From his fears in this respect arises a proportion of our natural meadows.

Yet it is not owing to this timid caution, that several patches of meadow land are left untouched by the plough.— Many of the small farmers, who depend on their little dairies for payment of their rent, are very careful in preserving their meadows. The old grass that grows on these narrow strips, or small patches of land, if not kept too wet, is, by the frequent overflowing of the water, incomparably more nourishing than the sown grasses, of only one or two years old, which they raise artificially. This nourishing quality is partly occasioned by *age*, for old grass in this respect is superior to young; partly by its situation in the neighbourhood of the stream, which keeps it always moist, while other places are frequently parched by drought in the summer months, and partly by its overflowing, as already mentioned; and especially by the reflux of the stream, (provincially called back water,) when a considerable sediment is left, that enriches the grass. In consequence of all these causes, black cattle are found to thrive extremely well, when fed on these natural meadows; and the milk of cows in particular is richer, fuller of cream, and contains a greater proportion of curd, when made into cheese, than the more copious, but also more watery milk, which is yielded by a cow, when fed on broad clover only, or in pasture ground that is only one or two years old. The superior value of old meadow grass deserves the attention of both the proprietor and of the farmer; and should induce them—not to attempt to force meadows, where the soil and situation of the ground are unfavourable, but to study to have a proportion of *old grass*, where they cannot have a water meadow; and where they can have this, to make out one with great care, and not to be tempted for the sake of present emolument to break up land that has been left out as a meadow; the preserving of which will be attended with great and permanent advantage.

The quantity of meadow land however is inconsiderable. While the county of Aberdeen contains 400,000 acres of heath, either on its bounding mountains, or scattered through its different divisions; and while it has not less than, 40,000 acres of benty moor, provincially termed riesk, and producing only the coarsest kinds of grass, it does not probably contain 4000 acres of good and well laid out meadow. Most of the landed proprietors are so far sensible of this defect, that they have inserted clauses in their tenant's leases prohibiting the breaking up of green-sward, or meadow; but few of them have done much to get part of their farms laid out in old grass, or formed into regular meadows, which are not to be broken up for the future. Mr. FERGUSON of Pitfour, as has already been observed, has established as a rule on his extensive property, that his tenants shall be obliged, if required, to lay out the extent of one-seventh part of their arable lands in meadow or old grass. Not that this proportion will be uniformly demanded, but it must be so laid out, if required. In fact, there are many farms which have only a small portion of their land adapted to this purpose; and there are others, on which the land on the banks of the rivulets, if left to nature alone, and still more, if properly dressed and laid out as meadows, would prove of the utmost im-

importance in supporting the arable lands, with a great proportion of excellent manure from the cattle fed upon them; and be of inestimable value to the dairy of Aberdeenshire.— Without attempting to force meadows, it will not be advancing too much to say, that ten times as much meadow land as we have at present, might be laid out for this purpose; and would both add much to the food for black cattle, and by increasing the quantity of putrescent matter, would tend greatly to enrich our arable land. At the same time it may be remarked, that wherever the *finer grasses* disappear, and top-dressing, from peat-moss, shelly sand, and lime, have not the effect of *renewing them*, the meadows should be broken up; and either laid down again after a proper dressing, or cultivated along with the arable land. The dung of animals should always be ploughed in, never left on the surface; and when one field of old grass fails, another should always be left in its room.

Meadow grass, in general, is not made into hay by the farmers, and it is not let by itself, but for the most part along with arable land.

The above remarks regard meadow grass in general.— When we come to the subject or irrigation, the value of water meadows will be particularly considered. At present we have been treating

> Of the bonny green sward knows,
> Where the bee bums, and the burnie rows.

Of the other kinds of natural pasture, viz. heath and moor, there is by far too much still remaining, though several thousand acres of these have been broken up, and more been improved within the last forty years. In general, the effect of lime, in the lower parts of the county, has been to extirpate heath: And wherever the plough has been introduced,

duced, and the land first fallowed for a year, and then limed, it has always been found to produce good pasture, not only when grass seeds were sown, but even when left out, after bearing injudiciously several crops of oats. Wherever calcareous matter is applied, heath no longer appears; but the poas, and other grasses come up, and even the natural pastures become of much superior value.

There are some small patches, not in the vallies, but on the declivities, or sides of hills, which the cottagers convert into excellent grass, by covering them with water. A spring well, instead of being allowed to flow whither it pleases, is confined by a small dyke of turf, provincially called a dam, built across its current. It overflows this, and at eight or ten yards distance, another dam is formed; and a second small piece of ground is overflowed. These small patches, formed by their industry, produce a great weight of valuable grass, which the cottagers (taking off their shoes and stockings,) cut down with the sickle, and carry off to their cows, twice or thrice a day during the summer months. The expence of these dams may be, at first, two or three days work in forming, and sometimes a few hours in repairing them.—But in the driest seasons, they afford their own cows abundance of good grass, if there be only a spring well of such power as to cover the surface; and indeed a small fountain is generally sufficient for this purpose.

On the sides of the mountains, the Reporter observed with pleasure, that wherever a small stream could be directed properly, and not allowed to fall too rapidly, the heath was effectually destroyed, and fine pasture was obtained by a little labour properly directed.

But towards the coast of Buchan, though there are some fine downs near Fraserburgh, there is a considerable extent of long and coarse bent, interspersed with sand, which cannot

not be rendered valuable pasture, nor yet be broken up with the plough, without exposing the lands in the neighbourhood to be overflowed by the German Ocean. The greater part of the parish of Forvie, near the mouth of the Ythan, was blown over with sand, and affords only a scanty pasture.

Besides natural meadows, or uninclosed pasture lands of different descriptions, several of the landed proprietors have inclosed extensive farms, and have let them as pasture lands, sometimes without improving them farther, than by an inclosure for keeping the cattle from wandering away. These inclosures, though they do not let nearly at so high a rent, as when the land has been improved, and laid out in good order, still pay very well—seldom under thirty shillings per acre, often above two pounds.

In the neighbourhood of Aberdeen, and in a few other districts occasionally, horses and black cattle are kept in inclosed fields of grass, at a certain sum per night, i. e. for 24 hours. The rate is different, according to the nature and situation of the pasture. Twenty years ago, a horse paid fourpence a-day; now he pays eighteenpence, where the grass is good.

In different places and districts, the farmers graze young cattle to the cottagers and artificers, for the season. The rate varies with the quality of the pasture. In the highest part of the county, Mr. CHARLES MACHARDY, who rents the extensive farm on Glen Geaullie, charges only twenty shillings for a horse, and six shillings for each head of black cattle, for the season, i. e. about five months. In the lower districts black cattle only are taken in to grass for the summer months at different prices, from five to twenty-five shillings, according to the age of the animal, and quality of the pasture.— This mode is particularly convenient for those who have provender in winter; but need grass during the summer. And

by

by this means they are enabled to raise a few young cattle, without being obliged to tend them during the busy season in which they act as day-labourers.

The mode of using the meadows and natural grass has been much improved within these forty years. Formerly the long, or taller grasses were cut down with the sickle, (the scythe at that time being unknown to many of our farmers, and used only by a few of them.) And the horses and cows, after collecting for themselves what they could gather through the day, were fed at night with these natural grasses, and with the thistles that were picked out of the corn fields. The best grass on the richer outfields was depastured by the horses, cows, and labouring oxen—the inferior natural grass on the outfields and faughs, was given to the farmer's young cattle, and the cottagers cows; while that on the barren hills, and benty riesks, as they were called, was given to the sheep. By this mode, none of the animals were well fed; and every farmer required *three*, and many of the farmers employed *four keepers* to attend these different charges. Now the meadows and taller grasses are cut down by the scythe. The horses and larger stock are fed principally with sown grass. The young cattle, during a part of the day, are sent to pasture on the natural grass, in the more barren places of the farm; but are always allowed better food during an hour or two every day.

SECT. II.—THE ARTIFICIAL GRASSES.

These demand the greatest attention in a county in which the rearing of black cattle is of so great importance. But the subject of the comparative value of different artificial grasses, is, as yet, not so well understood as could be wished, as we have hitherto been contented with raising only a few kinds of them.

The principal artificial grasses now in use are, red, white, and yellow clovers, plantain or rib grass, and perennial rye grass. A few pounds of lucern, and a few bushels of sainfoin, have occasionally been sown, in the way of experiment, by some of the principal landed proprietors, but have neither been long continued in one place, or generally distributed over the county, nor do they seem adapted to the soil or climate of Aberdeenshire. There are a few places, which, owing to their being sheltered by hills or plantations, or by hanging on sloping banks which ly to the south, have eight or ten degrees of more local heat, than is possessed by their neighbours. In these, when the soil is good, and more especially if the subsoil be pervious to the roots of lucern, or sainfoin, by the attention of the cultivator, either of these valuable grasses may be raised with success.— But though lucern grows to great advantage among the rocks in Switzerland, (where its roots find a passage through the fissures, or loose soil,) the granite, or the hard tilly subsoil of Aberdeenshire, which can be pierced only by the spade and mattock, are impervious to the long tap roots of lucern.— In this county also we have no Cotswold downs, or soil incumbent on a bed of chalk, in which sainfoin can be raised to advantage, if sown in our fields. What may be done by cutting the tap roots of these grasses, and transplanting them from the seed bed, it may be too hazardous to anticipate, till

proper

proper experiments are fairly made on a variety of soils.—But it should be the first object of the farmer and grazier, to be at great pains to cultivate these grasses, which we know, by experience, to succeed in our soil and climate; and his second inquiry should be, whether among the native grasses, which we find already in our fields, some are deserving of being selected, and sown either alternately with those now in use, or substituted in the room of any of them. For it is suspected, that by too frequent repetition, the soil becomes tired of raising the same vegetable, whether corn, root, or grass.

After these remarks, the artificial grasses, which are now generally used, may be separately noticed.

1. CLOVER.

Of this we have three kinds, the red, white, and yellow. The red is properly a biennial plant; and the white and yellow are perennial. None of these are raised for the purpose of preserving their seed; although in an early season this could be done with success. And none of them are sown separately, but they are either mixed together, or with a quantity of rye grass, or plaintain, before they are sown.

Red clover is sometimes sown for the purpose of soiling; and in this case is mixed with only a small proportion of rye grass. It is most generally sown for hay, with a mixture both of rye grass, and white clover. In this way we sow out all our fields that have been employed in raising turnips, potatoes, beans, or cabbage, the preceding year. The land is always in good order.

It was formerly the practice to sow out grass seeds with a crop of bear or bigg; but from the high proportion of malt tax, on that species of barley, and the great demand for oats, at at least one half of our sown grasses are now laid out with that kind of corn, viz. either potatoe or red oats, which have been both imported into this county. This no doubt is not so beneficial to the young plants of clover, as when they were sown out with bear or bigg.

Clover and other grass seeds are sown out occasionally after a complete fallow; though from the nature of our soil, there is occasion for that kind of fallow only in strong clays.—When wheat is sown in such land, clover and rye grass seeds are harrowed in, some time in spring, as soon as the ground is dry, and the weather favourable.

The land is usually manured, for the turnips, or dunged on the bare fallow; but sometimes it gets a top-dressing when the clover and grass seeds are sown; and at other times it receives in autumn a top-dressing of peat ashes, shelly-sand, or compost manure. (When shelly sand has been ultimately mixed with peat moss, it is well adapted to this purpose.)

The quantity of clover sown on an acre depends on several circumstances. When it is intended for soiling cattle, from 10 to 14 pounds of clover, with half a bushel of rye grass, are commonly allowed to the Scots acre. If it be meant to be cut for hay, for one year, and then to remain some years in pasture, *eight* pounds of red, and *four* pounds of white clover, with at least a bushel of rye grass, (some indeed use above two bushels) are generally sown on an acre by the better sort of farmers, along with bear or oats, or in the spring after wheat. Some add to this mixture three or four pounds of rib grass, or plantain. Where there is too little red clover, there is a greater weight of coarse hay.—When too great a proportion of clover is sown, the hay is finer in quality, but the crop is lighter, and is supposed not to be so serviceable for horses that are severely wrought.

Where the land is thought too weak to bear red clover,

and

and intended only for pasture, either no broad clover is sown, or four or five pounds of it, with an equal weight of rib-grass, and double the quantity of white clover is sown, along with two bushels of rye grass.

Red and white clover are either first sown by themselves, and then the rye grass sown separately on the same land; or the whole seeds are mixed in certain proportions, previously determined, and then all sown at once. When the weather is perfectly calm, either of these methods may answer. But if there be a gentle breeze of wind, it is better to sow them separately. When it blows pretty hard, they must either be sown in this way, or deferred till the weather become moderate. The most correct practice among our farmers is to sow them separately; and, except the air be perfectly calm, either very early in the morning, or late in the evening, when there is the greatest chance of their being evenly spread on the land.

As clover seed is sometimes very dear, the most economical practice of sowing it, is to harrow, and slightly roll the land after the corn is sown; as by this means the surface is rendered perfectly or nearly flat. Grass seeds are then sown; and the ground gets a second harrowing with a close-toothed harrow, of the improved construction, (by which every teeth has a particular rut;) and then the whole is rolled a second time. As by this mode the grasses are all near the surface, a field may be as well sown with seven or eight pounds of clover to the acre, as it is in the common way, when it gets double that quantity.

The produce of hay *from red clover*, with a small proportion of rye grass, seldom exceeds two tuns per Scotch acre.— But if a proper quantity of rye grass be sown, it will sometimes amount to three tuns, and sometimes, as will be afterwards mentioned, it will even exceed that quantity. *Yellow* clover is now seldom used.

2. RAY,

THE ARTIFICIAL GRASSES.

2. RAY,

(Or provincially Rye Grass.)

This is used in *very considerable quantities* by all our farmers, except when it is mixed with red clover for soiling, in which case, only a small proportion of ray grass is allowed. When intended for hay, it is seldom sown so thin as it ought to be. At least two bushels to the Scots acre, and frequently two bushels and a half to that measure (i. e. two bushels to the English acre) are allowed in sowing out ray grass.—There can be no doubt that red clover is choked, or at least injured, when more than one bushel of ray grass is sown on an acre. And it cannot be denied, that when ray grass is allowed to perfect its seed, or even to be nearly ripe, it exhausts the ground very much. The landholders, instead of stipulating with their tenants, in their leases, that they should always sow a certain proportion of ray-grass with their clover, should stipulate, that they shall never exceed a bushel and a half of ray grass to the Scots acre. It seems highly probable, that when other grasses are better known, a less proportion of ray grass will be used by our farmers.

It was already observed, that the land tired of a too frequent repetition of the same crop; and it is generally believed that turnip and red clover do not thrive so well in Norfolk as they did sixty years ago. It is also true, that even in this county our crops of sown grass, in some places, are not so weighty as they were, when we laid out a small proportion of our arable land with clover and grass seeds. But whatever be the cause of the deficiency in Norfolk, the *greater quantity of manure laid on an acre* by our first improvers, will account for the greater weight of hay which was occasionally obtained thirty years ago in Aberdeenshire. That this weight
was

was very great on some of *our old croft land*, is fully established by the following passage in Dr. ANDERSON's Original Report.

"Those who have never seen it can form no idea of the
" possible luxuriance of a crop of rye grass in this county:
" and will not perhaps be able to give credit to what I am
" going to state.

"A man who farmed a small patch of rich land in the
" neighbourhood of Aberdeen, boasted in my hearing, that
" he had obtained at one cutting, 500 stones of hay from an
" acre of ground. (The stone is sixteen pounds, each pound
" containing twenty-one ounces averdupois.) I did not give
" credit to the account, because it so far exceeded any crop
" I had ever seen produced in the neighbourhood of Edin-
" burgh. Next year I had a field in grass, of the soil above
" described, (i. e. old infield or croft land) which had been
" laid down with the bear crop after turnips, the ground
" having been made very clean, and completely limed and
" dunged to the turnips. The rye grass on this field grew
" more luxuriantly than any I had ever seen before; and as
" the season happened to be very dry, I was enabled to al-
" low it to stand till the stalk had obtained a tolerable de-
" gree of firmness, before it was cut for hay. I thought it
" the greatest crop I had ever seen obtained at one cutting;
" and indeed, considering the circumstances, believed it to
" be a *maximum*. To satisfy myself of the weight, I mea-
" sured off, with great exactness, one quarter of an acre of
" the best of it. The weather continued very dry till the
" hay was thoroughly made. When it was completely made,
" and fit to be put into any stack, I had it exactly weighed,
" and found the crop amounted to 640 stones per acre. The
" rye grass in this part of the field measured in general about
" three feet and a half in length, and was so strong in the
" stalk

2. RAY,

(Or provincially Rye Grass.)

This is used in *very considerable quantities* by all our farmers, except when it is mixed with red clover for soiling, in which case, only a small proportion of ray grass is allowed. When intended for hay, it is seldom sown so thin as it ought to be. At least two bushels to the Scots acre, and frequently two bushels and a half to that measure (i. e. two bushels to the English acre) are allowed in sowing out ray grass.— There can be no doubt that red clover is choked, or at least injured, when more than one bushel of ray grass is sown on an acre. And it cannot be denied, that when ray grass is allowed to perfect its seed, or even to be nearly ripe, it exhausts the ground very much. The landholders, instead of stipulating with their tenants, in their leases, that they should always sow a certain proportion of ray-grass with their clover, should stipulate, that they shall never exceed a bushel and a half of ray grass to the Scots acre. It seems highly probable, that when other grasses are better known, a less proportion of ray grass will be used by our farmers.

It was already observed, that the land tired of a too frequent repetition of the same crop; and it is generally believed that turnip and red clover do not thrive so well in Norfolk as they did sixty years ago. It is also true, that even in this county our crops of sown grass, in some places, are not so weighty as they were when we laid out a small proportion of our arable land with clover and grass seeds. But whatever be the cause of the deficiency in Norfolk, the *greater quantity of manure laid on an acre* by our first improvers, will account for the greater weight of hay which was occasionally obtained thirty years ago in Aberdeenshire. That this weight was grass, equal to those three kinds of crop in any part of Great Britain.

It is not to be denied, however, that ray-grass, when raised in such quantity, as above mentioned, and allowed to stand till it be firm in the stalk, is a scourging crop, and tends very much to exhaust the soil. And we may presume, that there are other grasses, besides clover and ray grass, (particularly the latter) which would be found more valuable. A wide field is here open for experiments and useful investigation; and a very ingenious and well-informed gentleman of this county, whom the Writer of this Report has long known and esteemed, but whose name he is not at liberty to mention, has turned his particular attention to this subject, and favoured him with a communication, which will be found in the Appendix. In this place it is necessary to add, on the article of Ray Grass, and Clover, a few passages of Dr. ANDERSON's Original Report, which ought to be preserved.

"The seeds of rye grass are saved here with great care;
" and the annual kind so much complained of in other parts
" of the country has never found its way hither. No-
" where does rye-grass prosper better, than on the old in-
" town lands of this county, when properly cleaned and
" limed. It affords very weighty crops of hay; and upon
" rich soils is one of the best pasture grasses yet known. It
" rushes up early in the spring; produces a vast profusion
" of blade; is sweet, and so much liked by beasts of all
" kinds, that if it be eaten down early enough, it never
" shews its flower stems, but continues through the whole
" season to afford a deep bite of the most succulent herbage.

" But upon *poor soils* it is perhaps one of the worst grasses
" yet known. Its leaves there are not more abundant than
" those of the dog's tail grass; and so dry and rigid, that
" cattle

"cattle are not fond of it. Its stalks spire forth very early,
"and being unmixed with leaves, they are tough as wires,
"so as to be disrelished by all beasts; and are all allowed
"to get into seed, when they become brown and sapless,
"and good for nothing. On poor fields no practice can be
"so bad as that of sowing rye grass. It extirpates all other
"grasses; and this is worse than any of them.

"Broad clover is sometimes a very good crop; but it is
"only on the stiffest soils, when abundantly manured.—
"White clover is a surer and thriftier crop, on most of the
"limed soils of this county. A mixture of the two kinds
"does best. The broad clover succeeds the first year; the
"white continues, and thickens the sward, especially when
"it is intended for pasture.

"No person should ever sow broad clover with an inten-
"tion to be cut green, without sprinkling along with it a few
"seeds of rye grass. It keeps down the great white gowans,
"(chrysanthemum leucanthemum) which frequently, when
"this is neglected, spring up among the clover. It thickens
"the crop, and brings the first cutting at least a fortnight
"earlier than it would otherwise have been, and makes it
"generally more abundant. On the second and third cuttings
"it has less effect, but always does good. The crop, in con-
"sequence of this addition, will, in general, be bettered
"nearly one third: For it is to be observed, that if broad
"clover alone be sown very thick, the plants stint each
"other in their growth. If thin, the first cutting is scanty,
"and full of weeds. Half a peck of rye grass seeds to the
"acre will effectually cure both of these defects."

To conclude this long account of clover and rye grass, the Reporter would briefly mention, that he has distilled spirits both from clover hay, and the seeds of rye grass.—

If the British Distillery should embrace the distillation of roots, and other vegetables, the seeds of rye grass will be found to yield a considerable quantity of spirits, with a peculiar flavour, which, mixed with others, might be deemed very palatable. That from *clover hay,* being an extract chiefly from *the stalks,* is less abundant in quantity; and its flavour is inferior to that from several roots.

SECT. III.—PASTURING FROM ARTIFICIAL GRASSES.

As the rearing of black cattle is a principal object with the farmers in this county, the sowing of clover, rye-grass, and other artificial grasses, has added greatly to their food in summer. Formerly, on all the east coast of Scotland, the chief dependence of the farmer (who was also a grazier in a greater or a less degree) was on the regular falling of the rains. A cold and a dry spring, that prevented the rising of the natural grasses, damped his spirits; and both prevented the feeding, and injured the sale of his cattle. Now, though he must ever be dependent on Providence, he is less affected by the seasons. The artificial grasses now supply him much earlier with pasture, which continues to nourish his cattle through the summer; and the after-math, or second crop, as it is commonly called, of the land which he cuts for hay, carries him forward till the end of autumn.

Since the introduction of pasturing from the sown grasses, many of the landed proprietors, instead of continuing the arable lands of their personal farms in raising of white crops, have first improved them, then laid out the different fields in succession with clover and ray grass, and sometimes after taking a crop of hay, at other times without cutting their

their grass at all, have let them from year to year. At first, their own tenants, who had not at that time improved their farms, were the principal persons who rented, or held at public auctions, the inclosures thus let by the landed proprietors. Now the farmers, those excepted who are dealers in cattle, generally raise a sufficiency of sown grass, both for hay and pasture; and the butchers and cattle-dealers are the principal merchants for these inclosures, which are rented at various prices, according to the state of the fields, and the demand for pasture in different seasons; but seldom under two pounds, and frequently above five pounds per Scotch acre. Mr. FERGUSON of Pitfour, who has the largest personal farm of any proprietor in the county, (extending to above three thousand acres) last year let some of his *grass parks*, as they are called, as high as six pounds per acre. Mr. ROBERT WALKER, in Wester Fintray, also let a field to a butcher in Aberdeen, for two years, for the same price, viz. at eighty-one pounds for thirteen acres and a half. This was drawing nearly the half of the rent of his farm for the grass rent of a single field. A judicious farmer can best decide for himself, when he should let his grass for a high rent, and when he should purchase cattle, where his own stock is unable to eat all his pasture grass. But letting his personal farm, (or inclosed lands) in perpetual pasture, when laid down in sown grasses, is perhaps the most profitable, generally the most prudent way, in which a landed proprietor can dispose of them, or recover the money which he expended on their improvement. He seldom succeeds so well as a corn-farmer, for he cannot, for the most part, rise early in the morning, set his servants to work, and oversee them constantly. For all these things he must be dependent on his bailiff, or farm-overseer. Nor is *he* often a good judge of black cattle; but would be frequently overcharged if he bought and sold his cattle, trusting to his own skill. And in this respect, though he be an excellent conductor of his farming operations, he may also be deficient. But when his grass inclosures are let by public auction, he has nothing to do, but to get good security for the offered rent, and to demand and receive it when due.

When land is sown out in clover and grass seeds, and not cut for hay, it is well worth six pounds per acre, for night soiling or pasture, if it be laid out in good order. In this case, it may be worth five pounds for the second, and four pounds for the third year's grass. But if cut for hay the first year, these rates will be rather high in the interior parts of the county. In the vicinity of Aberdeen, grass is seldom let the first year under fourteen or fifteen pounds; nor for the the second under ten pounds per acre. But in this case, the person who lets it, has expended generally ten pounds for dung on every acre, at the time that it was sown with turnips. This reduces his profits on the subsequent years.

The weight of grass in the neighbourhood of Aberdeen when used for soiling, and when the land has been well dunged, is seldom under fourteen or fifteen tuns of green clover, in the different cuttings. And if a cow-feeder obtain an hundred weight of good clover for a shilling, which is twenty shillings per tun, he has no reason to complain. In the country, where the *same quantity of dung* cannot be had, and where the *manure* from the compost dunghill, however well *fermented* and *prepared*, is not so *strong or powerful* as the *night soil* of the city, pasture from sown grasses cannot be so valuable.

SECT. IV.—HAY HARVEST.

While the introduction of artificial grasses rendered the pasture more nourishing during the summer season, the laying in a quantity of hay for winter provender to horses, (especially after the busy season commences, when the ploughs go two journeys, or in the dialect of this country, go two yokings daily,) for food to fat cattle, whose flesh requires to be firmed, before they are finished off, and for cows who have newly dropped their calves, occasion the proper mode of harvesting hay a matter of great consequence to all these different animals.

This was formerly conducted in a very tedious manner, and with considerable danger of injuring the crop. It is now managed more expeditiously, at much less risk, and with far less labour and expence.

Till about twenty years ago, the hay, even when made from natural grass, was made up into small *cocks*, each of which would have contained about *seven* pounds Dutch weight, or the *third part of a stone* of hay. And these cocks were put up either in the afternoon of the day in which that part of the field was cut down, or in the morning of the succeeding day. After standing two days in this form, three of these cocks were mixed together, and formed into a small cole, which would contain nearly 22 pounds averdupois, or 20 pounds Dutch, being the provincial weight for hay used in this county. At the end of two or three days more, these were put into larger coles (by the hand, but not pressed) each of which would contain two or three stones, according as the state of the hay, and the appearance of the weather rendered prudent. At the end of four or five days more, these large *hand coles* were put into *tramp* coles, i. e. were pressed

pressed down by the weight of the person who built them, while another man forked the hay, and a third shaped the cole. Each of these tramp coles would contain from 50 to 60 stones, (at a medium, half a tun averdupois) or if the state of the hay, and appearance of the weather permitted it, into tramp coles of a tun weight, when they were considered out of danger, and where they remained for some weeks, till the hay rick, or stack, or as it is called the *hay sow*, was made up from the whole hay belonging to the farm. It is easy to see that this was both a tedious and an expensive process.— And what was still worse, if the rain fell in any quantity, while it was in any of these cocks, or small coles, they were generally completely drenched, were of necessity turned over repeatedly, and spread loose to admit the wind and influence of the sun, before they could be dried. This bleached and exhausted the hay, by frequently exposing it to the weather. If, on the other hand, the season continued dry and favourable, its juices were too much dried up by this tedious process, which exposed it so much to the influence of the sun, and of the dry atmosphere, as well as the blowing of the wind.

Now the system of hay-making is completely altered; and it is a just tribute to the memory of Dr. ANDERSON, to whose instructions the farmers in this county were much indebted, to mention, that the alteration of our system in this case, was in a great measure occasioned by his salutary advice and good example. Instead of turning over the cocks or coles, every other fair day, (as Dr. MOIR of Peterhead happily expresses it) to bask in the sun for several weeks, there are now many instances of large fields of hay being properly made, and put in the *tramp coles*, (when it is out of danger,) in the course of a week after beginning to mow the grass. The practice now most generally adopted is to allow the hay

hay to remain in the swath at least two days (and longer in hazy weather) till its surface, which has been exposed to the sun's rays, is sufficiently dry, and the moisture, of course, is considerably less. After this, by noon at farthest, and as much earlier as the dew of the preceding night has evaporated, the swath is turned over, exposed to the sun and air for a few hours, and made it up into coles in the afternoon. (Care is always taken to turn no more out of the swath, than what can be made into coles that day, before the dews begin to fall; because while in the swath, it can bear a good deal, but when once turned over, is easily soked, if rain should fall upon it in that state.) The coles contain from a stone to a stone and a half, or 28 lbs. English each, and the hay remains in these small coles for three days generally.— At the end of a week, at most, from the time of mowing, it is usually put into larger coles, of five or six stones, after being turned over for a few hours. If the weather is favourable this is done in five days; and after other two days it is put into the tramp coles, which vary in their size from 30 to 60 stones, as is judged most expedient. It is then out of danger, and remains there till it is stacked.

This is now the most general practice on the personal farms of the landed proprietors, and among the most considerable farmers in the county. It must be acknowledged, however, that there are several very sensible and judicious men, who still retain the old plan, of curing and harvesting their hay. They are unwilling, they say, to allow their grass, when cut, to be dried too much in the swath, or by being made too early into hay, to sweat, either in the tramp cole, or in the stack. Where a small farmer has a number of hands, and not much hay to harvest, he may still follow the old system. There is, however, one remarkable circumstance in regard to the making of hay, with which he is

un-

unacquainted, or to which he does not pay sufficient attention. It is, that *as men are affected by sympathy* for the situation of one another, so in a hay cole, or rick, the *moisture* of wet grass is attracted by that part of the swath, which was *too dry*, and the whole mass in a very short time becomes of one uniform or equal degree of dryness. Whether the practice introduced by Dr. ANDERSON be better or worse than the old and tedious method of hay-making, it is now generally adopted, except among the small farmers; and nine-tenths of the hay now made in this county, is cured in the expeditious method above-described.

It must be acknowledged at the same time, that there is an error in regard to the cutting of hay, which is too prevalent in Aberdeenshire. This is the delaying to cut it, till the grass seeds are formed; and even till the clover is sometimes peeping out, and shewing its seeds. There are many solid objections to this delay. 1. The hay is always harder, and of a worse quality, than if it had been cut down eight or ten days earlier, or as soon as the flowers of the ray grass are ready to fall off; and before the clover has begun to shew its seeds. 2. Sometimes the ray grass plants are so much injured, that they do not appear next season at all. 3. The land is severely scourged by this delay in mowing the grass, and loses more than all the difference of crop can recompense. In fact, ray grass, when allowed to ripen, exhausts the land as much as if it had carried a crop of corn. 4. There is by no means any considerable addition; often no addition at all to the *quantity* of hay, or first crop cut that season.— 5. The *aftermath*, or second crop of grass is *greatly retarded*, and sometimes is very inconsiderable, owing to this delay; and neither it, nor the following year's grass, are of nearly the value which they would have possessed, if the hay had been made early, and taken off the field only *ten or twelve days*

days earlier. 6. It is often good husbandry, when there is abundance of old straw, and when owing to a dry summer, the farmer dreads a scanty crop of *herbage*, (or *fodder*) for the cattle in winter, to cut the *aftermath* in the end of August or beginning of September; to mix this with a proportion of old straw, which tends, from its superabundant dryness, to make the aftermath soon ready to be put up into coles. By this means, when forage seems to be scarce, a considerable quantity of this mixture, of which the cattle are extremely fond, will be a great supply of food to the live stock, and raise a great quantity of putrescent matter for the manure of next year. But this is impossible, if the first crop is not cut down early, as the equinoctial rains set in before the aftermath, that has been mixed with straw, could be ready to be put up in either ricks or tramp coles. And it is to be observed, that if this mixture be exposed to much rain before it be out of danger, it is frequently rotted on the field, and is not only rendered useless, but injures the grass on which it was laid.

In every view in which it can be considered, nothing can be more improper, than, in a county that lies in the 58th degree of latitude, to allow any crop to be late, which can be early harvested. Many of the landed gentlemen, as well as of the principal farmers, are sensible of the impropriety of this delay, which they are careful to avoid: But the practice of delaying to cut down hay, till the grass seeds are half ripe, and then using these seeds for the next year's ray grass, is so common among the small farmers, that it was thought necessary to state the above objections. Where ray grass seeds are intended to be preserved, the hay should stand till the grass seeds are fully ready; when they may either be threshed out in the field, or what is better, bound up in sheaves, put up in ricks, and threshed out in the spring

spring. By this means *one* bushel of good and newly threshed rye grass will be worth *six* bushels of the shakings of hay, that has been too long on the ground to be good hay, and too short to make good rye grass.

It may also be proper to inform those who both pursue the old tedious method of hay-making, and also delay cutting their hay till the seeds are pretty ripe, that while the swath continues untouched, and the grass seeds are not nearly ripe, the hay will not be easily hurt by the dews, and will even endure a great deal of rain, as the greater part of this runs off in the sloping direction of the swath; but that after the swath is turned over, it both imbibes the rain, when it falls in great quantities, and the rye grass also rushes out when exposed to the weather.

It is not to be denied, that the green colour which so often appears in the English hay, is seldom to be seen in the hay stacks of Aberdeenshire; that our tramp coles, and hay ricks seldom sweat from fermentation, and are never burned, by the hay being stacked too early. But this advantage is more than counter-balanced by delaying to cut down our hay, till the ray grass has lost its green colour, and till from the ripening of its seeds, the stalk has assumed a whitish appearance. What has once become grey cannot be again made green: And both for improving the quality of the hay, and making this a more meliorating crop, than it is in many cases, the above strictures may be useful to some persons who may read this Report.

Nothing here is meant to censure the conduct of those prudent farmers, who in hazy weather, that seems to indicate the approach of rain, or when the rains have already fallen, delay cutting down the grass intended for hay. Such a situation not only excuses, but in fact requires them to cut with caution. Impatience in this case would be very injurious: It is only when there is no danger of rain, and when from a sordid economy the crop is injured, that the Writer of this Report finds it his duty to censure unnecessary and improper delays. And he can add with pleasure, that in general the hay harvest in this county is well conducted, and every year increasing, both in the quantity and quality of hay.

SECT. V.—FEEDING.

The county of Aberdeen exhibits all the extremes of good and bad natural pastures, of good and bad sown grass, of high and low rented lands, and consequently of good and bad feeding.

In the mountainous district of Marr, and in a dry season where the grass is stinted for want of rain, and the heath-flower has not begun to shew itself, the goat browsing among the rocks, may find a sufficiency of herbs; and the sheep thinly scattered over the surface, may collect daily such a quantity of grass, as suffices to support them. But the black cattle can pick up a very scanty herbage on the mountains (except by the side of a streamlet, where a green border is sometimes to be seen); and even in the glens or narrow vallies, the grass is very poor till the summer be pretty far advanced. Yet even here, when the heath-flower appears in the beginning of August, and the natural grasses have come up in tolerable quantities, the finest flavoured and the most juicy mutton, far exceeding what is raised by the artificial grasses, is fed by this apparently scanty herbage.— We must, however, be cautious of calculating the *produce in meat*

meat per acre; because several acres are necessary to the pasturing of a single sheep; and because there is a considerable difference in the fertility of different mountains. Lochnagar, for example, which is an immense collection of granite, has the *everan,* the best of the mountain berries, growing within a few yards of its summit, though nearly 3800 feet above the level of the sea; while Benaboard, or the *Table Hill,* exhibits a flat and nearly level surface for about two miles, with scarcely a single pile of grass, at only 140 feet of greater elevation. Not less range than *ten acres* on the highest mountains can be allowed for the pasture of a single sheep. But more than the half of their surface is altogether *barren;* and perhaps four or five acres is the highest allowance that can be fairly made for that part of the mountain on which the animal goes in quest of food. The horses and black cattle that are fed in the glens, or narrow vallies in this district, are small sized; and where the pasture rent of a horse is only twenty shillings, and that of each of the black cattle six shillings a-head, high feeding is not to be expected. Yet in the end of summer, the young highland cattle are generally in good order, and are eagerly bought as *winterings,* that is, to be reared during the winter, by the farmers in the lowlands, and put to good grass the ensuing summer. In these districts, in which the inhabitants depend entirely upon the natural pastures, as they raise no artificial grasses for their live stock, there is the greatest difference with respect to the profits on feeding cattle, in an excessively dry, a moderate, or a rainy season. But as the live-stock fed on this unproductive soil, is a mixture of various kinds of animals; as the extent of pasture is unknown, or undefined (for the tenants of several adjacent farms, and sometimes of different landed estates, have a common right of pasturage) it is obvious that no accurate calculation can be made

be-

between the number of animals, and the number of hundreds of acres, by the herbage of which they are nourished.

As we descend from these mountains to the highest arable lands, we find a number of small farmers, and small fields, which raise not only food for man, but herbage or forage, as winter provender for those animals that are partly fed in summer on the neighbouring hills, and partly sent away to be pastured on the distant mountains. As the extent of land, of such various qualities, and situated in different districts, cannot be connected with the number of horses, sheep, and black cattle—the only safe way of calculating the profits of a highland farmer, is by the *difference* between the *price* of these animals, when he *purchases*, or might sell them, in *spring*; and when he disposes of them in *the end of autumn*. That price indeed varies with the demand at the two periods. But the Writer of this Report, from the best information that he could obtain, would state the profits of a small sized sheep at three shillings, and that of a black-faced Linton sheep at six shillings, when the demand in autumn is considerable; or at half-a-crown for the former, and five shillings for the latter, in the general state of the market. In very brisk sales of the best sheep, the profits may occasionally amount to seven shillings and sixpence per sheep; but this happens very rarely. Those on black cattle, are, at an average, thrice as much as those on sheep; but the small size of the cattle in the mountainous districts prevents them from increasing in weight, though their meat is sometimes pretty fat, and always firm and juicy in the end of summer.

If we compare the number of cattle, in the interior districts, which are equally remote from the sea, and from the mountains, we will find that the *natural* grasses even of the arable lands, do not sustain, or feed, (as far as the summer grass can fatten them) above *one-third* of the number of animals

mals which are both supported and fed, on an equal measure of ground, that has been laid down *with sown grass,* after being properly cleaned. For even after the broad clover is gone, and the rye grass in a considerable degree decayed, the white clover, plantain, and the natural grasses which spring up, are far more valuable in land that has been well manured, and laid *down clean,* than on what was, according to the old system, first *exhausted* by white crops, then left out to *nature,* without either seed or manure; and for the first two years producing very little pasture grass, and indeed in the first year, little else except sorrel and other weeds. The profits of feeding cattle in summer are far greater in the middle, than in the higher districts, in which the turnip husbandry, and the laying down of land in good order with sown grass, is so little practised, as not to affect the general produce. But where there is a mixture, amounting to only a third or fourth part of artificial grass, and where along with this the cattle have a sufficient quantity of natural grass, that is three, four, or more years old, the short, but nourishing bite of this, when it is both old and good of its kind, with an hour's pasture twice a day, (provincially termed *baiting)* enables the farmers in Cromar, and other intermediate districts, both to support, and generally to feed very well during the summer months, a considerable number of cattle; and to pay pretty high rents for their farms, though situated from twenty to thirty miles from the sea-coast.

In the lower and maritime districts of the county, where lime is applied, and turnips, clover, and other sown grasses raised in considerable quantities, we can approximate pretty nearly in calculating the number of the different kinds of stock that can be maintained on a given measure of land.—But still we must specify the particular rotation of crops
which

which is adopted, or at least the proportion of turnips properly cultivated, and of sown grass used in pasture, as well as the quantity of hay, and both the number and size of the animals pastured or fed, before we can speak with any precision of the quantity of butcher meat raised on an acre.—And it must be added, that as Aberdeenshire in general, is rather a cattle-rearing, and grazing, than a *feeding* county, the same farmer who fattens four or five, commonly rears, or has on his farm, from *forty* to *fifty* black cattle. The pasture in summer, and the turnips in winter, are, in various proportions, used for both purposes. But taking a general estimate from the best farmers in the lower or maritime districts, a full-grown ox ought to pay from three to four guineas for his pasture in summer, and from five to six for his hay and turnips in winter. This, if the animal consumed the produce of two acres, and if his carcase has increased two hundred weight, with an improved *quality* of his meat, and quantity of tallow, equal to other two hundred in value, would give, along with the skin, offals, and manure obtained in feeding him, a pretty near average of the expence of fattening, and of the additional quantity, and of the improved quality of his meat; which added together would be equal to *two hundred weight* per acre.

In the neighbourhood of Aberdeen, the cow-feeders, who pay a very high rent per acre, require, owing to the excellent crops raised, the smallest measure of land. The common allowance for a cow, of about four cwt. is one-fourth of a Scotch acre, (or two hundred beds as they term it) of clover in summer, and an equal measure for winter provision.— To this is to be added, a portion of land in coleworts, and a quantity of brewer's grains, or of pollard, equal in price to another quarter of an acre. So that the produce of three quarters of a Scotch acre, or at most, a statute English acre, (for which they pay about fifteen pounds in money,) is very nearly the average measure and expence of keeping a cow.— This animal generally yields, when fattened, as much money, as when she was bought, a few days, or at most a few weeks, after dropping her calf. Here we see, that the keeping and feeding of a cow require two pounds more money, but only two-fifths of the measure of land, that the feeding an ox requires in the country; while she yields a considerable additional value of milk, on which the cow-feeder's profit depends, and from which the expence of tending the animal, and taking her milk to market, falls to be deducted. A more complete contrast can scarcely be imagined than between the extent of land consumed in pasturing or feeding stock in the highland districts, and that of pasturing and fattening milch cows in the vicinity of Aberdeen.

The usual calculation of one Scotch acre for pasturing a horse, and one for an ox or cow, is far from being accurate. It is a nearer guess (for it cannot be called any thing else,) to allow an acre of pasture to a three year old stot, two-thirds of an acre to one of two year's old, and half an acre to one of a year old. But where land is in good order, a less extent of surface is sufficient; and where it is in bad order, more will be found necessary. On the farm of Wester Fintray, it will appear, by looking back to the distribution of crops, and to the number of horses and black cattle, that 11 horses, and 85 black cattle, are maintained during the summer by 90 acres of pasture from sown grass, and by the aftermath, or second crop, of 27 acres, cut for hay. From this it appears, that eight acres of good pasture maintain seven black cattle and one horse, during the summer season. Perhaps the number of black cattle in this county is very nearly seven times as much as that of the horses; and the proportion on the above may be applied to those lands in the county which

which are in good order. But as this is by no means true of the whole arable lands, and as these form only a small proportion of the surface of Aberdeenshire, at least two acres of arable land (at an average of its quality and produce) will be necessary for this purpose. In the vicinity of Aberdeen, the cow-feeders, as already mentioned, in general require about one-fourth of a Scotch acre of their excellent grass for each cow, though some of them may use 300 beds, as they term it, or three-eighths of a Scotch acre. But they occasionally prefer a little of the brewer's grain, or of the miller's bran, to induce their cows, by a change of nourishment, to eat more food than what they would otherwise consume. It is a great object with them to add to the quantity, though they should not improve the richness or quality of their milk.

The extent of grass land required for feeding cattle in summer, depends in no small degree on the manner in which the animals are fed, or supplied with the grass. They are either fed in the house on grass cut for soil; or they are kept loose in the fields, and attended by a keeper, provincially a *herd*; or they are tied by a rope, and thus shifted from place to place, by what is called a tether. If they are fed in the house with grass cut for soil, half an acre of good clover, or of clover mixed with rye grass, will be sufficient for each of the horses and black cattle, at an average size of both. If allowed to go loose, attended by a keeper, nearly double that quantity will be requisite for every full-grown animal. And if tethered, or tied by a rope to a stake, or bound by a kind of iron chain, they will require an acre and a half, for this mode of tethering is by no means economical. If the stake, provincially termed a baikie, be not removed frequently, the cattle tread down a great proportion of the grass, and they do not eat it clean, or crop it evenly, but large tufts "the tufts of grass produced by one species of animals, are eat by those of another kind, and nothing is lost."

Though the method of tethering cows thus obtains the Doctor's warmest approbation, he seems to hesitate whether cutting the grass when green, and giving it to the cattle in the house, is not a still more economical practice. There can be no doubt that it is so; and that a less measure of grass cut by the scythe will serve a cow, than what she goes over by tethering. The Doctor, however, adds very properly in a note, as to the cow-feeders in Peterhead, that "in their situation the grass could not be cut, and be carried into the house to be there consumed green, as no horses nor servants could be kept for the purpose, and that this is one among a thousand of the instances, which prove that the economy of a farm must be in a great measure regulated by its size."

The reader will see from this, the various modes of feeding cattle with grass in summer, which are used in this county; and he will judge which of them has most recommendations. But it may be observed, that not only when broad clover is purposely sown for soiling, but also that when any particular field, laid out with clover and rye grass is very full of clover, and owing to its richness, or to the failure of the rye-grass, not so fit for being made into hay, the most prudent plan is to cut it green either to horses or black cattle, (as is commonly practised in the southern counties) and that where the grass is poor, and the field in bad order, it ought to be depastured with *black cattle,* mixed *occasionally,* at any rate, with *horses and sheep;* or rather by the two first; and after *they* have gone over the ground, to be eaten last of all by the sheep.

It is proper to add, that the large farmers and cattle dealers who rent inclosures for the season, shift their cattle from one

" shells, which gives it a tendency to produce white clover,
" pasture rye-grass, and the poa-grasses in greater perfec-
" tion than any where else; and the owners feeling the supe-
" rior benefit they derive from cows, above that which they
" could obtain from the plough, have wisely determined to
" let that land ly in grass for feeding cows. As no person
" there has more than two or three cows, they have adopted
" a very economical mode of applying the produce of these
" fields, to the feeding of milch cows, without the aid of in-
" closures, merely by tethering the cows in these fields, in a
" regular and systematic method; moving each tether for-
" ward in a straight line, not above one foot at a time, so as
" to prevent the cows from ever treading on the grass that is
" to be eaten: And as it is always fresh, and gives a deep
" bite to the cows, they feed upon it greedily, and eat it
" clean up without the smallest waste; care being always
" taken to move the tether forward at stated intervals, so as
" to advance regularly forward, like a person cutting clover
" with a scythe, from one end of the field to the other:—
" And by the time they have thus got to one end of the ridge,
" the grass on the other end of it is ready for being again eat
" in this way. I believe a greater number of cows can be
" kept on the same quantity of grass, than in any other way
" I have seen, unless perhaps where it is cut and given to
" them green in the house. One gentleman in particular, who
" would not perhaps wish to have his name publicly men-
" tioned here, has carried this system to great perfection,
" and derives a profit from his cows that is very uncommon.
" That no possible waste of grass may be incurred, he has a
" few sheep upon longer tethers, that follow the cows, to
" sweep up any refuse that may be left; and occasionally
" he makes his horses pasture upon the same fields, so that

‘ the

pear in many places. If they are tended by a keeper who knows his duty, and is careful both of his cattle and of the grass, he allows them only a small proportion of new pasture, and shifts them back on what has been gone over; thus whetting their appetite, and causing them to eat the grass very clean. But where the owner of the cattle can get them fed by *soiling*, that is by green clover cut with the scythe, there is no method that is so economical. The grass that is thus cut down seems to recover, and to grow up more quickly and more equally, than what has been eaten up by the cattle: And these animals consuming the grass in the house, are free from the excessive heat of summer, and from the stings of flies, which to them are very vexatious.— At the same time, both their dung and urine are of much greater value, when preserved in dunghills, and carried whither they are most wanted.

Each of these methods has its recommendation. The small farmer, who cannot afford the expence of a keeper to a few cattle, either soils them in the house, or tethers them, or does each of these alternately. The large farmer, for the most part, sets them loose to the field, under the care of a keeper. The only exceptions to this are, that his horses and cows in this county, receive a certain portion of their grass cut by the scythe, while they supply themselves with the rest from the fields of pasture grass. For soiling either horses, or cows altogether, is not practised in this county, except about Aberdeen, where grass rents so high.

It is proper to add, that the least economical of these three methods seems to meet the approbation of Dr. ANDERSON, as will appear from the following passage:—

"In the near vicinity of Peterhead are some of the finest "grass fields that are to be seen any where in this county.— "The land is a rich sandy loam, fully impregnated with "shells, one field to another, and sometimes indeed to a field at a considerable distance, both to allow the grass to grow up more quickly after being eaten, and to tempt the cattle from a change of place and food, to consume more grass, which occasions them to fatten more quickly. But though they thus shift them occasionally, they do not wish them to get too much exercise, but in general to feed at their ease.

One practice of the best farmers in this county, with respect to their grass fields, deserves general imitation. This is when the weather is cold in April, or in the beginning of May, and when the grass that was sown the former season, and is intended for hay, appears thin and brown in the top, (from the blowing winds, or coldness of the season) they send their cattle for a few days to crop this grass hastily, (not to eat it bare) and then remove them in a few days. The effect of this is to thicken the grass very much, and to make it shoot up with new verdure very quickly. By this means, the hay cutting may be a week longer delayed; but there will always be found a great addition to the quantity, and an improvement to the quality of the hay, which otherwise would have been thin, hard, stunted, and neither sweet, nor a weighty crop.

CHAPTER IX.

GARDENS AND ORCHARDS

SECT. I.—GARDENS.

Although horticulture is both a pleasant amusement, and a branch of agriculture, it is comparatively of less value than the labours of the ploughman, or the unceasing attention of the shepherd, and the feeder of black cattle. In some places of the South of Scotland, it is a common remark, That such a man is too good a farmer to be also a good gardener. This is saying, in other words, that he cannot attend both to great concerns and to small matters. Yet the best cultivated farm is that which approaches nearest to a garden; and the best farmer is a man, who, to a comprehensive mind, unites unwearied attention to every thing, from the cropping his arable land, the cleaning his turnips, and the feeding of his livestock, to the dressing of his garden. No doubt the spade, or some more humble instrument, preceded the plough; but after the latter was introduced, from its great expedition, and from employing less human labour, it soon became more general in the cultivation of the soil. Yet the most ancient occupation, that of a gardener, ought not to be despised; and
in

In an age of luxury, horticulture is an important object, with every man of taste, as well as of fortune.

The county of Aberdeen is distinguished beyond any other county in the island, for the preparatory branch of all good gardening, viz. *trenching the soil to a proper depth.* This however is a species of improvement that is not confined to land that is destined for gardens; and therefore will be treated of more fully under the head of Improvements. It is only necessary to mention in this place, that all garden grounds are trenched, when first set apart for this purpose; and are occasionally trenched thoroughly to the depth of 16 or 18 inches; or else they are half-trenched, provincially over-spaded, that is, narrow stitches, about 15 inches deep, and two feet wide, are laid upon an equal breadth of untilled land, and in that situation exposed to the winter's frost. When gardens have been incessantly cropped for several years, the trenching down of the surface, and bringing up new soil, is attended with the greatest advantage.

On the other hand it must be acknowledged, that this county is situated in so high a latitude, as not to be calculated for producing, without artificial heat, many articles which are raised on the continent of Europe, or even in the south of England; and that we have not so many hot walls, nor so many orchards as are found in the more cultivated counties of Scotland. But we have several hot houses belonging to different proprietors in the county, and about a dozen of these in the vicinity of Aberdeen; though it is unnecessary in this Report, to give any plans, or minute descriptions of those which are most remarkable.

The tax upon gardeners, who are employed by landed gentlemen, has deeply injured that profession, by lessening the number of apprentices, in consequence of limiting the employment of journeymen, for whom this tax is payable:
Yet

Yet we still have several gardeners scattered over the country; and we have a numerous class of gardeners in the vicinity of Aberdeen, who cultivate the lands in the neighbourhood of that city, and whose practice deserves to be generally known, and generally imitated.

They, in fact are kitchen gardeners, seedsmen, or nurserymen. They raise all sorts of roots for the inhabitants of the cities of New and Old Aberdeen, various seeds for the use of the county at large, and nurseries so extensive, and so carefully managed, that besides serving the landed proprietors in the county, and the owners of villas near the towns, they export considerable numbers of plants to England.

It was stated, as one of the preliminary observations, that by the intermixture, or alternate use of the spade and plough husbandry, the lands in the vicinity of Aberdeen, though of poor and thin soil, produced for the most part, very weighty crops, were much more highly rented than those in any other part of the island, excepting perhaps those near Penzance, in Cornwall, a small, but fertile district, where the rent is nearly equal to that around Aberdeen. The highest rents in this county are paid by gardeners, of different descriptions, chiefly by seedsmen and nurserymen. Without derogating from the opulent merchants and manufacturers, or the other affluent improvers of land near Aberdeen, the humble and industrious kitchen gardeners and seedsmen of that city, by their industry, their emulation, (or that competition which always take place, when the professors of any art, or the persons who follow any particular occupation, are numerous) and by their being contented to dig the ground with their own hands, and to live in that humble sphere, in which both economy and industry are requisite to procure a moderate competence, are enabled to pay much higher rents for their small extent of land, than what could be paid by those

who

who were better fed, better clothed, and rather dealers in garden stuffs, than operative gardeners. For it must here be remembered, that though large farmers, with great capitals, valuable teams of cattle, high-priced implements of husbandry, and renting extensive farms, either of arable or grass land, are very useful in improving a country, that has not been either generally or well cultivated, the small gardeners in the vicinity of Aberdeen, are more beneficial to the citizens, than a few great capitalists would be, who might easily conspire together, and both raise the price of their garden stuffs, and reduce the rents of their lands. These rents at present vary from six to twenty pounds per Scotch acre.

The garden stuffs raised by the numerous gardeners near Aberdeen, are chiefly cabbages, coleworts, potatoes, turnips, green pease, garden beans, spinage, carrots, parsnips, onions, lettuces, raddishes, and other sallads. These are raised in such plenty, within two miles of the city, that very few persons from the country bring any of these articles to the market for sale. (The cow-feeders indeed sometimes go nearly three miles from Aberdeen to purchase turnips for their cows.) There is no regular course of cropping; but the great object is to raise as many crops as they can, within as few years as is possible. For they could not pay their rents, except for a few of the most valuable roots, if they did not raise more than one crop every year.

The fruits which are raised by these gardeners are red, black, and white currants, gooseberries, raspberries, and strawberries. Apples, pears, cherries, and other stone-fruits, are chiefly raised by the gentlemen, who have their gardens inclosed with stone walls. For it must be observed, that the greater part of the lands in the vicinity of Aberdeen is inclosed by stone walls, whose height does not exceed three or four feet, or by hedges, which could not protect the higher fruits

from

from the depredations of idle or disorderly persons. Indeed a considerable portion of these lands is inclosed only partially along the great roads; and a few wooden pins annually placed by the land-measurer, denote the lines of partition between the different occupiers for a particular year. Of these different fruits, gooseberries and strawberries are raised to the greatest amount; although the caterpillar is often very destructive to the former.

The *produce* of these different articles is extremely various. The highest sum that can be drawn from an acre, is when sown with onions, which are sold by *beds*, that measure 48 square feet.—1153 of these are in a Scotch acre; and a bed of onions sells from a shilling to eighteenpence. At a shilling for each bed, the amount for an acre is L.57 13s.—. At fifteenpence, it amounts to L.72 1s. 3d. A bed of carrots varies from sixpence to ninepence, or half the price of onions. Coleworts are purchased by cow-feeders, at about L.2 or two guineas; potatoes and turnips, from L.2 5s. to L.3 for the *hundred beds*. But *seven score* are allowed to the hundred: and commonly an eighth part of an acre is *now* allowed for the hundred beds. In short, the value of potatoes is from L.18 to L.24; and turnips from L.16 to L.22.

The nursery grounds are kept remarkably clean; and as already mentioned, are in high repute. The best evidence of this is, that they fetch a very high rent. In old leases, they are seldom under L.7 or L.8; and the last acre of nursery let on a lease of 19 years, is rented at L.18 for the Scotch, which is above L.14 3s. per English acre. The seedsmen who have nurseries of small seeds or plants, are numerous, and have not much capital. But those who deal extensively in this branch, and raise the young plants of different trees, both require, and possess a greater capital, or considerable credit, to enable them to export young plants

of

of Scotch fir, larch, and various other trees, which are raised by their apprentices, or by other servants and labourers.—The first great dealers in nursery have been very successful; and one of them left a considerable fortune (L.10,000) at his death.

There are, however, two things which operate as a drawback to the nurseries of Aberdeen. First, owing to the seeds of turnips, ruta baga, and coleworts, being planted too near each other, the different kinds are apt to mix when in flower. Therefore the farmers ought to raise such seeds for themselves in corners of fields, or places considerably remote from each other. Secondly, Those young trees which are raised near Aberdeen, and sent to the southward, are generally hardy, and succeed well. Hence they are in great request for the English planters of wood. But on the other hand, those plants which come from the sister kingdom to Aberdeenshire, are not able to endure our cold weather, especially the alternate frosts and thaws of the spring months.—Nay in some exposed places, in the interior parts of the county, it is more adviseable to raise young plants of hard wood in nurseries at home, than to purchase them from the warmer nurseries near Aberdeen.

SECT. II.—HOT HOUSES.

It is not to be expected that the gardeners near Aberdeen, who are comparatively poor, or in but moderate circumstances, should be able to build *hot walls*, or *pineries*, or carry on works which afford more pleasure than profit. Many of the landed proprietors have more or less of thsoe accommodations

at-

attached to their gardens in the country; and there are a number of hot houses in the vicinity of Aberdeen. One of the best is the property of ALEXANDER YOUNG, Esq. of Cornhill, about a mile from the city. There are altogether, about a dozen of these accommodations to a villa, which in an age of luxury, are considered as articles of convenience, or indications of taste, to be found among our opulent merchants, and within an easy walk from the exchange, or from the counting-room. PATRICK MILNE, Esq. of Crimonmogate, one of the partners of two of the principal trading companies in Aberdeen, has a very excellent hot house within a hundred yards of Union Street. It is pleasant to see, in the 58th degree of north latitude, the front of an elegant house, where the massy pillars of granite, polished at a considerable expence, attract the eye of a stranger. It is no less agreeable to see at a small distance, the productions of a warmer climate here flourishing artificially. Here, without walking a great distance from his counting-room,

>Luctantem Icariis fluctibus Africum,
>Mercator metuens, etiam et oppidi
>Laudat rura sui——

SECT. II.—ORCHARDS.

THE gardens of the different landed proprietors are more or less stocked with fruit trees: But we have only a few orchards, properly so called, in this county. Sir ARCHIBALD GRANT, of Monymusk, grandfather of the present Baronet, paid particular attention to the raising of fruit, as well as forest trees; and had at one time 16 gardeners employed in his orchard, extensive nurseries, and garden. The tax on that useful body of men, as already mentioned, has contributed not a little to lessen the number of gardeners, and very few are now employed at Monymusk.

GEORGE SKENE, Esq. of Skene, grandfather of the present proprietor, laid out nearly two acres in an orchard, and stocked it with variety of fruit trees. But as the principal seat of this family is now at Carristown, in the county of Forfar, the same attention cannot be paid to it, as if the proprietor were generally resident in the county.

It has already been mentioned, that there is an excellent garden, with hot houses, and fruit trees of different kinds, at the Castle of Cluny. And it may be added, that at Keithhall, Fintray House, Logie Elphinstone, Castle Fraser, Abergeldie, Culter, Drum, Kemnay, and many other gentlemen's seats, there is abundance both of wall and other fruits in the gardens, though no place is set apart for the sole purpose of orchards.

The only modern orchard, of any considerable extent, is that which is now laying out by JAMES FERGUSON, Esq. of Pitfour, member for the county. Both on account of its antient and present condition it deserves to be particularly mentioned in this Report. The site of it, nearly 600 years ago, contained the Abbey of Deer, (which was founded in 1218), and the garden belonging to the Monastery.—We may talk of the luxury and refinements of the present age, and boast of our superior attainments in horticulture; but if we examine facts attentively, we shall find that our ancestors paid more attention to this than at first we are apt to suppose. At any rate, it is a striking proof of the luxury of the Romish clergy, and of their uncommon skill in the raising of fruit trees, that when Mr. FERGUSON was laying out his new orchard, he found in the Abbey garden, first rich soil, above three

feet

feet deep; secondly, a well paved causeway of granite; thirdly, a bed of pure sand one foot deep; fourthly, another causeway of granite; and below the whole a considerable depth of rich mould. No greater precaution could have been taken to hinder the roots of the fruit trees from being injured, by piercing into a cold or wet subsoil.

The ancient precincts of the Abbey contained about 11 Scotch, or 14 English acres. Mr. FERGUSON's new orchard contains, within its walls, 10 English acres, and fifteen hundredth parts of an acre. But in the middle of it the ruins are most carefully preserved; and these cover an acre and a quarter. The ground therefore laid out for an orchard is eight acres and nine-tenths.

It is situated on the north bank of the Ugie, nearly a mile above the village of Old Deer, and between that river and the turnpike road leading from Peterhead to Banff. Its exposure is excellent, as seen from the road to New Deer, on the opposite side of the Ugie, on a sloping bank towards the south, and its figure is nearly the segment of a large circle. Three lines of wall next the turnpike road appear as cords of the circle. The first of these is a wall 16 feet high, and 143 yards long. The second is intended for hot houses, and is 16 feet high, and 105 yards long. The third is a hot wall for fruit trees, with three flues, and is 16 feet high, and 146 yards long. These three walls extend to 394 yards along the north side of the orchard. The south wall, containing the segment of a larger circle, is 500 yards in length. Wall trees are to be nailed to the south side of this wall, between which and the river there is an open space intended for a walk and shrubbery. Beyond these, either a hedge, or a chevaux de frize, will be raised for protecting the wall fruit.

The expence will be very considerable, judging from what is already laid out. But its amount cannot at present be known.

CHAP.

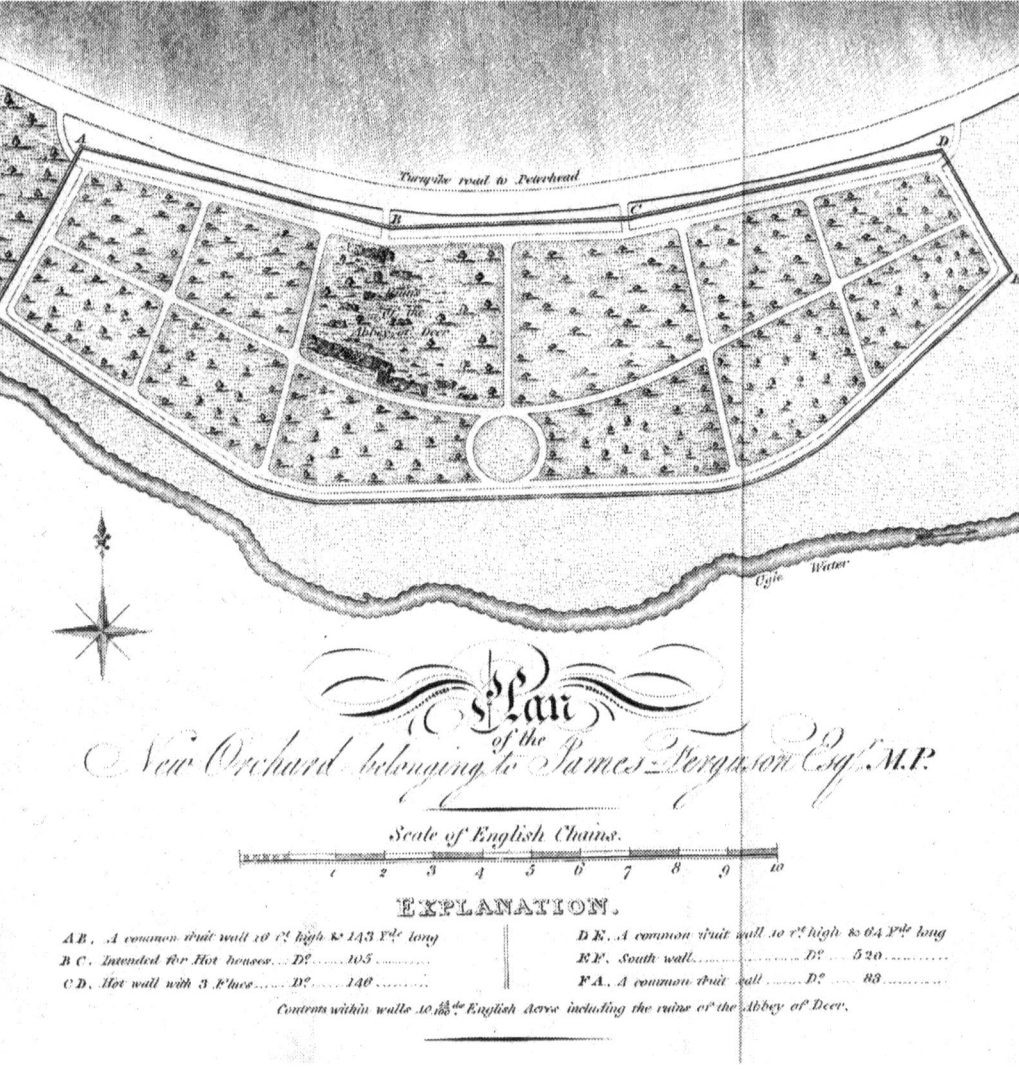

Pitfour's Orchard

CHAPTER X.

WOODS AND PLANTATIONS.

There is a very great extent of land in the county that is left out in natural woods, or inclosed for plantations in the highland districts. Inclosures of various dimensions, planted with various kinds of wood, are dispersed over the lower parts of the county, and are generally near the mansions of the different landed proprietors. Besides these plantations, small strips, or belts, from ten to sixty yards in breadth, are raised, both for shelter and for ornament, in places where the land is too valuable to be allotted for plantations. And single rows, with or without a hedge, are planted around the farmer's gardens; and in many places a few trees are found in the cottager's kail yard. The proportion of wood, however, is very different in the different divisions of the county. In Marr, especially in the higher and mountainous districts, there is a greater quantity of wood than in the other four divisions of the county.

This division, as already mentioned, contains about seven-fifteenth parts of the whole surface of Aberdeenshire.—And above one-half of this district is included in nine parishes, four of which are united to others, so that the whole are under the charge of five clergymen.* The population of this district is below 8000 persons, of all ages; and the arable lands do not raise food to maintain the inhabitants. But the woods and plantations (which in many cases greatly exceed the extent of the corn fields) are a source of wealth to the proprietor; and the cutting down, and disposing of these, in small parcels, in the more cultivated parts of the county, both employs, and liberally rewards the labour of a considerable number of the small highland farmers or artificers. These, in return for the wood which they sell, purchase oatmeal, and other articles for their families.

The woods and plantations in this higher division of Marr occupy nearly 100 square miles. But they grow very irregularly, being in some places very thickly planted, and in others raised by nature, at very different distances between the trees. Of these woods and plantations, nearly one-third has been both inclosed and planted; one-third has been raised by nature, without either inclosing or planting the ground; and the remainder has been surrounded by fences for keeping out the cattle, and then been stocked with wood, raised from seed either blown by the wind, or carried by the rooks, who, by some instinctive impulse, carry the cones of the Scots fir in their bills, to provide habitations for their offspring, at a remote period, when the seeds contained in these cones become trees, in which they may build their nests. In these higher districts wood grows so easily, that the

---

* These parishes are, 1. Birse—2. Aboyne and Glentanner—3. Glenmuick, Glengairn, and Tullich—4. Crathie and Braemar—5. Strathdon. Besides these, a detached part of Tarland divides the parish of Strathdon, and lies in this mountainous district. But to balance this, the lower part of Strathdon, intersected by Glenbucket, reaches down along the district of Cromar, and is not here included.

the proprietors need only inclose an extent of hilly ground, and thus shut out the cattle. The wind and the crows will in time supply him with seeds.—But when these natural woods grow very irregularly, it is found prudent to assist nature, by sowing, or occasionally scratching into the soil, a number of seeds of the trees which are wished to be reared in the vacant spaces.

A few examples will give the reader a more correct idea of the proportions of wood raised in this district.

The extensive estate of Invercauld, besides about 20,000 acres in Perthshire, contains 86,130 Scotch (or 109,044 English) acres, in the county of Aberdeen. Of this the manor place of Invercauld, including a few small possessions, contains 8964 Scotch acres—Of this only 274 are arable, and above 3300 are in woods and plantations. A considerable quantity of wood is raised in other parts of this property; but the above is adjoining to the mansion-house or manor place of Invercauld; and the whole has been correctly measured.

The estate of Abergeldie has also been measured, and abounds both in natural woods and artificial plantations.—Its total contents amount to 23242 English acres, of which 2120 are covered with wood of different kinds, and are more than twice the extent of the arable land.

WILLIAM FARQUHARSON, Esq. of Monaltry, on an estate of near 20,000 English acres, has a much greater proportion of wood than of land that has been subjected to the plough. In fact, from the irregularity of its surface, and the nature of its soil, this district is peculiarly adapted to the raising of wood.

The greater proportion of these woods consists of Scotch fir, a species of pine that grows slowly at first, but is very valuable. It is well known, that where the soil is most barren, and the trees grow very slowly, the wood of the Scotch pine is of the best quality.—A remarkable tree, at Invercauld, was cut down about 40 years ago; and the number of concentric circles near its root, viz. 229, shewed it to have grown and increased in size for 229 years, besides the time that it continued stationary. Its wood was declared, by all who saw it, to be much superior in quality to any that had ever been imported from the north of Europe. In 1804, Mr Burns of Haddington, the ingenious architect, who built the wooden bridge over the Don, opposite to Parkhill, on the turnpike road from Aberdeen to Oldmeldrum, brought all his fine wood from the forests of Braemar, at 60 miles distance. He informed the Writer of this Report, that it was far superior to any that ever came from Memel, Riga, or any port in Prussia, or of Norway.—Mr FARQUHARSON of Invercauld has sold to the amount of L.16,000 of this excellent wood, which was from 18 inches to 2 feet in diameter; and was sold at 14d. per cubic foot.—It will give some idea of the closeness of the texture, and fineness of the quality of the wood in the upper part of Aberdeenshire, to mention, that the Reporter cut a tree in the forest of Glentanner, and carried it home to the distance of 40 miles, the root of which, though only 15 inches diameter, contained 132 concentric circles when cut across; (consequently, the tree was above 132 years old;) and what was still more remarkable, where the diameter of the tree was only seven and a half inches, the number of circles was 110; and the quality of the wood was excellent, almost to the top. Yet the fir in Glentanner is considered inferior to that in Braemar, where the Reporter has counted from 150 to 160 concentric circles in different trees. It is this slow growth and closeness of texture which marks the difference between the woods in the higher, and those in the lower parts of the

county. And if the landed proprietors would only inclose all the barren grounds which are not worth the expence of cultivation, we should have no occasion for wood either from Prussia, Norway, or North America. There are thousands of fir trees in Braemar, some of which are nearly 6 feet in diameter, which are superior in point of quality to any wood of that denomination that was ever imported into any place in Great Britain.—And both the legislature and the Highland proprietors, each in their different spheres, should take care that this country be independent of any sort of ship timber. England can always raise oak and elm; and Scotland can raise fir on the mountains and hills; and beech and ash in the more cultivated districts.

Besides the very great extent of Scotch fir, considerable quantities of oak, beech, and aller, are found among the natural woods of this mountainous district; or growing in scattered groupes along the banks of the rivers: So that this district, though neither populous, nor productive of a great quantity of corn, may, both from the great quantity and excellent quality of its woods, be rendered very beneficial, not merely to the wealth, but to the strength of the empire. It is only necessary that turnpike roads, canals, railways, and every mode of facilitating the carriage from the forest to the port of Aberdeen, be encouraged by the legislature—on national grounds; for the district is too remote from Aberdeen to support the necessary expence from the fortunes of individuals.

The second or middle district, in the division of Marr, comprehends the whole parishes in the districts of Cromar, Kildrummy, and Alford, and reaches down to Monymusk, and the lower extremity of the parish of Midmar. Though it contains no natural woods of any considerable extent, yet a number of plantations have been made by the landed proprietors,

prietors, particularly by Sir ARCHIBALD GRANT, grandfather to the present Baronet of Monymusk, Mr. FARQUHARSON of Breda, Mr. FARQUHARSON of Haughton, Mr. BYRES of Tonley, Mr. MANSFIELD of Midmar, the late Mr. GORDON, Mr NIVEN LUMSDEN of Clova, Mr GORDON of Craig, and Mr. BREBNER of Learney. It is probable the whole plantations in this division will amount to thirty square miles. And it is deserving of notice that the woods of Monymusk, which were planted in the barren hills, from 60 to 80 years ago, though not equal in quality to the woods of Braemar and Glentanner, are superior to those in the low country which were raised in better soil, and both came sooner to maturity and have decayed earlier.—Slow growing fir is always the most durable.

The lower division of Marr (which 600 years ago was almost a continued forest, distinguished by the names of the forest of Drum, the forest of Kintore, and the forest of Stocket,) besides a great extent of well cultivated land, contains in the plantations of Drum, Culter, Countesswells, Hazlehead, and Murtle, on the south; Skene, Housedale, Castle Fraser, Cluny, Kintore, Caskieben, Pitmedden, Craibston, and various small proprietors near Aberdeen, other 20 square miles, so that the whole division of Marr contains 150 square miles of natural woods, or artificial plantations—or very nearly a sixth part of its whole surface.

In the division of Garioch the whole natural woods and plantations do not exceed 12 square miles, or 6000 Scotch acres. These are distributed chiefly in plantations (for there are few natural woods) near the mansions of the Earl of KINTORE, Sir WILLIAM FORBES of Craigievar, Mr. FORBES of Balbithan, Mr. ANDERSON of Bourty, Mr RAMSAY of Barra, Mr. URQUHART of Meldrum, Mr. M'KENZIE of Glack, Mr. HORN ELPHINSTON of Logie Elphinston, Major KNIGHT of Pit-

Pittodrie, Mr. LESLIE of Balquhain, Mr. LUMSDEN of Pitcaple, Mr. LEITH of Freefield, Mr. GORDON of Newton, Mr NIVEN LUMSDEN of Premnay, General HAY of Rannes, Mr. GORDON of Knockespock, Mr. GORDON of Wardhouse, and several smaller proprietors. A more considerable proportion of ash, elm, and other hard wood is raised in the Garioch than in the division of Marr.

In the division of Strathboggie there are not above 4 square miles, or 2000 Scotch acres, of wood of any description, and, except a small patch near Abbachy, these belong to the Duke of GORDON, and are situated near Huntly Lodge, the residence of the noble Marquis of HUNTLY.

In the division of Formartin there are considerable plantations belonging to the Earl of ABERDEEN, General GORDON of Fyvie, Mr. MORISON of Bognie, Major DUFF of Hatton, Mr. FULLARTON UDNY of Udny, Sir WILLIAM SETON of Pitmedden, Mr SKENE of Dyce, Mr PATON of Grandholm, Mr. RAMSAY of BARRA at Straloch, Mr. GORDON of Hallhead, Mr. FORBES of Blackford, and Mr LESLIE of Rothie, and other inferior proprietors. The whole may amount to 20 square miles; and contain a considerable proportion of hard wood, all in a thriving state, except near the sea coast, where they are injured by the spray.

In the division of Buchan, wood seldom is raised near the German Ocean; and where raised, is not in a thriving state. Yet in the interior parts, viz. at Pitfour, Strichen, Shivas, Kinmundy, Aden, and Auchry, there are considerable plantations of thriving trees; and near the sea coast several patches of wood have been raised with success, though many others have been blasted or stinted by the spray from the ocean. Perhaps there may be in this whole division ten square miles of woods and plantations, of which one half is in a thriving condition, and the other half is more or less injured by the German Ocean.

On the whole, owing to the greater quantity in the division of Marr, nearly one tenth part of the measure of the county is either in natural woods, or artificial plantations. Of the first class, those which grow naturally in the higher parts of Marr, are more unevenly, and on the whole, more thinly scattered over the surface, than those which have been planted artificially. The woods of Braemar, from the barrenness of the soil, are of slower growth than those which have been planted in the lower and more fertile districts; but this is more than balanced by their excellent quality.

It has been already mentioned that the lower parts of Marr, or the ground between the Dee and the Don, extending from the sea coast about fifteen miles backward, was one great forest in the time of King ROBERT BRUCE. It may be added, that Buchan also, even near the sea coast, was a well wooded district, and that in the peat-mosses both of Buchan and Formartin, large roots and trunks of different trees, particularly oak, fir, birch, and aller, are found in considerable quantities.—But two causes contributed to the destruction of the woods both in Buchan and Formartin—The first was a great storm, several centuries ago, from the north east, that appears to have laid thousands of them flat to the ground, where they rotted; and, from the stagnated waters, were gradually converted into, or covered with, peat moss: while the direction in which their trunks are found, shews the quarter from which the tempest assailed them. The second cause of the destruction is ascertained to have been the rebellion of the Cummines and their adherents, the ancient proprietors of these districts, which occasioned their lands to be burned, and their property to be destroyed, by King Ro-
. BERT

bert Bruce. In the ancient woody state of this county, the trees sheltered one another so much, that they grew very near to the sea shore, where they cannot be raised, or at least do not thrive at present: and experience, in this as well as in other cases, has shown, that it is easier to destroy than to raise or repair.

It would no doubt require a considerable time, and much labour and expence again to raise wood in any quantity on these coasts, now that they are bare or stripped of wood; for it is very difficult to get the better of the rigours of climate and peculiar disadvantages of exposure. But the task of renovating wood in Buchan, though difficult, may be accomplished. By planting hedge-rows and belts of laburnum, keeping these hoed for several years, and sowing ruta baga in the vacant spaces to prevent the hares from injuring the laburnum, the bark of which they prefer to every thing except ruta baga, shelter could be given both to oaks and larches, near the sea coast; and if once a breadth of 100 yards of wood, next the sea, could be raised to a tolerable size, other plantations would quickly follow. The Reporter shall only add, that on Dr. Anderson's farm at Monkshill, situated in sight of the German Ocean, and near 400 feet above the level of the sea, he was shewn by him a variety of trees, raised in a place that was much exposed, where the laburnum stood all the severity of the storm, when the wind blew from the sea, and where the oak, raised on the spot, appeared to resist the inclemency of the seasons, better than any other tree except the laburnum. It may be proper to insert here a quotation from Dr. Anderson's Original Report, and to suggest a few hints on the subject of planting.

"All along the sea coast, for many miles backward, Aber-
" deenshire till of late might be said to be perfectly destitute
" of wood; and Buchan, in particular, is even proverbially
" bare,

" bare, so that in many parts of it, CHURCHILL's description is
" literally verified,

"Far as the eye can reach no tree is seen."

" But this defect will soon be entirely removed; for almost
" every gentleman is now active in making plantations on
" his estate; and I am confident that in no part of the
" world have so many trees been planted in the same space
" of time, by an equal number of men of the same extent
" of property, as within the last thirty years in this county,
" unless perhaps in Angus, and some of the northern
" counties of Scotland. At first, these plantations were
" chiefly of the Scotch fir, which is not the tree that is best
" adapted to the lower part of the county; but many other
" trees have been interspersed among them; and of late the
" larch has become, with good reason, the favourite tree;
" and great quantities of them are now planted every year,
" so that in a short time, this, from being the barest coun-
" ty in Scotland, will become one of the best wooded dis-
" tricts in the kingdom. The effect of this change will be
" experienced in an astonishing degree, by the descendants
" of the present generation; for it will soon appear, that
" the want of wood has been one of the principal causes
" which retarded the improvement of the county.

" On account of the general barrenness of the county,
" and the impossibility of raising solitary trees in these
" circumstances, the plantations have all been made in large
" masses; to which, the diminution of expence, in thus
" planting, has contributed a great deal: so that the country
" still has a bare and naked look."

It no doubt is true, that the mode of planting in large
masses, or making square, round, or oblong plantations has
not the same effect in ornamenting a country as is produced by distributing belts of wood in all directions; although

though the quantity of wood planted were the same in both cases. Not only would the county obtain a better aspect, but more shelter, and several degrees of more local heat, by planting belts from 30 feet to at least as many yards in breadth (as much wider as can be done with propriety) on the boundaries, and most exposed places of estates or farms; so as to obtain shelter from the *north, north east,* and *north west*; and, near the coast especially, also from the *east*.— One of these belts should surround the greater part of every landed proprietor's personal farm. Several degrees of additional local heat may be obtained in many cases, by thus planting judiciously. But the kindly visits of the sun should not be excluded by planting too thick belts, on the *south east, south,* and *south west*. In many cases, especially near the sea coast, the rays of the morning sun may be with propriety excluded, by a belt of planting on the east. For though these are friendly to an early harvest, yet in an unfavourable or late season, if a mildew or rain in the evening be succeeded by frost at night; and if the sun dart his rays in the morning on the wet corn when in flower, or on the pease, beans, or potatoes, before they are fully ripe, the effects are generally fatal. The only remedy, viz. that of two persons going very early with a rope, between the furrows of a ridge, and shaking off the rain or dews, cannot be practised on a large scale; though it may save the potatoe crop of the industrious cottager that he know this, and put it in practice, when he sees the hoar frost, early in the morning. A more permanent remedy is to have a small belt of wood on the east, (which generally prevents the suns rays from injuring the crop), which is usually dry before the sun appears in the south east; for the danger is over, as soon as the moisture of the night is dried up.

It would also be very conducive to the improvement of the

the country, where two gentlemen's properties meet, in wet places, where draining is necessary, to throw up a ditch of 5 feet wide, and at least three deep, at the distance of 5 or 6 yards from the boundary, on each side; thus leaving a belt of ten or twelve yards in the middle, and planting this at their joint expence;—and then dividing the trees, when raised, or their produce when sold. This would soon defray the expence both of throwing up the ditches, and of planting the belts of wood: And, instead of going to law about a few acres of moor, and spending several hundred pounds in a lawsuit, which begins with irritation, is carried on with heat, and ends in heart-burnings, they might with the utmost harmony divide as many hundred pounds, at the sale of this wood, whether for cuttings, weedings, or grown trees: and, at the same time, the making two ditches, and planting the intervening ground, might cost little more money than they would pay for a fence or march dike, one half of which every proprietor is compelled to build, when called upon to do so. On every considerable estate the same kind of belts might be made as boundaries between the different farms.—This would add much both to the shelter and to the ornament of the county; which in some districts, particularly in Buchan, has rather a bare and bleak appearance.

It is a judicious remark, made by Mr. GEORGE ROBERTSON, in his very able survey of Mid Lothian, that, "forest "trees, planted in hedge-rows, seldom come to good; and," he adds, "that a thin belt of four or five rows will not " stand the severity of the climate in that county," where he says " there should be twenty rows at least." Hedge-rows, in this way, are neither good hedges nor good trees; and a thin belt is in many respects inferior to a thick or wide one. But the Reporter knows, from his own experience, that much less than twenty rows will answer the purpose of shelter, and

and protect each other while growing. By the advice of the late ROBERT BARCLAY, Esq. of Ury, he planted, and kept for five years under hoe, a belt of planting which is above two hundred yards in length, but has only seven rows in breadth. It is situated in one of the most exposed places of the county, 320 feet above the level of the sea, 80 feet above the level of a small brook only 150 yards distant, and 180 feet above that of the river Ury.—Yet it has for 28 years afforded ornament, and for the last 20 years, shelter to his house, garden, and part of his glebe. Those who can spare ground may very properly make belts of wood round their premises of 20 rows—But those who are only life-renters, or have but small properties, may be contented with 10, and may be sheltered with 6 or 7 rows. As many would be deterred by the expence, and by the quantity of land, from making belts of 20 rows, the Writer of this Report would make this trifling correction, on a work of great merit, merely for the sake of promoting the ornament and shelter of the more exposed parts of the county;—and he would, from his own experience, suggest, that neither spruce firs, nor Scotch firs should be used in those narrow stripes of planting, but oaks, ashes, elms, planes, birches, and especially larches, of which, after the first ten years, the greater part may be thinned out, and cut as weedings.

From having made different surveys of the county, with some attention, he would, in the most respectful manner, suggest another hint to the landed proprietors.

There are a multitude of streamlets, or burns, scattered over the county; and the irregular broken ground that is formed by their curvatures, may, in many cases, at a very moderate expence, be formed into small plantations, where the land is unfit to be left out as a meadow, in perpetual grass,

grass, or to be let either to a small farmer or a cottager.— A judicious proprietor will avail himself of such a situation.

On the whole, the difference that would be occasioned in the aspect and shelter of many places in Aberdeenshire, by planting either belts, or small patches of wood, will be an apology for these imperfect hints, from one who has a great partiality for planting.

But in order to encourage this as a general practice, two things appear to be necessary.

The first is, that the proprietors of entailed estates should be empowered by law to burden their property with the appretiate value, at their death, of all houses, inclosures, and plantations; and that the amount of this burden might be applied in the first place, for repayment of debts due to their creditors, and secondly as a provision for their younger children. This would give a stimulus and an energy to many a proprietor of lands that are strictly entailed, who possesses both a fine taste and public spirit, but looks at his younger children with a sigh, and complains that he cannot support his rank in life, and improve both the aspect and value of his estate, (or rather of that property of which he is the temporary possessor) without defrauding his creditors, or beggaring his younger children. It also would give great relief to such a man, when he has in his youth contracted large debts in embellishing this property; and in his old age is seized with the deepest distress, when he thinks that he has made but little provision for his blooming daughters, while the family estate, by the will of the entailer, goes to an heir male, who may sometimes be only a distant relative. An act of Parliament by which every heir of entail could burden his successor to the amount of the money which was thus expended, would injure no person, and would be attended with many advantages.

The

The second thing necessary is, that when farmers have long leases, and plant such belts of wood, or make plantations on the barren parts of their farms, they should be valued at the expiration of their leases; and the tenant at his removal should be allowed at least a moiety of their value. During the currency of the lease the farmer should also be at liberty to use what weedings he had occasion for—if he planted the wood himself. The Reporter is happy to remark, that it was from seeing a regulation of this kind, in the articles of agreement between GEORGE SKENE, Esq. of Skene, and Mr. ROBERT WALKER, in Wester Fintray, that this sentiment first occurred to him, not as a matter of speculation, but as what had been actually practised. And he has been lately informed, that Mr. ALLARDYCE of Boyndsmill has planted wood to a considerable extent, in consequence of an article in his lease, by which his landlord is bound to pay the full value of all such wood, at its expiration.

From the very different situation of the agriculture, and of the uses to which wood is applied in this county, compared to the counties of England, no copse woods are either raised or cut in Aberdeenshire.

When our fir woods are cut down, and when the ground is not planted again, but consigned over to the plough, the roots are grubbed up in various ways; and the land after repeated ploughings is sometimes sown with oats; at other times with turnips. The latter is considered as the preferable mode: and Mr. JAMES SCOTT, in Mains of Craigevar, has in this way improved several fields, at an expence of about L.10 per acre.—But this expence is so extremely different in different acres, that no general estimate would be at all agreeable to truth. In this situation the Writer declines giving any Estimate. For although Mathematical *calculation* is superior to

every

every sort of *logical argument*, it should be remembered, that a Mathematician is always vulnerable in *his data*.

The price of wood has risen very much within these 40 years. In the year 1770 the cubic foot of Scotch fir, which grew in the lower parts of the county, was only 3 pence.— Now it is a shilling; and sometimes has been as high as eighteen-pence. In 1772, wood from Lord FIFE's forest in Braemar, of the finest quality, was bought at 4d. per foot on the spot; and after being floated down the Dee to Aberdeen, was sold, *in retail*, at 8 pence. Now it is sold in the forest from a shilling to 14 pence; and particular trees at 18d. in the forest; and from two shillings to half a crown at Aberdeen. Ash in 1770 was sold at 6d. and is not now below 4s. per cubic-foot. In no way has money been laid out to more advantage than in raising plantations of forest trees. But both the prices and the profits have been too variable, owing to the occasional stoppage or opening of the ports in the Baltic, to admit of drawing up an account of profit derived from rearing wood.

Although American fir has been imported in considerable quantities, during the present war, it is well known that it is comparatively friable, and easily broken. A bowsprit of a vessel, when the tempest covers it with water, is very apt to break if it be of American fir. If brought from the forest of Braemar, it may dip in the water, but rises again unbroken by the waves. It is therefore incomparably superior to the fir, which is imported from America.

It may be added to this long account, that a ship wholly built of Oak, raised in Lord ABOYNE's woods in Inch Marnoch, and called the Countess of Aboyne, was built at Aberdeen; and the oak, though inferior in point of size to oak from Hampshire, was remarkably close in its texture, and both tough and durable.

SECT.

CHAPTER XI.

WASTES.

SECT. I.—MOORS—EXTENT.

A Great deal of moorish soil is found in the different divisions and districts of this county. In the mountainous part of the division of Marr, about two-thirds of the surface is of this description; and as we advance from the lower to the higher ground, or from the moor to the mountain, the line that separates them is only ideal. In Strathboggie, about one half of the surface is moor; and, towards the four hills of the Cabrach especially, wherever water stagnates in the moors, peatmoss of greater or less depth is formed. On the bounding hills of the Garioch a great proportion of this soil is found, adjoining to the arable lands of inferior quality; and even in the valley itself, though more fertile in general than any other division in the county, several patches of moor are interspersed with soil of superior quality. In Formartin, there is a still greater proportion of moor, which sometimes is covered with hard bent, but more generally with heath. And in Buchan, at least one half of the soil is either moor, or has been reclaimed from being moor, or, by the stagnating of water in its surface, has become peatmoss, of various depths and qualities. Nearly one half of the surface of the County is moory soil, or has been reclaimed from moor by the industry of the inhabitants.

The value of the moors is very different, according to the quantity of the stock that a given extent, suppose an hundred acres, of moor can maintain; and according as it is intended to be improved or left in a state of nature, from twopence an acre in Braemar, to two shillings in Formartin, or Buchan.

The most usual application of moor was to the feeding of the small sized sheep, who are natives of the county. But now sheep of a larger size, from Peebles or Lammermuir, are introduced in their stead; and sometimes young cattle are sent to depasture in the moors.

The most approved method now followed in the improvement of moors is first to surface-drain, and plough them—to allow them to remain in fallow for one or two years, then to sow them with turnips; and lastly to lay them down with oats and grass seeds.

Lime in this case is liberally applied; paring and burning are very seldom used, and only white clover, rib grass, and rye grass, are sown out along with the oats. When as much dung can be collected as to raise a tolerable crop of turnips, the profit is very considerable. And when the land is well surface drained, or the soil pretty dry, the laying on of lime pays abundantly. In the district of Formartin, Mr. BARCLAY at Mill of Knockleith, and Mr. ALLARDYCE, of Boyndsmill, have been very successful in the improvement of moors. And in all the districts of the county, large tracts of moor have been added to the arable land. A considerable portion of this has been inclosed, and after being

im-

improved, is allowed to remain some years in sown grass.—Afterwards it is broken up, and subjected to a regular rotation of crops.

SECT. II.—MOUNTAINS.

The higher districts in the division of Marr, which reach along the boundaries with the counties of Angus or Forfar, Perth, Inverness, and the uppermost part of Banffshire, contain mountains of greater altitude than what are to be found in any other part of the island, except in a single instance near Fort William; viz. Ben Nevis, which is the highest mountain in Great Britain.

These mountains had never been measured till the Writer of this Report made two surveys of them in the course of last summer—the particulars of which will be found in the Appendix. In this place it is necessary to mention only in general terms, that Ben MacDuie, the highest mountain in this county, is only 50 feet of less height than Ben Nevis, or very nearly 4,300 feet above the level of the sea, at half-flood; that Cairntowl and Breriach are 4,200 feet above that level, and that these two are exactly of equal altitude.—(The river Dee takes its rise between these two mountains; and its two highest wells are very nearly 4000 feet above the level of the sea.)—The summits of Ben MacDuie, Cairntowl, and Breriach, are not above two miles distant from each other.—And about three miles to the north east of Ben MacDuie, is Cairngorum, on the confines of Inverness and Banffshire, formerly accounted the second mountain in Great Britain; but now found to be 260 feet lower than Ben MacDuie and 180 less than either Cairntowl or Breriach. Along the boundary between Banff and Aberdeenshire are the mountains of Benabourd, and Benavon, nearly 3,900; and towards the boundary with Forfar and Perthshires, the most interesting of all the mountains in this county, viz. Lochnagar, which is 3800 feet above the level of the sea—Between it and the vertex of the county, both along the Geaullie, and near the Dee, there is a large assembly of mountains; all of which are more than 3000 feet above that level.

It is obvious, that in this Alpine district, especially near the tops of the mountains, there is a great extent of surface, that for the purpose either of cultivation or of pasture, must be denominated *waste*. On the summit of Ben MacDuie, and for a considerable way down the mountain, there is no food for any animal. On the top of Cairntowl, and for several hundred feet below it, there is nothing to be seen but one continued mass of rocks or large stones, very difficult of ascent. On Breriach, whose broad shoulder is equal to the peak of Cairntowl, the rocky summit hangs over a precipice above 2000 feet high, which is awfully sublime. Benabourd, or the Table Mountain, presents a dead level of near four miles in length, where one walks among flat rocks, without a pile of grass; and when we descend about 500 feet, a species of grass is found that is poisonous to horses, and must be carefully avoided. On the highest peak of Ben Avon, large masses of granite render it very difficult to climb to the summit; and no food, even for sheep, is found near the top of the mountain. At Lochnagar alone, that immense pile of granite, grass is found growing very near the top of the *white month*; and what is yet more remarkable the *Ca Cuim*, or highest peak, is decorated with the *averan*, or largest berry that is found on the mountains.—The immense quantity of granite, of which very little is decomposed, is the probable cause of the comparative fertility of Lochnagar;—for a

breast-

breast-work of nature, (that would be degraded by being compared to the most noted fortifications of Vauban) formed of solid rock, nearly perpendicular, and from 950 to 1300 feet high, extends above a mile and a half in length; and shews that this interesting mountain is composed of massy granite. It may also be remarked, that probably owing to the local heat occasioned by the greater quantity of stone in this than in the other mountains, there are trees found growing here in a higher elevation, than in any other place in the county. It must be acknowledged however, that in this Alpine district there is a great proportion of the surface that produces neither grass, nor corn, nor wood of any kind.

Let not the fastidious critic, however, consider these mountains as *mere wastes*. Nature produces nothing in vain; though we may not always see the use of her productions. It is worthy of our attention, that the granite which is found in such quantities near Aberdeen, is trenched by the spade and mattock, and is carried to pave the streets of London; and that topazes, beryls, and other precious stones, called from the name of two of our mountains, *Cairngorums*, are also dug up by the spade, or trenched by the mattock, in these Alps of Scotland. While the carriages of the great are driven along the streets of the metropolis, which are paved with Aberdeen granite, the heads of the finest ladies of the Court are adorned by the gems that are found in these lofty and apparently barren mountains. The writer of this Report, in the course of his survey, saw a number of persons digging for these precious stones; where some were adventuring for themselves, and others were employed as common labourers at 3s. 6d. per day, or a guinea a week, in search of them.— He saw an aqua marine, or species of beryl, which it was supposed would defray the whole summer expences, laid out by two undertakers, on the mountain of Ben Avon; and his guide

guide pointed out to him, in the highest Cairngorum (for two of the mountains bear that name) a hole of four or five yards diameter, and about five feet deep, out of which precious stones had been dug, last summer, (1809,) by the country people, for which L.75 had been paid them by a dealer; who would no doubt double that sum, when he sold them to the jewellers in London.

The extent of land, toward the tops of these mountains, that has been reckoned barren, reaching from Lochnagar on the south, and Cairneelar on the west, to Benavon on the north, is at least 100 square miles; and the moory or mossy soil on their ridges, or in the valleys or narrow glens between them, amounts to 200 square miles.

The only improvements, which occur to the Writer of this Report as proper to be attempted, are the following.

1. Where the Highlanders are in search of precious stones, after trenching, let them be obliged to build a Galloway-dike with the granite that is dug out of the ground. This appears to be the most proper quit-rent that they should pay for the privilege of digging for Cairngorums, or Scotch topazes.

2. Let the proprietor sow Alpine grasses in these small inclosures, strowing a little lime on the surface. These patches will afford both shelter and early food in the spring to the sheep, who are often in great need of this.

3. Let the mountain streamlet be conducted slowly, and by various windings along the sides of the mountains, by the shepherds who attend the sheep in summer. Let these men receive a gratuity from the proprietor, and also from their employer, for every acre, or either greater or less measure, which by properly conducting the water along the sides of the mountains, they convert from heath to sweet grass. This method of irrigating mountains is not a mere theoretical ad-vice.

vice. The Reporter saw the worthy and intelligent Mr. STUART, at Allanquoich, (Factor or Land Steward to Earl FIFE,) improving the grass of the mountains attached to his farm, by this simple process. But in order that the shepherds may find their interest in performing this work, which is no part of their proper employment, he thinks that both the proprietor of the lands, and the owner of the sheep, should bribe them to irrigate, in this way, as much of the surface of the land as they possibly could.

SECT. III.—BOGS.

WE have but a small proportion of the surface on the east-coast of Scotland, that may be properly denominated boggy, as unable to bear the weight of the human body in the summer months. But there is a great proportion of peat-earth, of various depths and qualities, in different parts of the county of Aberdeen, which is used for fuel by the inhabitants. Before the introduction of carts, (which were not generally used till about 40 years ago) the farmers spent the greater part of the summer in preparing and bringing home fuel from the different peat-mosses. Now much less time is spent in this way, as a greater quantity of fuel is conveyed by carts, than what could formerly be carried on horses backs in *peat creels*, a sort of panniers that is now entirely disused. But as our peat mosses are nearly exhausted in the lower districts of Marr, Garioch, and Formartin, no farmer would think of improving such soil for the purpose of agriculture, unless he have either a great quantity of this soil, or be situated near a sea port. Therefore there is not a great proportion of such land

land which, as yet, has been subjected to the plough.—Not less than 5,000 acres have been trenched out from barren moors or rocky soil, or otherwise torn in by the plough, within these fifty years: but not 500 acres of bogs, or of peat moss, in the whole county, have been added to the extent of arable land. The late Mr. CUMMING of Auchry, partly improved by the plough, and partly planted more of this soil than any other person in the county; and the late GEORGE BURNETT, Esq. of Kemnay, planted very successfully in peat mosses. ALEXANDER MOIR, Esq. of Scotstown, Sheriff-Depute of the county, has, by draining, liming, and manure, added considerably to the quantity of arable lands near his Mansion-house; and raised plantations on the ground that was of a moory soil, and was cut off from the bogs, or peat moss. The different modes of improving peat-moss in this county, are the following:

1. Draining.—Here large open drains, called *main drains*, are sunk in the hollow places; and smaller covered drains are made to communicate with these.

2. Burning was formerly used, with all the soils which were called burnt lands, once every 8 or 10 years: but now the ashes of the burnt soil are spread over the field, when turnips are sown, or the land laid out in sown grass for some years; at the end of which it is brought to a rotation of cropping.

3. Where the soil is not burnt (a practice that is generally disused,) hot dung, with a top dressing of lime, is successfully applied to raising turnips, bear, and corn crops, on such soils.

It must be acknowledged, however, that the improvement of bogs or peat mosses, in Aberdeenshire, has not kept pace with the cultivation of dry arable lands; that the peat soils of this county, in which a considerable proportion of *acid* is always present, are probably inferior to those in the west of Scotland, and certainty are not, like many of those in the county of Angus, incumbent on a bed of marl. The substances found in the bottom of our peat mosses, though analyzed carefully, do not hold out great encouragement to the farmer to pierce below the moss in order to mix the soil with the subsoil.

The rent of such soil is seldom above one half of what is paid for good arable land near the rivers; and in fact, the vicinity of the peat moss occasions a number of small farmers and cottagers to pay a higher rent for land of this inferior quality, than it is worth, compared to early and dry soil.

These remarks regard the divisions of Marr, Garioch, Formartin, and Strathboggie.

On the coast of Buchan, the immense quantities of shelly sand are an inexhaustible source of improvement to the peat mosses;—and here, by cutting drains in proper directions, a good deal is already done; but far more will probably be effected at no distant period.—And the instructions of Dr. RAINNIE and Mr. AITOUN will no doubt be found beneficial to both the proprietors and farmers in this division of the county.

SECT. IV.—FENS AND MARSHES.

THERE is very little of the surface of this county which approaches to a level; consequently there is little that can be called fenny or marsh-land. Sometimes at the confluence of two rivulets a proportion of marsh is to be found; but by a little draining, this may be easily removed. We have no grounds that are denominated *fens*, or used as such: but we have a kind of salt marsh, between Formartin and Buchan, that

that will gradually emerge out of the water, without any aid from human labour. It is thus described by Dr. ANDERSON, who resided within a few miles of it, and is called the Sleeks of Tartie.

"In the river Ythan, about a mile from its mouth, there is a capacious bason, which is filled with water every tide, that empties nearly dry at low water. This is called the Slitch, provincially, the *Sleeks* of Tartie; and affords a beautiful example of the gradual progress of water being converted into dry land, in similar circumstances. For there can be no doubt that in the course of years, the whole of this bason will be converted into dry land, when it will form a salt marsh similar in kind, though of much smaller extent, to Romney Marsh in Kent. Its progress is thus effected.

"When the river is in flood it carries down a great quantity of the richest mould from the corn-grounds, upon its banks, and makes it thick and muddy.—On account of the rapidity of current at that time, and the agitation it sustains, the mud is not deposited in quantities until it comes near the river mouth. But as the tide here meets it, and makes it overflow this bason, the muddy water is there allowed to stagnate; and the mud of course falls to the bottom; forming a soft slimy bottom, called *slitch* or sleeks, which gradually fills up the cavity, and rises nearer to the surface. It is easy to see that the greater proportion of the sediment must be deposited in the deepest places; and that, of course, as it fills up, the surface must approach nearer to a level, till at length it will emerge out of the water entirely. In the upper part of this bason, an island is already formed of considerable extent; and to the best of my observation, the slitch, in the

deep-

" deepest places, has risen from three to four feet, in the
" course of thirty years past."

SECT. V.—FORESTS.

The term forest implies either woods or plantations, or uncultivated land set apart for the purpose of preserving, and of hunting for game. In the upper part of the county, the Earl of FIFE has allotted about 10,000 acres for a deer-forest. No part of it, however, is inclosed, and the woods in it are all natural. No tenant is permitted to depasture either black cattle or sheep, in this ground; nor to kill any of the deer that roam in this forest. Mrs. FARQUHARSON of Invercauld, also has allotted a considerable extent of land for the pasture of both red deer and roes. At Abergeldie, Monaltry, and in the forest of Glentanar, this priviledged race are preserved by the proprietors; all of whom are attentive to the improvement of their personal farms, but are satisfied with the present state of their forests—or improve them by additional plantations.

In the lower parts of Marr, along the Dee, the forest of Drum has been more or less subjected to the plough for these six hundred years; and the forest of Stocket has been improved at an unparalleled expence by the Citizens of Aberdeen. In the intermediate district between the Dee and Don, the forest of Morven has been shared by the proprietors of land in Cromar, and the forest of Culblean, extending downwards to Skene, has chiefly belonged to the family of Skene, and the proprietors in the parishes in Midmarr, and Lumphanan. But the name forest occurs only in antient charters— no forest lands, or privileged deer-pastures, being allowed

in

in this district—though both roes and red deer occasionally visit the woods, as if they meant to preserve the prescriptive right to their former possessions. The mountain of Benochie, on the boundary between Marr and the Garioch, formerly a royal forest, has an extensive peat-moss near its summit, and considerable plantations of wood along its ridges, which belong to the proprietors of the adjacent lands. A deer park belonging to the Earl of ABERDEEN, at Haddo-house, near the confines of Formartin and Buchan, preserves these noble animals, who find no forests in the lower part of this county.

Taking a general view of the antient forests of Aberdeenshire, except in Braemar, they are all converted into private property, and more or less subjected to the plough.

SECT. VI.—HEATHS AND DOWNS.

ALTHOUGH a great proportion of the surface of the county of Aberdeen is covered with heath, and although there are several downs on the sea-coast, from Fraserburgh, where Lord SALTOUN has some excellent downs, to Peterhead, and a few along the Buchan coast southward, to the old parish of Forvie, the greater part of which was blown over with sand; yet sheep farming in those districts has been long on the decline, since the introduction of inclosures and the turnip husbandry. Lime however has been applied as a manure to a great part of our heaths, which are yearly decreasing in extent, and converted into corn fields. And though our former small breed of sheep cannot be retained, yet there are at present many examples of mixed breeds, and of the
South

South Down sheep, thriving on our downs, and in heath soils, which have been limed and improved, and changed into sweet and rich pasture lands.

General Observations on the above Descriptions of Waste Lands.

Independently of what may be afterwards cultivated as sheep farms, or brought into a regular rotation of cropping, a very considerable proportion of the barren moors, of the ridges of the mountains, of the ancient forests, and uncultivated heaths, might very properly be added to the numerous plantations that have been raised in this county.

In the division of Marr, there is still a great extent of ground that produces little food to man, or to either cattle or sheep, which might be planted with some species of wood, if the land was inclosed; and by being at a little pains to make small ditches in swampy grounds, oak, beech, elm, and birch, might be raised on many empty or waste places, where the Scotch fir does not succeed. But the expence of making roads to the more distant mountains, or even to remote places of the same mountain, is often so great, that the proprietor allows the ground to ly in an unproductive state, rather than plant these inaccessible places. And, indeed, without going to the mountains of Braemar, there are many hills in the division of Marr, which are so much neglected by their owners, as to be little better than wastes. These, for the most part, could be planted with some species of wood; and, from the great profit made by planting, they will probably soon be applied to this purpose.

On the bounding hills which divide the Garioch from Marr, Strathbogie, and Formartin, there is also a considerable quantity of barren ground, (part of which has been planted).

planted). The greater part of the remainder might be inclosed and formed into plantations, by dividing the hilly ground between different proprietors, in such a way that the boundaries might be straight lines, instead of the very irregular *marches* (as they were termed) between the different estates.

In the divisions of Formartin and Strathboggie, a considerable proportion of moor, at present covered with heath, might be subjected to the plough. And as no heath remains on land that is once thoroughly limed, very extensive fields of sown grass, in this event, would be found in places where the pasture, both for sheep and young cattle, is either heath, bent, or the coarsest natural grass.

In the extensive division of Buchan, there are at least 100,000 acres of land, over which the plough, several centuries ago, has had a partial authority, (namely *narrow ridges* and *broad baulks*, or wide interstices of barren land). From the spirited exertions of many of the landed proprietors and farmers, in this division, there is reason to hope that these long neglected fields will be restored to the dominion of the plough, with more extensive authority than it formerly possessed; and when the whole are properly cultivated, will raise a very great additional quantity both of corn and grass.

One unpleasant circumstance respecting the boundaries between the five divisions of the County, and even the boundaries of individual proprietors, where not exactly ascertained, deserves to be mentioned in this place. The *proving a right to a barren and unproductive waste* is sometimes attended with more expence, than would be occasioned by *improving it*, so as to render it highly fertile—A few acres of a barren mountain become the subject of litigation. To the *proprietors or claimants* they continue *waste*; but are most *productive to* *the lawyers*. Even a small patch of ground will become a great bone of contention; for there are no *wastes* in this county in the eye of *the law;* but the most barren is the property of some person, or other; although the proprietor's right may be dormant, or not well defined.—These disputes are best settled by arbiters mutually chosen*.

Dr. Anderson, on this point, has inadvertently fallen into an error, which requires to be corrected. Speaking of the calamities occasioned by the unfruitful seasons from 1693 to 1700, he says—" Of so little value was land in this county, " at that period, that there are instances of considerable " tracts of corn land being so totally abandoned, as to be " allowed to pass from one proprietor to another, merely by " a prescriptive title of occupancy,—for upwards of 40 " years, without a challenge." It is here necessary to remark

---

* A small rivulet, in the district of Garioch, from the accidental lodging of a few turfs in the middle of its channel, about which a quantity of sand and loose earth was collected, gradually formed a diminutive island, or what is called *an inch* in the provincial dialect of Aberdeenshire. In the course of a few years, this *inch* became the subject of litigation between the proprietors on each side of the rivulet. After some time spent in wrangling and contention, they agreed to meet at a public house, or country inn, about six miles distant,—along with two gentlemen who were to examine witnesses, and settle the matter as umpires or arbiters. A lawyer, who attended on the occasion, and who had more good sense than his client, requested a farmer in the neighbourhood of this *inch*, to take a spade and throw the patch of ground down the rivulet, and apprize him when this was accomplished.—The farmer did so; and apprized the Lawyer, as he was ordered. One of the parties being called out, and understanding what had been done, came into the meeting in a great rage, swearing that the *ground of dispute* was taken away. The arbiters, who were informed of this at the time he was called out, said they were happy that the matter was settled. " No," says he, " it is *not settled*—the ground " has been thrown down the water"—So may all trifling contentions end.

mark that by the Scotch law, a prescription of 40 years, on *a sasine*, or infeftment, constitutes a right; but no term of possession without any term in writing, can preclude a man, who has a written title, from ejecting the occupier, who has only prescription in his favour.—In general, landed estates have what is called bounding charters; so that in the legal sense there can be *no wastes*, or lands which have no owner.

Two very important questions on the subject of waste land may claim our notice? If we cannot answer them decisively, we may at least call the general attention of those persons who are able to answer them in a satisfactory manner.

1. What is the proportion of the waste or barren to the cultivated lands in this county? 2. How great a population could be maintained by all the lands that already are, or that can be, made arable?

Dr. ANDERSON's original report contains some information on these topics, which ought not to be suppressed, and also some opinions that ought carefully to be examined.

" *Proportion of unimproved Ground in this County.*

" A wish has been expressed that the proportion between
" the cultivated and unimproved parts of this district might
" be ascertained. But as there has been no map of Aber-
" deenshire since that of Gordon of Straloch, published in
" BLEAU's Atlas, in which this particular was wholly disre-
" garded, I find myself unable to say any thing that can be
" deemed satisfactory on this head. But I should suppose
" that in the district of Marr, not much above one-fourth
" part of the whole surface has ever been brought under the
" plough;—In Formartin, perhaps, two-thirds, in Buchan
" three-fourths, and in the Garioch about one-half, has been
" at one time or other under culture. Of the lands that may
" now

"now be reckoned waste, perhaps one-half is susceptible of
"being brought under the plough. The remainder, con-
"sisting of hills that are too high, or too steep and rugged,
"is not accessible to that implement.

"The inclosed ground does not exceed one twentieth
"part of the present arable; and not above one hundredth
"part of the arable ground possessed *by the farmer*, is in-
"closed.

"To the possible improvement of this county no limits
"can be assigned. It is certain there are many patches in
"it, which at present produce a thousand times more food
"for man and beast than they did twenty years ago; and
"there is not perhaps, in the whole district, a hundred acres
"that may be supposed to yield nearly its utmost maximum
"produce; though I doubt if there be another in the king-
"dom, under the plough, that approaches so near to it. The
"land is, in every part, susceptible of a high degree of im-
"provement; if canals shall be opened up through every
"part of the county, where these are practicable, and other
"measures adopted that the rising prosperity of the nation
"might suggest, I should think it by no means an over-
"strained computation to say, that this county might be
"brought to support within itself a hundred times the
"amount of its present population. But the reader will be
"pleased to advert, that my opinion of the possible improve-
"ment of land greatly exceeds that of most persons who
"will speak upon the same subject.—What is just stated
"above, that I have actually seen in Aberdeenshire, is a
"proof it is not unfounded; and every person in the coun-
"ty has it in his power to know if the facts be so or not.—
"What has been done in one case, it is certainly possible
"may be done in another; though many are the bars that
"remain to be removed, and long may be the time before
"the

" the minds of men shall become sufficiently enlightened to
" admit of any thing approaching to what I have here hint-
" ed at."

On this long quotation from a very able writer, it is necessary to make a few remarks, both on the proportion of cultivated and inclosed land; on its probable or possible increase; and on the population which this county can be ever able to support.

1. As to the proportion of arable land in the division of Marr.—When Dr. ANDERSON supposed that about one-fourth of this division had at any time been under the plough, he did not consider, or probably did not know, that about 280 square miles of this district, which included all the lands on the south of the Dee belonging to Aberdeenshire, between this river and the Grampian mountains, along the boundaries of Kincardine, Angus, and Perth, had been overlooked altogether in his Survey; and that above 240 miles of this extent, being almost entirely mountainous, never could be cultivated or rendered arable by the plough. With the extremely small proportion of arable land on the north side of the Dee, in the upper parts of Marr, he also seems to be unacquainted.—Instead, therefore, of saying that one fourth part of Marr in general had at any time been brought under the plough, it would have been nearer the true proportion of such arable to the waste or barren lands to have said,—" Of " nine hundred, or nine hundred and twenty square miles, in " this extensive division, probably one-sixth part, or one " hundred and fifty square miles, was the highest computa-" tion of the proportion of land that had ever been plough-" ed.*

2. As

---

* The great estate of Invercauld has only 2858 arable lands out of 86130 Scotch acres; that of Abergeldie has only 843 English acres arable.

2. As to the proportion that has at any time been under the plough in the division of Formartin, which contains a number of hills, and a great extent both of peat moss and of moorish soil, and barren heath, or coarse *benty riesks*, one half of its surface, or 130 square miles, is the highest proportion of land under the plough at any period.—Dr. ANDERSON says only *perhaps* two-thirds, which implies a doubt whether there was so much.

3. As to the division of Buchan, which contains a very great proportion of peat moss, moorish grounds, one extensive mountain, a number of small hills, and nearly one whole parish blown over with sand, all the lands under the plough, or that have ever been subjected to its operations, (deducting those which have been blown over with sand), cannot be estimated above two-thirds of the whole extent, or 300 square miles.

4. The division of the Garioch, though the valley itself has a much higher proportion of arable land, has so little of this on its bounding hills, that it is very fairly estimated at one half of its surface under the plough, or 75 square miles.

5. The division of Strathboggie, though its whole surface be nearly equal to that of the Garioch, does not contain above 55 square miles, or three-fourths of the arable lands belonging to that division.

On a general review of what may be called *the waste*, or unarable lands of Aberdeeenshire, that is, those which have been at no time under the plough, they amount to nearly 1250 square miles, and the whole arable to 700 miles in

arable, from 23240 of surface; that of Earl FIFE, of 100,000 acres, has not above 2000 arable.—The woods and grazings are more valuable than the arable lands in these mountainous districts.

in round numbers—Or taking the whole surface of the county at a million of Scotch acres, which it is extremely near, the arable lands may be stated pretty correctly, at 360,000 Scotch, or 457,500 English. Dr. ANDERSON, however, seems to be very well founded in supposing that nearly as much more as has ever been under the plough could be added to the arable land;—in other words, that double of the former quantity might be brought into cultivation. By this means, 720,000 Scotch, or 915,000 English acres, might be made arable. One fifth of the whole area of the county (or 200,000 acres) consists of lofty mountains; and 80,000 Scotch acres must yet be allowed for rivers, roads, canals, or waste lands that cannot be cultivated in the lower districts of the county.

From considering the proportion of the arable to that of the waste lands, and also what addition could be made to the former, it may be proper to enquire how great a population could be supported, if the above 720,000 Scotch acres were all subjected to the plough.

The population of the county, being 123,000, we shall, to avoid fractions, call it 120,000, or one-sixth part of the number of acres that are supposed capable of cultivation— At this rate, every individual at present has the produce of six Scotch, or seven and a half English acres. How many persons could be contained, if the whole 720,000 acres were all in the highest state of cultivation?—The question, tho' apparently speculative, is of very great importance; because, when men aim at *the highest objects,* they certainly acquire a higher degree of excellence than if their desires had been more moderate.—Dr. ANDERSON, a man of genius and literature, and enthusiastically fond of agriculture, supposes that the arable lands of Aberdeenshire could maintain, in their highest possible degree of improvement, 100 times their present

sent population. These must be considered as *jactantia verba*, or words spoken at random, and not intended to be literally or strictly interpreted. But let the reader be pleased to read, not a computation made by fancy, but a short calculation in figures, where the Writer certainly wishes to simplify knowledge, and convey useful information in the plainest language.

As this county is situated in a high northern latitude, the Writer of this Report certainly considers it as an overstrained hypothesis, (not to speak of any computation), to suppose that it can support one hundred times its present population, or above 12 millions, nearly a million more than are at present in Great Britain. He doubts whether such a population could be supported even in China, where a crop of rice is obtained twice every year: and he is certain that such a population could not be supported by any crop which can be raised in this county.—But he hopes to satisfy every candid and attentive person, that ten times our present population could be permanently supported by the produce of 720,000 Scotch acres, such as we have in this county, and producing crops that are suited to our soil and climate; and not estimated beyond a moderate and reasonable produce.

He assumes as data the following articles, which he hopes may be allowed him:

1. That one-sixth part of the said extent of land, or 120,000 Scotch acres, may be brought to endure a four shift course of 1st, turnips, potatoes, or naked fallow—2d, bear, or big, or wheat after fallow—3d, hay, or clover, for soil—4th, oats.

2. That ———— of this extent of land, as being of secondary quality, is fit only to endure the five-shift course,—1st, turnips or potatoes, one-half of each green crop—2d, bear

bear or bigg, with grass seeds—3d, hay—4th, pasture—and 5th, oats.

3d. That other two-fifths of it, as being of inferior quality, should be cultivated in a sixth-shift course, consisting of 1st, turnips or potatoes, one half of each—2d, oats, with grass seeds—3d, hay—4th and 5th, potatoes—6th, oats out of lea.

4. That a crop of wheat in the best soils, after a bare fallow, will for every acre supply four persons with food, during a year, i. e. a man, woman, and two children, supposed to be in opulent circumstances.

5. That a crop of oats, besides sowing the ground, after potatoes, turnip, or lea, will, on every two acres, raise food for a man, woman, and two children, in moderate circumstances.

6. That a crop of bear, or bigg, after allowing for seed, will, on every two acres, also afford nourishment for a man, woman, and two children, in poor circumstances.

7. That an acre of potatoes will amount to ten tuns avoirdupois, and would supply food for two men, two women, and four children.

8. That an acre of turnips employed in rearing or feeding cattle, will amount to at least thirty tuns; and will raise butcher-meat equal to the support of a man or a woman, with a child, or only one-fourth of the potatoe crop.

9. That all the grass, and pease and beans, and vegetables, except as above, are supposed to maintain and feed the horses, sheep, and black cattle, and to raise milk, butter, and cheese, for the community.

10. That while the rich eat wheaten bread, they exchange part of it for butcher-meat and vegetables; that those in moderate circumstances exchange oats or bear with the poorer classes, for potatoes; and on the contrary, that the poorest

est classes, though they use potatoes, and other vegetables, share in oat-meal and bear.

On the above data we will find that 1,220,000 persons, or nearly 10 times our present population, could be supported by the produce of the lands that might be made arable, without overstraining our computation, or calculating from too high a produce per acre. The particulars may be concisely stated under the following articles—beginning with the 120,000 acres in the four-shift course:—

1st year, fallow for wheat in the four-shift course, 10,000 acres, no produce at all.

	Persons.
2d do. turnips on other 10,000 acres of this land, will produce butcher-meat equal to the yearly maintenance of persons of all ages, at two for the Scotch acre,	20,000
Do. potatoes on other 10,000 acres, food for 8 persons per acre,	80,000
2d. The wheat after 10,000 acres of fallow, food for four persons per acre,	40,000
3d. The barley after potatoes and turnips, 20,000 acres, food for two per acre,	40,000
4th. The oats after lea, 30,000 acres, food for two persons per acre,	60,000

(The hay and night soil, other 30,000 acres, allowed to the inferior animals)

The number of persons maintained by 120,000 acres of best land, is two per acre in this four-shift course, or, - - - - - - - - 240,000

On the 300,000 acres in a five-shift course, the number stands thus:—

1st year, potatoes, 30,000 acres, food for eight persons per acre, - - - - - - - - - 240,000

Carry over, 240,000

Brought over,	240,000

Do. turnips, 30,000 acres, butcher-meat, for two persons per acre, - - - - - - - - 60,000
2d. Oats, with grass seeds after potatoes, 30,000 acres, food for two persons per acre, - - - - 60,000
Do. barley or bear with do. do. - - - - - - 60,000
3d and 4th. The hay and pasture of 120,000 acres allowed to the lower animals.
5th, oats after lea, 60,000 acres, food for two per acre, as before, - - - - - - - - - - - 120,000

On this five-shift course, every 10 acres maintains 18 persons.—In all, - - - - - - - - 540,000
On the 300,000 acres of land in a six-shift course, the number is as follows.

1st year, turnips, 20,000 acres, food for 2 persons in butcher-meat, - - - - - - - - - 50,000
Do. potatoes, 20,000 acres, food for eight persons per acre, - - - - - - - - - - - - 160,000
Do. pease, beans, or coleworts, on wet or poorer soils, 20,000 acres given to cattle.
2d. Oats, with grass seeds, after these three kinds of crop, 60,000 acres, food for two per acre, - - 100,000
3d, 4th, and 5th. The hay, and two years' pasture given to the lower animals.
6th. Oats after lea, food for two persons as before, 120,000

On this six-shift course 10 acres *maintains 15 history* —In all, - - - - - - 450,000
The five-shift course as above, for 300,000 acres, - 540,000
The four-shifts, on 120,000 Scotch acres, - - - 240,000

The total number that may be maintained by 720,000 Scotch acres, nearly 10 times our present population, - - - - - - - - - - 1,220,000

But

But after the lands have been 20 or 30 years in these rotations, a great proportion of wheat could be sown; and a large share of the five-shift course could be moved forward into the four-shift; and also a considerable extent of the six-shift could be moved into the five-shift, which on the whole is best adapted to the soil and situation of this county. By this means an English acre might support as many persons as a Scotch acre is here supposed to do. And 12 times the present population could be maintained by 720,000 Scotch, or 915,000 English acres, the utmost extent of arable land in Aberdeenshire—Supposing the population of Great Britain increased to 100 millions, what nation could do us any injury?

Perhaps it would have been more accurate to have valued the butcher-meat contained in the produce of 270,000 acres of hay and pasture, and to have discounted all the barley or big, that is consumed in malt and ardent spirits; but this would have rendered the calculations more complicated; and in fact, a considerable quantity of beer or bigg is made into meal by the peasants, and even by persons of better condition.—It will, however, be remembered, that after allowing hay and pasture for horses, and for rearing sheep and black cattle, the butcher-meat obtained from grass would more than balance all the quantity of barley or bear that is made into malt or ardent spirits. What is manufactured either into meal, or into pot-barley, goes directly to the food of man.

On the whole, the Writer of this Report has the highest veneration for the memory of the late Dr. ANDERSON: but notwithstanding the respect which he has for his genius, learning, and for both that speculative and practical knowledge, and that zeal for the interests of agriculture, which Dr. ANDERSON undoubtedly possessed, he must repeat, that the computation that this county, if cultivated to the highest pos-

possible degree, could support an hundred times its present population, has been overstrained, or made hastily. At the same time he thinks that the lands of this county could (if part of that mercantile capital, which has been unprofitably directed to speculations in commerce, were laid out in improving our waste lands, and improving more highly those which are already under the plough) be easily made to support ten times, and, in the course of 20 or 30 years, might support twelve times the number of its present inhabitants.

An excessive population, as MALTHUS has shown incontrovertibly, is no advantage to a nation.—But where this population is supported by agriculture, where green crops can be raised in abundance, to preserve the lives of men in calamitous seasons, and where the manners of a numerous peasantry are simple, their morals pure, and where, from the love of liberty, every man should be willing to die for his country, the number of the people may be increased in a ten-fold proportion; and a manly character, and warlike spirit, by proper institutions, may be infused into the breast of every man, in this free and happy country.

The reader will forgive this long disquisition, which it is hoped will be found a pardonable digression from the subject of waste lands, as the writer's object is to rouze the nation to improve those which are still in that condition, and to cultivate, as highly as is possible, (by regular approximations, and by courses of cropping adapted to the nature of the soil and climate,) every acre in the kingdom. Could his feeble voice be heard by men in power, the tyrant of France should soon feel that we are not a nation of shop-keepers only; but with ships, colonies, and commerce, and without attacking learning, destroying trade, or violating the liberty of the press, can unite the professions of agriculture and arms.

CHAP.

CHAPTER XII.

IMPROVEMENTS.

SECT. I.—TRENCHING.

ALTHOUGH the plan pointed out by the Board of Agriculture, even with all its minute divisions and subdivisions, has omitted to introduce this section under the chapter of improvements, or under any other chapter or section, yet, as trenching is really the most correct method of improving barren ground that is unmanageable by the plough, and as it tends very much to meliorate the crops of arable land, it has been judged proper to introduce it, as the first section of this chapter on improvements, in the Report of the agriculture of a county in which it has been carried on both more extensively, and more successfully, than in any other part of the island. In describing this species of improvement particularly, and even minutely, although not included in the plan of the Reports, as amended by the Secretary of the Board of Agriculture, the Writer of this Report is acting agreeably to the concluding observation of the President, which is very happily expressed in the following words:—

"One point is of peculiar importance, that the Surveyors
" should

"should enquire into new or peculiar practices; and should
"ascertain as minutely as possible, the nature and effects
"thereof: For a single practice, discovered, by means of
"these Surveys, in a narrow district, or even on a single
"farm, if spread, through the medium of the Board of Agri-
"culture, over the whole kingdom, may add more to the
"national wealth than the possession of the Indies."

It has already been mentioned that the greater part of the land in the vicinity of Aberdeen has, from the most barren and unproductive state, been thoroughly improved by trenching. Not less than 3000 acres have been trenched within three miles of Aberdeen; and in all places of the county considerable additions have been made to the arable, by trenching the barren lands. On the personal farm of PETER GORDON, Esq. of Abergeldy, the Writer of this Report saw a large field undergoing the operation of trenching, at a great expence, not less than twenty pounds per acre; and he was informed that forty-two acres of barren and very stony ground, had by this means been added to the arable lands of this single farm. Several thousand loads of stones, taken out of the soil, had been carried to the river Dee, and laid as a sloping bank of granite to prevent its farther encroachments when swelled by the rains in summer, or by the melted snows and more dangerous shoals of ice in winter. These operations were conducted 46 miles from Aberdeen;—and two miles farther up the Dee, at Crathienaird, belonging to the Rev. THOMAS GORDON of Aboyne, the Reporter saw lately, (in July 1810), a large field of twelve Scotch, or fifteen English acres, undergoing the same operation of trenching, at the expence of L.16 per acre, besides the expence of carting off the stones, which will amount to at least L.5 per acre; and after these are taken off, the expence of manuring and dressing a crop of turnips will probably amount to L.6

more

more. At Allerg, in Curgarff, near the source of the Don, he saw another, but smaller field, also trenching out of moor, at the expence of the Rev. ROBERT FARQUHARSON. It was not so full of stones; and may cost about L.20 per acre, when the first crop of turnips is manured and ready for use. From these specimens in the highland districts, the reader will judge of the spirited exertions in this favourite method of thoroughly subduing barren and stony land, that prevails in Aberdeenshire.—No species of improvement is equal to that of trenching such unproductive and unmanageable soil. The spade and mattock with a quantity of gun-powder, and a few tools for boreing, will subdue a rocky soil which no violent effort of the power of horses, nor patient exertion of a team of black cattle could be able to accomplish. And it deserves to be attended to, that in those highland districts, where there is so great a proportion of waste lands, with very little arable, every acre added to the latter supports, during the winter months, the cattle that depasture the natural grasses in the glens, or distant mountains, in the summer season. A proprietor of an estate in Braemar, or Curgarff, can therefore improve land at an expence which a landholder in Surrey or Middlesex would be afraid of incurring.

The expence, indeed, owing to the rise in the money price of labour, has greatly increased within these 20 years: and the following account both of this practice, and of the expence of carrying it on, deserves to be quoted from Dr. ANDERSON's Original Survey.

" The only practice that can be accounted peculiar to this
" county among the improving farmers, is that of trenching
" ground by the spade. This practice was originally adopted
" for the purpose of getting the stones taken clean out of the
" ground, where they frequently prove troublesome to the
" plough; and it has been found to be beneficial in many
" other

" other cases. For by thus burying a coat of moory earth,
" which originally occupies the surface of many fields, and
" which being more difficult to fertilize, than the clay on
" which it lies, this trenching turns out to be a great im-
" provement. Indeed, on another account, that of deepening
" the soil, so as to allow the water to sink below the roots of
" the plants, wherever the subsoil is retentive, I am inclined
" to believe that this will be found in all cases, upon such
" soils, a very beneficial practice. For it is difficult to say,
" how much such a soil is benefited, by a considerable depth
" of loose earth below the furrow. The expence of trenching
" ground, that has never before undergone that operation,
" to the depth of from 12 to 14 inches, is from L.2 13s.
" to L.4 per Scotch acre. Ground that has been formerly
" trenched, is sometimes done as low as *two-pence* per fall,
" or twenty six shillings and eight-pence per acre."

Before the Reporter makes any remark of his own, he feels it his duty to quote the following observation which has been made by a Gentleman on the broad margin of the Original Report.

" It has long been a problem to me, whether or not the
" modern mode of improvement, by many ploughings, drag-
" harrowings, &c. (and perhaps after all, losing two years
" crops,) could not be cheaper accomplished by trenching
" alone. I am certain that the work would be more effectu-
" ally done; and where the soil is ill to reduce, and full of
" stones, I will venture to say, the odds will not be so ma-
" terial as people may conceive, all things considered. A
" marginal note would not contain a tenth-part of what
" might be advanced on this head. The burying of the
" moory earth, mentioned by Dr. ANDERSON, and bringing up
" good mould in its place, is of the utmost consequence; and
" what ploughing, even trench ploughing, could not so re-
" gularly

"gularly perform—the field will be regularly levelled, and
" drains put where necessary, as the trenching goes on, which
" saves a double work."

As trenching has been practised to a greater extent in Aberdeenshire, than in any other county in Scotland, or perhaps in Great Britain; and as the ground is both more thoroughly improved, and a greater quantity of human food is raised by this, than by any other method, it is proper to state the different ways in which it is practised, as well as the peculiar advantages that attend the different methods of trenching.

It is practised in barren land which abounds in stones of different dimensions; sometimes, where the soil is dry, and in other cases, where it is wet—united with draining. It is practised when the object is to deepen the soil; or to mix a portion of the subsoil along with it—it is practised when the subsoil is tilly or very tenacious, as well as when that next the surface is unproductive, moory, or exhausted by over-cropping—And lastly, it is practised when the land is foul, and when stronger or cleaner soil can be brought up to the surface.

1. It is practised in barren land, which abounds in stones of different dimensions. Used for this purpose, it is the most complete method of rendering such land arable. In this case, the surface is cut into square or oblong pieces, and thrown in the bottom of the trench, except when forming the first trench, which of necessity, is thrown on the barren land. The labourer cuts off a breadth generally of three feet, and throws out both soil and subsoil to the depth of about fourteen or fifteen inches, and as long as he finds it expedient.—He throws all the stones on the surface, and generally requires *two stamps and two shealings*, as he terms it, to fill up his trench. The first stamp, or spit dug by the spade, is about eight inches deep, if he be able to pierce so far into the ground—this is thrown into the open trench; then with a

shov-

shovel or spade, he throws the loose earth above that spit; after which he digs a second spit, which he lays above the former, and uses the shovel or spade a second time, to level the bottom, and throw the earth, sand, or clay, which he finds there, on the top of the trench. If these two spits and two gatherings by the shovel, amount to the depth required, he goes backward, and shapes off another trench:—if not, he must, in hard soil, dig a third spit, and then apply the shovel a third time. The stones are always picked out, and thrown on the surface.

But it often happens that the spade cannot pierce the stony ground. In this case, a kind of mattock, provincially termed a *pick*, is applied. By means of this, the stones and earth are loosened to the depth of several inches; and they are then thrown out with the spade—a second, third, and fourth picking are given, shealing the trench with the spade, between every picking, till the requisite depth is obtained.— Where the pick or mattock is not sufficient to raise some of the large stones, crows, or levers of iron, from three to five feet, are used; and wherever these fail, a long wooden lever, of several yards long, is applied. When no lever can remove them, the labourer digs around the stone and loosens the earth; after which the boring irons are used; and then gun-powder poured into the hole, which is afterwards jammed in with bits of stone, is lighted by a match, and the stone is blown in pieces; by which it is broken into such fragments as can be raised out of the ground. The labourer is generally allowed a penny for every inch depth of this bore, independently of the expence of trenching.

When the ground is wet, the labourer shews particular accuracy and dexterity in laying the bottom of the trench.— The grassy surface is most carefully laid on its back, and every sod or spit compactly joined together, that the water may

may form a new *pan* or channel, between the soil and subsoil. The bottom of the trench is cleared of all dirt or loose earth, and formed into a slope or level, with a small descent if possible for the water, before the surface is inverted, or compactly laid on its back. It is the care used in scouring the bottom of the trench, and laying the inverted surface flat in the bottom, and joining all the sods together, that renders trenching a permanent improvement in wet soils.

If draining be necessary, as well as trenching, the drains are cut as much deeper than the bottom of the trench, as that the whole water contained in the drain shall be below the level of the bottom of the trench; and the breadth and depth of the drain are proportioned to the quantity of water:—only, a small part of the bottom of the trench is made to slope gently into the drain, that no water may remain among the grassy sods, but that the whole moisture may fall into the drain. Conducted in this manner, trenching has been found beneficial to wet soils; when the inverted grassy substance has answered the purpose of filtering the moisture, and carrying off the otherwise stagnant water.

When the object is merely to deepen the soil, the ground in the subsoil is chiefly loosened by the pick or mattock, and then thrown by a spade above the spit of soil which was thrown into the bottom of one trench, after being taken off the surface of another. By this means the soil can be deepened is 16 or 18 inches, if necessary, where it was formerly only 5 or 6 inches deep. But in this case it is necessary that the bottom of the trench be completely picked up, and that no interstices of hard subsoil be left between the trenches. For this purpose the pick or mattock should take up the bottom by a perpendicular stroke; and the labourer must *clean the teeth* of the trench, so as to prevent any space being left untouched or unloosened by the spade. Negligence

gence here is very apt to break a plough: for after trenching, when the soil is loose like that of a garden, the horses go on rapidly, and somewhat carelessly; so that when they meet with a sudden obstruction from an *ill-picked* trench, a broken plough is the first indication of a ridge in the bottom of it, which was left by an unskilful or dishonest labourer. When properly executed, as is generally is, the deepening of the soil is a permanent improvement, that more than doubles the value of the land.

It often happens that the soil and subsoil are of different qualities, and by trenching them they are mixed; and a better soil than either is produced by the mixture. The surface no doubt is turned into the bottom; but by using the *shovel* twice in what are called the *two shealings*, the soil and subsoil are considerably mixed; and in the course of two ploughings the mixture of the two is completed. Mixing of opposite soils is thus a great improvement.

Trenching is also practised with great success when the subsoil is tilly and very tenacious. It is very usual to have five or six inches deep of moory, black, unproductive soil, next the surface of our moors, and clay or till below this.— When the moory soil is thrown into the subsoil of a trench, and a mixture of moor with the till in the bottom is laid on the top, the superabundant moisture is carried off, and the ground becomes dry and very productive.

When land has been once cropped, (especially after being limed, or injured by getting too great a quantity of lime), by trenching, and bringing up a few inches of soil from the bottom, it is completely renovated; and either dung or lime, or both, can be applied to the greatest advantage. Where additional depth can be obtained by the plough, as in deep clay soils, trenching is not necessary; even after over-cropping. But in this county such soils are not common; and

by

by piercing the pan and bringing up new soil, dung and lime are always applied with success. The Secretary to the Board of Agriculture mentions that the farmers in Norfolk are afraid of piercing the *pan* or taking up the hard subsoil, lest this should produce *charlock*. An Aberdeenshire gardener or day-labourer, who is as well acquainted with trenching as the other is with ploughing, has no fear of breaking the pan, or of making out a new channel. He knows that trenching destroys weeds; and if a few remain in a foul subsoil the hoe will soon extirpate them. But if the Norfolk farmer, either by the plough or spade, brought up soil that had never been cropped, or rested for many years, he would not, if he dunged it sufficiently, complain so much of its tiring of clover or of turnips.

Trenching is practised when the land is foul, or when either stronger or cleaner soil can be brought to the surface. This can be always done by the spade, when the common plough would not answer at all, and when the trench plough could act very imperfectly:—and when the price of labour was moderate, it was much used for this purpose; though it is now chiefly used in breaking up barren and rocky soils.

While the Writer of this Report felt it his duty to describe the process of trenching in different cases; and its beneficial consequences to the soil of Aberdeenshire, he thinks it but fair to subjoin a few observations.

1. What is certainly the most eligible and expedient method of improvement in one county may not be so in another. In Aberdeenshire, where granite is found in detached masses, or in small insulated quarries, and where the subsoil, whether gravelly or moorish, is often within a few inches of the surface, trenching has many and great recommendations, as above detailed. It deepens the soil, and it forms a regular

nel for water to pass, between the soil and subsoil, at a proper depth from the surface. The water which formerly oozed out in small springs, or chilled the soil, where it was retained, in winter, and rendered it unproductive, no longer meets with any impediment from the irregularity of the strata; but is filtered by the grassy surface in the bottom of the trench, and a new channel is formed before this grass is rotted. The subsoil is better and more evenly loosened than by Mr. ECCLESTINE's Miner, or is brought up to the surface from a greater depth than it could be by DUCKET's Trench-Plough. Nay, where it is too hard for any plough to pierce a bed of granite, this can be accomplished by the spade and mattock; and the ground that was formerly broken and irregular, may both be laid dry, and of an uniform depth, which produces both an equality of weight of crop, and of ripening in harvest. Where this trenched land has been exposed to a winter's frost, and then *well dunged* before it is sown with turnips, the expence of cleaning or hoeing out the turnips is a mere trifle; and as there are no weeds in the soil, a weightier crop, first of turnips, then of barley, and thirdly of hay or broad clover, is obtained, than could be raised by the plough. But though all these things are proper in the thin lands of Aberdeenshire, where granite so much abounds, it might be more expedient in the moory soils of Mid-Lothian (Gogar moor for example) first to plough for a crop of oats, then summer fallow, drain, dung, and afterwards take a crop of wheat. Nothing can be worse founded than to prescribe one uniform mode of improvement in all places, on all soils, and in all situations.

In the immediate neighbourhood of populous cities, where rents are high, and where dung can generally be procured in any quantity, by those who choose to pay the price of night soil, trenching is often very eligible; because horses cannot al-

always be hired for frequent ploughings, and because day-labourers in the winter months will engage to trench by the piece, i. e. at so much per fall, at little more than half the sum they would ask in summer. For this they have two reasons. First, the moisture in the earth enables them to trench the land more easily; secondly, they are glad to get constant employment in winter, though at inferior wages.—But let it be remembered always, that—

Every thing to which labour can be applied has its price. More dung must be used after trenching by the spade and mattock, than after cultivating by the plough. The soil brought up from the bottom is not for some time in a state of vegetation, till it be exposed to the frost, and to the influence of the weather; and though incomparably cleaner than that at the surface, stands in need of much more manure than the other would have required. The expence of trenching is greatly altered within these 30 years. In 1778 the Reporter was settled in his present benefice; and he has at different periods trenched about twelve Scotch acres of his glebe, or parsonage lands; a considerable part of which was taken out of barren moor. In 1781 he paid *three-pence half-penny* per *fall*, or L.2. 6s. 8d. per Scotch acre—In 1793, for a piece of stiff moory soil, he paid L.4 per acre—For the old croft land, in 1799 and 1800, he paid at the rate of L.6 per acre, for nearly four acres. He could not now trench the barren moory soil under L.16 per acre, or four times as much as in 1793.—And for what he paid L.6 per acre in 1800, he would now pay L.12. In short, it is evident that trenching by the spade is incomparably the most perfect of all modes of improvement; but between that and other approximations to perfect culture by common ploughing, trench-ploughing, and using a miner for loosening the subsoil, the command of dung, of labourers, and of ready money, produces such questions of expediency, as the relative situation of individuals only can solve. It is no doubt true that the greatest population can be maintained by the labour of men, who trench with the spade and mattock, on lands that are able to bear this expence, in those situations in which there is a superabundance of manure: But while both the citizens of Aberdeen, the gardeners in its immediate vicinity, and the proprietors of the lands of this county, in improving their personal farms, have set an example worthy of being followed by all persons who are in a similar situation; it is not pretended that this pattern ought to be universally imitated.— Excessively high pretensions are always improper. Infallible systems of theology, catholicons in medicine, precedents in law, which are pretended to be universally applicable;—and practices in agriculture to be followed in all soils, and in all situations; are the offspring of ignorance, illiberality, and inexperience.

There is indeed one case, in which trenching with the spade (and with the mattock also when necessary) ought to be generally imitated by other counties. This is when land of inferior quality happens to ly contiguous to better soil, and is surrounded by the same inclosure. It is observed by Mr. ROBERTSON, in his very able Account of the Agriculture of Mid-Lothian, that "inclosures are too often
" formed by straight lines into regular figures; and that
" there is an absurdity in this practice. For if the land is
" irregular, and also wet, the ditches, which constitute a chief
" part of the fence should be laid out in such a direction as
" to carry off the water at the same time; and even when
" it is dry and lying regular, the nature of the soil should
" be adverted to; that land of opposite qualities may not be
" included in the same close." This observation holds true in many cases. But, perhaps, in some situations, *draining,*
trench-

trenching, and *levelling* the wet or irregular grounds, and carting, from the too deep old croft land, a portion of its superabundant soil to the thin dry ground, would preserve the uniformity of division, and occasion a great improvement to the land of inferior quality, that happened to lie in one corner of the inclosure. The *line of beauty* no doubt is a curve line, and should be followed in laying out the *policy* as it is called, or the pleasure grounds, of a great landed proprietor. But the *line of business* is a straight line, which is the nearest distance between two points; and it is most convenient for the operation of the plough. Besides, the method here pointed out of trenching and improving such corners of an inclosure, is only restoring to every one his own. For *muck feal, and divot*, or turf carried to the dunghill, was at that time the order of the day, in every part of Scotland; and the *croft lands* were the great oppressors of the poor earth in their neighbourhood.

The importance of this article, and its being a peculiar practice in the agriculture of Aberdeenshire, will be an apology for the length to which it has swelled.

SECT. II.—DRAINING.

This is a species of improvement, which, when properly planned and executed, is both beneficial and permanent.—For it has been found that wet lands, when properly drained, carry very weighty crops. But if the drain is not made in the proper place, if it be either too shallow or too narrow, its good effects are but partial, and it is soon choked up, or rendered useless.

The landed proprietors, and also all the better sort of farmers, have for more than 40 years paid great attention to the draining of wet lands; and it ought to be stated in justice to the memory of Dr. ANDERSON, that he had not only *executed a number* of drains on his farm at Monkshill; but that he had *publicly* recommended the *tapping* of wet ground, before Mr. ELKINGTON's practice was known. This, however, does not derogate in the least from the merit of Mr. ELKINGTON, who first reduced draining into a system; and who well deserved the reward he obtained.

Many of the landed proprietors of Aberdeenshire have employed Mr JOHNSTON, a disciple of Mr ELKINGTON, to survey not only their personal farms, but those of several of their tenants, for the sake of obtaining directions from him, with respect to draining off the superabundant water. Some of the landholders have also agreed to allow the expence of drains as a melioration, to be paid for at the end of their tenants leases; and others pay the whole expence of draining, charging the tenant with the interest of the money so expended. Yet the greater part of the drains which have been made, on lands that are not in the immediate possession of the proprietor, have been made by the farmers.

The drains used in this county are of three kinds, which are distinguished by the names of *rumbling syres, laid* or covered *drains*, and *open drains*.

The first is made in the following manner:—A narrow ditch, from two feet to two feet six inches at the surface, from 18 inches to two feet in the bottom, and of such a depth as shall be found necessary to obtain a solid foundation on which the stones may rest, is cut through a field, in such a direction, or in two or three such lines as are judged most proper for carrying off the water. The loose earth, clay, and sand are most carefully thrown out of this with a shovel; and two rows of stones, set obliquely, and leaning to each other at the top, are laid along the bottom of the drain

drain, like the couples or roofing of a house. If the drain be narrow, and the stones picked out of the field are of a tolerable size, the whole water in the drain passes through between these two rows. If the stones be small sized, there are two and sometimes three rows of these couples, with a flat stone between every two rows. By this means a more considerable stream of water may pass through, in the open spaces. In short, both by rows of stones set like couples, and by flat stones placed on the top of these, as many small drains as are judged necessary are made in the bottom of the ditch. Above these, a number of loose stones of a small size, picked off the fields (which generally abound with these stones) are thrown promiscuously on the others, till the top of the drain is as high as the subsoil reaches, or from 14 to 16 inches below the surface—(The object here is to be below the reach of the plough, which might injure the drain or its cover; and to be near enough the surface to convey away the moisture of the land when ploughed.)—When the stones are all levelled by a spade on the top of the drain, they are covered with a quantity of weeds taken off the field, or with a coat of turf, pared by the breast-plough, (provincially, *flaughter-feal*,) which is laid with the grassy surface inverted, and placed very compactly above the stones. After this, the earth that was taken out of the ditch, or as much of it as can be used, is replaced, and the rest carted away,— then the ground is ploughed over.—This kind of drain is found to answer very well when there is only a quantity of superabundant moisture, provincially a *sink*, without any fountain of running water, except what is occasioned by making the drain. Sometimes if there be but a little water, the stones are thrown promiscuously without being coupled; but this is considered as very incorrect draining. Where the stones are coupled, or made to incline to each other at

the

top, and where the two rows fill the bottom of the drain, leaving an opening of from eight to ten inches in the middle, a spring-well is sometimes conducted through a field, when the stones are pretty large, by this coupled drain.—But

2. If a stream of running water, even though only a spring well, or small fountain, enters at the top, and runs along the whole course of the drain, it is generally found adviseable to use a *laid drain*, i. e. a row of stones laid on each side, with an opening of from six to ten inches between them, and a course of flat stones laid above these; and then to fill up with loose stones, and cover as in the former case.— The opening between the rows is made as wide as is necessary to contain all the water; and the flat stones above them, if they do not cover the whole, at least cover the edges of the rows. This drain being *nearly square*, is much more capacious than the coupled drain, whose openings are *triangular*; and by this method of draining, a considerable fountain, or collection of water in wet land, is carried off; and the whole field is ploughed over, without any appearance of drains in any part of it.

3. When there are several drains in a field, and when the streamlets of water collected by them are so considerable, that it is found exceedingly difficult, or from the want of long stones, impossible, to cover it properly, then an open ditch, or, as it is provincially called, a *main drain*, is left without any covering. In this case, the various smaller drains are built very carefully near the termination, or junction with the main drain; and the opening near the latter is made as wide as that a hoe or spade can be pushed up a few feet, to clear off any impurity when the main drain is cleaned or *scoured* occasionally. The bottom of the large open drain is also sunk a few inches below the bottom of the covered drain. Near Aberdeen, where the ground is

va-

valuable, and where splinters from the stone quarries can be obtained for covering pretty wide drains, even the main drain is covered: but in this case, it is commonly built with stones on each side, at least a foot high; and the bridges or flat stones which cover these are proportionally large; so that the opening in the bottom of the drain is at all times sufficient to hold the greatest quantity of water that can be expected to run through it.

In making out the two first kinds of drain, it is a rule strictly observed by the judicious improver, to sink the stones and bottom of the drain so deep in the *subsoil*, as that all the water can be contained without rising so high as to touch the soil; and also to fill up the drain so far that the top of it shall be able to receive any water from the soil, by being an inch higher than the *hard* pan or *subsoil*; or as the labourers express it, getting a *grip* both of the *land* and of the *hard pan*, or channel. When the drain is not sunk in the subsoil, a few inches at least, so as to contain the whole water, it does not dry the land; and where the top of it does not communicate with the soil, if this be very tenacious, it does not receive the surface-water, which stagnates between the ridges. Hence in stiff soils a medium depth of drain is preferred; viz. such as shall be always beneath the reach of the plough, but not so deep as to have no communication with the soil. It is however, a matter of great consequence, and therefore is particularly attended to, that the bottom of the drain be firm, and that all the *springs* or fountains be cut, or tapped, in the making of drains.

When it is found that a drain has not answered the purpose intended, a few more branches are made and connected with the former ones. But to prevent this second operation, a judicious drainer always cuts his drain, at least two or three yards higher up the field, that the spot where the
mois-

moisture appears. This was the practice of the late JOHN RAMSAY, Esq. of Barra, in draining some very wet fields near Straloch, above 35 years ago, and before the system of tapping was known in this county. In a sloping field in which moisture is found, it is evident that the fountain must be higher than the place in which the water appears.

Where the declivity of land is gentle, it is found that drains stand better, or are more durable, than where the water in the drain descends very rapidly. Wherever the current is very quick in its motion, quantities of sand are always collected, (especially near an angle or turning in which one drain is joined to another) and this collection of sand frequently chokes or fills up the drain. To prevent this, an experienced drainer is very careful that the angle which one drain makes in running into another, shall not be sharp or acute, but either obtuse, or else made with a sweep, or in a curved form, immediately before its junction. Where several branches meet, near the head of a principal drain, which are provincially named its *toes* or *taes*, (from some resemblance to the letter T),—these branches generally enter it at an obtuse angle.

In covering drains, when turf or a quantity of weeds cannot be had, rushes, broom, furze, (or whins as they are called) are used for covers: and the farmer in general is very careful, that no earth fall into a drain, or be mixed with the stones, when the drain is laid, and before it be covered. It is of no consequence, though the cover of the drain should rot in a few years; because the earth laid immediately above it soon acquires a degree of consistency, and is in no danger of being mixed with the stones, unless the land be ploughed by far too deep, or the top of the drain has been improperly filled up too high, i. e. too near to the surface of the soil.

The practice of keeping drains open, or of having a number

ber of open ditches in wet lands, though very common in England, is attended with a great loss of ground. Though stones cannot be had in many places, so readily as in this county, yet wherever there is abundance of brush wood, small drains may be made at a moderate expence, into the lower places of any field, and the main drains only, in such a case, require to be kept open. Indeed, wherever the current of water is either very rapid or pretty considerable, an open main drain is most adviseable,

SECT. III.—PARING AND BURNING.

This not only is very little practised, but in most leases or covenants with tenants, is expressly prohibited. Burning the furrow made by a plough, was formerly very common on breaking up those lands which are still named *burnt-lands*. The practice was to set fire to the furrows, when they were so dry as to be capable of burning; but that mode of consuming peat-earth, as a preparation for tillage, is now abandoned. The cause of giving it up was, that owing to the treatment which it received, (viz, bearing three successive crops of oats, after burning the surface), such land was completely exhausted. Burning six or eight inches deep, or as deep as the plough could reach, and then cropping severely, was wretched husbandry. But where dung cannot be obtained, it has been doubted by many, whether paring off one or two inches with the breast-plough, (provincially, *casting thin flaughter-feal*), would not be beneficial; provided a crop of turnips were raised on this mossy soil, from the ashes of the surface, and the land, which had been drained before burning, were, after the turnips, laid down with bear

or

or oats, and sown grass. There may be cases in which this experiment may be tried for once; but it certainly ought never to be repeated on the same land.

SECT. IV.—MANURING.

The various kinds of manure, which are used in Aberdeenshire, in such quantities as deserve to be named in this Report, are the following, viz.

1. *Night-soil,* or street-dung, is used on the lands in the vicinity of Aberdeen, and is sometimes carried several miles into the country, by the landed proprietor or farmer's carts. It is also occasionally carried up the canal to Inverury, a distance of 18 miles by water-carriage. It is not a strong or durable manure; but it operates very quickly; and minutely divides the particles of the soil on which it is laid. The Commissioners of Police, to whom the charge of cleaning the streets is, among other things, committed by Act of Parliament, employ a number of scavengers to collect this species of manure, which is a nuisance if allowed to remain in the streets. Night-soil is very unequal in point of strength. It is formed into dunghills in different places in the suburbs, where the dung is mixed, laid out in square plats, and sold by auction—not according to any determined measure or weight, but by lump, generally 10 or 12 loads in a hill. It sells at all times very high—And if laid on the side of the canal, toward the approach of turnip season, when the *demand is greatest,* its price is from seven to nine shillings per tun. In the cheapest sales it is seldom below five shillings for about 15 cwt. or six shillings and eightpence per ton. The Writer of this

this Report paid six guineas for 16 tuns; and the expence of carrying it to Inverury was two shillings per tun, to the canal company, besides paying for the boat, horses, and servants; so that a tun of night-soil, carried from Aberdeen to Inverury, would cost altogether about 12 shillings. Some years ago, night-soil was carried from London to Aberdeen, and laid on lands in the parish of Belhelvie, about 7 miles from that city. The price of dung is one of the best criterions of the spirit, or zeal for agriculture, among the farmers of this county. The Commissioners of Police, in the course of the year 1809, received for night soil L.1158 17s. and the quantity sold by private families, or scavengers not employed by the Commissioners, would amount to at least one-third of that sum; so that the night-soil of Aberdeen sells for L.1500 yearly. From fifteen to thirty tuns are laid on the acre.—It pays best when laid on a field of ruta baga.

2. *Stable-dung* belonging to the inn-keepers, stablers, and others in Aberdeen, is of various qualities, according to the feed of the horses. That which is got from the poorer stablers, where the horses are fed chiefly on straw, with a proportion of oats, sells from two to three shillings per load, or from three to four shillings and sixpence per ton. But at the stables kept by the principal inn-keepers, and owners of carriage and saddle horses, where hay and oats are given liberally, stable-yard dung is very high; and can scarcely be had at any price, without being particularly acquainted with the owner of the horses. It sells from four to six shillings per load, or from six to nine shillings per tun. It pays best when laid on a field intended for potatoes—And from 20 to 30 loads are laid on the acre.

5. *Soap-lees* are also used as a manure, in the neighbourhood of Aberdeen, and operate very powerfully and quickly; but they can seldom be had in any quantity, as the soap-boil-

ers in Aberdeen do not carry on such extensive works as to make this a general object to the Farmers. The soap-lees sell from three to five shillings per load;—or from five to eight shillings per tun.

3. *Cow-dung.*—The Cow-feeders in the neighbourhood of Aberdeen, frequently apply their dung to the fields which they rent. When they sell it, they get from four to eight shillings per tun, according to the supposed value of the dung, and the cow-feeder's distance from the city.—From 20 to 25 loads are commonly allowed to an acre of turnip-lands, where cow-dung pays very well.

4. The Dung of the butcher-market is usually sold in the lump by auction;—But is esteemed as the most valuable of any, and is supposed to sell from eight to ten shillings per tun, according to the demand at the time it is sold.

6. The refuse of whale blubber, at a particular season of the year (i. e. after the Greenland ships arrive) is also used as a manure. It is not powerful or durable; but it minutely divides the particles of the soil, raising a very rich crop for the first year, and a tolerable one for the second. It is improper to lay it on grass lands, as the animals who depasture the grass, or eat the hay, do not relish its flavour; but it may be successfully applied either to corn-lands or to a field intended for turnips. Its price is various, according to its quality. For there are two kinds of it, one of which contains much more oil than the other. Those who purchase it generally mix some stable-yard dung with it, to prevent its overflowing the casks in which it is carried: for owing to its liquid state, it cannot be carried in a cart like other manure.— Yet the Reporter has seen it carried above six miles from the shore, by that spirited farmer, Mr Pirie, in Mains of Watertown; who purchases it by the boat-load, not by the tun weight.—It was mixed with dung from the stable-yards, and

con-

conveyed in large casks, which seemed to be old puncheons.

7. Near the sea coast, especially of Buchan and Formartin, immense numbers of sea dogs, after the fishers have taken out their livers for oil, are used as a manure. In some seasons they are caught in greater quantities than in others.—Their price has risen, within these 30 years, from ninepence to three shillings per hundred.—But in some places six, and in others, seven score are allowed to the hundred. It has already been mentioned that PATRICK MILNE, Esq. of Crimonmogate, has used seventy thousand of them in one year. It may be added, as a specimen of the quantity taken by the different fishing towns, that in the fishing village of Cairnbulg, the population of which consists of 186 persons, the average number of sea dogs for the last three years, is 51,000. It is probable, that there are at least a million of them taken in a favourable season, in the whole fishing towns in the county; and supposing the average price only *two* shillings per hundred, or 20 shillings per thousand, their total amount would be L.1000. They are a most powerful manure for poor soil that has been for the first time brought into cultivation.—They last only for two years; and like the refuse of whale blubber are of great service in minutely dividing the particles of the soil—Mixed with peat-moss, in such quantities as to prevent the worms, which form in their bodies, from escaping, they are a very cheap manure to poor soil that is near the shore. But it is both unpleasant and expensive to carry them to a great distance.

8. The dunghills near fishing villages are also valuable, from the refuse which they contain. But their price is extremely variable, according to the situation of the village—perhaps the average is three shillings per tun.

9. *Sea-weed*, provincially *sea-ware*, is also used in considerable quantities on land near the sea-coast. It is usually

applied to lands intended for a crop of bear or bigg, and sometimes it raises very great crops. But it is very remarkable that the bear raised from sea-weed on the coast of Aberdeenshire, is of inferior quality to other bear raised in the county; and therefore, when the Sheriff strikes the fiars, it is always from a shilling to two shillings cheaper than farm or market bear. Yet the barley on the coast of Dunbar and East Lothian, raised from sea-weed, (perhaps from their soils being a heavy clay, while that of this county is a light loam, or gravelly soil), is always of the best quality.

These are the principal kinds of manure used in the neighbourhood of Aberdeen, and of the sea-coast.

10. Peat ashes—are used in as great quantities as peat is consumed in the county:—And principally from the quantity of salt contained in them, they are extremely useful in pushing forward the young plants of turnip. But they endure only for one year; and therefore a more powerful manure ought to be applied along with them. It may be doubted whether *peat* is not a more expensive kind of fuel than coals, where a man resides within ten miles of the sea coast—But the value of peat ashes will induce the farmers in general to use peat in preference to coals, and even the landholders to consume a considerable proportion of peat in their kitchens. Six loads of them to a Scotch acre will have a great effect in the early stages of a turnip crop,—after which, other manure, slower in its operation, but more durable, will carry on what the other had begun.—Peat-ashes should be kept in ash houses, by every farmer; for if they get rain, they soon become very insipid. Therefore, they should be covered, or laid in a hollow spot, where the rain cannot carry off any moisture. They were used as a top dressing to bear, before turnips were commonly raised. But

But now they are chiefly applied to the turnip drills, where they are most beneficial in the early part of the season.

11. Lime is used in very great quantities, and in all proportions, from five or six bolls per acre, as a top dressing, on land that has been repeatedly limed, to 25 or 30 bolls of shells, of 128 Scots pints per boll. It was at first applied by the farmer to their outfields, and other lands of inferior quality, in a most improper manner. For it was laid on the surface, and after lying some time exposed to the rain and frost, was ploughed, and then five crops of oats were taken in succession: Nay, sometimes eight or ten successive crops, (without any manure, except the first liming, and with the intervention of a crop of pease every third or fourth year), were taken, on lands that had either been infield, or were esteemed to be of good quality. But the ground that was treated in this manner was destroyed, or rendered unable to bear the expence of cultivation. Lime is now generally applied, either after a complete fallow, or along with turnips, or as a top dressing to the land, that is newly laid out with grass seeds. The finest or purest lime is the lightest. Some of the coarser lime shells, imported from Lord ELGIN's works, as well as those which are burnt in the division of Marr, and in some places of Buchan and Formartin, weigh 28 stones Dutch, or 490 pounds, neat avoirdupois, per boll. Some of the finest lime imported from Sunderland, or carried from Ardonnel in the division of Strathboggie, will not weigh above 24 stones Dutch, or 400 neat pounds avoirdupois; yet will yield from two bolls and a half to three bolls of slacked lime, from the boll of shells; while the coarser or heavier will not produce two bolls. The farmers therefore, proportion the quantity of lime, according to their opinion of its quality, as well as the condition and strength of the soil on which it is laid. For a slight top dressing, five or six

six bolls of shells, or 16 of slacked lime,—for a moderate liming, 12 or 16 bolls of shells,—and for a thorough manuring with lime on new or strong land, from 24 to 30 bolls are used. The measure of a boll of lime shells contains 128 Aberdeen corn pints of 60 ounces and 12 drams avoirdupois of water, or 105 cubic inches for each pint. This boll is almost exactly six and one-fourth Winchester bushels, and the chalder or sixteen bolls, 100 such bushels. The price of the boll at present is four shillings and eight-pence for the best lime-shells. So that the expence of liming varies from 25 or 30 shillings per acre, for the very slightest top dressing, to four, five, and six pounds, for the price of lime only, besides carriage from the lime-kiln or sea port. The expence by land carriage is proportioned to the distance; but when carried in the canal boats to Inverury, is only one fourth part of the price of the lime, when laid down at Port Elphinstone, at 16 miles by land, or 18 by water, from the quay at Aberdeen.

12. In the neighbourhood of that city, old *lime walls*, or rather the lime mortar and small stones, along with old plaister, are also used as a manure; and sell from eighteen-pence to three shillings per ton, according to the supposed strength of the mixture. This manure operates more quickly than lime shells, but not so quickly as night-soil, though it is more permanent than this, especially on thin shallow soils.

13. The *rubbish of old houses*, whether consisting in the rotten covers of turf and straw, or the mixture of clay and lime mortar of old walls, is also used as a manure in the inland parts of the county. It is sometimes mixed with animal dung in a compost dung-hill; and is seldom sold, as it generally belongs to a person who rents more or less land. When sold, it fetches rather less than half the price that is paid

paid for the lime rubbish, provincially *lime redd,* of Aberdeen.

14. *Soot* is also used as a manure for turnips, or a top dressing for corn that is sown along with grass seeds. The boll of it costs about two shillings, varying from 20d. to half a crown, according to the season of the year and demand for this manure. It is cheap, but not durable.—It is very light, so that 6 bolls of it can be put on a cart; and one can get as much of it for thirty shillings as will answer for a top dressing to an acre of turnips or bear. It should not be laid on grass land; but on turnips or on land that is sown with bear and grass seeds: for the cattle do not relish the grass till the soot be all washed into the soil.

15. *Compost dung-hills,* made up of the earthen walls of old inclosures, peat moss, the scourings of ditches, weeds that have been previously rotted, and the dung of horses, black cattle and sheep, are made up by the judicious farmers with great care; and laid on the turnip lands in very considerable quantities; from 20 to 40 loads of a single cart per acre. In this compost dunghill, the dung of black cattle is preferred to that of horses in general, as it produces more lasting effects. But good stable dung is considered as the best manure for potatoes. The dung of sheep is most generally used by itself; and produces a powerful effect on the shallow soil of the outfields or faughs, though it is not durable.

16. For the gardener, the contents of the necessary, the dung of swine, poultry, and pigeons, are found very beneficial in raising onions, and various garden roots. These are found chiefly near the houses of the landed proprietors, and are never sold.

It has been already noticed in the preliminary observations, that there is too little attention paid either to the making

ing of dung, though this is the moving power of the farmer. The following quotation from Dr. ANDERSON's Original Report, will shew that the account already given is not exaggerated.

"I know no such instance of barbarous management in "regard to agriculture in Aberdeenshire, as the general "practice that prevails there respecting dung-hills. In "every particular it is slovenly in the extreme, and calcu- "lated to occasion an unthrifty waste in a thousand ways. "The dung is usually thrown out through a hole in "the wall, made for that purpose; and there allowed to "wither in the sun and wind, with a large irregular sur- "face, exposed just as it chances to fall, while the moisture "is suffered to drain from it, as if its presence would prove "nearly as hurtful as the juice of the cassava root, which "the natives squeeze from it with care. That juice is "allowed to run upon the highway, or into rills or rivu- "lets of water, or any where that chance may lead it to, "without the smallest attempt to preserve it. While this "operation is going forward at the home stead, the farmer "is busied digging up some sods in any waste corner, usually "the worst that he can find; and for the most part, moory "earth that proves a poison to any soil on which it can be "applied, as far as its influence can extend. This he calls "*muck feal*; and is at a great expence in casting his dried "dung and it together into some corner, where he mixes "them together in the proportion of three or four parts of "earth to one of dung. This mixture he turns over perhaps "once; and after lying some months in this heap, the whole "is carted out to his infields at a very considerable expence. "And this kind of earth has no doubt contributed in part to "confer upon that soil, the light puffy foulness, for which it "is so remarkable."

There

There certainly is too little attention paid by many of our farmers, especially by the small farmers and their subtenants, in the making up of dunghills.—Yet a number of spirited and improving cultivators of the soil have shewn a better example; which in the gradual and silent method of correction, (without either reproving or reproaching), has tended very much to reform the slovenliness in the management of dunghills, that too generally prevailed. Instead of *muck feal*, or thick turfs cut from barren soil, the greater part of the farmers now lay in the bottom of their dunghills, either the turf walls of old houses, or dikes, (i. e. inclosures of turf,) or rich earth taken from the deep soil of a head ridge, or a quantity of peat earth if the soil be heavy, or of clay, that has for some time been exposed to the weather if their ground be of a sandy or gravelly nature. They lay on horses dung next above this, then another coat of earth, afterwards a coat of the dung of black cattle;—and so on alternately till they have raised the dunghill from four to five feet high at most, including a light covering of earth. This is turned over and carefully mixed, till the whole has undergone the putrid fermentation, in the end of May or beginning of June, when it is carted to the turnip field.—Or the summer-made dung is first carted to the corner of that field about Christmas; and then regularly formed into a dunghill, by mixing what additional earth or other compost seems most necessary. The dunghill though at first from four to five feet high, falls to about three feet, when the first fermentation has subsided. After being turned over, when the outside is thrown into the middle, and the whole is completely mixed, and the larger lumps broken, a second fermentation takes place, by which the particles of the dung are rendered more completely divisible. Some farmers, (running from the extreme of ill prepared *mack feal*, imperfectly mix-

mixed with dung, to the opposite one of keeping their dung over year, and of breaking the cloddy parts, and wasting its strength by too frequent turnings), found to their cost, that this method reduced the quantity, and injured the quality of their manure. It is now admitted, that animal dung, newly dropped, is as good a manure for turnip as can be applied; and it is no longer kept over year, but used as soon as the compost in the dunghill has been completely rotted, and has undergone the putrid fermentation.

In the neighbourhood of peat-moss, or where peat earth can be got in boggy places of a farm, many of the farmers have adopted Lord MEADOWBANK's method of making up compost dunghills, with a mixture of animal dung and peat earth.— But it must be acknowledged, that there is by no means such attention paid to the collecting, preserving, and preparing of dung, in too many cases, as the writer of this Report could have wished, and as is certainly practicable in most situations.

On the coast of Buchan, shelly sand is much used as a manure. It is not *equal* in purity, or in the quantity of calcareous matter, in all places where it is found; but for several miles along the coast, it is extremely pure; And as shell marle has improved the lands in the *valley of Strathmore*, shelly sand must in a few years add much to the fertility of *Buchan*. Mr. FERGUSON's *canal*, which extends nearly 7 miles in length, will be of great service in transporting this valuable manure into the *interior parts* of that extensive division of Aberdeenshire.

There is extremely little marle in this county.—The fresh-water wilk, that fertile source of marle, is found in none of our lakes.—But a species of clay-marle, and of rich unctuous clay, has been used very successfully, in mixing up both mossy, and sandy soils. This method of improving ground by a mixture of opposite soils is a permanent one: and

and there are few farms of any considerable extent, which do not contain either soil, or subsoil, of so opposite qualities, as that they would be much benefited by their mixture.—By this method alone, several industrious small farmers, when they had a little spare time, have with the aid of their horse and cart, very much improved their small possessions.

On a general view of the quantity and quality of the manure now annually laid upon the lands of this county, it is certain that both are in a progressive degree of increase; that in 1810 much more than double, probably not less than triple, of the former value (not the money price only) of manure is now used; and the different applications of calcareous matter have operated, both as stimulants and alteratives, in fertilizing the soil. In some districts, particularly on the thin gravelly soil of Dee-side, (where several successive crops have been taken after liming), the injudicious farmer has no doubt injured his grounds by overcropping.—In other districts of the county, where the soil is not so shallow, nor the subsoil so gravelly, the same erroneous application of lime and tillage, though no doubt hurtful, has not been so injurious, merely because the land was not previously cleaned by turnip crops, immediately after being limed, and the soil being both stronger and fouler, or more abundant in weeds, could not be so much exhausted. But after all the deduction for the improper application of manure, and of subsequent scourging crops, by far the greater proportion of lime now used, is well managed; the animal dung is better fermented with the other ingredients in the compost dunghill; and every species of manure is laid on with more skill. Hence both in regard to the raising of grass and corn of every kind, with one single exception, the county is in a state of progressive improvement.

The exception alluded to, in this summary, is the raising of of bear or bigg; which has been checked by an excessively high and ill proportioned tax, on malt from it, compared with malt from barley.

SECT. II.—WEEDING.

On the Infields, or old croft lands, there is a great multitude of weeds. The outfields and inferior lands are comparatively cleaner, because they have received less manure.— Under the old husbandry, the *only weeds* which the farmers were at pains *to pull up, (for they hoed none,)* were *thistles* and *mugworts.* Even those were pulled up, not because they were weeds, but because they were useful. The thistles, in the scarcity of grass in summer, were given to the farmer's horses; and the mugworts were either given to the cattle, or used as greens by the farmers' servants and cottagers. Now the hoeing of the turnip and potatoe crops has occasioned the destruction, or at least a great diminution, of the weeds which grew formerly so abundant. A complete summer fallow, on strong and foul lands, is sometimes resorted to, for destroying wild oats, wild raddish, and wild mustard, and also for extirpating the various dock weeds, marygold, and couch grass. Two successive crops of turnip, potatoes, or carrots, properly horse and hand hoed, require to be only once manured, are very valuable, and have been found by the Writer of this Report, more efficacious in cleansing foul land, than a bare fallow, with any number of ploughings and harrowings, in a single year. But in soils of a very loose texture, the yarr, or spergula, is increased a hundred fold, by summer-fallowing, or too frequent ploughings. Liming, or covering with shelly sand, has the effect

effect of destroying sorrel, heath, and several other weeds of *an acid disposition,* which cannot thrive in soil that is impregnated with calcareous matter. And, on the other hand, several weeds appear after ground is limed, that were not observed, or did not thrive in the soil before it was limed.

Women and children are employed in weeding the small patches of lint, that are now raised. But it must be acknowledged they are too seldom seen in the turnip or potatoe fields; and are never employed in hoeing the drills of a corn field; or even in picking out the principal weeds in such land, where weeding would be of great service.—Thistles are sometimes pulled up by the farmers' servants; but too often they are allowed to grow undisturbed, now that they are not necessary for *supper to the horses;* and the farmer purchases coarse gloves, at sixpence each, to his shearers in harvest, to save their hands from being pierced by the thistles. It would be wiser for him to give each of his servants half a crown, to purchase fine gloves for Sunday, as a reward for pulling up the thistles when green.

It may be incidentally remarked, that one class of these weeds, the *dock,* is never found to prosper in bad or unfertile soils. About 40 years ago, a man took a small farm in the division of Marr. When he entered to it, at the usual term, viz. Whitsunday, he found that there was not a weed of the dock kind on his farm. At Candlemas, or nine months after, he called on the proprietor; and apprized him that he was to leave it. The gentleman asked him, " why he gave up a " farm, before he saw what crop he could raise on it ?"—He was answered, " Sir, there was not a dockan" (the provincial name for dock) " on it at Whitsunday. I brought dock-" ins from different places and have planted them; but they " have not answered at all; and I know that what will not
" grow

"grow docks cannot grow corn." This self-taught botanist was perfectly in the right; for the farm was really a bad one.

SECT. VI.—WATERING OR IRRIGATION.

Even under the old husbandry, irrigation was frequently practised; though it was not often conducted in the most proper manner. Sometimes one of the old farmers made with the plough a triangular rut along the head ridge of a field, and a few other *curved furrows* in an irregular direction, so as to water a considerable part of the ridges below. At other times, those who were more attentive, formed with a breast plough, (provincially the *flaughter spade,*) a multitude of small curves, which when filled with water, became *rills* that were very neatly conducted over as many ridges, as the spring or fountain, or different streams of which they had the command, could irrigate; and after these ridges were sufficiently saturated, the same implement formed other curves, which were in like manner covered with water, until the whole field was irrigated. Irrigation was seldom attempted till the natural grass was three years old, and never till it was two. The water was continued for the greater part of two years upon the same field; and the ground was commonly broken up at five years old.

The purposes for which irrigation was at that time practised, were chiefly two:—1st, to supply the want of manure, and fertilize the land—2d, to destroy the immense quantities of wild oats, which at that time were so common, and for the destruction of which no other means were known, previous to the introduction of fallowing and green crops.

Though

Though a few individuals, in various districts of the county, continue to water their ground, in one or other of these methods, yet it is not above *twenty* years since irrigation was conducted in a regular manner. At present, it is practised very successfully by an Englishman, named JOHN BOULTON, who was brought from England for that special purpose, by JAMES FERGUSON, Esq. of Pitfour. An account of the quantity of land flooded by him, in different places, and of the produce of the lands irrigated, is here inserted from a paper communicated by him; both as stating the fact, and as giving an example of irrigation.

"*An account of the expence and produce of the water meadows,*
"*belonging to* JAMES FERGUSON, *Esq. of Pitfour, M. P. for*
"*the county of Aberdeen; and situated in the parishes of Old*
"*Deer and Longside.*

"No. I.

"The first water meadow, in Old Deer parish, consists of ten
"English acres; and was made out in the years 1801,
"1802, and 1803.—The expence, including hatches and
"bridges, was L.62 in whole, or L.6 4s. per acre.—The
"interest of this, at five per cent, is nearly six shillings and
"fourpence per acre, or for the whole 10 acres, L.3 2 0
"Attendance in keeping in yearly repair, is 10
"shillings per acre, or for the whole, - - 5 0 0
"Rent of the land originally, ten shillings per
"acre, or in all - - - - - - - - 5 0 0
"The expence of cutting and making the hay,
"L.1 5s. per acre, or in all, - - - - 12 10 0
"Total annual charge nearly L.2 11s. 2d. per
"acre, or in all, - - - - - - - - L.25 12 0

" The produce per acre is 125 stones, of 22lb.
"avoirdupois each, or 1250 stones in all. The
" value of this at eight pence per stone, 22
" English pounds, is L.4. 3s. 4d. per acre.—

" In all, - - - - - - - - - L.41 13 4
" The after math, worth 10 shillings per acre, or
" in all, - - - - - - - - - - 5 0 0
" Value per acre, L.4. 13s. 4d. or in all, - - L.46 13 4
" Deduct the charges as above, L.2. 11s. 2d. per
" acre, or in all, - - - - - - - - 25 12 0
" Profit per acre, of L.2. 2s. 1¼d, or on ten Eng-
" lish acres, - - - - - - - - - L.21 1 4

" N.B. This meadow is on a cold soil, and subsoil, full of
" iron ore; and is not regularly supplied with water, on ac-
" count of a corn-mill, which must first be served by the
" stream.

No. II.

" The next water meadow is in Longside parish, was begun
" in 1801, carried on in 1802 and 1803, and finished in
" 1804. It consists of nearly forty-one English acres;
" and cost, including hatches and bridges, made by stone
" masons and carpenters, nine pounds per acre, or in
" all, - - - - - - - - - - - - L.369 0 0
" The interest of this at five per cent. is fifteen
" shillings per acre, or - - - - - - 18 9 0
" The expence and attendance in keeping it in
" repair, is 10s. per acre, or - - - - 20 10 0
" The original value of the land, 10s. per acre, or
" in all, - - - - - - - - - - 20 10 0

Carry over, L.59 9 0

Brought over,	L.59 9 0
" The expence of cutting down, and making the "hay, 22s. per acre, or in all,	45 2 0
" The total charge is	L.105 11 0
" The produce 160 stones, of 22 English pounds "each per acre; or in all, 6560 stones, at "eightpence per said stone of this excellent "meadow hay, L.5. 6s. 8d. per acre, or in all,	L.218 13 4
" The after math, at only five shillings, per acre,	10 5 0
" Total produce, L.5. 11s. 7d. per acre, or in all,	L228 18 4
" Deduct expences as above,	105 11 0
" Gain cleared per acre, L.2. 15s. 3d.—in all,	L.113 7 4

"N.B. The soil of this meadow is partly *yellow*, partly "*black*, and partly *gravelly*. The subsoil is clay, gravelly "and sandy.

No. III.

"The 3d water meadow is also situated in Longside pa-"rish, was made out in the years 1804, 1805, and 1806. It "consists of eighteen and a half English acres; and the ex-"pence of making it, including the work done by stone "masons and house carpenters, was twenty pounds per acre, "or in all, - - - - - - - - - L.370 0 0

" 1. The expence and attendance in keeping it in "repair, is per acre, 12 shillings—in all,	11 2 0
" 2. Original value of the land eight shillings per "acre—in all,	7 8 0
" 1. The interest of the money at twenty shil-"lings per acre,	18 10 0
" The total charge amounts to	L.37 0 0

" It was rouped, or let by auction ; and the first
"   year it was sold for   -   -   -   -   -   -   L.27   8   0

" Being but *newly finished*, there was a *loss* this
"   year of   -   -   -   -   -   -   -   -   -   L.9   12   0
" But next year it rose to   -   -   -   -   -   L.78   4   0
" Deduct from this the annual charge and former
"   years' loss, viz.   -   -   -   -   -   -   -   46   12   0
" There remains a profit at the end of the second
"   year, of   -   -   -   -   -   -   -   -   L.31   12   0
" The third year 1808, it sold for   -   -   -   L.66   11   6
" Deduct the annual charge of   -   -   -   -   37   0   0
" Remains what may be supposed to continue a
"   profit of   -   -   -   -   -   -   -   -   L.29   11   6

' N.B. The soil of this meadow, yellowish, with some iron
" ore, and a little moss in the subsoil.

No. IV.

" The next water meadow is likewise situated in Longside
" parish, was made out in the years 1804, 1805, and 1806,
" and consists of six English acres.—It cost L.10 per
" acre, or   -   -   -   -   -   -   -   -   -   L.60   0   0

" The interest of this, ten shillings per acre, or in
"   all,   -   -   -   -   -   -   -   -   -   L.3   0   0
" Expence of keeping it in repair, and attendance,
"   fourteen shillings per acre, or in all.   -   -   4   4   0
" Original value of the land—six shillings,—or
"   in all,   -   -   -   -   -   -   -   -   1   16   0
                            " Carry over,   L.9   0   0

"   Brought over,   L. 9 0 0
" Expence of mowing, and making hay, L.1 per
"   acre—in all,   - - - - - - - -   6 0 0

" Total charge is L.2. 10s. per acre—in all   L.15 0 0

" The quantity or produce of hay, 100 stones per
"   acre, or 600 stones in all, worth at the above
"   price, L.3. 6s. 8d. per acre—or in all - L.20 0 0
" The aftermath at only 3s. per acre, is   -   0 18 0

" Total produce, or value, L.3. 9s. 8d. per acre—
"   or in all   - - - - - - - - -   L.20 18 0
" Deduct the charge, as above, L.2. 10s. per acre 15 0 0

" The profit is 19s. 8d. per acre—or in all   L.5 18 0
" N.B. The above meadow is of poor gravelly soil and
" subsoil.

                        No. V.

" This is a piece of catch work at Pitfour, in the parish of
"   Old Deer; was made out in 1802 and 1805, and consists
"   of about three English acres. It cost only L.5. per acre,
"   or in all   L.15 0 0

" The interest of this, 5s. per acre, or in all L.0 15 0
" Keeping it in repair is L.1 per acre, owing to
"   a long water-course—or in all   - - - -   3 0 0
" Original value of the ground L.2 per acre—or
"   in all   - - - - - - - - - - -   6 0 0

" Total charge L.3. 5s. per acre—or in all   - L.9 15 0
                        r f   " Its

"Its produce cut for soiling is well worth L.8 per
"   acre—or in all - - - - - - - - L.24 0 0

"Profit L.4. 15s. per acre—or in all - - - L.14 5 0

No. VI.

"The last piece of water meadow, also situated near Pitfour,
"   was done in 1802, and consists of two and three-fourths
"   English acres, which was not worth 2s. per acre; being
"   full of stones, with a mossy soil; the subsoil a stiff iron
"   ore, and stoney poor bottom. It cost L.21 per acre, or
"   in all L.57. 15s.
"The interest of this money 21s. per acre—or
"   in all - - - - - - - - - - - - - L.2 17 9
"The original value, not above 2s. per acre—or
"   in all - - - - - - - - - - - - -   0 5 6
"Attendance, and expence of keeping in repair,
"   15s. per acre—or in all - - - - -   2 1 3
'The total charge is L.1. 18s. per acre—or in all 5 4 6

"It is now worth L.2. 10s. per acre—or in all L.6 17 6
"Its value is yearly increasing.
"Deduct expence, as above, - - - - -   5 4 6

"Profit, 12s. per acre—or in all - - - - L.1 13 0

"For the two first fields Mr. Ferguson obtained the silver
" medal of the Board of Agriculture in 1805. But they have
" now risen in value considerably since that time."

"It is to be observed, that except the last piece, which had
" been totally neglected, all the land that had been made into
" meadow, as above described, was formerly ploughed, and
" left out very rough. That circumstance, added to the ex-
"   " pence

" pence of large feeders, and large flood-gates, or flood-
" hatches, and various cuts on turns of the river, makes the
" expence per acre, greater or less, according to circumstan-
" ces: all the expences of which are included in the forego-
" ing statement.

" It may also be observed, that the above six water mea-
" dows, consisting in all of 81 English acres and 1-4th, al-
" ready yield a profit of L.185. 16s. 2d. sterling; (and besides
" the profit arising from them, this is an improvement that
" will never waste or wear out so long as water runs;) that it
" always insures a good supply of winter food for cattle,
" which in this climate is very useful; that it makes an ex-
" tra quantity of dung for improving other land; and that
" the hay which grows on water-meadows is very little infe-
" rior to hay from sown grass to horses, and for black cattle,
" is at least equal.—Wherever good water can be copiously
" employed, it will make any kind of land worth from L.3 to
" L.4 per acre; and in some cases, L.6, L.7, or L.8. That
" is to say, when the water runs through a fertile country,
" or when it receives the washings of large towns and villages,
" or when it abounds with good springs, either of these will
" do; two of them will answer very well; and when all the
" three are combined, they will answer best of all.

" In this statement I have not under-rated the expence,
" nor over-rated the produce; but I would conclude, with
" remarking, that when water meadows come to be properly
" understood in this county, they will be much cheaper
" made, and also much cheaper kept in repair than they can
" be at present.

" Mr. FERGUSON was put to a considerable expence in
" bringing me from England, and in paying me a salary,
" when I could not at first be constantly employed. But I
" could not charge this, except when engaged in working or

" super-

" superintending labourers; and I consider the loss of my
" time as balanced by teaching his men to work; although
" the above 81 acres of meadow had cost him L.1200.

"JOHN BOULTON,

" A Gloucestershire Flooder."

On a general view of this valuable communication, it appears that Mr. FERGUSON, at an expence of L.933 15s. has made out 81¼ acres, at the average expence of very nearly eleven guineas per acre, (a general average which has escaped the notice of this honest Englishman) and instead of drawing only 5 per cent. for the money so laid out, viz. L.46 13s. 9d. draws already L.139 2s. 5d. more than that interest; or L.185 16s. 2d,; or very nearly 20 per cent. for the money so expended. When land sells commonly above 30 year's purchase, whenever a landed proprietor can make but 6 per cent. by a permanent improvement, he is a great gainer. And it is much more prudent even in a *merchant* to invest a proportion of his wealth in the purchase of an estate, by which he at first draws but 3 per cent. (but by the improvement of which he can soon make 6 per cent. both of the original purchase-money, and subsequent expences,) than to embark all his capital in foreign speculations, where he is at the mercy of a Tyrant, and where his country derives comparatively little benefit, even if he is successful. For it should ever be remembered, that in foreign trade, all that the nation can gain is the *difference* between the *real* value of British materials, and labour, in the articles which we export, and that of a foreign nation, whose rude produce or manufactures we import—but that in cultivating the soil, in improving the corn lands, or making a water-meadow, the land, the water, the grass, or corn raised, the cattle fed, the labour employed, the permanent wealth accumulated, *are all British*—

British—all belong to that free and happy country; which, would it study its own true interest, and attend to its own resources, would bid defiance to all the tyrants of the earth, by raising a hardy race of men, both for ploughing the land, and ploughing the ocean; by discouraging every manufacture, or article of commerce, that renders us effeminate, or dependent in any shape on other nations, and by encouraging agriculture, which adds to the health, strength, and virtue, as well as to the wealth of the empire.

The other parts of the plan laid down by the Board of Agriculture, respecting Irrigation, are not applicable to this county; and one of them, entitled Water Mills, a dreadful nuisance, is not founded in fact, so far as regards Aberdeenshire. *Thirlages* to mills are a heavy servitude, but watermills are no nuisance at all, if the stream be properly attended to; and every streamlet in a short time will be attached to a threshing mill; nay many rivulets may by this means be brought to higher levels, and promote irrigation more extensively. In a *champaign country*, where the water runs slow, a water mill may often be a *grievance*; but water, while it is *clean*, can be no nuisance. Our navigable Canal from Inverury to Aberdeen, has sometimes been injurious to those in its neighbourhood, to whom the damages paid were not always a recompence. But it also has afforded in the summer months, a supply of water to the citizens of Old Aberdeen, when they would otherwise have been in great scarcity. It is bad water, as well as a bad wind, that brings good to nobody. Every thing of this nature has its advantages, as well as its disadvantages.

CHAP.

CHAPTER XIII.

EMBANKMENTS.

There is no embankment against the sea in this county, though the blowing of sand, as already mentioned, has nearly destroyed the parish of Forvie. But a bank made by the sea has formed the lake or loch of Strabeg, within the last 160 years; and there is now no probability of the burn of Rattray being able to force its way in its old channel, nor of the landed proprietors again attempting to lay open its access to the ocean.

On the river Dee there are two embankments, one made by a mass of small stones at Abergeldy, which was noticed under the article *trenching;* and another made by the late Captain HENDERSON, at Newton of Murthile, or Dee Bank, now the property of ARTHUR ANDERSON, Esq. This has been serviceable in repelling that rapid river from carrying off a pretty large meadow on the estates of Murthile and Cults, within five miles of Aberdeen.

On the Don there are several embankments, which have been very useful, on a small scale. One adjoining to the bridge at Inverury, prevents the Don from overflowing the adjacent meadow; and from running into its old course about 600 yards above its confluence with the Ury.

A se-

A second has been made by the farmer of Kinkell, and has protected about 40 acres of rich haugh or meadow from the overflowing of the Don.

A third, and the most considerable, has been made by Mr. ROBERT WALKER, in Wester Fintray; who has embanked about 1000 yards, and repelled the direct overflowing of the river, which sometimes overflowed about 100 acres of haugh, and in 1768 carried of 2000 threaves of corn.

A fourth has been made by the farmer of Bidliestown, in the parish of Dyce, by which he has checked the encroachment of the river, which threatened to be very destructive.

These three last embankments deserve to be noticed, as being done by farmers, on their own charges. Mr WALKER alone gets at the end of his lease about two thirds of the expence incurred—by his agreement with GEORGE SKENE, Esq. of Skene.

A small embankment has also been made on the Ury, which protects several acres of ground from the overflowings of that river, and has been done by the tenant at Mill of Keith-hall.

All these embankments are made with a gentle slope to the river, consequently with a broad base. The expence is from a shilling to two shillings per yard—in both making the embankment, and laying the turf in a proper angle to the river. They are carefully beat by the back of a spade or shovel, when the turf or feal is laid on; and are exposed to great danger the first season before the grass has formed a kind of matting on the surface, and before the joinings of the turf become firmly united.—After a year or two, they are in no danger from the water, but may be hurt a little by the breaking of the ice, or by large shoals of ice, when the river is swelled excessively by a sudden thaw.

It would be a matter of the greatest consequence to the

pro-

produce of about a thousand acres of the finest land in the kingdom; that a new bed were made for the river Don, for seven english miles, extending from a little below its confluence with the Ury, to the church of Dyce. The river does not fall above forty feet in all that distance by a straight line, and in about 10 miles by its present windings.—The quantity of ground lost by these windings, and rendered useless by the river shifting its course, is very considerable, probably 200 acres; and the injury done by the overflowing of the Don in some years has been very great.* The river Eden in Fife, and Kelvin in Stirlingshire, have been straighted in this way; the latter at the expence of only L.600 for a course of four miles.—As the Don is much larger than either of these, it might probably cost L.3000 to make a new course for seven miles; yet the additional quantity of land gained, and the greater produce of the whole *haughs,* or meadows on each side, would soon repay all the expence.—But unfortunately the greater part of the lands are strictly entailed; and the present proprietors cannot injure their families in the improvement of an estate, which may soon go to the heirs male of the entailers, or to their eldest sons, without benefiting their younger children.—Until the proprietors of entailed estates are entitled to burden their lands with the expence of such useful works, the straighting of this river cannot be attempted

* A cut of less than one mile of the lower part of the Ury, for a sum not exceeding L.500, would add more than L.100 yearly to the produce of the burgh roods on the west, and the Earl of Kintore's on the east bank of that river.

CHAPTER XIV.

LIVE STOCK.

SECT. I.—CATTLE.

As Aberdeenshire is a breeding county, which raises a greater number and value of black cattle, than perhaps any other in Scotland, the history of our cattle trade, an account of our different breeds, and of the number consumed at home, and carried away to other places, will occasion this article to be more fully and minutely discussed, than has been necessary with respect to any other section in this Report.

The price of every article in commerce is regulated by the demand for it, compared to the quantity in the market: and wherever a merchant or a farmer can make money, by raising a different commodity, he will gradually relinquish a former practice, which he finds to be less beneficial.

In Great Britain, different in this respect from that of ancient Rome, *tillage* pays better than *pasture*, or *wheat* and the other kinds of *corn* are more valuable than the rearing of *cattle*. Hence in the progress of luxury in this kingdom, which has been accompanied by improvements in agriculture, a county or province, which formerly reared horses or black cattle, becomes a corn raising district, and cultivates *wheat*,

or

or two-rowed barley, the most valuable species of grain.—
The county or province bordering upon this, becomes a nursery for black cattle; and the sheep farmer removes to the more distant and uncultivated districts.

About 120 years ago, before the Union of the two kingdoms, when the disputes ran high between the two nations, the English parliament, in order to overawe that of Scotland, prohibited the introduction of Scotch cattle into England.—The stroke was aimed against the nation in general; but it fell principally on the southern counties.—At that time, Berwickshire and the Lothians were the great cattle rearing counties on the east coast. Even Fifeshire, though it had an excellent breed of cattle, was more dependant on its sheep and wool:—And in all the counties north of the Tay, the high price of Scotch wool in the market of Europe, and the distance from England, added to the small size of their black cattle, which at that time were seldom carried to the southern counties, as the demands were limited, occasioned the prohibition of Scotch cattle from entering England to be little regarded. The rearing of sheep for the sake of their wool, which at that time bore a high price on the continent, was then the great object of the farmers in the north of Scotland.

The Union of the two kingdoms put an end to the disputes between the two nations; and was equally beneficial to both. The wars during the reign of Queen Anne, occasioned an increasing demand for black cattle. But as the prices were still low, and as wheat and barley were supposed to pay better, the farmers in Berwickshire and the Lothians, began to sow wheat as one of their regular crops, and barley more generally than bigg. The introduction of the turnip husbandry greatly improved the condition of these farmers, and before 1760 they began to turn their attention more to the raising of corn, than to the rearing of black cattle.

In the mean time the limitation of the Scotch wool to the market of Great Britain, in consequence of the Union of the two kingdoms, produced a great fall of its price; and the farmers in Fifeshire, and the greater part of Angus—and even in the south part of Kincardineshire, or How of the Mearns, began to turn their minds to the rearing of cattle. The demand for work oxen to the plough by the farmers in the northern parts of Angus, and in the counties of Kincardine and Aberdeen (who formerly got their cattle from the Lothians) was principally supplied from the county of Fife, or from the Falkland market.

The native breeds of this county during that period, were diminutive in point of size, and used in the plough only by the poorer farmers. The cattle dealers who bought the best working oxen, when young, also purchased them from our farmers at 11 to 12 years old, (the Reporter once saw an old ox kept till 17 years old), and gave them young cattle in exchange, on what terms they pleased.

At last, the landed proprietors of the county saw the great loss which their tenants sustained; and in order to improve the native breeds, and to raise a large kind of cattle within the county, they purchased both bulls and cows from other places; partly from England, partly from Holland, and partly from the southern counties. This, however, produced but a partial effect, till the principal farmers procured bulls from Fife, or at least from the Falkland markets.

It is proper here to mention, that the Falkland breed, which at this time was one of the best in Scotland, had originally been raised from some English cows, which that sagacious Prince, Henry VII. had 300 years ago sent in a present to his eldest daughter, the Queen of Scotland; who had
been

been married to King James IV. In reporting facts nothing should be suppressed; and that pride is unmanly which would induce a man to conceal the truth. In the contentions between the two nations, the breeds of cattle were mixed, by carrying them off as plunder: But in this case, the cattle of the south of Scotland were improved in their size by a father's present to his daughter. To return from this digression.

Hitherto there had been very little demand for Scotch cattle in England: But in 1762, a disease among the horned cattle in that country, occasioned salt beef to be in request for the navy. In 1763 there was no demand for cattle: but salted beef was purchased at the average of one penny per lb. In 1764, a few dealers from Galloway, and from the west of Scotland, began the carrying of Scotch cattle to England. In 1765, owing to a sudden fall in the price of black cattle, the dealers in Aberdeenshire universally stopped payments; and the necessity of rearing within the county, cattle of a proper size for the plough, became more evident to the farmers themselves; and in the course of a few years the native breeds were considerably increased in size by crosses with the Fifeshire bulls, and by purchasing both bulls and cows from different places in the south of Scotland. Several of the landed gentlemen endeavoured to raise the Lancashire breed of cattle, either by themselves, or by crosses between them and the native breeds; but neither of the schemes have been successful. Attempts to raise the Holderness, and other English breeds also became abortive, as we had not food to support these large sized animals. Besides, the mixture of the Falkland and native breeds were found to have more *points*, as the cattle dealers term it, or more eligible qualities, than any of the English breeds, or of the crosses made with them. The Fifeshire or Falkland were a little inferior in respect of size to what had formerly been brought from the

the Lothians; but they possessed *far more points* than the large but mixed breed, that is *now generally found* in the county of Fife. By carefully avoiding all crosses, with the too large and long-legged breeds, which some of the proprietors had brought from England, and by raising well shaped bulls at home, or purchasing them from Fife, the present largest breed of Aberdeenshire, (when five years old, weighing from five to seven cwt.—at seven years old, from seven to eight cwt.—and when full fed, from ten to twelve cwt.) was raised to great perfection.

In 1766, cattle dealers *came* from England in the end of June and beginning of July, to purchase Scotch live-stock for the English market; and the low prices of the preceding year were succeeded by a reasonable price for all our old cattle that could be spared from the plough, and for all others that were suited to the English market. This gave our farmers room to hope that they might not only rear cattle to serve themselves, instead of depending on Fifeshire, but also might raise them to a size adapted to the English market. But a second and sudden fall of the price of black cattle towards the end of 1767, again damped the spirits of the farmers, and ruined all the cattle-dealers. This was succeeded by a gradual decline in the market, till they fell to their old prices in 1770.

In the meantime, the landed proprietors, soon after 1760, and a very few before that period, had made a number of inclosures round their personal farms, which they had laid down with grass seeds, and employed chiefly in grazing black cattle. Many of them also began to cultivate turnips and potatoes in the fields, which were hoed in narrow drills by the hand, (or horse-hoed) and also laid down with bear and grasses. A few of the better sort of farmers began to imitate the example of their landlords, so far, as to have an acre of

tur-

turnips and potatoes, and an acre or two of sown grass.—
From these small beginnings, an important alteration of our
agriculture took place; and the breed of cattle raised by the
landed proprietors, and by a few of the better sort of farmers
was greatly improved, by the additional quantity, and better quality of their grass. For the turnips at first were not
employed in the rearing of cattle, but were given only to such
of the oxen or cows as were to be sold to the butcher.

Fortunately, the prices began to rise in 1771 and 1772;
and with the exception of a few months in the end of 1775,
this rise continued till the commencement of the war with
America. During that war the demand was steady, but the
prices moderate.

The calamitous season of 1782 compelled such of our
farmers as were not ruined by it, to abandon their old
wretched system of husbandry; to introduce turnips and
sown grass, as a part at least of their distribution of crop;
and to extend this gradually as far as their capital or credit
permitted them to go. Also where the old farmers had been
ruined, or rendered unable to carry on their business, the
landed proprietors saw that it was necessary both to select
good tenants, and either to induce or oblige them to improve
their farms.

In 1783, 1784, and 1785, black cattle sold well; and
there was great demand, owing to the thriving state of our
commerce and manufactures on the restoration of peace.—
This encouraged the raising of the breed of cattle to their
present size and quality.

From 1786 to 1793, the demand was steady, though the
prices were not very high. But an Act of Parliament in
favour of the licensed distillers, enabled the farmer to go on
rapidly in extending the size of his turnip fields; and to sow
bear or bigg with grass seeds, in great quantities. This gave
ad

additional food to our cattle, both in rearing for the English market, and in fattening for our internal consumpt.

From 1793 to 1801 the prices were so very great, owing to the demand, that they might be called exorbitant. This demand, though first occasioned by the war, was increased by the scarcity of wheat in England in 1796, and of all kinds of corn in Britain in 1799 and 1800.

The short-lived peace that was patched up in October 1801, occasioned a temporary fall of 25 per cent. in the price of cattle; and for *the third time* in the present Reign, *ruined most of the cattle dealers.* But in 1802, though the prices were lower, the demand revived. (An ox worth L.20 in September 1801, fell to L.16; and next year rose to L.17) *They rose again in 1804 and 1805*; but though the demand was steady they fell in 1806 and 1807, and with some variations of price, but a steady demand, they have again risen in value; and are now very high.

The present state of the cattle trade, compared with what it was 60, or only 40 years ago, is as follows. The *grandsons* of the Lothian farmers, who supplied us with working oxen, and the *sons* of those in Fifeshire, who succeeded them in rearing cattle for ploughing our lands, are now become great corn farmers, who raise wheat and two rowed barley; and they now purchase, for different uses, the cattle bred in this county, that are now in high estimation, not only in the south of Scotland and north of England, but find their way to St. Faith's Fair in Norfolk, and when fed in that county, to Smithfield market. Sixty years ago, when 10 and sometimes 12 oxen were used in one plough, *the greater part of our working oxen came from the Lothians.* Forty years ago they came from Fifeshire, but not in so great a proportion.— For when the keep of cattle came to be improved, and the breed raised, nearly one-half of the oxen, who worked chief-

ly

ly on the *land-side*, were raised by *our farmers* from either the *mixed* or *native* breeds. Afterwards, three or four, and next only *two* of the whole number were bought from the cattle dealer. And since 1782, with very few exceptions, they have all been raised on the farm. And now they work either in pairs, like horses, or in strong lands four are yoked in the plough. In a very few instances, where six are used in ploughing very rough ground, they work with patience and steadiness in tearing up barren land, which the more spirited animals, the horses, cannot so easily be brought to render arable.

Every thing human is transitory.—While the Aberdeenshire cattle are so much improved, those of the Lothians are little attended to; and even in Fifeshire, the Falkland breed has been much injured by improper crossings. In these southern counties corn chiefly is raised, and rearing of cattle is disused. In fact, the rents there are so high, that they could not be paid by any thing but raising of wheat.

If the rent of land continues to increase, and improvements in agriculture to go on as rapidly as they have done for the last 30 years, Aberdeenshire may become a corn country, and the rearing of black cattle for the market of England, may, in the course of a century, remove from the Pentland hills to the Pentland firth; and Caithness and Sutherland come in the place of Berwickshire, and the Lothians. But when, in the progress of luxury, the feeding of cattle becomes of more value than the raising of corn; the cultivation of the soil by the plough will make a retrograde motion. The Lothians and Berwickshire will become grazing counties; and from Northumberland and Durham, to Kent and Middlesex, the soil will be occupied by the grazier, instead of being cultivated by the farmer. May luxury never produce the effects in Britain which it did in imperial Rome.

There it was said, What produces the most certain profit? *Feeding well?* What is the next? *Feeding moderately?* What is the third? *To feed even badly?* The least profitable of all was *tillage*.* Let our Legislators remember, that an improper system of corn laws, established to appease the clamours of noisy manufacturers, who know not their true interest, may produce this effect in Britain, which, to the ruin of Italian agriculture, was occasioned by improper largesses of corn to the Roman populace. To return to the cattle trade of Aberdeenshire.

Though the rearing of cattle does not, in Great Britain, pay so well as the raising of wheat, yet from the late increase of their price, and the steady demand, the county of Aberdeen at present draws from cattle, sent either to England, or the south of Scotland, - - L.150,000
viz. 5000 best cattle, from L.16 to L.17 each, at

at an average - - - - L.82,500
3000 young cattle, from L.11 to L.13,
   at an average of L.12 each - 36,000
1000 ditto, from L.9 to L.11 each, at
   an average of L.10 each - - 10,000
2000 cows or heifers, from L.7 to L.9,
   at an average of L.8 each - - 16,000
1000 highland cattle, small-sized, at
   an average of L.5 10s. each - 5,500
                                    L.150,000

The Writer of this Report is indebted to Mr. George Williamson,

* " Quis est certissimus quæstus? Si bene pascas.—Quis proximus? " Si mediocriter pascas.—Quid tertium in agricolatione quæstuosum " est? Si quis vel male pasceret." These were the opinions of Cato, as recorded by Pliny and Columella.

WILLIAMSON, farmer, at St. John's Wells, in the parish of Fyvie, (who is the principal cattle dealer in the north of Scotland,) for the history of the cattle trade since 1762, and its present annual amount. He has known him for above 42 years, and can depend both on his extensive and intimate knowledge of the trade; and upon the fidelity and accuracy of his information. Mr. WILLIAMSON, and his two brothers, JAMES and ROBERT, generally sell about 8000 cattle yearly in the markets of England, and of the south of Scotland, of which two-thirds are raised in this county. They rent above 2000 Scotch acres of land, besides paying L.500 of grass rent within the county. They have at present about 200 acres of turnips, employed in feeding as many black cattle, and in rearing 400 young cattle or winterers. As they are cattle dealers, the Writer of this Report did not think it fair to quote them as examples of good farming, though their lands are in excellent order; but he submitted to them the following table of the different kinds of black cattle raised in Aberdeenshire, which both received their corrections, and met their approbation.

They decidedly prefer the true native breed, unmixed, and raised by good keeping, to the mixture of the Falkland, or Fifeshire breed, with that of this county; and consider both these to be much superior to the English, or to any foreign breeds. They justly remark, that the *food, or keep should be always above the breed, and not the breed above the keep.* They consider the *small* highland cattle, which are generally bought by inferior dealers, as *too restless and impatient* for feeding well. They prefer the native low country breed to the larger ones, as they are most easily maintained, more hardy in work, have flesh of the finest grain, and pay better in proportion to the goodness of their keep.

A GENERAL VIEW OF THE DIFFERENT BREEDS OF BLACK CATTLE IN THE COUNTY OF ABERDEEN.

Name of the Breed.	Character of the Breed.	Ploughed when, and how long.	Age when put to grass a certain age.	Weight at that age.	Expence of grazing.	Weight after 6 months grass.	Expence of Hay and Turnip.	Weight when full fed.	Age when killed 40 years ago.	Price at that time.
1. Largest English, or Foreign Breed.	Slow feeders, Generally not hardy, nor easily kept, but very docile.	Commonly for 2 years, but sometimes for 4 years.	At 5 years, 5 years bull old, average 6 cwt.	This largest breed L.4	At 5 years 6 months is 7 cwt. 2 qrs.	At 6 years old, is from L.6 to L.7.	At 6 years old, from L.6 to 9 cwt.	Seldom killed here 40 years ago.	Not sold here
2. Largest Scotch or Fifeshire, mixed with native	Good feeders Also generally docile, and pretty hardy	Also generally ploughed only 5 but for 2, or for 4 years	At 5 years, 4 cwt. 2 qrs. L.13 to 15.	This mixed breed L.3	At 7 years, Do. Do.	At 7 years 6 months, 9 cwt. L.7.	At 8 years old, 11 cwt.	Do. Do.	
3. Native and unmixed lowland docile, or Aberdeenshire	Great feeders sufficiently docile, and not at all very hardy	Ploughed one year, or 4 years old	sometimes 7 years old, about L.1 to 18.	Do. Do.	At 7 years and 6 months 8 cwt. L.6.	At 8 years old, 9 cwt. 2 qrs. to 10 cwt.	Do.		
			Usually at 4 years old com. 4 cwt. L.11 to 13.	At 4 years, breed L.2 16s.	At 4 years and 6 months, 5 cwt. 2 qrs. to 6 cwt.	At 5 years old, 7 cwt.	Do.	From L.1 5s. to L.1 15s.		
4. Native and unmixed highland docile, but sometimes used to good for winters feeding	Extremely hardy but not chiefly sold breed	Ploughed 4 years old old, aver. L.12.	At 4 years, only 2 old from lands, 6s. to L.1	In the highlands from 3 cwt. 2 qrs.	Not fed with turnips, but sold to the dral-south country.	At 6 years old, 7 cwt. 2 qrs. to 8 cwt. 2 qrs.	At 8 years old, 9 cwt. 2 qrs. above that age	Seldom under 12, often above that age	From 15s. to L.1 1s.	
						Not full fed till Do. carried to the				

N.B. The different Breeds of Cows are not included in this Table, as the time of feeding them is so extremely variable. A number of them are purchased by the Cow-feeders in Aberdeen, who sell them when their milk is gone off, and when they are fit for the Butcher.

To this account the Messrs. WILLIAMSONS add, that every succeeding generation, for the last 30 years, has increased in size: and that by good keeping, the native breed is double its former size (i. e. weighs at least double its former weight) since the introduction of the turnip husbandry.

They are also decidedly of opinion, that wherever a landed proprietor feeds more than one year, for family use, the stot should not be tied up, but allowed to feed loose, in order to get gentle exercise along with his food; that the second year he may be put to high feeding, and be tied up, and may be continued with this high feeding as long as he seems to thrive; but that he ought to be killed whenever he loathes his food, or appears to be sickly, or not thriving. So much for the different kinds, or breeds of cattle.

FATTENING.

The feeding with hay and turnips in winter, in this county, is less an object than the *rearing* of black cattle. And few persons, except butchers, make a *separate* business of this branch. Some farmers feed with turnips, and a proportion of straw; others only litter with straw, and feed, first with turnips, and then with a proportion of clover hay.— Their profits are proportioned to the richness of the food given to the cattle: and some of our farmers feed as thoroughly as any in the island. To feed an ox of 600 cwt. besides hay and straw for litter, half an acre of turnips is generally found necessary: And the farmer expects to receive from five to six pounds in addition to the price of the ox for his expence and trouble. But in fact, the profits of feeding cattle are very uncertain, when live-stock is purchased at a high rate, and a greater number are fed, than what are wanted for the supply of the Aberdeen market, and that of the few towns in the

the county. The price of feeding cattle, however, has risen considerably. About twenty-five years ago, the late Mr. BARCLAY of Ury, was in the practice of contracting with a butcher in Aberdeen, to feed a certain number at four pounds per head. They were carried off weekly in a fixed proportion, from December to the end of April. At least six guineas would now be expected for such plentiful feeding as Mr. BARCLAY's was. In fact, the price of feeding ought to be regulated by the measure and quality of the turnips, and the quantity of hay and straw consumed; and when the two last are very dear, owing to a dry summer, the farmer charges from the butcher, or cattle dealer, a much higher price for the acre of turnips. Some years ago, Mr. DAVID WALKER in Blair, brother to Mr. ROBERT WALKER in Wester Fintray, set two acres of excellent turnips, which fattened four bullocks of at least nine hundred weight each, for only ten guineas, affording litter for the cattle. But in 1808, when hay and straw were very dear, eleven guineas were refused for the acre of turnips, along with straw or litter.

The following singular instance of good management in feeding cattle, deserves the reader's attention. Mr. JAMES GORDON,* at Orrok, begins with keeping his *calves* in excellent order; and by uncommon attention from the *calf* to the *fattened ox*, has both increased the size of his cattle, and sold

---

* Mr. GORDON entered to *this farm* only seven years ago. The first year, from the dung left on the farm, he could sow only *three-fourths of an acre*, in turnips; but by paying unwearied attention to raising of manure, and buying night soil in Aberdeen, nine miles distant, he has now 26 acres in turnips, ruta-baga, and potatoes; and has six horses and fifty two black cattle on a farm of 92 Scots, or 117 English acres of arable land. No farmer in the county supports and feeds so many cattle, on so small an extent of land. He has only 32 Scotch acres in white crop.

sold them at very high prices. Last year he sold his fat calves to the butcher, at six weeks old, from L.2 10s. to L.3; and at nine weeks old, for L.4. And he bought a young stot (the provincial name of a young ox) in September, at L.3 16s. 6d. fed him with globe turnips till January,—with yellow turnips till the middle of March,—with ruta baga and potatoes in the end of spring,—and lastly, with early grass, cut for soiling, till the beginning of June; when he sold him for L.15, having cleared L.11 3s. 6d, for nine months keep. The animal was of the small sized native breed. He also raises more corn, and feeds a greater number of cattle, than any one who possesses the same extent of ground. On a farm of 92 Scotch acres of arable land, he had in 1809, 20 acres in green crop, viz. 15 in turnips, 3 in ruta baga, and 2 in potatoes, (besides 3 acres in bare fallow.)—This year, (1810,) he has 20 acres of turnips, 3 in ruta baga, and 3 in potatoes. And he is fattening for the butcher 21 black cattle, who were put to turnips towards the end of August; while he is rearing 31 cattle, which are all in excellent condition.

The statement of these facts will serve to introduce the following communication from Mr. GORDON. It was originally sent to Sir JOHN SINCLAIR, President of the Board of Agriculture, in answer to certain Queries respecting the Aberdeenshire Breeds of Cattle, and the value of them when fattened.

(Among the remarkable examples of good feeding, it may also be remarked, that Mr. WALKER in Wester Fintray, has received L.50 each for two bullocks reared on his farm, and killed at seven years old;—that he received L.35 each, for other two only four years old;—and that he has frequently received L.30 for young stots either sold to the cattle-dealer, or fed to the butcher.)

SHORT

SHORT ACCOUNT OF ABERDEENSHIRE CATTLE.

1. STOTS, OR YOUNG OXEN.

" Few stots in Aberdeenshire are put to high keeping to
" fatten for the butcher, till they are four years old. The
" practice of weighing live animals has not been introduced
" into this county; so that nothing certain can be said of
" their *weight* when lean. The *value* of such stots is very va-
" riable in different seasons. Nay, frequently in the same
" season, every other fair brings about a rise or fall. The
" ordinary weight of middle sized stots, from *four* to *five*
" years old, when fed with rich grass for one summer, and
" with turnips the following winter, is from 32 to 38 stones
" Dutch, or from 40 to 48 English, at 14 lb. per stone.

" In summer 1806, taking the medium of the season, stots
" of the above weight could have been purchased, when lean,
" at from L.10 to L.12; and taking the medium of 1807,
" from L.8 10s. to L.10 10s. each The value of a high fi-
" nished (or thoroughly fattened) bullock at Aberdeen, at
" an average of the years 1806 and 1807, was from 6s. 3d.
" to 7s. 10½d. per English stone of 14 lbs.; or from 50s. to
" 55s. per cwt. sinking offals.

" At no time has the value of butcher meat been more un-
" steady than since the year 1799. In 1800, 1801, and 1802,
" a well fed bullock was worth 8s. 9d. per English stone, of
" 14 lb. or 70s. per cwt. sinking offals. In the course of
" the last mentioned years, *lean* stots, of the above weight,
" sold from L.15 to L.18 each, to the cattle dealers, to be
" driven to England to fatten. In the year 1803, the best
" fat bullock killed at Aberdeen, sold at 65 shillings per cwt.
" and in 1804 and 1805, at 5s. less, or 60s. per cwt. sinking
" offals.

" offals. In 1806 and 1807, the price was still lower, being
" from 50s. to 55s. per cwt.

" But there are numberless instances of stots bred in
" Aberdeenshire, which are below, and others which are
" greatly above the medium weight above-mentioned.—
" There are several stots killed at Aberdeen every year, of
" the above age, which do not exceed 26 or 27 stones Dutch,
" or 32 to 34 English. On the other hand, there have been
" instances of Aberdeenshire stots, brought to the slaughter
" in a high-finished state, from the age of between 3 and 4
" years, which have been found to weigh 64 stones Dutch,
" or 80 stones English, for the four quarters (or sinking of-
" fals). There have also been instances of middle-sized cows,
" when crossed with small, but well formed bulls, that pro-
" duced stots which weighed 8 cwt. or 64 stones English, at
" the age of between *three* and *four* years, and which rose to
" 80 stones English, when only one year older, (besides up-
" wards of 8 stones Dutch, of tallow, or 10 stones English, in
" each stot), and their flesh was as richly marbled as that of
" an ox of the first quality, of six or seven years old, and
" fattened to the highest possible state. There have been
" some stots killed at Aberdeen, between the age of four and
" five years, of 8 cwt. or 64 stones English, which produced
" upwards of 14 English stones of tallow in each of them.
" There have been a few stots killed at Aberdeen, at the ear-
" ly age of between *two* and *three* years old, which weighed
" upwards of 48 stones English, for the four quarters, and
" whose flesh was of excellent quality, besides upwards of 8
" stones of tallow, raised from cottager's cows of a small
" size, (not above 24 stone weight when fed.) In all the
" above instances the stots were bred and fattened by profes-
" sional farmers, and not by the landed proprietors.

It

"It is well understood that young stots, even when very
fat, do not tallow so well as full-grown bullocks.

2. COWS.

"The cow is an animal that can seldom be fattened to a
"very great degree, as every tolerably well kept cow brings
"forth a calf once a year. In that case, from the time the
"milk goes off, till the season of calving, sufficient space
"cannot be left for the purpose of fattening; so that most
"of the cows are killed in a half fat state, by inferior but-
"chers, who retail the flesh of these animals at a reduced
"price, to the lower class of the people. Of course, the
"weight of these half-fat beasts cannot be very considerable.
"Taking the medium of cows, the ordinary weight is from
"24 to 34 stones English, for the four quarters: they were
"sold in the years 1806 and 1807, at from 35s. to 40s. per
"cwt. sinking offals, in that half-fat condition.

"Cows that do not take the male in the course of the season
"after calving, (provincially named farrow cows) are purchas-
"ed by some graziers in the following spring; and then are
"put to grass. Such stock usually consists of small cows that
"have been ill kept, but which, after being laid on rich
"pasture, take on fat very quickly. After being put to
"grass, they generally soon take the male; consequently,
"cannot be allowed much longer than six months to fatten.
"The average price of that kind of stock, when lean, in
"1806, was from L.4 10s. to L.6 10s. each; and in the
"year 1807, from L.3 10s. to L.5 10s. each; and, when fat,
"was sold at from 30s. to 40s. per cwt. sinking offals. The
"average weight is from 24 to 34 stones English. In the
"years 1800, 1801, and 1802, such cows could not have been
"purchased when lean, under from L.6 to L.8 10s.

"In

"In some instances, lately, there have been cows of the small Scotch breed, which suckled their calves till the end of October; and the cows themselves were killed in the following February, when from four to five years old.— They weighed 40, 48, and 56 stones English, for the four quarters, besides about 9 stones of tallow each.

"There were *two cows* of very small bone and size, but extremely beautiful, and of admirable figure, killed at Aberdeen in July 1804; the one was six, and the other seven years old. They had been kept as store beasts, chiefly on straw, with a very small allowance of turnips during winter; and were put to high keeping for the short space of two months, by soiling with grass in the house, when it was found that they had missed calf. The one, which was six years old, weighed 56 stones English; and the other of seven years old, upwards of 64 stones English.— They produced about 10 stones Dutch, above 12 English, of tallow each: and were so fat, that the butcher would not risk the danger of carrying them to Aberdeen. They were sold for L.52 10s. sterling, being at the rate of 70s. per cwt.

"It is necessary here to state, that all the above-mentioned cows had brought forth calf at the early age of two years; and that they were fattened by men whose sole dependance was on the profession of agriculture.

"Passing from cows to their calves.—On the 25th of September 1805, a calf of *five months old*, of the small Aberdeenshire breed, happening to be put into an inclosure among other cattle, admitted a male, who was only one year old. In the month of June following, at the age of fourteen months, she brought forth a very fine calf; and in summer, 1807, she brought forth another equally good. The first calf, after *working in winter, spring and summer,* 1809, was
"killed

" killed the 30th January, 1810, aged three years and seven
" months, weighed 772 lbs. neat Avoirdupois, (or 12 lbs.
" less than 7 cwt.) and was sold for upwards of L.24. The
" second calf, at the age of three years and six months, was
" killed the 16th December, 1810, and weighed 56 stones,
" or 7 cwt. Avoirdupois, and was sold for L.25. It may be
" added, that on the 30th of December, 1807, the mother was
" slaughtered at Aberdeen, at the age of *two years and eight*
" *months,* after having brought up the *above calves.* She was
" found to weigh upwards of 34 English stones the four
" quarters (or sinking offals) and was well tallowed. The
" butcher, a man of character, declared, that a finer beast
" could not have been killed. She was sold for L.13 10s.
" beef being then *at a low price.*

" Some years previous to this, a male calf, of the small
" Aberdeenshire breed, happening to meet with an accident,
" was obliged to be fattened in the stall with turnips; and
" when slaughtered at the age of twelve months, was found
" to weigh 4 cwt. or 32 stones English, the four quarters.
" His beef was sold at the highest price.

" With regard to the time required to fatten a bullock to
" the state which is properly termed high-finished, few gra-
" ziers in Aberdeenshire think of continuing any bullock
" longer than 12 months at high keeping, provided such bul-
" lock is put to fatten *in good condition.* Most graziers say,
" that a full-grown ox of *six* or *seven* years old, is apt to lose
" rather than gain, after being well fed and cared for, under
" favourable circumstances, during 12 months. They are of
" opinion, that a bullock of this description has then reached
" the acme of perfection; and in many cases this fact has
" been fully ascertained. Some bullocks, however, fatten as
" much in six months, as others do in double that time :—
" Much depends on the nature of the animal, the quality of
                                                    " the

" the food, the state of the weather, and the care and atten-
" tion bestowed during the time of fattening.

" No branch of the agricultural art requires more skill,
" experience, and close attention, than that of fattening live
" stock to the utmost advantage.

3. Breeds.

" There are many varieties of different breeds of cattle in
" Aberdeenshire, which are produced both by crossing and
" by difference in the mode of keeping. These breeds also
" have been much improved of late, by giving the most
" beautiful and best formed females to the most beautiful and
" best formed males of the same breed.

" The Aberdeenshire cattle are held in high estimation
" among the English graziers, who fatten for the Smithfield
" market. Of late, some cattle were driven from Aberdeen-
" shire to England. On being inspected by some English
" dealers, and graziers of respectability of character and
" great experience, they were declared *to be the best that had
" been ever driven from Scotland to* England.

" Perhaps to point out any material difference with regard
" to the superiority of the Aberdeenshire cattle, compared
" to the breeds of the neighbouring county of Angus, might
" give offence to many of the eminent breeders and graziers
" of that county. However it must appear evident that these
" gentlemen pay much respect to the Aberdeenshire cattle,
" from the great number of them who attend our fairs, par-
" ticularly the great Aiky Fair at Old Deer, where they pur-
" chase many cattle to raise for the English market, and also
" to fatten for the Edinburgh and Glasgow markets, &c. As
" examples, of the excellent cattle brought to Aiky Fair, it
" may be mentioned, that Francis Garden Campbell, Esq. of
" Troup,

"Troup, several years ago, and before cattle rose so high, sold
"a lot of stots, at L.25 each, to Mr Fiddes, cattle-dealer; and
"that Mr. Watson, farmer in North Essie, a tenant of Mr.
"Ferguson's of Pitfour, sold another lot of stots to the Messrs.
"Williamsons, cattle-dealers, to drive to the English market,
"to be fattened, at the same price. They would now fetch a
"greater price if they were to be sold.

"It would be tedious to give a particular detail of all the
"fine full grown oxen of the large breed, that have been
"both bred and fattened in this county, and killed at Aber-
"deen within these few years past. A bullock of this des-
"cription, weighing from 90 to 100 stones English, is scarce-
"ly ever taken notice of. There was an ox, of about eight
"years old, killed at Aberdeen, (in the year 1805,) which
"weighed 92 stones Dutch, or 115 stones English, the four
"quarters, i. e. sinking offals. In the month of April 1793,
"a stot was killed at Aberdeen, at the age of *three years and*
"*four months*, bred and fattened in this county, whose four
"quarters weighed upwards of 88 stones. And in December,
"1807, there was another stot about the same age, killed in
"Aberdeen, of the small, but improved Aberdeenshire breed,
"which weighed upwards of 73 stones English, the four quar-
"ters, and proportionally tallowed. His fat cut uncommonly
"thick, his flesh was as fine grained and marbled as a kyloe,
"and as tender as a heifer. There was a peculiar circumstance
"in this case, that no doubt would surprize both graziers and
"butchers; namely, that this stot had been two years a
"bull. Yet the beef was considered by the best judges, who
"eat of it, as of excellent quality.

"These, and a few similar cases, which are well known,
"shew the value of the Aberdeenshire breed of cattle, and
"the importance of good feeding in fattening them for the
"butcher.

"To

"To conclude, no market in this kingdom can boast of finer beef than Aberdeen. At the same time it is to be regretted, that a number of cattle, not nearly half fat, are killed weekly at Aberdeen, to the great loss of the feeder, and with no advantage to the consumer; because it is clear that a pound of good meat must go much farther than a pound of bad.

"In the above statement, the weight of the four quarters only has been considered, or what is termed the neat weight of beef, sinking offals.*

"JAMES GORDON."

This valuable paper, drawn up by an experienced and attentive feeder of cattle, contains a number of facts, and judicious observations, which will require no comment. Few men, however, have paid so much attention as Mr. GORDON has done to the feeding of cattle; and no man who has a family, could devote so much of his time and care, as Mr. GORDON has devoted to every branch of agriculture.

The landed proprietors, in all the districts of the county, who cultivate their personal farms, pay much attention to the feeding of cattle. A few examples of their feeding deserve to be mentioned.

JAMES FERGUSON, Esq. of Pitfour, among his fed cattle, had two free-martins, i. e. two heifers, who never produced any calves. They began working in the plough, the one rising three, and the other rising four years old. And they continued to work for six years. In the last of these, viz. in spring 1808, they wrought moderately till the 8th of May when they were put to pasture grass till the middle of October; but the spring being extremely cold, and vegetation com-

---

* Since the above paper was drawn up, the price of cattle has risen considerably.—J. G.

consequently late, they derived little benefit from the grass till June. In October they were housed, and got meadow hay and globe turnips, till the 10th of January, when they got yellow turnips, and ruta baga; and as the hay was of indifferent quality, a small sheaf of corn for supper, till about the first of March, when the allowance of corn was doubled, till slaughtered, the 5th of April, 1809, at the age of rising 10 and 11 years. Their weight was then taken by the butcher who killed them, whose account is subjoined.

ABERDEEN, 7th *April*, 1809.

Weight of Free-Martin Heifers, bought from JAMES FERGUSON, Esq. M. P. is as follows:

Gray Quey, fore leg, Dutch pounds	298
hind do.	311
One side is	609
The other side is also	609
Four quarters of the Grey Heifer	1218
Brindled Quey, fore-leg	233
hind do.	282
Weight of one side	515
The other side	515
Four quarters of the Brindled Quey	1030

Tallow in the Grey Quey, 7 stone 5 lbs.
Do. in Brindled Quey, 10 stone 17 lbs.
N.B. The stone of tallow is 26 lbs. Dutch.
Sold one-half of the whole weight, at one shilling per lb. and the other half at eightpence; the tallow at twenty shillings per stone of 26 lb. Dutch.

Hide

Hide, Grey, 112 lbs. Dutch.
Brindled, 80
———
Both 192. at threepence per pound.
[Signed] WILLIAM REID.

Upon inspecting the above weight, the butcher's receipts, or money paid him, may be thus arranged:—

1124 lbs. Dutch, at 1s. per lb.	- L.56	4 0
1124 lbs. Do. at 8d. per lb.	- 37	9 4
17 stones 22 lbs. tallow, at 20s. per 26 lbs. Dutch	- 17	16 11
Hide or skins of the heifers, 192 lbs. at 3d.	2	8 0
Amount of the whole was	- L.113	18 3
They were sold at a public auction for	100	1 0
Yielding a profit to the butcher of	- L.13	17 3

It may be observed, that when we reduce the Dutch or Scotch Troy weight to English Avoirdupois, the four quarters of the first heifer weighed 1218 lbs. Dutch, or 1332 lbs. English, neat, or only 12 lbs. less than 12 cwt. or 95 stones English, at 14 lbs. or 166½ Smithfield stones, of 8 lbs. each.

When we also reduce the weight of the other heifer from 1030 lbs. Dutch, we find that it amounts to 10 cwt. and about 6¼ pounds, or 80 stones and 6¼ pounds English, or nearly 113 Smithfield stones.

It may likewise merit the reader's attention, that the largest of these heifers was of an English breed, introduced about 60 years ago by the late commissioner UDNY; and that the other was of the native breed of the county, raised by good keeping.—It was remarked by the best judges, that the flesh of the smallest or native heifer was greatly superior in quality to the other. This they attributed to the superior quality of the Aberdeenshire breed; but it might have been, partly at least, occasioned by the smallest heifer being the fattest. For although she had 188 Dutch pounds, (equal to above 205 English) of *less beef*, she had 90 Dutch, (above 98 pounds or 7 stone English) of *more tallow*.—At the same time it must be observed that both were well tallowed.—For the largest had above 204 and the smallest 303 English pounds of tallow.

The largest ox ever killed in the city of Aberdeen, was the property of Francis Garden Campbell Esq. of Troup; and was reared at Delgaty, in the parish of Turriff. He was fed for three years; and weighed upwards of 115 stone English, as stated in the above communication from Mr. JAMES GORDON. But it was believed that after the end of the second year's feeding, this animal fell off instead of improving; and that if he had been slaughtered a year earlier he would have yielded both more beef, and considerably more tallow. There is a certain acme beyond which no animal can be fattened; and every judicious farmer should slaughter every feeding ox when he observes that he becomes reluctant to feed, or ceases to fatten. *

At the same time, where any ox or sheep continues to thrive, it does not admit of a doubt that beef or mutton which is feed for two or three years, is superior in point of quality to what has been fattened in a single season, and still more to what has been at good feeding for only two or three months. Some of the landed proprietors are sensible of this; and feed the animals intended for family use, for at least two seasons. For this purpose they purchase a lot of young cattle, and either kill or sell what they don't find their interest in feeding; while

---

* Mr GARDEN lately sold eight full fed oxen, at L.40 each, or L.320 for the whole; all of which, as well as the above-mentioned ox, were slaughtered by Mr. JAMES WILLIAMSON, butcher in Aberdeen, who corroborates the above statement.

while they carefully fatten those that continue to feed well, and to relish their high keeping. The flesh of such is always better marbled, or mixed with fat—is more delicate—more economical, and keeps better when salted.

As it is of great consequence to every landed proprietor, that the meat which he uses in his family be of the best quality; the following examples of the advantages of feeding black cattle, from one to two years, deserve to be inserted in this report.

ROBERT D. HORNE ELPHINSTONE, Esq. of Logie Elphinstone, is in the practice of feeding for family use, a number of black cattle. His farm overseer (or bailiff) sells in different states of fattening, such as he does not judge it prudent to keep over year; and retains those only who continue to fatten till they are slaughtered.

On the 14th of September, 1808, he purchased three stots at Keith for L.10 15s. and one for L.3 5s. *Each of the four* at that time would *have weighed, when lean,* 250 lb. Dutch, or 273 *neat* Averdupois. But the *fourth*, which was the cheapest, did not fatten so well as the others; and was sold on the 20th of May, 1809, for L.11. Mr Horne's bailiff informed the Writer of this report, that he considered the *aftermath, and turnips,* which this small stot consumed in the course of these 8 months, as worth from L.4 10s. to L.5. He received L.7 15s. more than the price of the stot; so that there remained a profit of from L.2 15s. to L.3 5s.; at a medium L.3. He considered that the animal would weigh nearly 360 pounds Dutch, when he was sold.

The other three, after being fed all the winter and spring, with turnips and rutabaga, were put to good pasture, (on land that had been cleaned with turnips, and laid down, some years before, with clover and rye grass seeds); and they all continued to fatten very much. On the 10th of October, 1809,

the first of them was slaughtered; and found to weigh, for
the four quarters 514 lbs. Dutch; or 5 cwt. 2 lbs. English.—
His beef was excellent; and estimating the value at 8d. per
Dutch pound, it would have sold for - - L17 2 8
The skin, tallow, and offals, at the usual rate of
  one-fourth of the beef - - - 4 5 8

The value of the animal to Mr. HORNE - L.21 8 4
His prime cost, one-third part of L.10 15s. 3 11 8

The recompence for keeping him nearly thirteen
  months - - - - L.17 16 8

The second was killed on the 10th of January 1810, when
the four quarters were found to weigh, independently of the
tallow and offals, 537 lbs. Dutch, or 5 cwt 27½ lbs Avoirdupoise. His beef was uncommonly fine, and well marbled,
or mixed with tallow; and estimating its value at 8½d. per
Dutch pound, the value of it was to Mr. HORNE's family
  L.19 0 4½
The tallow, skin, and offals, one fourth of this 4 15 1

The total value of the animal amounted to L.23 15 5½
Deduct the prime cost, as before - 3 11 8

Recompence for keeping him sixteen months L.20 3 9½

The last was killed on the 4th of May, 1810, and was
found to weigh 532 lbs. Dutch, or 5 cwt. and nearly 22 lbs.
English. The beef was certainly the finest that the Writer
of this Report had ever seen; and he fortunately saw it on
the 5th of May, the day after the animal was killed. Estimating it at only 9d. per Dutch pound, (which is below 4s. 6d.
for

for the Smithfield stone of eight pounds English,) the value of the beef was - - - - L.19 16 0
And the skin, tallow, and offals, at one-fourth of this - - - - 4 19 0

Total value of the animal - - - - L.24 15 0
Deduct the prime cost, as before - - 3 11 8

The recompence of keeping from nineteen to twenty months - - - - L.21 3 4

By looking to this last sum, it will be seen that the recompence for keeping this last ox is less in proportion than that for the two former. And Mr HORNE remarked, very properly, that he was not certain that fattening *for so long a period* was advantageous; but that he considered that meat which was fed for *two seasons* was richer, and more economical, than what was fattened in *one season*: especially what was fattened *in a few months*. Of this there can be no doubt, with any man who has studied the subject. The beef of the second bullock, (which the Reporter ate on the 3d of May, after being *nearly five months salted*,) had very little of the salt taste which inferior beef acquires in *as many weeks*. And that of the last bullock, as being killed in the beginning of May, was more valuable, even if intended for the market, because butcher meat at that season was both scarce and high priced.

Another example of feeding, by Mr. HORNE's bailiff, was lately communicated, and goes to establish the same thing; though the profits of feeding are less at present, from the high price of lean stock.

On the 13th of June, 1809, he bought in Sliach Fair, two young stots, each of which weighed, as nearly as he could judge, 270 lbs. Dutch, above 295 lbs. English, for

L.15,

L.15, or L.7 10s. each. They were put to grass, along with the young cattle, only of a secondary quality, during summer and autumn, and fed with turnips till the 15th March, 1810, when one of them, that did not seem to feed so well, was sold for - - - - - - L.16 0 0
Leaving a profit, including the expence of keeping
  for nine months - - - - L.8 10 0

The other was killed on the 12th of October, after getting good grass all summer; and the four quarters were found to weigh 570 lbs. Dutch, above 623 lbs. neat Avoirdupois.— This beef was uncommonly fine, and at only 8d. per lb. the price of it amounted to - - - - L.19 0 0
And the skin, tallow, and offals, at one-fourth
  of this - - - - - 4 15 0

Total value of the animal - - - L.23 15 0
Deduct the prime cost, viz. - - - 7 10 0

Leaving for sixteen months keeping - L.16 5 0

At 9d. per pound for this excellent beef, this
  would amount to - - - - L.18 4 7¼

This, even in *an unfavourable year*, when lean stock was very dear, is above L.14 yearly for feeding a stot of the small Aberdeenshire breed, whose weight was more than doubled in 16 months. These examples will shew the value of progressive feeding, beginning with grass, and ending with turnips, from 10 to 12 months. Beyond the last-mentioned period, no feeding of black cattle will pay the expence; although a fat animal costs less than when beginning to feed.

As a proof of this, it may be mentioned, that Mr. JAMES GORDON, when residing at Mains of Logie, in Buchan, in order to ascertain how voracious an animal is, in the first stages of feeding, weighed 48 stones Dutch, or 840 nett Avoirdupois pounds of *white globe turnips,* and gave them to a small cow beginning to feed, and that she ate up the whole in three days. She did not at that time weigh, sinking offals, above half the weight of her three days provisions of turnips. In the last stage of feeding, *the same weight of yellow turnip* would have sufficed for *above a fortnight,* and an equal weight of *rutabaga* would have served for *nearly three weeks.*

The dairy of this county is not so important as the trade in oxen. For the greater part of the milk, butter, and cheese is *consumed by the farmers,* and other inhabitants of the country.

The whole number of cows is very nearly, in round numbers, 28,000.

Of that number, at least 1000 of the best cows in the neighbourhood of Aberdeen, yield butter or cheese to the value of L.20 each, or   L.20,000
2000 might be rated at L.15 each, or   30,000
5000 may be estimated at L.10 each   50,000
10,000 of farmers cows at L.8 each   80,000
5000 of cottagers, or villagers, at L.6 each   30,000
5000 small highland cows, at L.4 each   20,000

Total annual produce of the dairy, is, in round numbers   L.230,000

What part of this is exported, is principally from the division of Buchan, and will be stated in the proper place.

The number of cattle killed in the city of Aberdeen, from October 1, 1809, to October 1, 1810, was considerably less than usual, being only 3680.

CATTLE.

Of that number, 300 at L.30 each, or the whole	L.9,000
Of do. at L.25 each, 600—price of them	15,000
Of do. at L.20 each, 800—price of these is	16,000
Of do. at L.16 each, 800—price of these is	12,800
Of inferior cattle, at L.12 10s. each, 800—amounting to	10,000
Of cows, at L.10 each, 380	3,800
Probable value of the whole	L.66,600
Calves killed in Aberdeen, 1621, worth at least	3,400
Total black cattle killed in Aberdeen, 5301—value	L.70,000
Killed in Peterhead, Oldmeldrum, Huntly, Fraserburgh, Turriff, and other inferior towns in the county, and by private families, probably near an equal number of cattle, but chiefly cows, or inferior cattle, not exceeding the value of	50,000
The number killed at least 10,000, and the price	L.100,000
The number sold to dealers, as formerly estimated, 12,000, and their price	150,000
Total value of black cattle killed or sold,	L.250,000

The whole number in the county of Aberdeen, which are almost all raised in it, according to the best information which the Writer of this Report could procure, is about 110,000. And their present value, in round numbers, may be sated at L.720,000.

That number and their value may be more particularly seen in the following abstract:—

28,000

28,000 cows, at L.7 each - - - L.196,000
22,000 calves, reared, (besides those which die, or
  are killed,) at L.2. - - - - 44,000
20,000 year-olds, at L.3 15s. each - - 75,000
19,000 two year-olds, at L.7 10s. each - 142,500
21,000 three years, or upwards, at L.12 10s. each 262,500

Total, 110,000, as above - - - L.720,000

The live stock kept, and the annual sales, are nearly a million—or at least L.970,000.*

It would have been frittering down this account of the black cattle of Aberdeenshire, to have adopted the minute divisions of the plan laid down by the Board of Agriculture.—But after discussing this subject in the most comprehensive and useful way that he could, (which he is certain is complying with the spirit of his instructions), the Writer of this Report shall now concisely go over the particular articles, which are mentioned in that plan.

The following are considered as the best rules for breeding.

1. For *beef*—the animal should be handsome, well formed, short legged, with a smart or keen eye, and a rough ear.

2. For *milk*—A small neck and head, broad in the hind quarter, her bag, or udder, lying well forward on her belly; and her paps or teats, well spread, or pretty distant from each other.

3. For *work*—a well shaped thick animal, well spread, both

---

* In so extensive a county, and where, from the constant trade in black cattle, the number is changing every week, it was impossible for me to get them exactly numbered, as Mr ROBERTSON has done in the small county of Kincardine; and owing to the proportion of highland cattle, the average prices of cows and young cattle are lower than he has very properly stated them in that county.—GEORGE SKENE KEITH.

both in the back and hook, deep in his bosom, small in the mouth, smart in his eye, clean in his throat, and rough in the ear.

4. In rearing young cattle for any purpose, feed well from the calf, but do not keep them either too warm, or too confined.—Let them be loose the first year under a shade.—In the autumn, give them always a quantity of turnips, especially of the leaves or tops, to prevent the aftermath of clover from thickening their blood too much, which produces the disease that is so fatal to young cattle, called the quarter-ill or spald.

5. Endeavour to raise the size of black cattle of the native breed by good keeping, rather than by introducing large foreign breeds, or by putting too large males to small cows.

6. As to the best form of black cattle, Dr. COVENTRY, Professor of Agriculture, has quoted a paragraph from Mr. CLINE, which deserves the particular attention of every breeder of cattle. " A compact round made, not flat, ribbed body, a " deep chest, a broad loin, full limbs, and bones not coarse " and large; and a soft but not thick skin, with hair neither " stairing nor hard, are among the chief marks of a good kind. " The shorter legged animals, too are generally preferable, " those of an opposite description being found to be the least " hardy, and the most difficult to rear or fatten."

7. The most hardy is always the best constitution; and the rules above quoted are tests of this, as well as of the form of the animals.

8. The colours which are considered as good, are *brown, black, brindled,* and *dun,* if not too *white.*

9. In crossing, avoid great and sudden changes; and be content with gradually approximating to perfection.

For *food* in winter, straw, hay, with a proportion of turnips, is given to all cattle, and coleworts to milch cows, for which

last

last boiled chaff is occasionally used; but no steaming of potatoes is practised in this county. In summer both natural and artificial grass are generally, and soiling with clover, occasionally, used. And water, from the diversified surface of the county, is generally abundant and of good quality; except when it rises in a peat moss, or is hard, owing to clay soil.

Salt is not used in mixing with food to cattle in this county, and is now too high priced to be applied to that purpose; nor is oil-cake, or corn, generally used. A few weeks before being killed, the landed proprietors, or better sort of farmers, allow a little corn to an animal intended for family use.—The cow-feeders in the neighbourhood of Aberdeen, use both the millers bran, and the brewers grains occasionally.

Rule for Fattening—First scour or purge with globe or red topped turnip, next apply the yellow Aberdeenshire, (not the yellow ox of Northumberland), afterwards ruta baga; and lastly potatoes, with good hay. The animal should be kept clean and dry.

Rule for the Dairy.—Old grass wherever it can be got for pasture, the leaves of cabbages, coleworts, with good clover, or boiled chaff, when the cow is in the house, are well paid, and the proverbial expression is—" a cow is a cow by the mou'—or mouth".—i. e. according as she is well fed.

Rule for Working—Never work an ox beyond his strength, but let him increase in size, and be kept always in good condition while working. One pound will keep him in this state, when two pounds will not bring him back to it, if he be rendered too lean by bad keeping or over working.

The stall should be roomy, and the crib frequently cleaned. The yards or sheds should be dry and airy, but well sheltered.

It is not the practice in this county to weigh live cattle; but most of the cattle dealers and butchers, and even many of of the farmers, having acquired a practised eye, will guess an ox within 5 per cent. of his weight.

Several distempers occur among the black cattle. The quarter-ill, or black spald, is very fatal, unless cured in time by copious bleeding. The tail-worm is also cured by cutting off a few inches of the tail, which bleeds pretty freely. And swelling from wet clover, or turnip tops, is usually removed by pouring oil, or tar, down the throat of the animal. A quantity of ardent spirits is also used to remove a cholic or gravel, as it is called. And a rope is forced down the throat of any animal, when in danger of choaking from a piece of turnip, or sometimes an entire, but small turnip, sticking in its throat.

The worked oxen are not one-fifth part of the number kept before 1782, nor one-tenth of the number that was kept 50 years ago. They increase both in size and value, perform only one journey, or yoking daily, or only half the work of horses; but are more steady, much easier kept, and seldom require shoeing, as they rarely go on the turnpike roads.— But very large farms require two pair of work oxen, with a number of horses proportioned to their work.

It has long been disputed in the southern counties, whether oxen or horses should be used exclusively on a farm. In Aberdeenshire, a mixture of the two kinds of ploughs has been found the best. And in tearing up barren ground, two oxen next the plough, with two horses before them, have by their patience broke up land, where four horses would have foetted and broke their tackle.

The importance of this article will be a sufficient apology for its length.

SECT. II.—SHEEP.

While the county of Aberdeen has so much improved its breed of black cattle, it does not at present rear one-fourth part of its former number of sheep.

It has been already mentioned, that the Union of the two kingdoms, which has been attended with many advantages to both countries, deeply injured the rearing of sheep, by limiting the Scotch wool to the British market. At that period the native breed of this county consisted entirely of the small white-faced sheep, weighing from seven to nine English pounds per quarter, and producing from twenty-one to twenty-eight ounces, or from three quarters of a pound to a whole pound, of twenty-eight English ounces, which is the provincial weight, of fleece wool. The sheep were fed on the mountains, hills, or barren muirs, and other inferior and uncultivated ground. And their wool, though deficient in point of length or quantity, was of most excellent quality, and not inferior to any Spanish wool. Stockings made from it were worn by persons of the first rank in Britain, and exported to the continent at very high prices. One lady belonging to this county knitted them of so fine a texture, that they were sold at three guineas a pair; and several pairs of them were commissioned for by the Empress of Russia. (They were so fine, that a pair of them could have been drawn through a ring, that was taken off the finger of the fair manufacturer.) But when fine wool declined in price, owing to its being limited to the British market, an increase of the size of our sheep became an object to the farmer. Hence, improper crosses with the coarse woolled, but larger breeds, took place; the quality of the wool was less attended to; and the introduction of turnips and sown grass, had the effect of banishing the

the sheep, especially the native breeds, from the lower parts of the county. In these districts a few only are kept for family use by the landed proprietors; and the better sort of farmers rear a few of the English, or mixed breed, that are more easily kept from injuring the inclosures, and are allowed good grass in summer, and a proportion of hay and turnips in winter.

In the higher districts, viz. in Birse, Braemar, Strathdon, and Strathboggie, a considerable number of sheep, partly native, partly of the black-faced, and partly of the mixed breeds, still remain. But though their size has been raised, the quality of their wool is much inferior to what it was sixty, forty, or only thirty years ago. This has been occasioned either by crosses between the native and the Linton sheep, or by bringing great numbers of the mixed breeds of the south of Scotland to this county.

Calculating from the best data, which the Writer of this Report could obtain, there were in 1690, or 120 years ago, at least 600,000 sheep in this county—60 years ago, two-thirds of that number, or 400,000—and only 30 years ago, 300,000. At present, the total number of sheep is less than that of black cattle, and cannot be estimated above 100,000. About two-thirds of that number is in the mountainous division of the county above-mentioned. In the lower districts, the number of sheep has decreased very much since 1778.— That year there were in the Reporter's parish 4500 sheep; and in 1809, there were only 141, of which 103 belonged to the landed proprietors, and 38 to the farmers and subtenants.

It deserves to be attended to, that sheep-farming in this county, which 120 years ago was of so great value, is at present at its lowest ebb, and will probably soon increase very considerably. In the seven unfruitful seasons in the end of of the seventeenth century, a number of sheep were killed for food in these years of famine. Their number, at the Union of the kingdoms, had been nearly filled up, when the limiting of the Scotch wool to the British market, by reducing the price, injured the quality of our wool; and the introduction of the turnip husbandry has been fatal to our ancient native breeds, which were too restless, in point of character, to be kept from turnip and sown grass, and too small in point of size, to pay the expence of good keeping.—But while our native breeds, from all these causes, have decreased so much, a number of other breeds, particularly of the Merino, and the South-down, have been brought into the county by different landed proprietors; and will probably, in a few years, be more generally reared, not only by them, but by the sheep farmers, as a valuable species of live-stock.

Even the native sheep, when properly cared for, improve much in size within a short period. The Writer of this Report, in 1809, saw some sheep, which in nine months time had been raised from 30 pounds when lean, to 63 pounds the four quarters. They were fattened by the Rev. WILLIAM PATERSON, at Logie Buchan.

The small-sized highland mutton is also considered as the finest eating. It certainly is more highly flavoured than that of the low-country, owing to the herbs and natural grasses, that are found on the mountains; and such of the landed proprietors as wish to have mutton of a superior quality, purchase highland sheep, and fatten them for family use. These animals, however, that have been accustomed to a wide range on the mountains, seldom thrive when stall-fed. Therefore it is found necessary either to throw turnips to them in the fields, or to enclose them in a small patch of ground, surrounded by a paling of wood, from which there is

ac-

access to the hay and turnips that are put in their cribs or mangers. While their mutton is fine flavoured, the wool of these small and highland sheep is generally of excellent quality.

As the sheep is an animal that has four distinct qualities, viz. the goodness and quantity of mutton, and fineness and weight of wool, all these things merit the attention of the sheep farmer; and a few examples may be selected from the practice of a man, who feeds every species of live-stock with the greatest attention.

Mr James Gordon sold his one-year-old sheep from L.2 2s. to L.2 5s. besides shearing seven pounds of excellent wool.—In 1794, when residing at Muiresk, he had a large kind of English sheep, from each of which he obtained 14 lbs. Dutch, or 14¼ lbs. English, when they were full grown, and 10 lbs. Dutch, or 11 lbs. English, when only a year old; but he sold them all off, because they were not so good feeders, nor had wool of so good a quality, as his present breed.—By proper crossing, and good keeping, he has raised the mixed, or half-Scotch breed, from below 9 to above 18 lbs. Dutch, or nearly 20 lbs. English, per quarter. And by high feeding a small ewe, crossed by a larger sized male, he obtained two excellent female lambs. The first of these took the male, and also had twins. The other had no lamb, but was killed at eighteen months old, when the four quarters weighed 60 lbs. Dutch, and the tallow, even at that early age, weighed 17 lbs. 8 oz. Where such a disposition to fatten is joined to a fine quality of wool, a breed of sheep is most valuable, and ought to be preserved.

No animal pays better than a sheep, when well fed, and kept dry and clean. From the attention paid to this subject, within the last ten years, it is probable that the time is not distant, when the rearing and feeding of sheep will again become general in this county. In the present state of this

branch

branch of live-stock, it is probable that the total value of sheep is L.100,000 in round numbers; and that the number killed annually is worth nearly L.30,000. Of this number, there were killed at Aberdeen, from October 1, 1809, to October 1, 1810, 13,480, which were supposed to be worth L.20,000. At least an equal number was killed in the county, in the small towns or villages, or by private families;—but as part of the sheep killed in Aberdeenshire, are reared in the counties of Kincardine and Banff, the annual value belonging to this county, is stated only at L.30,000.

There is at present *no regular system* observed with respect to sheep. In the highlands, where they are fed on the mountains, they roam at large, and ly down to rest where they please. In the low country, under the old husbandry, they were folded, when every farmer, who had a plough of land, had a flock of sheep. Now they are too few in number, to afford the expence of a shepherd, except when kept by the landed proprietors. No rules are generally observed with regard to their crosses. Their carcases have increased in size, but their wool has rather declined in quality.

Their food in the highlands, is at all seasons chiefly what they can collect for themselves. In the low country they get hay and turnips in winter, and pasture from natural and artificial grasses in summer. But it is found to be highly improper to allow the sheep access to young grass, as they destroy its roots; and they are commonly pastured on old grass, or on a field which is intended to be broken up next year.

In the mountainous districts they cannot be folded, as their pasture is many miles distant from any arable land.—In the low country their number is now too small to render their manure an object of the farmer's attention. Formerly, a farmer who had forty black cattle, and two hundred sheep, considered the dung of both to be sufficient for tathing a field

of

of Scotch acres, in twenty or twenty-one weeks. But at present there are no data for making a calculation of the value of the dung of this useful animal, which was formerly raised in so great numbers. Sheep farming, however, must soon be more an object than it has been for some time past in the county of Aberdeen.

SECT. III.—GOATS.

These are also on the decrease; there not being above a thousand of them in the county. They are chiefly reared in the higher parts of Marr, where several goat-whey quarters are kept in the summer months for consumptive persons. But it is impossible to state exactly the value of goat milk, which is sold at 4d. per Scotch pint, after keeping a sufficiency for the kids. Perhaps the flesh of goats killed, and the milk sold annually, may amount to L.300. Formerly, cheese was made both from goat milk and from ewe milk, to a considerable annual amount. But it is now used only by the poorer peasants; and is too inconsiderable to be deserving of notice in this Report.

SECT. IV.—HORSES.

There are various breeds of horses used in this county.— The smallest is that of the common highland poney, from ten to twelve hands high, remarkably hardy, and surefooted in climbing the mountains. They are adapted only for the saddle. The largest is the West-country, or Ruglen breed; which is very strong and docile, and draws in a single cart a much greater weight than any other breed in the island.—The native breed of horses, reared in the lower parts of the county, though not so large as that of the Southern or Western counties, hold an intermediate rank between the highland poney and the Ruglen cart-horse; and is distinguished for their spirit, agility, and hardiness. They have a considerable portion of blood, and are fitter for the saddle, or the plough, than for the cart or waggon. This breed is said to have been greatly improved in 1558 by a number of Spanish stallions, which were landed on our coasts, when many of the ships belonging to the Armada were wrecked by a storm, in attempting to sail round the island, after the Armada was defeated by the English. But since carts come to be generally used, and our horses to be employed in pairs in the plough, attempts have been pretty successfully made to improve the breed, by stallions from the southern counties. There has also for many years been a regular exchange of our light, active horses, (which are fittest for the saddle, and are sold in the counties of Angus, Kincardine, and Perth,) for horses of a heavier make, and higher price, suited to the single horse cart, or fit to drive in carriages. From the very great attention that is paid to breeding young horses, both by the landholders and principal farmers, it is not improbable that we may, in other forty years, become as much distinguished in that line, as we are now in rearing black cattle. But at present, it is believed, that in bartering horses with all other counties, we lose at least L.20,000 annually. This is owing to the great increase of cart, carriage, and saddle horses in the city of Aberdeen.

The number of horses kept on a given extent of land is
ex-

extremely various,—from the small farmer who keeps two poor horses to plough 15 Scotch acres, to the large farmer, who ploughs at least 50 acres, with a pair of stout horses, besides the land which he has lying in grass. Nothing can be more clear than this, that it is not profitable to keep a horse where one has less than 20, or two horses, where he has less than 30 acres of land. The Reporter knows, that a small glebe, (or parsonage lands) of 4 or 5 acres arable, and 2 or 3 acres of grass land, is a losing bargain to every clergyman who keeps a horse, and often two horses, with so small a proportion of land in cultivation. Indeed no man, who is not either a horse-dealer, or a common carrier, can derive any advantage from keeping a pair of horses, except he has as much work as will find employment to himself and his horses round the year. There are by far too many small farmers in Aberdeenshire, who have not employment for the horses which they keep on their lands.

It is not, however, to be denied, that some of those who are horse-dealers, or who keep a breeding mare for raising young horses, derive very considerable advantages from those employments. But it should also be considered, that their profits are owing to their skill in horses, and attention in rearing them, not owing to the small size of their farms, which is certainly a disadvantage.

The work performed is in proportion to the strength of the horses, the extent and situation of the different farms, and the number of hours which they employ in labour.— In ploughing with oxen, they take only one yoking or journey of about five hours per day; and the quantity of ground seldom exceeds one-fourth of a Scotch acre. But on large farms, the same ploughman has two pair of oxen, which he uses in succession. Where horses are used, and the ground is light, and nearly level, a pair of horses can plough an

English

English acre in three journies, or yokings, of four hours each; but the average of work done, by a pair of ordinary horses, cannot be stated at more than a Scotch acre in four yokings. Where the horses are good, the season pressing, and the farm extensive, it is not unusual to plough an English acre at two journies of 5 hours each: And indeed some farmers make it a rule, in engaging their farm servants, that they are to work their horses ten hours a-day, except in the winter months. But where the ground is stiff, and has a considerable declivity, their horses are soon worn out by this severe labour, when continued through the whole year.—During the season of seed time, a great exertion is often made, without any injury to the horses, who are well fed while they are hard wrought. And when the winter has been unfavourable, from great rains in November, and severe frosts in December, January, and part of February, it is absolutely necessary to make a very great exertion, in order to overtake the ploughing, and finish the sowing in due time. But in this case the horses must be better fed, to enable them to endure severe labour, from the time that they begin ploughing, after a long continued storm, till the turnip seed is finished. On the whole, the horses in this country, excepting during the winter months, (when, from the shortness of the day, they can work for only one journey, or when they are stopped occasionally by excessive rains, or for a considerable time by intense frosts,) work eight hours a day, and plough from half a Scotch, to two-thirds of an English acre. From the irregular surface of the county in general, they cannot accomplish nearly as much work as is performed in Norfolk, where the land is flat, the soil light, and the furrow is comparatively shallow. And even on the coast of Buchan and Formartin, where the surface is pretty level, the soil is generally so stiff and tenacious, that the

horses

horses cannot plough more than is done in the sloping fields, but lighter soils, of the other districts. More work no doubt could be done than is usually performed by our oxen: but the farmer does not choose to work these hard, as they are generally young; and as he wishes to make a profit by selling them in good condition.

The *food of horses* is extremely various—from that of the highland poney that is fed on the mountains, or glens, in summer, and on straw in winter, without getting any oats, or any other kind of corn—to that of the high fed horse for the saddle or carriage.

The herbage that is collected in the glens or narrow vallies, or along the sides of the mountain torrents, is found sufficient for supporting the small horses of the highland farmer during the summer months. And although the *crop* raised in the higher districts, where turnips are very little used, is neither so clean, nor nearly so weighty as in the more cultivated parts of the county; yet the *straw* is much better, as it contains a great proportion of *natural grass*, that grows up along with the corn. By this means, a small-sized, but hardy breed of horses is maintained in much better condition than they could otherwise be, without getting a proportion of oats regularly every day, if the straw were coarse, as it is where the crop is cleaner and more valuable. Yet even in these highland districts, the farmer gives his horses the lighter oats, provincially *shillocks*, and also part of the chaff, and light grains of his bear, which last are boiled, and partly given to the horses, and partly to the cows.

In the intermediate districts, between the mountains and the lower parts of the county, the straw is not so good, because turnips are partially used: But the horses are fed with much better grass in summer, owing to the proportion of grasses sown; and in winter, they get not only the refuse of the different kinds of corn, but a proportion of oats regularly from the large farmer, and after the seed time begins, a proportion of hay between the yokings, or journies for two hours in the middle of the day. And when there is a sufficient quantity of it, they are fed with hay constantly from the commencement of the seed time.

Where turnips, potatoes, and sown grasses are generally used, the horses are fed with good grass in summer; and get generally a lippie, about 4 English lbs. of oats, before they go to work, and an equal measure of boiled bear at night.—They are fed in the winter months also with hay, by the landed proprietors, and some of the best farmers, or they get a proportion of hay during that season. But after they begin to go two journies daily, they are well fed with hay, and oats. Potatoes and yams are also given by many farmers to their horses, and save half of the allowance of oats. And both rutabaga and carrots are given by those who have not the other kinds of food. Carrots are much relished by horses, and are inferior only to potatoes and ruta baga. Potatoes are generally given raw, and steaming has not been introduced. But the Writer of this Report knows, that when boiled, and when mixed with *bear meal*, or even with bearsids (i. e. the hulls of bigg with a small portion of meal adhering to them) they are much more valuable than when given unboiled.

The expence of maintaining a small horse in the highland glens in summer, as already mentioned, is only twenty shillings for about twenty weeks. And in winter, the oat or bear straw, and other food given by the highland farmer, may be about L.3. So that a horse is maintained there yearly for L.4. In the intermediate districts, bordering on these, the expence of keeping a horse is at least L.8. Where oats and hay are given three months in the spring, it is not less than L.12. And

And where a stout draught horse is well fed, on hay, suiting, and oats, carrots, and other good keeping, it cannot be estimated at less than L.20 yearly. In short, the expence of keeping a horse may be estimated at L.4 in the highlands; at L.8 in the districts next adjoining to these; at L.12 where he gets hay and oats only occasionally during three months; and at L.20 when at full keep, as a draught horse, employed in the plough or cart. Carriage and saddle horses according to their size, and rate of keeping, cost the inn-keeper, or *horse-hirer*, as he is called, from L.26 annually, or 10s. weekly, to L.32 10s. yearly, and the landed proprietor, who feeds them highly, 14s. weekly, or L.36 10s. per annum.

The shoeing of horses has, from various causes, also increased in value. Thirty years ago, it was not above 5s.; it now amounts to 13s. annually for work horses; and where the horses have often occasion to go on turnpike roads, it is not below 15s. To common carriers, and inn-keepers, or posting horses, it is at least 20s.—often 26s. annually.

But the *decline in value* is the most serious article respecting this useful animal. Before the great rise in the price of horses, this was estimated at 20s. It is now above 40s.; perhaps on good farm horses is not below L.2 12s. yearly.— After 6 years old, a horse falls rapidly in price; and the most prudent method for a farmer, who is not a horse-dealer, is to buy at 3, 4, or 5 years old, and to work them as long as they are able to labour.

The harness of horses in this county is extremely various. In the highlands, ropes of straw, and of hair, with very little of leather are used. The whole harness does not cost as many *shillings* there, as it does *pounds* on the sea coast. But there, owing to the want of good keeping, and of good roads, except near the highways, no good carts could be used.— The harness of the small highland horses, which are used only occasionally in the cart, does not cost above 4s.; and of a full-mounted *shaft* horse at Aberdeen is L.4. But taking into the account the different ranks and situations of the farmers, and that one-third part of the horses *work in the traces*, the general average may be stated at L.2 for each horse, taxed for husbandry.

The number of work horses, per the Collectors' books, amounts to 8184

That of saddle and carriage horses is 421

Total, 8605

The distempers of horses form a subject for the farriers, of whom we have a considerable number; many of whom are very illiterate. The Writer of this Report is no veterinary surgeon, nor does he pretend to know the diseases of horses. But though there are a few persons whom he esteems, and who follow this profession with great success, he has, in general, no great confidence in a science, where the *afflicted* animal, or *patient*, cannot *inform the Doctor* what complaint he feels. He knows, that in 1799, a poor carrier's horse, when bringing out some coals for him to Oldmeldrum, was so much lacerated by an accident, that his life was despaired of; and that a subscription was made to get another horse to the poor man, who could not afford to pay a farrier in a desperate case, but covered the torn chest of the animal with a coat of tar, and a bit of old canvas. The poor horse, *turned out to the fields to die*, and left to nature, without any hopes of reviving, completely recovered in about six weeks. In all probability he must have died, if one of those who pretend to be farriers had been employed to heal him.

It may be proper to add, that Mr. FERGUSON of Pitfour has brought a stallion and two mares lately from Suffolk, to improve our breed of horses.

SECT. V.—ASSES.

Asses are not used for any sort of work in this county, except by a few poor men in carrying fish to the inland districts. And it does not seem adviseable to use them; as our horses, many of whom have a good deal of blood, cannot endure to see them. She-asses are kept for consumptive patients, and their milk is sometimes of service to such.

The highland poney can be used to greater advantage than the ass, where a small foot, and light animals, are necessary, in hoeing narrow drills.

SECT. VI.—MULES.

There are not perhaps twenty mules in the whole county. And there is a prejudice against using an animal that is generally very indocile.

SECT. VII.—HOGS.

Forty years ago, these were very numerous; and the salted pork of Aberdeen was in high repute. At every corn mill

mill there was a boar, a sow, and a number of pigs annually reared. Many of the landed proprietors and farmers kept a hog for eating up the offals. But at that time the pound of pork sold one-third higher than the pound of beef or mutton; and lately it sold at one-third less, sometimes at only half the price of butcher meat. Hence the rearing of hogs has been found unprofitable; and of course has been laid aside. A number of the mill-masters apply the mill-ring, (i. e. the corn that remains about the mill-stones,) to the feeding of horses. Few of the farmers now either rear or feed any hogs. But several of the landed proprietors, and the inferior class of millers still keep a few, though there are not above 1000 kept over year in the whole county. They are not now a profitable concern, except to a miller who has no horse, or to private families, who feed them with the offals of the kitchen. Before the licensed distillers were stopped, in 1799, a number of hogs were fed with the grains and spirit-wash; and at that period they paid better than any other kind of stock. From the dislike to pork, which is generally entertained by the country people, hogs are too seldom fed for private families; otherwise the whey of the dairy, and other offals which are thrown to the dunghill, could be profitably applied to the feeding of this species of animals. The rise in the price of pork, which has lately taken place, will probably occasion more hogs to be reared.

SECT. VIII.—RABBITS.

There are no rabbits raised for sale; and only a few for amusement—not a hundred in the whole county.

SECT.

SECT. IX.—POULTRY.

The number of these is very great; and the constant demand for eggs, not only for Aberdeen, but even for the London market, has raised the price, and increased the real value of poultry. Most of the landed proprietors, and many of the better sort of farmers rear geese and turkies; ducks, and barn-yard fowls, (especially the latter,) are kept by almost every cottager. In 1778, the Writer of this Report purchased eggs at 1d. or 1¼d. per dozen, and chickens at the same price each. Now eggs cost from 6d. to 9d. per dozen; frequently 1s.; and chickens from 5d. to 9d. each. A good barn-yard fowl, in 1778, cost 5d. or 6d. and now generally from 18d. to 2s. 6d. The price of eggs has increased in the greatest proportion, owing to the demand from London, and the increase of the inhabitants of Aberdeen. It is impossible to calculate the annual amount of poultry of all kinds, and of eggs that are sold, or the value of what are consumed by the country people. But the Writer of this Report believes, that at the advanced price of these articles, they are not below L.20,000. In the harvest, the poultry is very destructive to the corn. But in the seed time they seldom do much injury; and by their scratching the soil, the crop is often more abundant, after a considerable portion of the seed has been picked up by the hens, than where the clods have not been properly reduced by the harrows.

It needs only be added, that pea and guinea fowls are kept by a few of the principal landholders. But none of these are sold. They are kept merely for pleasure.

SECT. X.—PIGEONS.

There are fewer pigeon-houses, and of course fewer pigeons, than there were 40 years ago. But in many places the children keep a few tame pigeons. The whole are too inconsiderable to do much hurt, even in their immediate neighbourhood; or to be accounted of much value to the proprietors. For in stormy winters, the feeding of pigeons for a few months, far exceeds their total value.

SECT. XI.—BEES.

This source of national wealth is in many districts of the county too much neglected: In others, it is both carefully and successfully managed. The profits of bees are no doubt fluctuating; and in late, or rainy seasons, are often inconsiderable; while a severe winter destroys a multitude of hives. But when an apiary is properly attended to, it pays very well, especially on the banks of the Dee, and in the higher districts of the county, where there is abundance of heath-flower, and other articles of summer provision. In the parishes of Birse, Crathie, Glenmuick, and Strathdon, and in the adjoining district of Cromar, considerable sums of money are made by the sale either of bee-hives or of honey. One man, (Donald Simpson) in the district of Cromar, who pays only L.20 of rent, sold in one season, bee-hives to the amount of L.41. And last year he and his sister-in-law received for honey the sum of L.42. No doubt great attention is required in the management of bees; but that attention, indedependently of the amusement it affords, is amply rewarded. The value of honey sold in Aberdeen, or consumed within the county, is not probably above L.2000 annually. But it might easily be rendered L.20,000. For the quantity raised in a few parishes in the upper division of the county is nearly one-half of the whole.

The great objection to the keeping of bees, is the expence of feeding them in an unfavourable spring. An ingenious Friend of the Reporter's has contrived to keep them in an *icehouse*, in a state of insensibility, which is *a saving of their winter provision*.

To this catalogue of Live Stock we may add

SECT. XII.—GAME.

1.—THE Red Deer are found in great numbers, sometimes 300 in a flock, in the mountains of Braemar. The roes, or smaller kind of deer, are also seen in the hills of Cromar, Glentanar, and occasionally as low down as Monymusk.

In the upper parts of the county, the *stalking* of deer is practised with great caution. The huntsman must always keep to the *leeward* of these animals; otherwise, from their acute sense of smell, they always know of his approach.— Both Earl FIFE, and the proprietor of Invercauld, have allotted a great extent of land for deer forests.

The deer are very destructive to the highland farmer, in winter; and the poor people are not allowed to kill them, even when in stormy weather they eat the corn in their stackyards. About 70 years ago, a highland farmer complained of this grievance to the landholder, who desired him to poind them, when he found them offending in this manner. The man, who had some humour, took the opportunity of decoying

ing 16 of them in a severe storm, into his barn, the door of which was purposely left open; and then gave notice to his landlord to relieve his poinded cattle. The gentleman sent for his bailiff, or farm overseer, in great haste, thinking that some of his oxen or cows had been poinded by the farmer: but on discovering that it was not any of his *black cattle*, but a number of his *red deer*, that had been laid hold of by the farmer, he was pleased with his good humour, and permitted him to kill one occasionally, when they harassed him in winter.

2.—*Foxes.* These about 30 years ago were much less numerous than formerly: But the extensive plantations of wood in all the districts of the county, (the sea-coast of Buchan only excepted) has given them more shelter; and though fox-hounds are employed pretty successfully, the number of foxes is increasing.

3. *Hares.* The different fields of broom, furze, and turnips, all afford protection to the timid hare. And though exposed to the attacks of men, dogs, and foxes, this prolific race is rapidly increasing. The farmer does not grudge her either her grass or a little corn, but is provoked when she nibbles off part from a number of his turnips, as the frost quickly destroys what she had left. Rutabaga alone remains uninjured by the season, after its rhind and part of its bulb are eaten by the hare. In the mountainous districts the *white hare* is seen, and killed occasionally.

4.—*Wild fowls*, or *winged game.*—In the mountains of Mar the *eagle* still retains her ancient habitation among the rocks, and now and then carries off a lamb to supply her young ones with food.

5.—The *Ptarmigan*, a most beautiful bird, perfectly white in winter, and of a light-blue colour in summer, is also found near the summits of the highest mountains, and is eagerly sought after by the sportsman.

The

The numerous tribe of *moor game* suffering occasionally from the inclemency of the season, are regularly beset on the 12th of August with a multitude of enemies; and the sportsmen are sometimes fatigued with killing them in the beginning of the season. Also in the lower districts of the county, the partridges are for several months exposed to the landed proprietors, to their game-keepers, and frequently to the poacher. Another foe, who has *a license from nature*, viz. the *hawk*, is also very destructive to the partridges. And the Writer of this Report was much pleased to see one of his neighbours take out of his shot-bag, a hawk that he had killed at the precise moment that his bill touched the back of a partridge. This was a prompt execution of justice on the criminal, though the partridge died at the same instant.

It would swell this article too much to take particular notice of all the wild animals, and birds, that are found in this county;—we may just remark in general, that

The *Otter*, so destructive to the salmon, is sometimes killed near the banks of some of our rivers; that the *badger*, and the *polecat*, or *fumart*, are occasionally destroyed, the latter sometimes in the act of killing our poultry; and that the *weazel* is killed by the herd boys for the sake of its skin, which is used as a purse. Also that of the winged animals,

The *heron* visits our lakes and streams of water in quest of small fishes; that the *wild-geese* visit us in the beginning of winter, and by their flight either to the sea or towards the hills, are supposed to prognosticate bad or good weather; that the *wild ducks* commonly frequent our streams; and the *snipes* our marshes; that the *woodcock*, like the wild-geese, stays with us during winter, and the *cuckow* and *swallow* during summer; that *rooks*, provincially *crows* (though the real crow is seldom seen with us) are very numerous, and also very useful in destroying insects in spring and summer, though very alert at dig-

digging potatoes in autumn; that the loquacious *jack-daw*, the *hooded crow*, and occasionally the *raven*, with the *plover*, the *linnet*, *goldfinch*, and *blackbird*, are found in the lower parts of the county. And that the *sea-gull*, and other noisy birds, regularly visit Lochnagar, and bring up their young at Lochanyeun, or the *Bird's Lake*, 40 miles from the ocean.

The only noxious *reptile* in the county is the *adder*. It is seen pretty frequently in the mountainous district. One of these bit a cow in the beginning of last July (1810) belonging to Mrs. FARQUHARSON of Finzean, but the wound was not dangerous. The Writer of this Report saw another the same day, while he was taking the elevation of Mount-battock.— It was a beautiful speckled, or rather striped, animal, nearly a foot in length, and about three quarters of an inch in diameter at the middle. It escaped very quickly among the heath, and the Reporter had no desire to molest it. The sight of it rather excited curiosity, than inspired fear.

Moles are frequently found in our richest fields; and the mole-catcher has risen in his demands for catching them.— His allowance, at first a halfpenny, afterwards a penny, is now three halfpence for each. By catching them at stated hours, cutting with a spade in their line of journeying—stamping the rising mole-hill with the heel, and digging it up instantly, some persons are very dextrous in destroying moles. These reptiles feed upon worms, and it is surprising to see a number of the latter throwing themselves out of the ground, when they perceive the mole is approaching. Traps of different kinds are set for catching the hare by the sportsman, and the mole by the farmer. And both traps and poison are used for destroying rats and mice. So much for the different species of animals found in the county.

CHAP.

CHAPTER XV.

RURAL ECONOMY.

SECT. I.—LABOUR.

Since the turnip husbandry was introduced, the quantity and value of labour employed by the farmer have much increased. In many districts it has become difficult to procure labourers, even at very advanced wages. About 40 years ago, the labour of the farmer, and of his servants and dependants, were much less considerable, in point of either value, or of the time spent, and the exertion made by the labourers, than these are at present. The cutting and preparing of peat-moss, and of turf from the moors, and carrying these different kinds of fuel on the horses' backs, or in a kind of wooden panniers, before carts were used, employed a great part of the summer: and the digging with the spade, or cutting with the breast plough, a quantity of turf, provincially *muck feal*, the carriage of this to the dunghill, and mixing it with the dung, and then carrying the whole on horses' backs to the field, occupied that part of the autumn which remained after harvest, and also that portion of the spring months which was not employed in ploughing the fields, or in sowing the different kinds of grain. The hoe was scarcely known, far less used, except in the form of a coarse and blunt instrument, which served both for hoeing the coleworts, scraping the dung behind the cattle, and collecting it into small heaps, where it was thrown out at a hole in the back-wall, by a three-pronged fork, provincially a *grape*. And the pulling of weeds was chiefly confined to thistles and mugworts. Now the labour both of men and of cattle is far greater, and directed to much better purposes. The pulverizing of the soil by frequent ploughings, the hoeing of the turnip and potatoes, and the union of hand-weeding with horse-hoeing, give employment both to the farm-servants and day-labourers. The cutting down and harvesting of clover hay, the sinking of ditches and drains, and the raising of stone fences, and other inclosures, have given rise to a more valuable species of labour, than what was formerly employed in pulling thistles or mugworts, in cutting the coarse meadow-grasses, or in building the *feal dykes*, (or inclosures of turf) which were intended to stand for only one year. And both farm servants and day labourers are well fed and well paid by their masters.

Farm servants are generally unmarried in this county. Not above one-sixth part of the men are married, and these reside chiefly with large farmers; some of whom keep always one, and sometimes two married servants, to whom they allow a cow's milk as part of their wages. Persons of this description, besides their cow, and fuel brought home by the farmer's horses, are allowed six bolls of meal, and from L.8 to L.10 in money annually. Unmarried servants board generally in their master's houses, and according to their comparative merits, receive from L.7 to L.10 every half-year, (only a few of them exceed L.9) and young lads get from L.4 to L.7 of wages; the last sum when they assist in cutting down the corns. Boys for keeping cattle generally go to school for three months, from Martinmas to Candlemas; and their wages for the three spring months, are from 15s. to 20s.; and for the summer half year from L.1 10s. to L.3. Maid servants are from L.2 10s. to L.3 every half-year; generally 10s. more in summer than in winter, on account of the harvest. They have their food always in the farmer's house.

Many of the small farmers engage a married servant for nine months, allowing him three months in the summer to provide for his fuel, and to hire as a day-labourer. In this case he has a small croft, and a little grass for his cow at night, when she is brought home from the farmer's cattle. He assists in cutting down and harvesting the corns, and ploughs and performs other farm-work till Whitsunday, when he is again released for other three months. During the nine months that he is a servant, he gets his breakfast and dinner in the farmer's house, and has an allowance in meal for his supper.

Some of the landed proprietors, and large farmers, build a small house called the *bothy*, and sometimes the *men's house*, in which their men servants eat and prepare their food. In a few instances a woman is allowed to cook their victuals, and besides her wages, receives a peck and a half of oatmeal weekly. The allowance to the men is from 9d. to 1s. for *sup*, (i. e. for milk or ale) or a cow to every three of them, with two pecks, or 16 lbs. Dutch, equal to 17¼ lbs. English of oatmeal.

The principal article of the food of both the married and unmarried servants, is oatmeal, or rather oat-flour. For in fact, the oats of Aberdeenshire are ground so small, and sifted so completely, to get clear of the hulls, that they look like a species of flour, when compared to the coarse oatmeal of the Lothians. The farm servants, only 30 years ago, used one-third part of bear or bigg meal; but this is no longer permitted to be mixed with the oatmeal, which in general is very nourishing. It makes very good porridge, a kind of oatmeal pudding, that is constantly used for breakfast, and sometimes for supper. When baked with water, on a girdle, or thin iron-plate, it also makes excellent cakes, a species of plain short bread, that is baked without leaven. Both porridge and cakes are commonly eaten along with more or less milk, according to the season; except occasionally in winter, when ale is substituted in the place of milk. Sometimes porridge is made of oatmeal and milk; but this is considered as a kind of feast, or given only when working in the peat moss, or engaged in other hard work.

There is always a proportion of the oatmeal that adheres to the hull of oats, when they are ground on a corn mill. And while the pollard, or hulls of wheat, are accounted of little value, those of oats are accounted very valuable. They are used by the name of Sids; and a kind of pudding, termed *sowens*, (or oatmeal flummery) is made of them.* It is remarkably light, and easily digested, but not so nourishing as it is palatable. From its slight degree of acidity, it is used for a dinner dish, as it prevents the servants from being thirsty. It is commonly eaten with milk, seldom with ale.

Pot Barley made with milk, provincially *milk broth*, is used occasionally; at other times barley mixed with coleworts and

car-

---

* The method of preparing sowens is—first, a quantity of oatmeal sids is steeped in water, about 70° of heat in summer, and 80° in winter. When the liquor has become a little acid, the water on the top, which is termed *pourings*, is poured off; the oat hulls are passed through a sieve, and the raw-sowens is then a thick liquor, and is either half-boiled, and drunk by the name of *knotting sowens*, or completely boiled, and eaten with sweet milk.

carrots, either with a little butter, or a small bit of butcher meat, is used on Sunday for dinner. Sometimes *milk brose* is given occasionally to the farm servants; but more commonly *kail brose*, i. e. coleworts and water are boiled, and the juice poured into a dish in which a small quantity of oatmeal has been put, and the whole is stirred, adding more boiling liquor, till it be found of a proper consistency.

At other times, when setting out on a journey in a cold morning, or when returning suddenly from any market, *water brose*,* or hasty-fare, are given to the men-servants. A quantity of oatmeal is put in a *wooden bowl*, or *dish*, and boiling water is gradually poured on, and mixed by stirring, till it is of a proper thickness. Then a quantity of sweet milk, (i. e. of milk newly from the cow) is poured on the top, to render it palatable.

In times of scarcity, when oatmeal is bad in point of quality, and high priced, meal made of ground bear or bigg, is used by such of the cottagers as cannot afford to purchase oatmeal. And even by persons of better condition, porridge, made of bearmeal is given along with milk to children; or when first made into pot-barley, is used at table in thin cakes, or *sconns*. These, when baked with hot water, are very light and palatable; and when baked with new milk, are preferred by many to wheaten bread.

The Writer of this Report bestowed considerable attention to the most economical way of using all the different kinds

---

* The difference between *porridge* and *brose* is this. When the meal is gradually mixed with the boiling milk or water, and boiled while mixing, the dish is called *porridge*. When a quantity of meal is put into a bowl, and boiling liquor, whether water, milk, or the juice of coleworts, is poured into the bowl, and mixed, by stirring it with the meal, it is called *brose*. It is of *Greek* derivation.

kinds of food in calamitous seasons. And after trying a multitude of experiments, he came to the following conclusions:

1st, That the most economical way of using bear or barley, is when it is ground on a barley mill, and boiled as pot barley, either with a little butter, and a few vegetables, (in which case it is provincially called *barefoot broth*) or with a bit of meat, where this can be had, or with milk, when it is called *milk broth*. A boll of this kind of corn, made into pot barley, and boiled for three hours every day, will go as far in nourishing persons who are not hard wrought, (i. e. sedentary persons, women and children) as *three bolls* made into meal, or *six bolls malted*, and *brewed into ale*.

2d, That the most economical method of using oats, is to make them into meal, and mix them along with milk, slowly and long boiled. This is called milk-porridge, and is allowed to stand some time to cool before it is eaten. It is not so pro-

---

* In 1799, the oatmeal was remarkably bad; and the Writer of this Report found it necessary to purchase 176 lbs. of rice, which was used along with an equal quantity of pot barley. Six ounces and a half of each, after repeated trials, with three Scotch pints (one and one-half gallons) of liquid, viz. three parts of milk, and one of water, served for dinner to five grown up persons, and six children. But on a complaint that they became soon hungry after eating of this light food, a small quantity of oatmeal was made into brose, out of the boiling pot; when the rice completely cured the bad taste of the oatmeal, and rendered the dish both palatable and salutary. On making this discovery, he sent several pounds of rice to such of his parishioners as had very bad meal, which was completely cured by the mixture. About four months after this was known, and practised in his parish, a clergyman of much higher rank than this Writer, (viz. the Bishop of Durham) pointed out the good effects of using a mixture of rice along with damaged flour. These facts ought not to have been omitted, when treating of Rural Economy; as they may be of service to mankind when the Bishop and the Reporter are both laid in the dust.

proper to make them into brose, except for persons who are at hard work; and even then, though made hastily, the dish should be *cool*, not cold, before it is eaten. When the oats are injured by a rainy season, an ounce of rice mixed in a pound of oat meal, will have a great effect in rendering them both more palatable and more wholesome food.

3d, That in such times of scarcity, (and indeed at all times when it can conveniently be done) no oats should be given to *horses* till they are ground, or at least bruised.— Two pecks of ground or bruised oats are nearly as valuable as three pecks given in the common way. They are more thoroughly digested, and go more directly to the nourishment of the animal.

4th, That after the bones of butcher meat *appear* to have yielded all the juice that can be extracted by boiling water and pot barley, the *more penetrating liquid*, in which *coleworts are boiled*, has the effect of extracting still more nourishment to the poor cottager, who by this *only process known to him*, obtains all the *nourishment* that *he* can be supposed to extract from *a shin of beef*, the *only part of his ox*, to which his light purse can extend in years of scarcity.

5th, That butcher meat should be *boiled* with pot barley and other vegetables, even by the opulent, in such calamitous seasons, and not *roasted* in any considerable quantity, as this is a great waste of provisions. Baking into puddings or pies, with flour or other articles, is not liable to this objection;— though there is most frugality in pot barley, and boiled meat.

It has been stated, on the authority of Sir JOHN PRINGLE, that *a pound of butcher meat* contains as much nourishment as is given by *twelve pounds of bread*. The Writer of this Report entertains great doubt of the truth of this assertion.—

But he has no doubt that one hundred weight of beef, worth
9d. per pound, or - - - - L.4 4 0
With pot barley, potatoes, and turnips, worth
  only - - - - - - - 1 1 0
And with either household, wheaten, or even oaten
  bread, worth - - - - - 1 1 0

All amounting only to - - - - L.6 6 0
will go as far in supporting a family with nourishing food, as *four times* that sum expended on roast beef, and fine flour, in years of scarcity. And he is also of opinion, that in the country, where milk can be had in abundance, the farmer's servants, and the cottager's family can be supported at one-half of the expence, even of *boiled meat*, where pot barley, oatmeal, and potatoes, are used with *milk*, and a proportion of *rice*, where the oats have been damaged. He can speak here from the experience of his own family, and that of many other families in his neighbourhood, in five different years of scarcity, viz. in 1783, and 1784, (in consequence of the bad season of 1782, and bad seed of 1783,) and in 1796, (in consequence of a bad crop in England in 1795,) and of 1800 and 1801, in consequence of the bad season of 1799, and of bad seed in 1800,

2. *Labourers.* It has already been mentioned, that it is often difficult to procure labourers, even at the present advanced wages; and that the employing women to hoe turnips, and to gather the stones and weeds off the turnip fields, has, as yet, by no means become general. When by giving high wages, and even premiums to the best hoer of turnips, this practice is established, the turnip husbandry will acquire great advantage. For a woman, though not so *strong*, is *more alert*, and generally *more neat* in picking the young
                                                                tur-

turnips with her fingers, when they are so close that the hoe cannot separate them.

The *price of labour* in summer varies from 1s. to 2s. a-day, besides victuals. In the peat-moss, the men, who dig up peats, earn 16d. and the women who carry them off in barrows, receive 1s. with their breakfast and dinner, and as much oat cakes for supper as is now worth 3d. (This they take home to their own houses.) In mowing hay, a man gets frequently 2s. and his board. In harvest he is hired by the season; but his fee, which varies from L.3 to L.4 is equal to about 3s. for every good day that he can work in the field. A good builder of stone dykes, or inclosures built of stone, gets 20d. with, or 2s. 6d. without his victuals. A woman shearer earns from L.2 to L.2 5s. for harvest work; and 9d. or 10d. a-day when hoeing turnips; she *ought* to get at least 1s. In winter, a day labourer, for the most part, gets 1s. and his breakfast and dinner. He gets his supper only when he can work till six at night. His year's earnings are now more than doubled; and he makes L.26 in an average of seasons, where he formerly did not earn above L.10 or L.12. His year may be said to consist of 24 weeks, at summer wages—of 5 weeks at high wages in harvest, (including a few rainy days)—of 6 weeks at autumn and spring wages—of 10 weeks at winter wages—and 7 weeks, during which, (owing to bad weather, or to his going to public places of resort,) he either earns no wages, or cannot work at all. This, at a medium of labour, may be stated as follows:—

24 weeks, (on full wages) at 2s. per day	L.14 8 0
5 weeks during harvest	3 10 0
6 weeks, at 1s. 8d. per day, in autumn and spring	3 0 0
10 weeks at 1s. 4d. daily	4 0 0
7 weeks no wages.	
Total, besides his board in harvest	L.24 18 0

An inferior labourer will earn only about *twenty guineas*; but a good stone-dyker will receive above L.30 of wages; often L.40 when he gets employment, and works by the piece.

The rise both of the *money*, and of the *real price of labour*, has been very great since 1782, when the turnip husbandry began to be generally studied. Previous to that period, a farm servant's wages seldom exceeded L.4 yearly, besides his victuals. Now they are at least four times as much, or L.16 a year, often L.18, and sometimes L.20. A day labourer, who got no food from his employer, did not earn above 5s. a week, where now he has 12s. At that time he was supposed to have high wages, if he earned the price of a peck of meal daily, even at building stone fences. Now he earns at least the price of a peck and a half of meal at that profitable employment. The rise in the *money* price of labour is therefore as *five* to *two*; and the rise in its *real price*, is as *three* to *two*. The decrease of the value of money within 28 years is consequently as five to three; and the rise in in the real price of labour, which is as three to two, is a proof of our increasing industry, and of the national prosperity, in spite of wars and taxes.

The hours of work are from 6 in the morning to 6 at night, with one hour's interval for breakfast, and another for dinner, to day-labourers. In a busy season, however, there is not above half an hour allowed for breakfast, and three quarters of an hour for dinner to their servants, by some of the farmers, who in seed time cause their horses to be kept at work for ten hours a-day. In autumn and spring, the farm servants work from seven to eight hours, and for two months in the middle of winter, only six hours in the field. But they thresh generally two hours in the morning before breakfast, except on those farms where threshing mills are erected.

Piece

Piece Work is also very common in making out ditches, or in building inclosures, as well as in trenching barren lands. A ditch, which formerly cost 1d. per Scotch ell, of 37 1-5th inches, now costs 2¼d. A stone wall, of four feet high, and three feet wide in the bottom, and eighteen inches at the top, formerly cost 2d. per Scotch ell, and now costs 6d.; or three times its former money price. This higher proportion of both the money and the real price of that species of labour, is owing to the great number of stone fences that are carrying on in all parts of the county. A stone mason also, owing to a similar cause, gets L.3 10s. for the rood of work, of 36 square ells, or above thrice as much as he got thirty years ago, which was from L.1 to L.1 8s. 4d. for coarse rubble work of one storey.

3. *Cottages attached to farms.*—Many of the better sort of farmers have one married servant, with a *cottage*, and a small garden, or *kail-yard*. Some of these *keep a cow* round the the year to such a person, who is generally preferred, as being more steady and attentive than the younger, or unmarried servants are. This method is certainly preferable, in *most cases*, to giving a few acres of ground to a married servant, as it leaves him, and his wife and children free of attending to a single cow, whose milk is not worth the expence of attending to her alone. But others, who have *detached* pieces of *broken ground*, give two or three acres to a faithful servant; who has a family of children, that can care for his cow, and occasionally for two cows in summer, and only one in winter. As long as the ground that is in cultivation, *is ploughed by the farmer*, it will often be prudent thus to accommodate *one* married servant; and the extent of his small croft must be determined by the quality and situation of the ground, and other circumstances. No general rule can be laid down for cases of expediency.

4. *Expence proportioned to the extent of land.*—This is very different in various situations. Where the land is flat, or lies pretty near to a level, and where there is a great proportion in grass, the expence of labour is much less on the same extent of land, that where the ground has a considerable degree of *acclivity*, or is rough, and in the provincial dialect of this county, *bank-set*, and when at the same time a great proportion of it is under the plough. By looking back to the expence of labour in the farm of Wester Fintray, in page 204, it appears that the charge on 276 Scotch, or 350 English acres, for labour, amounts to L.480, viz.

To 4 married servants, for board and wages	L.100
7 unmarried men, and 4 women servants, for wages	130
Articles of provision for ditto	80
Labourers, and implements of husbandry	170
Total for labour, or labourers, and for implements	L.480

This is above 27s. per English, and nearly 35s. per Scotch acre. But when land is rough, with a considerable declivity, so that two horses or two oxen cannot easily plough it, the expence of labour may be stated at a guinea and a half per English, or L.2 per Scotch acre. In small farms also it is greater than on large, except when the farmer holds the plough himself—ploughs sparingly, and depends chiefly on profits from his live-stock. The consideration of the great rise in the price of labour (to which that in the price of corn, which ought to regulate that of labour, does not correspond) clearly proves, that fertile arable land, which lies flat, or with a very gentle declivity, is worth much more rent, than rough and high lying farms, which both require more labour, and yield much less returns.

SECT. II.—PRICE OF PROVISIONS.

It is a decisive proof of the advantages which are derived from the cultivation of turnips and of the sown grasses, that though the price of wheat has risen more than that of every other kind of corn, yet provisions, as far as regards corn, or butcher meat, have not increased in proportion to the rise in the money price of labour. Two hundred years ago, or in 1609, the quarter of 9 Winchester bushels of the best wheat cost L.2 10s. and in 1608, L.2 16s. 8d. And at the medium of 100 years, from 1595 to 1694 inclusive, a quarter of wheat of 9 bushels cost L.2 7s. 8½d. Consequently, a Winchester quarter of 8 bushels cost L.2 2s. 5d. From 1697 to 1796 inclusive, a quarter of wheat cost only L.1 14s. 10½d. or 7s. 6¼d. less than in the former century. Since 1794, the prices of corn have been higher, owing to various causes, but not so high in proportion as the increase of the price of labour.

The price of oats and of oatmeal in Scotland, has not risen so much of late years as that of wheat, and still less than that of labour. In 1641, when the best wheat in East Lothian was 16s. 8d. per boll, of 85 Scots pints, the best oats was 11s. 8d. per boll of 124 pints. In 1811, the best wheat is at 40s. per boll, Linlithgow measure, while the best oats are only 22s. per boll of corn measure. They should be 28s. in order to hold the same proportion to wheat. In Aberdeenshire, the fiars of oatmeal, of a middling quality, called *Farm Meal,* may be considered as a better proof of the real price of provisions, than the price of any species of *corn.* For all kinds of *corn* are different in their value in different years, according to the state of the weather, when

fil-

filling in the ear, the earliness of the harvest, and the goodness of the season, when they are cut down and harvested. The following short table will shew the price of a boll of farm meal, of 128 stones of Scots Troy, commonly called Dutch weight, which is nearly equal to 140 lbs. neat Avoirdupois, or half the weight of a sack of flour.

	£	s.	d.
From 1705 to 1714 inclusive	0	6	$4\frac{1}{10}$
From 1715 to 1724 do.	0	6	$1\frac{1}{3}$
From 1725 to 1734 do.	0	7	$6\frac{2}{3}$
From 1735 to 1744 do.	0	8	$1\frac{1}{10}$
From 1745 to 1754 do.	0	8	$8\frac{3}{15}$
From 1755 to 1764 do.	0	9	$9\frac{1}{3}$
From 1765 to 1774 do.	0	12	9
From 1775 to 1784 do.	0	12	$1\frac{12}{13}$
From 1785 to 1794 do.	0	12	$8\frac{1}{3}$
From 1795 to 1804 do.	0	18	$3\frac{1}{18}$
For the first 50 years the average is nearly	0	7	6
For the next 50 years the average is nearly	0	9	$9\frac{1}{4}$
The *highest* price in the whole was in 1800	1	17	0
The *lowest* price was in 1706 (not one-eighth part of the highest price) viz.	0	4	$5\frac{2}{3}$

On looking at these prices, it appears, that for the first 25 years of the last century, the price of oatmeal was only 6s. 6d. per boll; the next 25 years it rose to 8s.; the third 25 years it rose to 12s.; and the last 25 years of the century it rose to 14s. 2d. It is very remarkable, that the prices of the 10 years, from 1775 to 1784, though they included those of 1782 and 1783, are not quite *twelve shillings and two-pence*, while in the 10 preceding years, in which there was no scarcity, the average price of oatmeal was *twelve shillings and ninepence;* and the 10 subsequent years, in which also there was no calamitous season, it was *twelve shillings and eightpence and one-fifth of a penny.* It is no less worthy of observation, that the average price of oatmeal from 1795 to 1804, (though it included the price of 1795, which was chiefly occasioned by the scarcity of wheat in England, and also the extremely high prices of 1799 and 1800,) was *only eighteen shillings and threepence seven-tenths,* while the price of meal since 1800, *has been nearly nineteen shillings,* exactly *eighteen shillings and elevenpence two-ninths,* as appears by the following table, which contains the fiar price of farm bear and of farm meal.

Farm Bear, 1801	L.1 1 6	Farm Meal,	L.0 15 6
1802	0 16 0		0 14 9
1803	0 16 0		0 16 0
1804	1 2 6		0 17 0
1805	0 18 6		0 16 8
1806	1 1 6		0 18 6
1807	1 4 0		1 5 0
1808	1 4 0		1 6 0
1809	1 2 0		1 1 0

Total 1 boll for 9 yrs. L.9 6 0   L.8 10 5
Average price, 1 0 8   0 18 11¾

This increase of the price of *oatmeal* must be attributed chiefly to the increase of paper money. That of the price of wheat (of which much more is raised in Scotland than was raised twenty years ago) must be chiefly ascribed to the progress of luxury. The comparative decrease of the price of bear or bigg, which formerly sold for one-fourth part more than a boll of the best oatmeal, is undoubtedly occasioned by an excessively high tax on malt from bigg. A direct tax on the landed property of the county, which raised *ten times* as much free duty to Government, would not so deeply affect our agriculture. The fiar prices of oatmeal, at a medium, of the years 1807, 1808, and 1809, was per boll, of 140 English lbs. - - - L.1 4 0

Those of farm or market bear, at the same aver-

age, only - - - - 1 3 4

When it is considered, that the Aberdeenshire boll is exactly five-sixths of the English quarter, and one-ninth part larger than the Linlithgow boll, it will be found that the Winchester quarter of bear is *only twenty-eight shillings*, and that the standard Scotch boll of bear costs *only a guinea*, while the boll of oatmeal costs *twenty-three shillings and fourpence:* hence the decrease of the quantity of bear sown, and the injury done to our agriculture, may be clearly understood. In one respect, however, this high malt tax has been an advantage: The farmer's servants, who are boarded in the house, get a much less allowance of beer than formerly. When the demand for bear or bigg was great, and the tax on malt was moderate, the farmers raised much greater quantities of this kind of corn; sold a considerable proportion of this to the brewers and distillers, and gave a liberal allowance of beer to their servants. Now they give them much less, as they raise less of this species of grain, and as the malt tax is now five times as much as it was formerly.

The quantity of ardent spirits used within the county is also much less than it was twenty years ago. The stoppage of the licensed distillers has produced this decrease. A number of poor people, who had nothing to lose, engaged in illicit distillation, and were frequently fined in small sums by the Justices of the Peace: but since the late Act of Parliament obliged the Justices to imprison for a space of at least three months, the number of illicit distillers, who make any considerable quantity, has greatly decreased. A few of the delinquents, who were convicted after this time, were imprisoned, in terms of the act; and were supported while in the jail,

jail, by charitable contributions. It is creditable to the landed proprietors to state, that in a general county meeting, finding that the Excise would give no *aliment*, (the Scotch term for an allowance to a prisoner) to the first offender * who was imprisoned for illicit distillation; and not choosing that such delinquents should be supported from the *rogue-money*, (a tax imposed on account of offenders against the *criminal laws of the land*) put their hands into their pockets; and contributed a sum, which was more than equal to what the man could claim for aliment. In consequence of this alteration in the mode of punishing such offenders, (many of whom are *now* brought before the *Court of Exchequer*,) illicit distillation of whisky has in a great measure been given up: and smuggled spirits, viz. Geneva and brandy, are more generally used, than any friend to the interests of either British agriculture, or British commerce, would wish. On the whole, not above two-thirds of the ale or beer, nor one-third of the spirits is used by the inhabitants of the country, that was used fifteen or twenty years ago. The decrease in the consumpt of ardent spirits is not to be regretted: But every friend to humanity would wish that the farm servants, and the industrious day labourers, who are providing food for the community, should not, by the operation of the malt tax, be stinted in their allowance of beer, while cultivating the earth by the sweat of their brows; as if they were so many mariners cooped up in a ship, and reduced to a short allowance of provisions, until they arrived at some friendly harbour.—In the city of Aberdeen, and in the different maritime towns of the county, rum, from the West Indies, which pays a duty to Government, is used by the more opulent inhabitants;

---

* This delinquent had been for some years in the army, and was then an old man.

tants; but in spite of all the regulations of the Legislature, a considerable quantity of smuggled spirits is consumed, in preference to the spirits of the lowland distiller, which are not all relished by the inhabitants of Aberdeenshire. It is not possible, by any Acts of Parliament, or regulations of the Board of Excise, to render the *corn* spirits of the south acceptable to our northern stomachs. For, *de gustibus non disputandum est*, (there is no disputing about matters of taste.)— The great distillers no doubt have made their *corn* spirits purer than they were formerly; but they cannot cure them of their peculiar flavour, or *haut gout*. And every man, who was accustomed to drink our *malt* spirits, while we had licensed distilleries, will say, while he tastes, and *only tastes*, the corn spirits of the great distillers,—

> I do not love thee, Doctor Fell:
> The reason why, I cannot tell;
> But I don't love thee, Doctor Fell.

Even when ardent spirits are of good quality, they are certainly more hurtful than malt liquor to the constitution.— If an allusion may be borrowed from agriculture, in this Report, the Writer would say, that ardent *spirits*, like *lime*, or calcareous matter, are only *stimulants*, but that *beer*, or fermented liquor, like rich *manure*, nourishes and invigorates the constitution.

Fuel.—Forty years ago, the great article of fuel was peat, dug from the mosses. This was a very expensive article; not so much from the labour of digging, as from that of carrying home the fuel. The city of Aberdeen, at that period paid at least L.3000 a year for peats, frequently L.4000.— The principal part of this sum was for fuel from the Causeway, and other peat mosses in the county of Kincardine, within seven miles of Aberdeen. Now pit coals, both from New-

Newcastle and from the Firth of Forth, are generally used in that city; and more or less in all places of the county, within twenty miles of the sea coast. The Canal, on the Don, and the different turnpike roads, have facilitated the carriage of coals into the interior of the county. But it must be acknowledged, that the want of coals and lime are a great drawback on the exertions of the Aberdeenshire farmer; and that the difference between having both within a few miles of his house, as in the southern counties, and getting both from a distance, as in this county, is equal to not less than 30 per cent. of the rent of a farm. Peats are now become more expensive, as our peat mosses are nearly exhausted in many places; and the only consolation for this expence is, that our people are learning how to manage coals. In many cases it would be an advantage, that the farmers brought all their fuel from Aberdeen, and spent the summer months in attending to the horse-hoeing and fallowing of their lands. To save *five* guineas on coals, they often expend *twenty* in misapplying the labour of horses.

In the highlands, considerable quantities of wood are used as fuel, though they are generally mixed with either peat from the moss, or turf from the moors. In the middle, and even in the lower districts, aged and fallen wood is frequently sold at an under value, and used as fuel. Wood is a very wholesome and exhilarating species of fuel; and in the capacious chimneys of the hospitable mansion of a highland proprietor, unites with the landlord, the table, and the bowl, in giving the traveller a warm reception. In the peat mosses, a species of fir, that is highly resinous, and inflammatory when dried, is found in many places; and, when peeled with a large knife, is used either as a faggot or a taper. It burns with a clear and steady light, and instead of being snuffed, is kept clean and bright, by breaking off from time to time the *part* that is burnt. Small round pieces of moss, strongly impregnated with sulphur, (called *creeshy clods*) are also found in the most *hollow place* of some of our mosses; and are used both for kindling the fire, and giving light to the poorer peasants. The *creeshy clod* is the *most*—and the thick *turf* of the moor—(full of sand and small stones, and incumbent on a bed of clay or till,) is the *least, inflammable article of fuel* known in the soil or subsoil of this county.— The *fir* is also the *most*, and the *oak* is the *least valuable* article of fuel found in our peat mosses.

The different kinds of fuel consumed in this county, cost at least L 50,000 a year for their price and carriage.

Circumstance of fuel meriting particular notice. This (in a county in which fuel is so expensive, and the peat mosses so much exhausted,) is the mixing of coals with turf, or of peat that has been imperfectly dried, or is of inferior quality. An unfavourable season, (viz. 1782) first taught the bakers of Aberdeen to heat their ovens, and prepare their bread at one-fourth of the former expence; and another unfavourable season, viz. 1809, taught our farmers, when their peats were very ill dried, and their turf of a bad quality, to mix a small proportion of coals, which had a very great effect in rendering their otherwise bad fuel, both comfortable and serviceable. The Writer of this Report saw lately a proportion of coals burning in the kitchen fire of a farmer, whose farm bordered on one of the largest peat mosses in the county. He was informed by the man, who is sufficiently careful of his money, that though residing in the immediate neighbourhood of an extensive peat moss, he found his interest in buying a few bolls of coals, and using them along with a greater proportion of turf and peats.

The mode of using them is this—The hearth is clean swept and the small coals are put above the half-burnt peats, then the

the turf, or imperfectly dried peat, is laid above the coals. After this has begun to cake, and is kindled completely, the coal is gently stirred, when it inflames and consumes the fuel that is of inferior quality.

It would be an affectation of knowledge, which indicates neither a *comprehensive mind*, nor *a spirit of minute research*, if the Writer of this Report pretended to make any *calculation of the total income*, and *of the general expenditure*, (arranged under particular articles of detail) of the whole inhabitants of this extensive county, in number above 123,000, and forming all the various classes of society; from the great landholder, who dwells in a stately mansion, and is proprietor of 100,000 acres, to the poor cottager, who rents a mean hut and a small kail yard, both comprising less than one-tenth of an acre; and from the opulent merchant, who has amassed a capital of L.50,000, to the miserable mendicant, who can only collect a few pence, either by unceasing importunity, or by that speechless modesty, which generally indicates deeper distress, and more powerfully affects a feeling heart. In a small county, which was possessed chiefly by gentlemen of moderate fortunes, residing on their own estates, and by farmers and their dependants, some guess may be made of the probable amount of these different articles; and a very able and ingenious writer, Mr. Robertson, in his Survey of Kincardineshire, has made the Rural Economy of that neighbouring county both an entertaining and instructive subject of minute discussion. To attempt this in a Survey of Aberdeenshire, would argue an equal want of modesty and of knowledge. A few general estimates, founded upon the best data, which the Writer of this Report could procure, (and he knows that he will not be accused of indolence, or of carelessness in seeking information) shall be given in a subsequent part of the Report, after considering the

the Trade and Manufactures of the county. In this place it is sufficient to say, that by far the greatest proportion of the land-rents in this county is spent within the limits of Aberdeenshire, (not above one-tenth part of it is spent in England); that great additions are yearly made to the plantations, inclosures, and the improvement of the antient arable fields, while a laudable competition is carried on between the plough and the spade in breaking up barren lands; that a considerable part of the landholder's rent is now allowed for building durable farm-houses; that the farmers, on the faith of their leases, have invested a reasonable share of their capital in the improvement of their farms; that the farm-servants and day-labourers are more alert, and better fed, (though their allowance of beer has been lessened by a tax which it is hoped is only temporary); that by far the greatest number of persons live upon their incomes, and though a few have suffered from farming injudiciously, and more have been injured by the failure of a few cattle dealers; yet, on the whole, bankruptcies have seldom happened to any considerable amount, in the agricultural part of the community; and both the fixed and the floating capital of the farmer, whether consisting in the permanent improvement of the lands, or the addition to the number and value of live-stock, or to the yearly produce of corn sold in the market, are constantly increasing.

Let this much suffice for a general outline of the Rural Economy of Aberdeenshire, as arranged according to the corrected plan of the Agricultural Reports.

CHAP.

CHAPTER XVI.

POLITICAL ECONOMY.

SECT. I.—ROADS.

No county in Great Britain has expended so much money on roads of all descriptions, as has been laid out in Aberdeenshire. Above 300 miles of turnpike have been made out since 1796, or will be completed in the course of a few months, at an average expence of L.350 per mile. And commutation roads, that is, roads which have been made by commuting the *statute labour* of the farmers and other householders, into a certain sum of money, (generally 30s. the L.100 Scots of valued rent) have been made in all directions for opening the communication with the turnpike roads, and other high-ways in every part of the county. The money laid out in roads of all the various denominations, amounts to nearly L.120,000; and L.30,000 more will soon be expended.

I. The turnpike roads shall first be mentioned, as being both the most durable, and most expensive.

1st. The great north post road from Aberdeen to Edinburgh is carried for two miles to the Bridge of Dee, the place where it meets with the county of Kincardine. The expence of these two miles was above L.1000 [*].

Another branch is also carried along the north bank of the Dee, to Charlestown of Aboyne, for 31 miles. This has been made

---

[*] It was made by the commutation funds; and only becomes turnpike if the Magistrates fail to keep it in repair.

made out at an expence of only L.8000. The comparative cheapness of this road is owing to the purity of the gravel, and other excellent materials. It deserves to be mentioned, however, that about nine miles of this turnpike road, through the parish of Upper Banchory, and part of Drumoak parish, is situated in that insulated district of the county of Kincardine, which lies on the north bank of the Dee. (This is that part of the division of Marr, which above 500 years ago, was taken from this county, and added to Kincardineshire.)— Yet the whole 31 miles of the Deeside turnpike road belong to the turnpike trustees of the county of Aberdeen.

2d. A turnpike road is carried in a direction nearly due west from Aberdeen for 7 miles, when it divides into two branches. Both, however, are turnpike. One of them is 19 miles in length, and goes through part of Echt and Midmar, in a direction nearly in the middle between the rivers Dee and Don, until it reach Drumlasie, in the north of the parish of Kincardine O'Neil. It is proposed to be carried to Tarland, or 10 miles farther. The other, which is 21 miles in length, is carried on in a line nearly westerly, to the church of Alford. This turnpike, (which is called the *Skene Road*,) and its two branches, already extend to 47, and will be 57 miles.

3d. The great turnpike road from Aberdeen to the northwest, at 4 miles from that city, also diverges into two branches. The most southerly extends through Kintore and Inverury, to Huntly, and to the boundaries of the county of Banff;—a distance of 34 miles from the point where the two roads separate, or nearly 38 from the Exchange, or middle of the city of Aberdeen. The most northerly goes through Oldmeldrum and Turriff, in a direction almost due north to the town of Banff, at a distance of 42 miles from the point of separation, or 46 from Aberdeen. The whole length of turnpike, on both roads, is 80 English miles.

4th

4th. The turnpike from Aberdeen to Buchan, extends from the county town to Ellon, 17 miles, and from thence to Birness, other 5 miles. At that toll-bar it is divided into two branches; the most westerly of which goes through the village of Mintly, near Old Deer, and is carried on to Fraserburgh, a farther distance of 24 miles. And the most easterly is presently carrying on to Peterhead, a distance of 12 miles; and also extends from thence to Fraserburgh, a distance of 17 miles, in a more circuitous direction from Aberdeen. This great turnpike road to Buchan, and its two branches, extend in all 75 miles.

Besides all these much frequented roads from Aberdeen, the county town, there is a turnpike road from Peterhead to Banff, (a small branch of which is carried to Old Deer) and this road, including the Old Deer branch, extends nearly 35 miles. Another cross-road, also turnpike, has been made from Newburgh to Udny, and is to be carried to Oldmeldrum, an extent of 13 miles. So that these two cross-roads amount to 48 English miles. A road from Oldmeldrum to Inverury, a distance of 6 miles, is also proposed to be made turnpike, for facilitating the access to the Canal at that place. Excepting the last, and a few miles of the road from Ellon to Peterhead, which is now making, the whole turnpike roads have been completed in the short space of 14 years.

The *method* adopted in making out these roads was the following:—

After the different lines of road were marked out by a Civil Engineer, and approved of by the county meeting, (on application from the landed proprietors of the district) they were first *formed* by undertakers, by either carrying off soil, where the ground was too high, or by adding quantities of earth till it was brought to a proper height, where it was too low. An open space of 14 feet wide, by as many inches deep,

deep, was left in the middle of the road. This was filled up with granite, which was broken into small pieces of about two inches square, and is called Metalling. When this was properly broken, and laid in the middle division of the road, a quantity of loose gravel, or very sandy soil, (where pure gravel could not be got,) was laid on the top, about an inch deep, and was called Blinding. The rain gradually washes a considerable part of this down among the stones, or metalling, and in a short time the road becomes very dry and pleasant. The breadth of the road is generally 40 feet in all, viz. 13 feet on each side, of earth or gravel, formed into a slight degree of convexity, and 14 feet of pounded stones, or metalling, in the middle.

The *materials*, as already mentioned, of that part of the road, on which horses and carriages are drawn, are of that species of granite which is so common in this county. And it is not till after the carts or carriages have in some degree pounded the upper part of the stones, or metalling, that the road becomes agreeable to a traveller. When any part of the road is sunk lower than the rest, an additional quantity of small stones, or metalling, is laid on it from time to time, by men employed for that purpose. It is generally six months after a new road is made out, and used by travellers, before it becomes at once dry and agreeable.

The *expence* of forming and metalling a turnpike road, at an average, of the whole roads, including sewers and bridges, has been, at least, L.350 per English mile; in some cases, where the ground was pretty level, and materials at hand, below L.250; but in many other cases, where the ground was low, and broken, it exceeded L.500.

Roads of Communication, which give access to these turnpikes, and have been made with the *commutation money*, (or by trustees, who on the faith of that annual fund, have borrowed

rowed money to make them) are all regularly formed; but only a few of them are metalled. They generally, however, get a cover of good gravel. It has always been found adviseable to metal, or cover, with stones broken very small in the middle part of the road; but where the funds are not adequate, this is only covered with a thick coat of gravel. Some of the landed proprietors, who have a well founded partiality for metalling, make it a rule to apply the parochial money, or commutation assessments, in this way only as far as it will go, instead of giving an imperfect repair to a greater extent of road. This has been done particularly by Sir WILLIAM FORBES, Bart. of Craigievar; who by this means has completely cured many bad passes, and in a few years will have metalled the road from Fintray House to Aberdeen, as far as his property reaches. The commutation roads, and highways, not turnpike, exceed 1000 miles in length; have cost already L. 20,000; and when the road across the county, from Alford to Kincardineshire, including the two bridges over the Dee and Don, are completed, will cost L. 10,000 more.

4. *Farm Ways* in this county are not made at the public expence. Yet many of the farmers make out for themselves such private roads as they judge to be necessary. And the Writer of this Report is acquainted with cases, in which the farmers have offered to be at half the expence of making out a communication road to the nearest highway. The people in general, have imbibed a spirit for road making, that must be attended with great advantage.

5. *Concave Roads* are not found in this county, except in places where a stream of water has made a cavity in the middle of the road, in running down from a higher ground.— But,

6. *Convex Roads* are generally used in Aberdeenshire.— The convexity, however is not great, not exceeding one inch in a foot, or 10 inches in a parochial road of 20 feet wide.

7. *Streets.*

7. *Streets.*—While so much money has been laid out in making roads in all parts of the county, (and even in Braemar, where there are no turnpikes, the roads are kept in excellent condition, *fifty miles* from the sea) the city of Aberdeen has been adorned by two new streets; and a bridge has been thrown over a small brook, called the Denburn, that makes the access from the south, formerly so difficult and unpleasant, both easy and agreeable. One of the streets, called King Street, extends above a mile in length from the centre of the city of Aberdeen, to that of Old Aberdeen. The other, called Union Street, has united the high grounds in Castle-street, (in which the Exchange is situated,) with a kind of new town, that is built on the south-west, by one of the finest bridges in the island. This bridge is of one arch, considerably less than a semi-circle, the cord of which is 132 feet.—It is built of excellent granite, and attracts most powerfully the attention of every stranger. It cost above L.10,000.

The opening of these two streets, and purchasing the ground from the former proprietors, have cost the Magistrates of Aberdeen above L.100,000; but they will, in the course of years, be partly, (if not wholly) indemnified by the great rents which they will receive for the sites of houses.

8. With the same spirit for improving the harbour, that appears in opening streets and making turnpike roads, the Magistrates of Aberdeen have obtained an Act of Parliament, empowering them to borrow L.120,000, for carrying out piers, or quays, on each side of the Dee, and deepening the harbour. There is some difference in opinion about a part of the proposed improvements; but from the excellent checks on the expenditure of the money, there is no doubt that the carrying out the piers will be honourably conducted, and the money on that branch well laid out. When this is accomplished, the depth of the harbour will be much greater than at present.

Before

Before concluding this subject, the Writer of this Report would state, that he received from the President of the Board of Agriculture, the Report of the Committee of the House of Commons, respecting Highways and Wheel Carriages. In return he sent to the worthy Baronet a short Account of the Roman Method of making Roads, which he was desired to insert in this Report, and which he now subjoins in the most concise manner in which he can express it.

These masters of the world expended immense sums in making the great roads of communication from Rome to the most distant provinces. The expence of the Appian and Flaminian ways was unequalled by any thing in modern Europe; as will be easily seen, by attending to the method they took in making their principal roads. They first levelled the bottom of the proposed road, and beat the earth very hard, by machines adapted to that purpose, that it might not sink by the weight of the stones laid upon it. Then they covered the line of road with a soft kind of stone, which they called *tophus*, which was a foot thick, and regularly paved.— Last of all, they shaped the narrow stones very neatly to a determinate thickness, and laid them on so compactly above the *tophus*, that the point of a sword, in many places, cannot be thrust between many of the stones, which have stood for above 2000 years, as monuments of the Political Economy of antient Rome. Were the same mode adopted in first beating the earth hard, then laying Portland stone, and lastly paving with Aberdeen granite, the streets of London— and a similar method adopted in all great cities, there would be less reason to dispute about the form of carriage wheels; although the Committee of the House of Commons was certainly right, in the different Resolutions which were proposed on this important subject.

II. *Iron Railways.*—We have none of these as yet in this county:

county: and the competition that subsists between the Canal and the turnpike roads, on the banks of the Don, forbids the attempting an iron rail way in that direction. Should the Legislature, however, see the importance of the *forests of Marr* in the proper light, an iron rail way might, at no distant period, be seen on the banks of the Dee.

III. As repeatedly mentioned, we have a Canal, which is carried from the harbour of Aberdeen to the burgh of Inverury. This Canal extends from the river Don, to the shore-lands of Aberdeen, being in length nearly 18¼ miles; and the altitude surmounted from low water-mark in the harbour to the summit level at Stoneywood, is 168 feet.

The principal works constructed upon it are, 17 canal locks—5 aqueduct bridges over considerable streams of water—56 accommodation bridges—20 culverts, for conveying smaller streams from the higher grounds, under the canal.—Besides basons at Port Elphinston, at the top of the canal; at Bridge of Dyce and Kintore; a temporary bason at Aberdeen; and many other pieces of masonry, of lesser magnitude.

The expenditure on the Canal previous to the 31st of December, 1808, was

For surveys, plans, and Acts of Parliament	L.2154 10	4
Law expences - - - - -	251 10	0
Land, and land rents, and damages to grounds	2134 3	8
The execution of the earthen part of the Canal, including boats and utensils -	19,917 0	1
Constructing 17 locks, 5 aqueduct, and 56 accommodation bridges, 20 culverts, and other works of masonry, including superintendence, management, and incidental expences - - - - -	19,438 14	9

Total expenditure Dec. 31, 1808,   L.43,895 18 10

It

It was opened in 1807, and the returns from it have been yearly increasing. Besides the tonnage on goods of various kinds, on wood, slates, stones, dung, coals, and lime, the rent of two fly boats for passengers has produced a considerable and increasing revenue. The tonnage conveyed during the last three years, is as follows:—

	bolls of lime.	bolls of coals.	tuns of dung.	other articles.
1808	19423	4335	468	5144
1809	25673	5521	528	2526
1810	25525	6192	736	1950
	70621	16048	1732	9620

The fly boats in 1808 were let at L.105. In 1809 they were let for L.186. In 1810 they were let for L.231. And they are now (1811) let at L.348.

There are already 12 boats employed upon the Canal for the carriage of coals, lime, and other weighty goods, besides one boat for carrying provisions. And there is no doubt that the opening of this Canal, though it will not for a considerable time prove advantageous to the proprietors, must be highly beneficial to the country. The quantity of *lime* in the course of the last three years is, in round numbers, 70,000 bolls, or 14,000 tuns of *shells*; which, independently of what has been used in building, has added much to the fertility of 5000 Scotch, or above 6300 English acres; and in the course of 20 years will occasion a great proportion of uncultivated land to be rendered arable. The quantity of coals, though less than that of lime, also has a considerable effect in adding to the improvement of the land; because the farmer's time is much less occupied in providing fuel from the peat-moss. And the quantity of dung, or night soil, carried from Aberdeen to the inland parts of the county, amounting to 1732 tuns, is a proof of the spirit of enterprize that animates

mates the farmers of the district that is washed by the Canal. It has already been mentioned, that night soil has been carried from London to the coast of Belhelvie. Perhaps it may, in the course of a few years, be brought by sea to Aberdeen, and by the Canal to the inland parts of the county. And that most valuable root (the only one as yet raised in great quantities, which will pay for distant water carriage), viz. potatoes, may be carried from the division of the Garioch, to the metropolis of Great Britain. It is a curious fact, that *onions* are cultivated in the low countries, *by the species of manure best adapted for raising them, which is imported from Stockholm.* Therefore it is not speculating too far to suppose, that the same kind of manure may be carried partly by sea, and partly by this Canal, to the inland districts of Aberdeenshire.

The Canal was originally only 17 feet wide by three feet deep. It is now, in the greater part of its course, 3 feet 9 or 10 inches deep, and from 21 to 23 feet wide. (This has been done by subsequent operations). It is to be regretted, however, that the *plan* of that excellent engineer, Mr. RENNIE, who proposed that the Canal should be 27 feet wide, and 4 feet deep, was not at first carried into effect; and that *some proper person appointed by him*, was not paid for superintending the whole work. For, excepting the first three locks, which were executed in a substantial manner, all the others were built so very slight and insufficient, that although the undertaker paid a considerable sum of damages, the proprietors of the Canal sustained great loss, and the trade has been repeatedly interrupted by some of the locks giving way. This is now in a great measure remedied; and as the undertaking was new, and the work of a nature so little known to our artificers, it was not to be wondered, that several errors were at first committed.

If a Canal were proposed for conveying the valuable forests

rests of Marr to Aberdeen, it would be adviseable to use no locks, but to carry it only to Hazlehead, within two miles of that city. It would be cheaper to cart the wood for these two miles, than to build locks. An aqueduct bridge, however, would be necessary over the burn of Culter; as the Canal should be carried on a high level, at least 200 feet above the sea at half flood.

To this list of public works, we may add the expence of building a Bridewell, which has cost L.10,000.

Besides what has been expended *within the county*, the Magistrates of Aberdeen advanced also L.12,000 in making out the turnpike to Stonehaven, in Kincardineshire.

The whole sums already expended, and what in a few years will be laid out on roads, streets, harbour, canals, and Bridewell, cannot be estimated below L.400,000; and very probably will amount to half a million.

It is a circumstance highly deserving of notice, that when the whole lands in the county of Aberdeen were valued in 1674, their annual amount was below L.20,000; and all the lands in the county, and houses in the city, would not, at that time, (if sold at the highest prices at which houses and land were then sold) have produced so much money as has been laid out on these public works since 1796, or will be laid out in a few years. It is to the credit of the landed gentlemen, to mention that (with very few exceptions) they freely gave up the ground for turnpikes, commutation roads, and also for the canal; otherwise the greater part of these useful works could not have been executed.

It needs only be added to this account, that the Bridewell has not as yet had many inmates, partly owing to the great lenity of the Justices, and partly owing to the good morals of the people.

SECT. IV.—FAIRS.

Of these we have a great number, of two descriptions.—The first are properly called Fairs, and the right of holding them is by a charter or grant from the King.—The second are called Trysts, at which no custom is paid for cattle for the first three years. The most considerable are the following:—

JANUARY.—OLD STILE.

Drumblade, 2d Tuesday—Turriff, last do. and Wednesday.—Cumineston, 1st Thursday—Old Deer, 3d Thursday.

NEW STILE.

St. Nauchlan's, Oldmeldrum, 1st Tuesday after the 18th—Yule market, Strichen, 1st Tuesday.

FEBRUARY.—OLD STILE.

Rattray, 1st Tuesday—Tarves, 2d do.—Old Deer, 3d Thursday—Huntly and Alford, last Tuesday—Strichen, last Tuesday and Wednesday—Fyvie, Fasten's Even day, (or Shrove Tuesday,) and Tarland, last Wednesday—New Pitsligo, 3d Tuesday and Wednesday.

NEW STILE.

Monymusk, 2d Wednesday—Abergeldy, last Friday.

MARCH.—OLD STILE.

Turriff, Saturday before Fasten's Even—Woodhead, Fetterletter, 1st Tuesday—Tarves, 2d Tuesday and Wednesday—Kirk of Leochell, Inverury, Finzean, and Kirk of Migvie, all on 2d Tuesday—Lenabo, 3d Tuesday and Wednesday—Fraserburgh, 2d Wednesday—Old Deer, 2d Thursday—Turriff, last Tuesday and Wednesday—Huntly, last Tuesday.

NEW STILE.

Longside, 2d Friday.

APRIL.—OLD STILE.

New Deer, 1st Tuesday and Wednesday—Cruden, 1st Tuesday—Auchterless, 2d do.—Hawkhall, 3d do.—Tarves, last Tuesday and Wednesday—Kepple Tryst, and Auchindore, last Tuesday—Cumineston, 3d Thursday—Old Deer, 2d do.—Inverury, 3d Wednesday.

NEW STILE.

Old Aberdeen, last Thursday.

MAY.—OLD STILE.

Ellon and Kildrummy, 1st Tuesday—Ballater, 2d Monday and Tuesday—Greenburn and Arnadge, 2d Tuesday—Strichen, 2d do. and Wednesday—Peterhead, Tarves, and Insch, 3d Tuesday—Byth, 4th Tuesday and Wednesday—Udny, 4th Tuesday—Huntly last do. and Wednesday—Hawkhall, Thursday before last Tuesday—Auchterless, Friday before do.—Marquis fair, Huntly, and Newbyth, 1st Thursday—Drumblade, 2d Wednesday—Old Deer and Kincardine O'Neil, 2d Thursday—Cumineston, 3d do.—New Pitsligo, 3d Tuesday and Wednesday—Auchindore, last Friday—Turriff, last Saturday.

NEW STILE.

Meikle Wartle, Thursday before 26th.—Huntly, Thursday after do.—Oldmeldrum, Saturday after do.—Tarland, Wednesday before do.—Rora, 1st Wednesday after the 12th.

JUNE.—OLD STYLE.

Ellon, Lonmay, Daviot, and St. Colm's fair at Cairness, 1st

1st Tuesday, Lenabo, 3d Tuesday and Wednesday—New Deer, 2d Tuesday and Wednesday—Sliach do.—Turriff, Tuesday and Wednesday before last Tuesday—Balnakettle and Aboyne, 3d Wednesday—St. Sair's Fair, last Tuesday and Wednesday. The sheep market the Thursday before—Greenburn tryst, last Tuesday—Tarland, Friday after do.—Old Deer, 1st Thursday—Fraserburgh, and Greenburn market, 2d Thursday—Cumineston, last Thursday—Sliach, 2d Wednesday—Charlestown of Aboyne, 3d Wednesday.

NEW STILE.

Alford, Tuesday before 1st Wednesday—Inverury, Tuesday before 2d Wednesday—Rhynie, Thursday before do.

JULY—OLD STILE.

Peter-fair of Huntly, 1st Tuesday and Wednesday—Strichen, do.—Tyrebagger and Fyvie, 1st Tuesday—Rathen, 2d Tuesday—Aiky fair of Old Deer, do. and Wednesday for horses—Kepple Tryst, Tuesday before do.—Turriff, last Tuesday and Wednesday—Tarves, 3d Tuesday and Wednesday—Glass, 3d Wednesday—St. James' fair of Greenburn, last Thursday—Balnakettle, last Wednesday—Inverury, Thursday after Aiky fair—Charlestown of Aboyne, Friday after Pady-fair week.

NEW STILE.

Mortlach, 2d Thursday.

AUGUST—OLD STILE.

Lawrence fair, Old Rayne, 1st Tuesday and Wednesday; sheep and timber markets, Thursday and Friday before—Arnadge, 2d Tuesday—Strichen, and Mickle Sliach, do. and Wednesday—St. John's fair, Strathdon, Friday after do.—Ellon, last Tuesday and Wednesday—Bartle market, at Crimond,

mond, and Auchindore, 4th Tuesday—Bartle Chapel, Friday after do.—Kincardine O'Neil, Wednesday and Thursday after last Tuesday—Old Deer, 1st Thursday—Fraserburgh, 3d do.—Cumineston, last do.—New Pitsligo, 1st Tuesday and Wednesday.

NEW STILE.

Monymusk, last Wednesday—Timber market at Aberdeen, last Thursday—Muchals tryst, 1st Tuesday.

SEPTEMBER.—OLD STILE.

Inverury and Coldstone, 1st Tuesday—Charles fair, Huntly, 2d do. and Wednesday—Tarves do.—Alford, Friday after do.—Greenburn, St. Andrew's, in Braemar, and Michael fair, at Hawkhall, 3d Tuesday—Michael fair of Kinkell, last do. and Wednesday—Old Deer, 1st Thursday—Fraserburgh, last Friday—New Pitsligo, last Tuesday and Wednesday—Ballater, 2d Monday and Tuesday—Greenburn, 3d Tuesday and Wednesday.

OCTOBER—OLD STILE.

Turriff, 1st Tuesday and Wednesday—Charlestown of Aboyne, 1st Tuesday—Marywell of Birse, 1st Thursday after do.—New Deer, 2d Tuesday and Wednesday—Kinnethmont, and Kepple tryst, 2d Tuesday—Rhynie, the day after do.—Culfork of Breda, Monday before do.—Byth, Old Aberdeen, and Daviot, 3d Tuesday and Wednesday—Turriff, Thursday after do.—Insch, 3d Tuesday—Greenburn, do.—Tarves, 4th Tuesday and Wednesday—Inverury, Wednesday after do.—Old Deer, 3d Thursday.

NOVEMBER—OLD STILE.

Martinmas fair, Huntly, 1st Tuesday—Oldmeldrum, Saturday after do.—Strichen and Ellon, 1st Tuesday and Wednesday

nesday—Ruthven and Peterhead, 2d Tuesday—Methlick, do. and Wednesday—Udny and Lenabo, 3d do.—Andersmas fair Rayne, 4th Tuesday—Old Deer, 2d Thursday—Fraserburgh, 2d Friday—Cumineston, 3d Thursday—Tarland, Tuesday and Wednesday after Old Martinmas.

NEW STILE.

Aboyne, 2d Wednesday—Inverury, 25th day, for hiring servants.

DECEMBER.—OLD STILE.

Andersmas fair of Huntly, and Ellon, 1st Tuesday—Turriff, do. and Wednesday—Old Deer, 2d do.—Andersmas fair of Fraserburgh, 3d do.—Tarves, 4th do.—Turriff, Saturday before Old Christmas—Cumineston, 3d Thursday.

NEW STYLE.

St. Andrew's, Oldmeldrum, 1st Tuesday after Dec. 4th—Strathdon, 1st Tuesday—Arnadge, 4th Wednesday.

N.B.—What is called old style, should have been 12 days, ever since 1800, but is still accounted only 11 days later than new style, i. e. the 1st of a month old style is the 12th of the new.

Besides the above *annual* fairs and trysts, there are markets for fat cattle at the Bridge of Don once a month; at Oldmeldrum once a fortnight in winter, and once a month in summer; and at Inverury, once a fortnight from December to the end of March.

SECT. V.—MARKETS.

The only considerable market town is Aberdeen, which has a well frequented weekly market on Friday. It also has a market for butcher meat on Wednesday. But excepting in the summer months, when killing twice a week is necessary, it is very little frequented. There is also a weekly market at Peterhead, Fraserburgh, Huntly, and Oldmeldrum, but none of them are considerable. That of Peterhead, however, has increased with the population of the town, and its flourishing commerce.

SECT. VI.—WEIGHTS AND MEASURES.

1. *Land Measure.*—This is almost universally done by the Scotch ell of $37\frac{1}{7}$ English inches, and its denominations are

6 ells Scotch (equal to $18\frac{6}{10}$ feet) make a lineal fall.
36 square ells (equal to $38\frac{44}{100}$ yards) 1 square fall.
40 square falls make one rood.
4 roods make one acre.

This statute Scotch acre contains 5760 Scotch ells, or $6150\frac{4}{10}$ square English yards. Among the gardeners in the vicinity of Aberdeen, land was formerly let by the 100 beds; but it is now the practice to measure by the Scotch chain of $74\frac{4}{10}$ feet. Some land-measurers, very improperly, use a chain of only 74 feet. This makes the acre too little by 593.6 square feet, or 1-93d part of the whole. This is a very dishonest practice: because the Scotch inch, though not now in use, was 1-185th part longer than the English inch, and the Scotch ell, of 37 Scotch inches, was found at the Union of the two kingdoms in 1707 to be 37 1-5th inches, or one-fifth part of an inch longer on the 37 English inches.— The Scotch acre is to the English as 38.44 to 30.25, being the proportion of the number of square feet in a Scotch or English *fall* or *perch*. This is very nearly as 61 to 48, as formerly mentioned.

2. *Corn Measure.*—The provincial or county measure of oats and bear, or barley, is by the boll of 136 Aberdeen pints, or 4 firlots, of 34 pints each, commonly, but erroneously, called Stirling Juggs; each of which juggs, when filled with rain water, of about 55° of heat, weighs 60 ounces and 12 drams Avoirdupois. This is above 26,585 Troy grains.

The real Stirling jugg, as found at the Union of the two kingdoms, weighed only 26,180 grains.

The same, by a measure of the late Professor ROBINSON's, of Edinburgh, made only 20 years ago, was 26,200 grains.

The cause of this difference seems to be the following:— By an Act of Parliament, establishing the Stirling jugg, as the unit of measure through all Scotland, 3 measures of this standard Scotch pint were appointed to be sent, one to Edinburgh, one to Perth, and a third to Aberdeen. Our forefathers were not very accurate in making such measures. And the pint measure sent to Aberdeen, was 1½ per cent. too large. It appears that the Perth measure was also larger than the standard one. Hence the Aberdeen boll of oats, barley, or bear, of 136 *Aberdeen* corn pints, is one-ninth larger than the standard, or the Linlithgow boll, of 124 *Stirling* pints.—

For pease, beans, and anciently for meal, which before 1694, was sold in Scotland by measure only, the boll formerly used was 104 Aberdeen pints, or 4 firlots of 26 pints each. And both malt and sids, (i. e. oat-hulls, with the meal adhering

hering to them) were sold by the heap of the meal peck.—But as there was no wheat measure used in this county, it became necessary to fix upon one, now that wheat begins to be more generally cultivated. On maturely considering the matter, the Magistrates of Aberdeen have appointed the Linlithgow wheat firlot (the standard measure of wheat, rye, and pease, in Scotland,) to be used for the future, as the standard for these kinds of grain. And GEORGE HOGARTH, Esq. present Dean of Guild of that city, a very intelligent and active Magistrate, caused a wheat firlot to be made very accurately, for the purpose of being used in correcting other wheat measures.—(It may be noted incidentally, that the difference between filling slowly with a small shovel, and quickly with a large one, was found to be 3 per cent. Where fraud is intended, it is more than 6 per cent.)

The practice of *heaping* being equally illegal and improper to be continued, the Magistrates of Aberdeen, several years ago, appointed all oatmeal *sids* to be sold by weight, and 12 lbs. Dutch, or Scotch Troy, to be given in place of the heap of the peck. The same rule has been adopted for malt, which is now commonly *sold ground*, in direct opposition to the city's charter, which expressly forbids *ground malt* to be sold in Aberdeen. This alteration in the mode of selling will prevent the frauds which were sometimes committed in filling ground malt by the heap of the peck. For a peck of malt, from Aberdeenshire bear or bigg, ground on a corn mill, will weigh 14 lbs. Dutch. And less than 12 lbs. of that weight, is a peck, when ground, or rather bruised, between rollers.

Potatoes also were sold by the meal peck; but it has been appointed very properly that the peck shall weigh 32 lbs. Dutch, or 35 lbs. English.

The city of Aberdeen has also a standard Winchester bushel, which is made of cast brass, and is marked *Anna Regina,*

Regina, 1707. It was sent to Aberdeen at the Union of the kingdoms in 1707; but by a strange liberty taken by the Magistrates of Linlithgow, who have a title to keep the Linlithgow standard, the *arms* of that burgh, with the word *Linlithgow*, as if it had been a *Scotch* firlot, have been added to the original measure. The Writer of this Report, in 1794, was employed by Thomas Bannerman, Esq. then Dean of Guild, to examine the standard weights and measures of Aberdeen, when he found that this standard measure contained only 77 lbs. 2 oz. 7 grains of spring or fountain water, in a heat of 50°. And since this sheet went to press, he was asked by Mr. Hogarth, to examine and compare different corn measures, all of which were reckoned Winchester bushels. He did so accordingly, when the brass bushel, above-mentioned, was found to contain of rain water, in 44° of heat, exactly 1124¼ oz. of English Troy, (or very nearly 7 Troy ounces less than the brass bushel of King Henry VII. was found to contain in 1696,) or nearly 10 oz. 7 drs. of less water, than is contained in the gauger's bushel of 2150.42 inches. This is a difference of nearly 1 per cent. between two standard measures. At the same time, a bushel belonging to the county was produced, which contained 21 oz. more than this brass bushel. That standard is kept in the Record Hall, and was got from Guildhall; and is nearly 2 per cent. larger than the brass bushel belonging to the city. Two bushels were also procured from the Custom House; one of these being very entire, was found to contain 4 oz. more than the brass bushel. The other had received some injury, and contained 2 lbs. 2¼ oz. less than the best bushel belonging to the Custom-House. It will no longer be used as a measure. A sealed bushel from London, belonging to a company of brewers, was 4 oz. less than the best Custom House bushel, and 4 oz. more than the brass bushel. This difference either way was but trifling; but a difference of 3 lbs. 3¼ oz. upon a bushel of wheat,

wheat, (of only 60 lbs.) between the largest and the smallest standard measures, ought not to have been passed over by the Reporter; and it shews not only that one correct standard of weight and measure should be universally established, but that all kinds of corn should be sold by weight, and only so far corrected by measure as to ascertain their quality.

Liquid Measures.—The measure for ale, spirits, and all other liquids in Aberdeen, till about twenty years ago, was another Aberdeen pint, that contained 63 oz. of water. It does not now exist; and the jug of 60 oz. and 12 drs. is now the standard both for corn and for liquid measure. It contains 105 cubic inches.

Butcher meat, meal, iron, and lately both malt and sids, are sold by the Scotch Troy, which is erroneously called Amsterdam weight. This is commonly computed to be $17\frac{1}{2}$ oz. Avoirdupois; but was found to be only 17 oz. and 7 drops, or $\frac{1}{140}$ part less.

Butter, cheese, fleece-wool, and rough tallow, are sold by the lb. of 26 Amsterdam, or 28 English Avoirdupois oz. The former is the largest of the two, by one-third of an Avoirdupois oz.; but the latter is now generally used in Aberdeen. In Peterhead, and in general in the division of Buchan, only 20 oz. Scotch Troy (probably the ancient Trone lb. used in this county) are given of butter and cheese; and in Oldmeldrum, 24 oz. are accounted a lb. by the dairy women.

Coals are sold by the boll of 36 stones Avoirdupois, or Scotch Troy, equal to 630 lbs. neat, or 5 cwt. 2 qrs. 14 lbs. English. Lime is sold by measure of 128 Aberdeen pints, of 105 inches each to the boll. For all groceries, the Avoirdupois weight is constantly used.

For measures of length, the English inch, foot, and yard, are used, except in the sale of a coarse kind of cloth called
plaiden

plaiden, which is sold by the plaiden ell of 38 and 5-12ths inches.

The Writer of this Report having for about 24 years paid particular attention to the subject of an equalization of weights and measures, would not have the approbation of his own mind, if he did not, in this place, take notice of the great number of weights and measures which are used in this county. But as he might be suspected of partiality for a favourite object, he shall, in the simplest language, and as concisely as possible, throw out some general ideas, which may be of use to mankind, when he has retired from the stage of life.

Before the late convulsions of Europe, which were occasioned by the French Revolution, three-fourths of the various commodities used among merchants, were sold by the ancient weight of Charlemagne, known by the names of French, Dutch, Brussels, and Scots Troy weight. It was a fortunate coincidence between this weight, which was so generally used, and a standard taken from nature, that a cubical vessel, whose length, height, and breadth was $39\frac{1}{4}$ inches, or equal to the length of the pendulum which vibrates seconds at London, contains very nearly a tun, or 2000 lbs. of Amsterdam weight. Hence, by the simple means of calling two pounds *one* pound, a decimal division could have been easily established; a thing that cannot be done with Avoirdupois or with Troy weight. The length would have been very convenient as a yard measure; and a correspondence could easily have been established between the weights, measures, and coins of all trading nations.

But as this weight is no longer in use, another standard taken from nature, may now deserve to be preferred to all others. This is a standard taken from the great circle of the earth, measured on the meridian. By a series of accurate measure-

measurements, the measure of the degrees of latitude between the Royal Observatory at Greenwich, and that of the French Academy, is ascertained within a few feet; and 3 degrees and 36 minutes of latitude, in the southern extremity of the island, is extremely nearly one-hundredth part of the circumference of our earth. It is a curious coincidence, that *one-billionth* part of this circumference corresponds very nearly with a Troy grain of distilled water. The pound derived from this would be above 9950 Troy grains. This weight could be easily preserved, and the philosophers of both Great Britain and France having been engaged in measuring the distance between the two national Observatories, it would be honourable to both nations to take a standard of weights and measure from that admeasurement. The *foot* taken from it would be above 15¾ English inches; the pound above 22 ounces Avoirdupois; and the coins, weights, and measures, could be adjusted to this standard.

No compulsion should be used to introduce it; at any rate for the first eight or ten years. Magistrates should be ordered merely to see that the weights and measures derived from it were carefully preserved; and others of the same kind adjusted with great care. The simplicity of its decimal division, and the circumstance of its pound being larger, would gradually reconcile men to it. Perhaps the recommendation and example of the merchant would go farther than the authority of the Legislator.

It is proper to notice an objection that has been stated by a very judicious person, and able writer, Mr. GEORGE ROBERTSON, who has been frequently mentioned in this Report, " Nothing seems hitherto to have retarded so much a more " general uniformity in weights and measures, as that absolute perfection (inconsistent with human attainments) " that has been aimed at by all projectors on this subject.— " What hinders any new standard that may be fixed on from
" being

"being preserved unaltered, as well as the present legal corn
"measure, which is still of the same capacity that was origi-
"nally determined 209 years ago?"

It is here necessary to observe, that if we are to adopt a new standard, it is certainly better to adopt one that is connected with something which is permanent in nature, and which can be had recourse to, if there be any necessity, than to establish a standard connected with nothing, but chosen merely by the arbitrary will of the Legislature. There is by no means such difficulty in selecting a standard that has many arguments to recommend, as to get men to adopt it. It is not the understanding that is to be addressed. It is the will that is to be influenced. It is not, however, certain that the present legal corn measure consists still of the same capacity that was originally determined on. The wheat measure of Scotland, in 1618, was no doubt intended to be made equal to the English measure. It was enacted, that the *wheat firlot* (the Scots name for bushel) should contain $19\frac{2}{5}$ *Scots inches* in the diameter of a cylinder and should be $7\frac{1}{4}$ inches in depth. Now let this be reduced to *English inches*, and the cubical contents of this firlot, as fixed by Act of Parliament, will be found, by correct calculation, to be 2150.345 English cubic inches, which is only *one thirty-three thousandth part less* than the present English bushel. Unfortunately our founders, who cast the *brass bushel* of Linlithgow, and the Stirling jug, could not execute the work so well as Baron NAPIER of Merchieston, (the celebrated inventor of the Logarithms) could calculate the contents of a cylinder; and there was a difference of nearly 50 cubic inches between the firlot, as containing $21\frac{1}{4}$ Stirling jugs, and as consisting of the dimensions that are expressed in the Act of Parliament. And it may be noticed incidentally, that Baron NAPIER lived to lay down the plan, but died before

fore it was put in execution. *He* was a *projector* of no mean talent. To this it may be added, that the standard pint measures sent to the principal towns of Scotland, were not of equal dimensions. And that in England, the Winchester gallon contained $272\frac{1}{4}$, but that the bushel of Henry VII. instead of being eight times as much as this, or 2178 cubic inches, contains only $2145\frac{6}{10}$, or $32\frac{4}{10}$ inches less.— The same disproportions are found in all our ancient weights and measures. Therefore the most prudent plan, in the opinion of the Reporter, (who is far from being *confident* in the success of the method which he proposes,) is first to connect a standard of weights and measures, with something in nature, not depending on the arbitrary will of man, but, if he may use the expression, something of God Almighty's making: and not to attempt to force men to adopt this standard, but to render it worthy of their adoption, and calculated to *compel their esteem*, by its simplicity, its universality, and extensive usefulness; to put it under the protection of the law for some time, not with the right of a *monopoly*, which would be opposed by the passions and prejudices of men, but with that recommendation, which in a competition for public favour, is often more valuable than an imperious mandate. Law is power. Riches are by power;— and honour is power. But let it be remembered, in the words of Lord Bacon, that *mind also is power*. And a system of equalization of weights and measures, that has the recommendation of *mind*, or of *great intellect*, will have the *better chance* to succeed the nearer it approaches to perfection in theory; if it be simple, and easily put in practice by those who do not understand the principles on which it is founded.

There can be no doubt, that the French, Dutch, or Scots Troy pound, which was 7616 Troy grains, by the mean of

the

the Edinburgh weights at the Union, in 1707, was anciently 16 oz. of the same weight of which the English Troy was 12 oz.; and therefore the former was 7680, while the latter was 5760 grains. These standards of France, Holland, and Scotland, that have been much used, have been worn down, and have decreased by friction. The English Troy weights being chiefly used for weighing gold and silver, have been seldom used, and are little worn; therefore they are more nearly of their original weight. But the standards of Avoirdupois weight, which are much used, have decreased by being employed in verifying other standards.* When the Writer of this Report examined the standard weights of the city of Aberdeen, 31st December, 1794, the Dutch stone weight was equal to 17 lb. 6 oz. 14 drops. In December, 1810, the same identical stone weight appears to be 17 lb. 7 oz. or 2 drops heavier; i. e. the standards of Avoirdupois weight, by being much used, have lost 2 drops, or 54 Troy grains of their weight. It is not, therefore, to be wondered, that in the course of several centuries the English Troy ounce, which is chiefly used for weighing the precious metals, should be 480, while the French, Dutch, and Scots Troy ounce should, by being used for most kinds of merchandize, be only 476 in Scotland, and in Amsterdam only 475 English Troy grains.

On all these accounts, weights and measures ought to be not only corrected from time to time, but also equalized and connected with a standard taken from nature.

* In the city of Aberdeen, the standards of English Troy, which amount to 512 Troy ounces, are in a high state of preservation. (The outward case, and a few of the smaller weights, have suffered a little by friction; the first, by being moved from place to place; the last, from being used.) And therefore it has been determined, that the other weights shall be yearly compared with these standards.

SECT. VII.—PRICE OF PRODUCTS, COMPARED WITH EXPENCES.

It is impossible to give a just, and, at the same time, a very general view of the various prices of the different articles of produce in this county, compared with the expence of raising them: the county is so much diversified in point of surface, situation, soil, and climate; and contains all the various modes of agriculture, from the most imperfect to the most correct practices in cultivation. On the other hand, it would be both tedious and prolix, to enter into a minute detail of all the various articles of produce and expence, in the different divisions, and even in the inferior districts of this extensive county; in which the style and manner of living of the inhabitants differ so widely, not only in the great divisions, but even in the same district. Pursuing, therefore, a middle course, between either being very general or very minute, the Writer of this Report shall endeavour to give as comprehensive a view as he can, of the Products of Aberdeenshire, compared with the Expence of raising them,—by including as many particulars, as will enable a stranger to form a pretty correct idea of these important articles of the agricultural economy of this county.

The least expensive, and to Aberdeenshire the least valuable product, is, where not only the landholder, but the farmers, their servants, and even their live-stock are, during the greater part of the year, non-resident. This is the situation of nearly twenty thousand acres of that mountainous district of Mars, which borders with the counties of Inverness and Banff. Owing to the great distance of any arable land in this county from these mountains, the farmers in Badenoch, (whose lands are situated much nearer to this district) pay a small yearly rent to Earl Fife, for the right of pasture in

Glen-garchary and *Glen-guisachan*, two long, but narrow vallies, on each side of the rivulets, whose united stream assumes the name of the *Dee*. There, for several months in summer, a motley assemblage of diminutive horses, (or highland ponies), of black cattle, and of sheep, depasture these long vallies, or climb for food along the ridges and mountain torrents of *Ben Macdouie*, *Braeriach*, *Cairntoul*, and *Benivrottan*. In the end of autumn, they are conducted back to their native county, or sold to the dealers, to be driven to the southern markets. None of these animals belong to Aberdeenshire; and their keepers reside only for a few months in summer, in a shealing, or earthen hut, to which their provisions are brought from time to time, during the pasturing season. The *rent* paid to the proprietor of these mountains and vallies is all that can be stated, as, in any shape, belonging to this county. The only expence, attending this least productive range of pasture, is the proportion of salary paid to the very attentive and intelligent land-steward, for superintending and letting the pastures, and collecting the rents. But the produce of two acres of highly cultivated land, in the vicinity of Aberdeen, is more valuable to this county, than the product of these twenty thousand acres is profitable to their noble and worthy Proprietor.

The highland farmers who reside within the county, and who to an extensive range of pasturage, unite a small farm of arable land, the extent of which they are gradually increasing, and which receives the benefit of manure from their cattle during the winter months, are much more productive labourers to this county, than the graziers from Badenoch are, although their system of husbandry is neither very productive, nor very expensive. For example, Mr. CHARLES McHARDY's farm at Delavorar, which the Reporter, (by measuring

suring it on ARROWSMITH's Map of Scotland, found to contain nearly 40,000 English acres,) has only about 18 Scotch, or nearly 23 English acres of arable land, while his whole rent is L.260. The produce of what is under the plough, even in that district, where corn sells at a very high price, cannot be estimated above L.130, or the half of his rent. The annual expence of his family and servants, including that of the shepherds who take care of his live-stock, on the mountains of Scarsoch, and in the valley of Glen-geaullie, is equal to all the produce of his arable land. It is, therefore, to the pasturage of horses and black cattle, and to the profits of sheep-farming, that he must look for every farthing of his rent, and for the interest or profit on his capital. The whole farmers, and most of the cottagers, in the upper districts of Marr, are, though on a smaller scale of rent, in a similar situation with respect to produce. Their comparatively small proportion of arable land cannot, in the most favourable season, raise food sufficient to support a population of above 7300 inhabitants during the whole year; and at an average of seasons, they must bring from the more fertile lands in Cromar and Towie, as much meal as will supply them for six weeks or two months.—The whole arable land, (at one time under the plough,) and its annual produce, are nearly the following:—

	ACRES.	
Oats, at L.5 per acre,	6000,	price L.30,000
Do. mixed with rye, at L.6 per acre,	100	600
Bear, at L.7 per acre,	1500	10,500
Wheat on the proprietor's farms at L.10,	10	100
Barley do. at L.8 6s. 8d.	12	100
Pease and Beans, at L.2 10s.	40	100
Total, Corn and Pulse,	7662	L.41,400

Po—

	ACRES.	
Potatoes, at L.10 per acre,	200,	price L.2,000
Turnips, at L.7 per acre,	200	1400
Cabbages and Coleworts, at do.	100	700
Gardens of proprietors at L.10 per acre	50	500
Total roots and garden produce,	550	L.4600
Hay or soiling, at L.6 per acre,	500	L.3000
Pasture from sown grass, or inclosures, at L.3 per acre,	1500	4500
Meadows, or good natural grasses, at L.1 10s.	1000	1500
Pasture on hills or mountains, above	200,000	5000
Produce of Hay, and Grass of all kinds,		L.14,000
Wood sold by the proprietors,	-	L.6000
The farmer's profit, on cutting down, and selling this in the low country,	-	L.1500
Total from the land, for corn, pulse, roots, grass, and wood,	-	L.67,500
The produce of live-stock, and wool sold,		25,000
Do. of bees, honey, eggs, and poultry,		1500
The total agricultural produce of this district is		L.94,000

From which the landholders' rent, viz. L.14,000
And the price of wood sold by them, viz. 6000
In all deduct   -   -   -   -   L.20,000

The remainders to the farmers and cottagers, for food, seed, servants wages, and maintenance; for implements of husbandry, family expences, household furniture, &c.   -   -   L.74,000

But in fact, it requires the whole corn, and green crops of all kinds, to supply the inhabitants, and support the live-stock;

stock; and a considerable quantity of meal or bear must be annually purchased from the low country. So that, in truth, the profits on cutting and carrying down wood to other places, and the produce on live-stock, wool, bees, and poultry, sold annually, amount to only   L.28,000
And from this is to be deducted their rent, viz.   14,000
Consequently there remains to them only other .   14,000

From this they must purchase meal, and pay servants, day-labourers, implements of husbandry, and all the other expences of cultivation and of their families.

It is proper, however, to state in this place, that the wages of farm-servants are not high, that day-labourers are commonly paid by work of the farmer's cattle, and that the implements of husbandry are frequently made by the farmer, (for almost all the highlanders are accustomed to act as artificers); also that the farmers generally assist in carrying on all servile work. Hence less *money* is expended either on labour or on implements of husbandry.

In addition to what was done by the *men*, the manufacturing of coarse cloth, and the knitting of stockings, by the *women*, formerly brought in as much money as paid the corn or meal that was procured from the low country; and till of late, the distillation of whisky was also a source of wealth to the highlander, or supplied him with what the money price of his live-stock would not purchase. But all these sources are either diminished, or entirely cut off. And in the present situation of this district, it would be improper to enter into a minute detail of either the farm or family expences of a high-spirited people, whose economy is almost equal to their hospitality. Would the rich and profuse, when they visit this Alpine district, in order to behold the varied scenery of its mountains, condescend to examine into the character and situation of its inhabitants, they would see that they were

were virtuous, contented, and even chearful, without enjoying many superfluities; and would also be taught this important lesson,

"Man wants but little here below."

It would not have been exhibiting a true picture of the county, to have omitted this statement of the Product, and general view of the Expence of a district, which contains a fourth part of its surface, and a seventeenth part of its population,—and which differs in many respects from both the intermediate and maritime districts of Aberdeenshire. In fact, all these are corn countries, in which the turnip husbandry has been long introduced, and is more or less generally practised. Therefore, instead of specifying the proportions of produce in each division, the following summary will shew the general average and amount of produce, in what may be esteemed the cultivated districts of Aberdeenshire.

	ACRES.	
Oats at L.6 per acre,*	110,000,	price L.660,000
Do. near Aberdeen, and sea coast, at L.10,	4,000	40,000
Bear or Bigg, at L.7	20,000	140,000
Do. near Aberdeen, or the sea coast, at L.10	2,000	20,000
Wheat at L.12 10s. per acre,	400	5,000
Barley, at L.7 10s. per acre,	360	2,700
Rye, at do.	40	300
Pease and Beans, at L.4 per acre,	1,500	6,000
Total amount of Corn and Pulse,	138,300	L.874,000

* The oats in many places are below L.4; but as it has become customary to lay down grass seeds with early oats, the average in the county is L.6; and near Aberdeen, L.10 per acre.

	ACRES.	PRICE.
Potatoes, at L.10 per acre, besides the expence of digging,	3000	L.30,000
Do. at L.20 near Aberdeen and Peterhead,	1000	20,000
Carrots, horse-hoed, in the county, at L.10,	100	1,000
Yams, at L.10 per acre,	100	1,000
Rutabaga, at do.	100	1,000
Turnips, at L.6 per acre,	11,000	66,000
Do. at L.8 per acre,	8,000	64,000
Do. near Aberdeen, at least L.15 for best turnips per acre	500	7,500
Do. do. of inferior quality, at L.10,	400	4,000
Cabbages and Coleworts in the county, at L.6 per acre,	2,000	12,000
Do. near Aberdeen, at L.15 per acre,	500	7,500
Gardens, and all garden roots, at L.12	2,000	24,000
Fruit of all kinds, hot-houses included, at L.15 per acre,	1,000	15,000
To Flax, Hay, and Hemp, Buck Wheat, Hops, and all other crops little cultivated,	200	2,000
Total roots, fruits, and gardens,	29,900	L.255,000
Hay in the country districts, at L.5.	20,000	100,000
Do. near Aberdeen, or sea coast, at L.10,	1,000	10,000
Clover cut for soiling in the county, at L.5 per acre	1,000	5,000
Do. near Aberdeen, at L.10 per acre,	1,000	10,000
Pasture from sown grass, at L.3 per acre,	45,000	135,000
Natural Grass of arable land, at 9s.	100,000	45,000
Natural Grass, from land deserted by the plough, at L.1, nearly	1300 acres,	1,300
Carry over,		L.305,000

Brought over,	L.305,000
Do. from moors, hills, and barren lands in the cultivated districts,	10,000
Wood of all kinds in the lower part of the county,	18,000
Total from Hay, Grass, Pasture, and Wood, in these districts,	L.333,300
Total produce of the cultivated districts of this county	L.1,462,300
Total produce of the mountainous district, as above,	67,500
Total produce of the land,	L.1,529,800
Add to this the price of Black Cattle, annually sold or killed,	250,000
Also the value of the Dairy,	230,000
Do. of Sheep and Goats	30,300
Do. of Asses, Mules, and Hogs,	700
Do. from Bees	2,000
Do. from Poultry, Pigeons, and tame Fowls, of all kinds, eggs included; also from Game, viz. Red Deer, Roes, Hares, and Wild Fowls,	20,000
Total produce from both the Land and Live-stock,	L.2,062,800
Deduct from this an annual loss on Horses bought from other places,	20,000
Neat annual produce,	L.2,042,800

As this table has been drawn up with considerable care, the neat annual produce may, as formerly stated, be estimated at two millions.

This is, in round numbers, two pounds per Scotch, or nearly a guinea and a half per English acre, for the annual produce of all the lands in the county, whether arable or inarable.

But,

But it deserves to be particularly remembered, that only 360,000 Scotch acres are supposed to have been at any time under the plough, so that their produce is, at an average, L.5 13s. 6d per Scotch acre; while nearly 640,000 acres, partly of wood, and partly of pasture, but all inarable, produce only L.39,000. This is only one shilling and twopence halfpenny for the acre of the uncultivated lands; and is a strong argument for bringing a greater proportion of the surface of the county under the dominion of the plough.

The annual produce of the county is considerably inferior to the whole capital employed in agriculture. Supposing that the whole farmers of Aberdeenshire, and all their dependants, were to remove at the term of Whitsunday first, and to be succeeded by another race of tenants and cottagers, they would not only have the whole annual produce of their lands, but also the melioration of their houses, (i. e. whatever sum their farm-houses, and offices, exceeded the landholders' inventory,) their household furniture, their implements of husbandry, and the whole value of their live-stock, deducting their year's rent, and the expence of harvesting the crop of 1811. Their accounts would stand nearly as follows:—

1. Their whole annual produce, as before stated, viz.

                L.2,042,800

2. The melioration of their houses, at least a year's rent, - - - - 200,000
3. Their household furniture of equal value, 200,000
4. Their implements of husbandry, at 15s. for every acre of arable land, - - 270,000
5. Their black cattle, as formerly stated, 720,000
6. Their sheep, do. - - - 100,000
7. Their horses, 8600, at L.20 each, - 172,000
8. Hogs, goats, bees, and poultry, - 28,000

Making a general aggregate sum of   L.3,732,800

Aggregate sum brought over, — L.3,732,800

From this there must be deducted,
1. A year's rent of the whole county, L.200,000
2. The want of a year's produce of the dairy, — — — 230,000
3. The wages due to servants, and expence of harvesting the crop, — 200,000

Total deductions, — — — 630,000

Total agricultural capital, — — L.3,102,800

But in calculating the *price of products*, compared with the *expence of raising them*, (if the object of this Section be to ascertain the *whole annual expenditure and receipt*,) in fact, all that can be done, is to estimate the whole *disposeable* produce, and to deduct from this the *necessary expences*. In this view of the matter, the amount, taken either way, is less considerable. For, in this cattle-rearing county, nearly a third part of the annual produce is applied to the maintenance or feeding of live-stock. Thus—

1. The whole straw, chaff, and light corns, supposed equal to one-fifth of the corn crops, — L.181,860
2. Also the whole pulse, except the seed, 5,000
3. Oats, worth L.2, and bear, worth 10s. for boiled meat occasionally, to 8000 horses, 20,000
4. Hay, soiling, and pasture to do. and to black cattle and sheep, being the whole produce of the hay and grass of the county, except L.30,000 to carriage, saddle, and posting horses 291,300
5. Turnips, rutabaga, yams, and carrots, given to live-stock, — — — 126,000
6. One-third of the amount of cabbage and coleworts, at least — — — 6,000

Carry over, L.630,160

Brought over,	L.630,160
7. One-sixth of the value of garden-roots given to cows,	4,000
8. One-tenth of the whole potatoes to horses and poultry,	5,000
Total disposed of to live-stock of all kinds,	L.639,160
9. To this add the whole expence of *seeds* of all kinds, viz. of 145,962 acres of corn and pulse crops, at 16s. per Scotch acre,	116,770
Grass seeds, at 20s per do.	23,500
Potatoes, turnips, carrots, and all other seeds for gardens, at	10,000
10. Add rent for one year,	200,000
Total for supporting live-stock, price of seeds, and rent,	L.989,430

Thus it appears that nearly one-half of the produce of the land is appropriated to these three necessary articles—Livestock, Seeds, and Rent. The other half only may be called disposeable by the farmers; and what they have no occasion to dispose of, goes to the remuneration of their labour and attention, or to the accumulation of capital. But the difference between having old or new leases, consequently moderate or high rents, and level or fertile, or rough, and comparatively unfruitful lands, occasions so great an inequality between different farms and farmers, that all that can be advanced on this branch of economy, should be written with diffidence, and read with the consideration that is due to a general outline, rather than to a correct statement.

1. The

1. The expence of farm-servants, day-labourers, and of harvest shearers, who are hired for that purpose, here forms a prominent article. That expence is proportionably less on large than on small farms, and on fertile than on inferior soils; and is proportionably much less on the rich, flat, and high rented lands in the south, than in the comparatively poorer, high lying, and lower rented lands of Aberdeenshire. Here it cannot be stated at less than two years of the land-rent, or in the whole county L.400,000. In order to come as close to it as he could, the Writer of this Report made an estimate of the whole sum paid for labour, in his parishes, charging for every able-bodied man L.22 10s.; for every maid servant L.9 annually, and the same sum for old men or boys who tended the cattle; and he found that the wages and maintenance of farm-servants alone exceeded two year's rent of the whole lands in his parishes. Where there was an *old* lease, and a large farm of near 400 English acres, with a considerable *degree of acclivity*, the expence of labour was nearly four times the rent. In those farms of less extent, and which could be managed by a pair of horses in the plough, it is a little more than one and a half year's rent.—This great rise in the expence of labour affects the farmers in this county much more than the rise of rent as yet has done. In the southern counties, where the price of wheat has raised the rents to L.5, L.6, and L.7 an acre, the price of labour, i. e. the money paid to farm-servants and day-labourers, is not above half of the rent, and often less than that.

2. The expence of manure, (which is greater in Aberdeenshire than perhaps in any part of the island,) consists in the price of various articles, formerly enumerated, of which the most considerable is that paid for lime.

Of this the lime quarries at Ardonnel, in the parish of Cairnie, and district of Strathboggie, yield 46,050 Aberdeenshire

deenshire bolls of shells, of 136 Aberdeen pints each, sold at the quarry for L.9390.

But as a part of this is sent to Banffshire, there may be only charged to this county	L.8000
The whole other lime quarries in the county amount, at least, to as much	8000
The quantity of lime brought up the Canal to Inverury, costs at Port Elphinstone	8000
That brought to Peterhead, Fraserburgh, and the coast of Buchan,	6000
That imported at Aberdeen, and Newburgh, besides what goes up the Canal,	23,000
Total for lime for the purpose of manure,	L.53,000
Night soil in Aberdeen, and dung of the butcher market,	L.2000
Dung belonging to the cow-feeders, inn-keepers, horse-hirers, and stablers,	1000
Night soil in other towns in the county, with cow and stable dung by inn-keepers,	500
Soap lees, whale blubber refuse, and dung of fishing villages	1000
Shelly sand, sea-weed and sea-dogs,	1200
Lime walls of old houses, and other rubbish from old plaister,	500——6200
Total paid for manure,	L.59,200
The expence of carriage, when hired, at an average,	2000
In all	L.61,200

3. The implements of husbandry, since the introduction of threshing mills and fanners, or winnowing machines, are now very expensive. In the whole county they cannot be estimated at less, for their prime cost, than 15s. for every arable Scotch acre, or - - - L.270,000

Sup-

Supposing only 10 per cent necessary to keep them
in repair, the sum is - - - L.27,000
4. The household furniture cannot in general be
estimated at less than a year's rent, and is probably more than this—But stating it at - 200,000
The keeping it in repair, at only 5 per cent. is 10,000

5. Loss on live-stock.—This consists of 3 branches:
First, the annual loss of buying and selling horses
  with other counties, - - L 20,000
Second, decay of horses by old age, at L.1 10s. each, 12,000
Thirdly, loss of live-stock, in general, by death, supposed 10,000

Total loss on live-stock annually - - L.42,000

6. Expence of turnpikes and of commutation roads:
The first of these cost yearly for tolls above L.6000
The other at L.30 in the L.100 Scots of valued
  rent, - - - 3600
Both cost - - - - L.9600

7. Expence of the farmer's horses and servants on these
roads. These can be only conjectured, but *it is not guessing
them too high*, to suppose they cost five times as much as the
tolls on the turnpikes, or - - L.30,000

8. The farmer's own expence in going to markets, and
fairs, paying custom, dining in the markets, &c. cannot be
estimated below the same, - - L.30,000

9. The expence of fuel is very great, and cannot be estimated at less than - - - L.45,000

10. The expence of the farmers own families. Here the
farmer is fairly entitled to charge as *an overseer*, in proportion to the extent of the farm, and his wife as superintending
the dairy, is also useful to the community. Where the farmer labours with his own hands, he then ranks with servants or labourers; but he is entitled to receive wages as an
over-

overseer, in proportion to the extent of his farm. Suppose that he has only four shillings per acre, and his wife only one shilling per acre, then a large farmer and his wife, who have 400 Scotch acres, would have L.100 for the maintenance of their family, independently of the interest of their capital;— one who had 200 acres, would have L.50;—and a small farmer and his wife, who had but 40 acres, would have L.10, besides what they could earn by their labour. On this supposition for 360,000 acres of arable land, the farmers ought to have for the support of their families,   L.90,000

11. The interest of capital is the last, but an important article of agricultural economy. In the county of Aberdeen, the interest of this, at only 5 per cent. is   L.155,140

On a review of the whole produce, and expence of raising that produce, in the county of Aberdeen, the following abstract may not be far from the truth, and may be satisfactory to the reader.

The whole annual produce of the county, in corn, pulse, potatoes, turnips, and all other garden roots, or fruits raised in the county; in hay, artificial or natural grass, and in wood annually cut down, or live-stock, either sold or slaughtered, is computed to be   - - L.2,042,800

Of this there is consumed by the live-stock yearly	L.639,160
The expence of seeds annually, -	150,270
The money paid for rent is - -	200,000
The wages and maintenance of labourers	400,000
The expence of manure, where not raised on the farm,	61,200
The keeping in repair the implements of husbandry	27,000
The keeping up of household furniture, -	10,000
The keeping up, or loss of live-stock, -	42,000
The expence of turnpike and commutation roads,	9,600
The farmer, his servants, and horses expence on these	60,000

Carry over,   L.1,599,230

Brought over,	L.1,599,230
The expences of fuel	45,000
The farmers and their wives' charge for overseeing the farm and the dairy	90,000
The interest of their capital, viz. (L 3,102,800)*	155,140
Total expenditure of the farmers,	L.1,889,370
General gain, to promote the accumulation of Capital,	153,730
Annual produce, as above stated,	L.2,042,809

From the above it appears that the profits on farming are not quite 10 per cent.; and that it is only from the progressive improvement of the soil, that the land-rents of the county can be increased. By employing the women more generally in hoeing and weeding of turnips, and other green crops, and by adding yearly to the extent of arable, and well cultivated lands, the same money price of labour that is now paid, on 360,000 arable acres, will be sufficient to the cultivators; and the landed proprietors can add to their rentals, only by granting long leases, with progressive rises of rent, and every encouragement to the farmers, that may both stimulate and reward their industry.

This general outline of Produce and Expenditure, according to the Plan of the Corrected Reports, is inserted as part of the Political Economy of this county. Perhaps it would have been as well to have arranged it under the Chapter of Rural Economy. But the Writer of this Report has taken the liberty of terming it the *Agricultural* Economy.

SECT.

* This sum is higher than the *ratio* of capital in my parish, in Chap. iv. Sect. because the price of cattle has risen since 1809; and because the value of household furniture, of the implements of husbandry, of hogs, bees, and poultry, and also the melioration of houses on a farmer's removal, did not form a part of that estimate.—G. S. K.

SECT. VIII.—MANUFACTURES.

The division of labour is one of the principal causes of the wealth of nations. Hence the farmer and the manufacturer supply each others wants, equally to the advantage of both, and to the general good of the community. And hence our happy country, from uniting a great landed territory to valuable manufactures and extensive commerce, is far superior, both in point of strength and opulence, to those nations which enjoy only one of these sources of national prosperity.

The county of Aberdeen has been long celebrated for its manufactures. Above 150 years ago, it manufactured a species of cloth called *fingrams*, for the foreign market, and both *seys* and *serges* for the use of the inhabitants. The lands of Gilcomston, in the vicinity of Aberdeen, which within the last 60 years have been so highly cultivated, were, at that period, an extensive sheep farm, rented by an opulent manufacturer, whose flocks depastured the suburbs which surround the city on the west, and the meadow of the Denburn, over which one of the finest bridges in the island has lately been erected. Uniting the characters of a sheep farmer and manufacturer, he acquired a considerable fortune; and, by giving employment to a number of labourers, was a great benefactor to the county. It is remarkable, that the lands which he possessed are now above *an hundred times their former rent*; and that several of *the first families in England* derive their descent from this manufacturer.* The particulars are marked in a note below.

---

* The person here alluded to was Mr. Thos. Garden, a younger son of Garden of Banchory, a respectable family, that possessed for several centuries a considerable estate, situated in the county of Kincardine,

The manufacture of fingrams was succeeded by a still more valuable one, the knitting of stockings. Fifty years ago, that manufacture brought L.100,000, and forty years ago L.120,000, into this county annually. It raised the rent of our lands above a third part; and in the then imperfect state of our agriculture, was particularly adapted to the situation of the county. For the small farmer or cottager's wife, or daughter, could attend to her cow, while pasturing on the *baulks*, or patches of unarable land between the ploughed ridges;

but within four miles of the city of Aberdeen. He was Deacon Convener of the incorporated trades of that city; and rented from Mr. MENZIES of Pitfoddels, the lands of Gilcomston, for 500 merks Scots, or L.27 15s. 6d. (They now yield above L.3000.) He left to his oldest daughter 27,000 merks Scots, or L.1500 sterling; a very considerable sum in those days. To the great mortification of her relatives, she married a Lieutenant CADOGAN, at that time a subaltern officer in Oliver Cromwell's army. But the marriage, though at first offensive, proved in the end both advantageous and honourable. Her husband rose to be a Colonel, a General, and lastly to be a Peer. Of him were descended the Lords and Earls of Cadogan, the Dukes of Richmond and Leinster, Earl Verney, Lord Holland, the Right Hon. Charles James Fox, and a number of persons of the first rank in England, in consequence of intermarriages with the Cadogan family. What is yet more to their honour, they did not neglect their Scotch relatives, who at first thought themselves affronted by Miss Garden's marriage. Lord Cadogan, in 1733, obtained a royal presentation to the church of Banchory, in which the paternal estate of the Garden family was situated, in favour of the Rev. James Nicolson, who was a grandson of Convener Garden's, by his second daughter. Mr. Nicolson named a daughter *Cadogan*, and afterwards a son *Charles*, out of respect to his Lordship. And it was in no small degree owing to the interest of Lord Cadogan, that Mr. Garden of Troup, the *male representative* of Convener Garden, obtained a very advantageous lease of the forfeited estates. A farm that has increased an hundred times its former value,—and a farmer who was the ancestor of so many Peers, deserved to be noticed in this Report.

ges; and could at the same time enjoy pure air, and carry on the knitting of her stockings. The improvement of our agriculture has rooted out these barren patches, and has completely established the dominion of the plough over every part of our arable lands. And both the introduction of *stocking frames*, which has rivalled that manufacture, (as carried on by the hand,) and the long continued wars in the north of Europe, which was our great market for this staple branch of our commerce, have deeply injured the trade in stockings, whether woven by machinery, or knitted by the hand.— From all these causes, the woollen manufactures of this county have been almost completely changed, where not destroyed.

While the knitting of stockings was generally and successfully practised in this county, particularly in the three divisions of Marr, the Garioch, and Formartin, a few spirited merchants in Aberdeen, and enterprising manufacturers in Huntly and Peterhead, introduced, and carried on, with various success, the linen manufactory, and the spinning and bleaching of threads, partly for domestic use, and partly for the foreign market. There was a great quantity of flax raised, and the rest, along with the most of the lintseed wanted for sowing, was imported. This manufacture spread over the two divisions of Buchan and Strathboggie. But for several years past, the want, or excessively high price of flax and of lintseed, occasioned by the war, has also deeply injured the manufacture, as carried on *by the wheel,* and *manual labour.*

These two manufactures of stockings and linen, which thirty years ago produced *from an hundred and fifty to an hundred and sixty thousand pounds,* do not now produce *one-half of that sum.*

But since the introduction of machinery, which abridges hu-

human labour, the manufactures of this county, (though they employ fewer hands) are carried on to a very considerable amount; and great companies have succeeded, who have erected spacious buildings, and employ a number of persons at their manufacturing works, besides those to whom employment is given in their own houses. At present, when this sheet is at press, (February 1st, 1811,) there is a considerable change in their situation, since the time when the Writer began drawing up this Report; and there is a little stagnation in our manufactures, owing to the peculiar situation of this country with regard to the Continent; and as this Report is properly an agricultural work, it would be improper to give, in their present state, any other than a very geral account of the manufactures of this county.

The three principal branches of these are the Woollen, the Linen, and the Cotton; the last of which was introduced only about thirty years ago.

The Woollen consists of two branches, the stocking, or hosiery, and the manufacturing of cloth.

The stocking manufacture, though less considerable than formerly, is in the hands of a number of persons. The principal house in this line, is that of HADDEN & Co. who have opened a new branch, besides that of stockings, viz. the knitting of frocks, mitts, and all sorts of hosiery, on a very extensive scale; and employ more hands than all the rest of the manufacturers in the stocking trade.

The manufacture of woollen cloth, (independently of seys, serges, plaidens, and other coarse cloths, which are made for family use in all parts of the county, by the farmer and cottagers' wives, daughters, or servants,) is carried on for public sale, by a number of active and intelligent persons.

Messrs. MIDDLER & Co. carry on a woollen manufacture in Aberdeen, and employ a carding mill and other machinery.

chinery. Their principal manufacture is that of coarse woollen cloth.

In the country, Messrs. CROMBIE, KNOWLES, & Co. have, for six years past, successfully carried on a manufactory at Cothal Mills, in the parish of Fintray, eight miles from Aberdeen. This is a rising, though a new branch of manufacture. Last year, this Company divided, with a manufacturer in Galashiels, both the first and second premiums which were given by the Trustees, for cloth under ten shillings and sixpence per yard.

In the division of Buchan there are two cloth manufactories.—The first, or oldest, is at Kinmundy, in the parish of Longside, belonging to the Messrs. KILGOURS, which has done a great deal of business in the course of about thirty years. The other, carried on by Messrs. DALGARNO, and Co. at Peterhead, is of a later date, but both are useful to the county.

The whole manufactories in woollen, hosiery, and cloth included, employ more or less constantly, above 7000 persons, and assist in supporting, where they do not entirely maintain, 20,000 individuals.

The manufacture of flax consists also of two branches, viz. the making of linen cloth, and the spinning or bleaching of thread, for various purposes.

1. The linen manufacture, strictly so called, has of late years varied considerably in its amount.—In 1808, the most favourable year for a considerable time past, the quantity manufactured was above 300,000 yards. By the official report of Mr. JAMES MILNE, Stampmaster in Aberdeen, (a very intelligent man, to whom the Writer of this Report has been much indebted) the quantity manufactured in 1808 was - - - - - 314,556 yards, worth L.31,000
In 1809 it fell to - - 102,297 yards, worth L.10,000
In 1810 it rose to - - 153,336 yards, worth L.14,000

The cause of its prosperity in 1808, was Buonaparte's Edicts for prohibiting the exportation from his dominions and all places under his controul, of any manufactures of flax or hemp,—and the cause of its decline, for the last two years, is, that he revoked these impolitic edicts, and that the price of flax has risen so high, that its manufacture is no longer so profitable a branch as that of cotton.—The remedy, viz. the prohibition of the importation of foreign manufacture, was early suggested by Mr. MILNE to the Trustees for Manufactures, &c. and application was made to men in power to get this prohibition enacted by Parliament—but nothing as yet has been done to remove the evil complained of, though the remedy is so simple and easily put in practice. It is hoped that Great Britain will never carry the *lex talionis* so far, as to burn all the manufactures of France, that are imported into this country; or imitate that insidious policy, which carries on war against men by *perfidy*, as well as by the *sword*, as in Spain, or against manufactures *by fire*, as in Germany and France. But it is very impolitic in our public Ministers to permit the importation of the manufactures of linen from those countries, in which the manufactures of Britain are confiscated and burnt.

A linen thread manufactory has been carried on at Peterhead since 1765, by the Miss PARKS, whose goods always find a ready sale, and a high price.

A considerable thread and linen manufactory has also been established at the same place by Messrs. JOHN ARBUTHNOT & Co.; and other manufacturers have carried on this branch with various success in the division of Strathboggie.

The manufacture of thread has been carried on to a much greater value than that of linen cloth amounts to at present. In 1808, by information from Mr. MILNE, it appears that upwards of 600,000 spindles of yarn were spun in the northern

thern counties, the greatest part of which belonged to manufacturers in Aberdeen, who paid the *women* alone, for spinning, upwards of L.35,000, besides giving bread to several hundred industrious men and boys, employed as flax-dressers, in preparing the flax for spinning, independently of the sums paid to ship-owners and carriers.

The greater part of the yarn was made into sewing thread, from 3s. 6d. to 4 guineas per pound, giving employment, in the manufacture of thread, to upwards of two thousand men, women, and children, in the city and neighbourhood, besides many thousands employed in spinning the flax, in all the northern counties. A great deal of money is distributed by the thread manufacturers of Aberdeen, in all the northern districts, for spinning yarn. These, and other manufacturers of flax, send their goods to Glasgow and Paisley, in the west, and to the manufacturers in Perth, Fife, and Forfarshires.

Besides the branches here enumerated, are the various kinds of linen yarn made by the spinning mills, lately introduced.

A very extensive building for spinning by machinery, has been erected at the Haugh of Grandholme, (a meadow on the banks of the Don,) at two miles distance from Aberdeen. The owners of this work have also expended L.1200 on a bridge over the Don, merely to facilitate their intercourse with the city and harbour; and they both spin flax by their machinery, and conduct the making both of thread and linen on a large scale, giving employment to above a thousand families, either at the manufactory on the Don, or at another in the city of Aberdeen. The produce of these works, the threads excepted, is made into cloth in the southern counties, particularly in Fife and Forfarshires.

The flax used by the spinning mills is all imported from Russia; from whence this great Company have imported above 600 tons in one year. By the other manufacturers, either

either Dutch, English, or Scotch flax is used indifferently, as it is found most expedient.

It is proper to mention, that the flax raised in Great Britain is fit for every purpose for which either Dutch or Russian flax can be used; and for many purposes the manufacturers prefer the former to the latter. It is well known that finer yarns are spun by the hand from flax of Scotch growth, than have been made from the finest Dutch flax. (In this case, however, the flax seed is imported.) Eight spindles of linen yarn have been spun out of a single pound of Scotch growth, which at present would fetch 12s. per spindle, or L.4 16s. for the produce of a pound of flax. It is well known that a sufficient quantity of flax can be raised within the British empire: and instead of permitting the importation of foreign linen, Great Britain ought to encourage, by the highest premiums, the raising of flax, and the linen manufacture, from native produce.

The whole manufacturers in flax within the county, give employment to above three thousand families, and bread to ten thousand individuals.

The Cotton Manufacture was first introduced into Aberdeen in 1779, and for about twenty years was carried on almost exclusively by one Company, by whose exertions it was spread over the county. It now employs a great number of men, women, and children, in the city and neighbourhood, in many of the villages in the interior parts of the county, and in the towns on the sea-coast. Within these twelve years several new companies have been formed, who are now prosecuting this manufacture, and whose united trade is said to be equal to that of the oldest company. The most considerable of these is the House of FORBES, Low, and Co. who have erected, close by the Quay of Aberdeen, a large cotton work,

work, with two steam engines,* (one of which was the first of this kind that was used in Aberdeen), and every sort of machinery requisite in that trade, except for printing cotton.— They give employment to above six hundred persons, and bread to two thousand individuals. Besides spinning and weaving cotton cloth, they manufacture threads for the sempstress, which have been found to rival those made from flax; and they carry on their manufactures on a large scale. Besides the original Company (which unites a printfield, bleachfield, and two cotton mills,) and this in which the mill goes by the power of steam, there are six companies of manufacturers, whose trade is limited to the weaving of cloth, who employ above seven hundred looms, and give bread to about two thousand individuals. And lastly, there are above an hundred looms belonging to the different weavers, (some of whom have six, eight, or ten each) and which support five hundred individuals.

The whole cotton trade employs above four thousand persons, of whom two thousand five hundred are grown up, and fifteen hundred are children, from 9 to 15 years old.

Earnings.—These are widely different, varying according to the state of the trade, particularly in the weaving department. In general, the men can earn from 9s. to 25s.; the women from 5s. to 10s.; and the children from 2s. to 5s. per week. And the printers from 3s. to 5s. a day, at job-work. It is doubtful whether the very high wages paid to many of the best workmen, by introducing habits of dissipation, be not a loss to the community.

The rents of houses, of gardens, and of grass land, and green crops, in the neighbourhood of these manufactories, are

---

* There are now 7 steam engines in the city and neighbourhood.

are considerably increased, because potatoes, milk, and vegetables, are indispensibly necessary. Poor's rates are not known in this county.

A manufacture of sail-cloth has been established for above twenty years, and bids fair to supply us with an article of the first necessity to a trading nation.

There is also an inkle manufactory at Aberdeen, which is of recent institution. It is carried on under the name of the Aberdeen Tape Company. Its linen tapes are equal in quality to those of any manufactory in Great Britain. Formerly, the tapes used in this country were made in Holland: but are now made in greater perfection in this city;—and in the present state of Europe, the importation of foreign tapes, as well as all linen cloth, ought to be prohibited.

There are three paper mills within this county, which carry on that manufacture to a considerable amount. The first, and oldest, is at Peterculter, eight miles from Aberdeen, where Mr. LEWIS SMITH, the owner of the work, uses patent machinery, by which paper is made like a web. It employs above fifty hands, and must require a large capital to carry it on, as it goes night and day, being only stopt occasionally to oil or wash different parts of the machinery. In the second, belonging to Mr. ALEXANDER PIRIE, at Stonywood, on the Don, (who has already been mentioned as an enterprising farmer) the old plan of vatts and wire frames is used; and a good deal of business is done. In the third, near Aberdeen, belonging to JOHN DINGWALL, Esq. sometime Provost of that city, vatts and wire frames are also used. These two employ a considerable number of hands. The three paper mills, last year, paid duties to Government to the amount of L.6178 12s. 1¼d.

Different rope-works are carried on in Aberdeen and Peterhead.

terhead. They employ 250 ropemakers, and give bread to 600 individuals.

There is also a nail manufactory in Aberdeen, which usually employs from twenty to thirty workmen. But the high premiums which they received from entering as substitutes into the Militia, having induced some of them to enlist, they drew off several of their companions; so that there are only a few employed at present.

There are also several breweries, tanneries, soap-works, and candleworks in Aberdeen and its neighbourhood, who pay considerable taxes to Government. The first brewery was established at Gilcomston in 1768, and is a very complete work. Another, called the Devanha Brewery, on the banks of the Dee, about a mile from the harbour, is conducted by Messrs. WILLIAM BLACK, & Co. It deserves to be noticed here, that its porter has acquired great celebrity not only in Aberdeen, Edinburgh, and Glasgow, but frequently finds its way to the city of London. There are several other breweries in the cities of Aberdeen, and of Old Aberdeen, and in their vicinity; also at Peterhead, Oldmeldrum, and Huntly, which are carried on with various success.— The brewing of ale, in private families in towns, and in the farmer's houses in the country, has much decreased; but that of the public breweries in Aberdeen, and in other towns, has increased with the trade of these places, and with the luxury of the age.

Ship-building is carried on both at Aberdeen and Peterhead, and employs a number of carpenters, who are at once useful in building and repairing ships and boats of all descriptions, and as a nursery for the public docks, and for carpenters to the ships of war.

A straw-hat manufactory has been for some time carried on with success at Aberdeen.

Even

Even *pin-making* has been introduced into this city. Mr. ROBERTS, a pin-maker, seems to go on successfully in an occupation which is promoted chiefly by the division of labour.

While the various denominations of manufacturers deserve to be mentioned in this Report, it should ever be remembered, that the shoemakers, taylors, house-carpenters, plough and cart-wrights, blacksmiths, and artificers of every description, who either reside in the different towns, or are scattered over the county, (though the earnings of each individual cannot be ascertained,) add much to the wealth and comfort of the community.

On a general review of all the articles of this Section, and without wishing, in the present state of the manufacturing interests, to distinguish the capital or profits of the different manufacturers, it may be remarked, that owing to the abundance of excellent granite, the buildings of the principal manufacturers are spacious, valuable, and durable—that we have comparatively few speculators—that our principal manufacturing companies are possessed of large capitals—and that the annual produce of the labour of their servants and apprentices, is probably above L.700,000, and in some years has exceeded L.800,000. The duties paid to the Excise on those articles which are exciseable, were last year L.51,231 7s. 7d.

Printing.—To this it may be added, that the city of Aberdeen was early distinguished in the art of printing. ED. RABAN, the first printer in any part of Scotland, except Edinburgh, flourished in the end of the 16th century, and died about 1620. There are now three printing offices in Aberdeen, which employ from 20 to 30 hands. There are two Newspapers printed weekly, viz. *The Aberdeen Journal* and *Chronicle*. The Journal was the first provincial paper established in Scotland.

Though

Though the Board of Agriculture has not, in the Plan of the Corrected Reports, included this article, yet it is proper to introduce a few paragraphs here, under the title of

SECT. IX.—FISHERIES.

Of these, the salmon fishery is the most considerable, particularly on the rivers Dee and Don. Those on the Dee, in the most favourable year, (viz. 1798) estimated in barrels of 4 cwt. each, were - - - 1890¼
And on the Don the same year were - 1667
In 1804, the most unfavourable year, were on the Dee, 652¼
The same year, on the Don, they were only 349

Formerly, they were sent in the spring months to London, and in the end of the season, were packed up in barrels when salted, and carried to the south of Europe, where they were used in the season of Lent. Now they are chiefly sold in London; and their price has increased in a five-fold proportion since 1768.

It is proper here *to correct an error in the amount of the rent of the Don fishings.* In fact, they are not all rented; but the proprietors carry on the fishing themselves. In such a situation, the Reporter would not choose to specify a rent.—But he would remark in this place, that the fishings of the river Dee were let, in 1399, for L.478 of our present money; and that those on the Don, then called *Aqua Borealis*, were much higher.

The fishings on the Ugie, the Ythan, and the sea coast, near the mouth of these rivers, vary from 200 to 600 barrels.

Not above one pound of salmon is consumed in the county for forty that are exported.

It is obvious that there must be a great inequality of the produce of the salmon fishing, in a year in which there is little snow in winter, and in which the rivers are seldom swollen in spring or summer, compared with that in which

the

the rivers are flooded by melted ice or snow in spring, or copious rains in the summer season.

The sea fishing, for haddocks, cod, ling, skate, turbot, halibut, &c. employs a number of fishers. Of these, some have pretty good boats, and others have only small yawls. (It is a great loss to the county, that we have no decked vessels, except for our very distant fisheries.) The ordinary fishing is carried on in different small creeks, or villages, on the coast of Buchan, and a few fishermen at Futtie, or Foot-Dee, near Aberdeen. It brings in from L.15,000 to L.30,000 yearly. Some years of late, when herrings frequented our coast, it exceeded that sum; besides from L.2000 to L.3000 more to a number of poor women, who carry the fish either to the market towns, or to the interior parts of the country.

The greater part of this fishery is consumed within the county. At Peterhead, however, there is a considerable export, generally 500 barrels of cod, and a considerable quantity of ling. The herring fishery has of late been very productive to the fishers on the coast of Buchan, though its produce is too unequal to afford any data for making an average calculation.

The Greenland whale fishery has also been carried on with great success, both at Aberdeen and at Peterhead.—At Aberdeen, during the last seven years, there were four ships, of 1044 tons, in all, which produced 248 whales, 3396 tons of oil, and 150 tons of whale fins.

At Peterhead it has increased rapidly. For seven years it had two ships engaged in this trade, which brought home 172 whales, producing tons of oil, 1835¼ do. of fins, 45

	tons oil.	do. fins.
A third ship last year had 17 whales, yielding	174	7¼
180 total whales in 7 years,	2009¼	52¼

This

FISHERIES.

This is a most thriving place, where an uncommon spirit of exertion has been shewn in the course of the last thirty years. A fourth ship is building there for the whale fishery, and will be ready next spring.

The whole fisheries connected with the county, yield from L.80,000 to L.100,000 annually. But both the whale and herring fishery are of too late a date, and too unequal in their produce, to admit of stating an annual average.

SECT. X.—COMMERCE.

ABERDEEN has long been considered as the metropolis of the north of Scotland; and its commerce has been greatly increased since the union of the two kingdoms. In 1712, its foreign exports stood as follows:—

Country.	No. of Ships.	Do. of Tons.	No. of Men.
Sweden,	2	100	13
Norway,	13	406	64
Germany.	1	40	4
Holland,	6	238	38
Portugal,	2	85	14
Spain,	2	150	17
Italy,	4	300	53
Total,	30	1329	203

Its exports to these places were—

1. Oatmeal, bear, pease, and malt—quarters	4132
2. Cod, or ling fish—barrels - - -	815
3. Herrings, do. - - -	228
4. Pork, do. - - - -	360
5. Hogslard, casks - - -	10

6. Stockings

6. Stockings and woollen cloths—trusses   156
7. Tobacco—hogsheads   -   -   17
8. Lead—bars   -   -   -   888
9. Coals—chaldrons   -   -   10

In 1788, it traded with Russia, Sweden, Norway, Denmark, Prussia, Poland, Germany, and Holland, in the north; and with Italy, Gibraltar, and Portugal, in the south of Europe; also with America and the West Indies. The number of ships belonging to the port of Aberdeen, in the foreign trade, was 63—tonnage 4964—men 319.

Its coasting trade, ships 93—tonnage 5520—men 396.

Fisheries, ships 12—tonnage 1236—men 202.

Total, ships 168—tonnage 11,720—men 917.

It is to be observed, however, that in all the above cases, not only the vessels belonging to the harbour of Aberdeen, but those at all the inferior ports, from Banff to Caterline, a line of an hundred miles of sea coast, are comprehended.

In 1810, the merchants and ship-owners of the town of Aberdeen alone had

Ships.	150	tonnage 17,131	men 932
Peterhead,	50	5,597	404
Newburgh,	7	662	37
	207	23,390	1473

The whole trade of the city of Aberdeen, as charged officially by the persons who imposed the municipal taxes, was, in 1794, only   -   -   - L.442,460
In 1809 it amounted to -   -   -   - 887,250
It is known that this is below the truth, and should
  be above   -   -   -   -   1,000,000
The trade of Peterhead, in 1794, was   -   100,000
Do. including the coast of Buchan, in 1809, was
  above   -   -   -   -   -   200,000

N.B.—The exportation of granite for the city of London, employs above 70 ships, of 7000 tons, and 400 men.

Taking

Taking a general view of the trade of the port of Aberdeen, and including the repeated voyages of the ships which have frequented that port for the last 16 years, the following summary, taken from the Custom House books, will shew the average amount for 5 years preceding 1801, the particular amount of that year, as a year of peace; also the average from that period to January 1811; and lastly, its amount for the last of these 16 years. (It is, however, to be observed, that no ships with lime or manure are entered in the Custom House books, or included in this summary.)

	Ships.	Tons.	Men.
Average of five years,			
Foreign imports,	74	7,298	443
Coasting do.	996	48,779	3999
Average of imports,	1040	56,077	4442
Foreign exports,	44	4,611	296
Coasting do.	752	34,440	2,999
Average exports	796	39,051	3,295
1801. Foreign imports,	109	11,746	678
Coasting do.	1,052	63,961	4,236
Total imports in 1801,	1,161	75,707	4,914
Foreign exports, do.	62	7,240	424
Coasting do.	741	37,878	3,144
Total exports in 1801,	803	45,118	3,568
From 1802 to 1811 inclusive			
Average Foreign imports,	69	8,674	559
Coasting do.	888	59,243	4,004
Average of 10 years,	957	67,917	4,563

Foreign exports, do.	58	8,570	547
Coasting do.	738	46,787	3,407
Total Exports,	796	55,357	3,954

From 1810 to January 1811,

Foreign imports,	63	9,017	646
Coasting do.	1,100	78,676	4,851
Total Imports of last year,	1,163	87,693	5,497
Foreign exports do.	81	13,424	895
Coasting do.	730	44,798	3,218
Total exports of last year,	811	58,122	4,113

From the above it appears that the trade of Aberdeen is increasing, and is now greater than it was in 1801, when we had peace with France.

The principal articles of import are,

Wheat, and wheat flour—quarters, average	4000
Barley do. do.	3500
Rum—gallons	18,000
Aquavitæ, or whisky—do.	12,000
Gin, or Geneva—do.	5000
Brandy—do.	2000
French wine—do.	1100
Portuguese, or Spanish do.	42,000
English coals—chaldrons	17,000
Scotch coals—tons	6000
Wool—cwt.	5000
Salt—bushels	24,000
Tobacco and snuff—lbs. above	200,000
Flax—cwt.	27,000

Exports.

Oats, or meal, from Buchan and Strathboggie.

bolls	18,000
Bear, or bigg, do.	5000

Salmon—barrels	2000
Cod, ling, &c. from Aberdeen	4000
Do. Peterhead and Buchan coast	4000
Stockings and woollen, worth	L.80,000
Cotton, variable, perhaps,	L.200,000
Linen and threads	L.100,000
Stones to the amount of	L.40,000

N. B. Mr. Sim, the very intelligent Clerk at the Custom House, informed the Reporter, that he supposed the import price of stones when delivered at London, was equal to all the tea and sugar imported into Aberdeen.

He also stated, that he considered that the taking stones instead of ballast, reduced the price of coals 15 per cent.

THE EFFECTS OF COMMERCE ON AGRICULTURE.

It can admit of no doubt, that Commerce, on the whole, is favourable to agriculture; and that in the neighbourhood of Aberdeen our extensive commerce has afforded the means of improving the most barren lands in the county. Where a man has acquired a fortune in trade, he generally wishes to employ a part of it in the improvement or purchase of land. The internal commerce of the city of Aberdeen, or towns on the sea coast, with the interior parts of the county, supplies the farmer with ready money, or with the means of improving the soil. The coasting trade supplies us with lime, coals, and other necessary articles; circulating commodities from one part of the country to another, and carrying off the superfluities of every district. The foreign trade, while it furnishes us with raw materials, or rude produce, which our own territory cannot raise (either at all, or to advantage), is not the opponent, but the auxiliary of the farmer. On the other hand, the high wages that are given to our sailors, our manufacturers, and to all persons connected with commerce, who would otherwise be employed as day-labourers, prevents the farmer from getting, on any terms, as many servants, or cottagers, as he would choose to have, and tends very much to raise the money price of labour, where he does employ such persons. These things produce great inconvenience, and occasion a temporary loss to the cultivators of the soil, until the price of corn rises in proportion to that of labour.

HARBOUR AND CUSTOM-HOUSE AT PETERHEAD.

The trade and shipping of this place are now much greater than those of Aberdeen were a century ago: and the inhabitants have got a plan and estimate of a new harbour, from that able civil engineer, Mr. RENNIE. His report is most favourable, and the expence of a harbour, which will enable vessels both to enter and sail with every tide and wind, is L.30,000. The merchants, (who possess an uncommon spirit of enterprise) state, that owing to the want of a Custom-House, they are injured to the extent of L.1000 annually; partly by loss of time and by seamens wages; partly by the expence of their shipmasters' journies to and from Aberdeen; and partly by the expence of bringing out either the Collector or Comptroller of the Custom-House from that city, when any of the Greenland ships sail or arrive, or of an inferior Custom-House Officer on the arrival or departure of their ships in the foreign trade, besides the loss incurred by coasting vessels, especially those which leave part of their cargo at Peterhead. The Writer of this Report, would here remark, that as BERNADOTTE is now the Crown-Prince of Sweden; as Peterhead is a *watch-tower appointed by nature* for the north-east part of the island; as ships in the Baltic trade frequent this port; the British Legislature are called upon to see that there be *a good harbour* established at Peterhead, with a Custom-House, which may give *every facility for the entrance, departure, and clearing out of ships*, not for the sake of the *trade of Peterhead*, but from a regard to the strength and security of the empire: For a few hours previous notice from a vessel landing at that port, may prevent or baffle an invasion of Britain.

SECT. XIV.—THE POOR.

It has already been mentioned that we have *no poors' rates* in this county. Yet the wants of the poor are supplied by the humanity of the higher and middle ranks of society, without the aid of compulsory laws; though we have laws for making assessments when necessary. Even the poor, or those who are by no means in affluent circumstances, are not deficient in supplying the wants of their poorer brethren.—Collections are made every Sunday in all the established churches, and occasionally in those of the Episcopal Communion, and in all dissenting congregations. Those who are really poor, *are found out*, (for they seldom *apply* for assistance) and their wants are generally supplied. In the extensive parishes in the *highland district*, the limited funds of the Church Sessions can afford but a small sum of money to each individual; but the benevolence of their neighbours supplies this defect; and very few beggars are seen who belong to the country parishes. It is in large towns, whither the poor flock from all quarters, (frequently from the remotest districts of the west highlands) that there is really the greatest penury. There, even if the laws of England were established, few of the mendicants would have any right to a settlement; and there, also, it is not always possible to discriminate properly between those who are really indigent, and those who are only clamorous, and cannot be satisfied.

On a general view, *the condition of poor women*, who can find little or no employment in the stocking manufacture, and who have not been accustomed to any other kind of employment, *is most to be pitied, in the present state of the country.* The *annual receipt and expenditure* of most women of this class, *are below four pounds sterling per head*; and yet many of them

them will not accept of aid from the parish. Those who are enrolled in the poor lists of the different parishes, receive from *ten shillings* to *five pounds yearly*, according to the state of the funds, the number of the poor, and their different degrees of poverty and of inability to supply their own wants.

In the city of Aberdeen there is a Poor's Hospital, in which young boys, who are either orphans, or the sons of poor inhabitants, are educated and maintained; but, strictly speaking, there is no work-house, nor house of industry, in this county, excepting the Bridewell, which was built, and is supported at a great expence, and could afford lodging and employment to several vagrants, who infest the streets of Aberdeen, or wander through the country as strolling beggars.

It may also be mentioned under this head, that there is an excellent Infirmary in Aberdeen, which generally admits a thousand patients annually into the house; and where advice and medicines are given freely to near double that number of out-patients. Also, that there are two Dispensaries in that city, where the physicians give not only advice and medicines, but also visit, gratis, the poorer people in their own houses.

Besides the sums which are given to the out-pensioners of the Poor's Hospital, or to the enrolled poor of the city, by the Church Session, the different corporations of tradesmen, who have ample funds, give large supplies to their own poor; and, the societies which have been long established, as well as those of recent institution, expend considerable sums annually; so that poors' rates are really not necessary in this county.

The Reporter feels it his duty to add some remarks of a more unpleasant nature.

Under the head of *Poor's Rates*, it was mentioned, that partly by *an oversight in the Property Tax Act*, and partly by the

the interpretation put upon that law by the *Commissioners for Special Purposes in London, the poor's funds in Scotland have been taxed—certainly without any intention on the part of the Legislature.* As the Writer of this Report has applied, in vain, for redress, it becomes necessary to state the particulars.

1. Owing to our distance from London, and the small capitals of our *Church Sessions* being insufficient for the purchase of stock in the public funds, the money belonging to the poor of the different parishes, is generally lent out, either on bills or bonds. From the ignorance of the framers of the Property Tax Act, or from their not attending to this circumstance, a clause, exempting from this tax, the interests, or annual rents of poor's money lent on such security, was neglected to be inserted in 1807. The present Chancellor of Exchequer, on being informed of this, acknowledged, in a very humane and condescending letter, that " this was an " omission in the Act of Parliament, and that all interests of " money belonging to the poor, in whatever way it was " lent, ought to have been exempted from taxation." Three years and a half have elapsed since this acknowledgment was made; yet owing to the pressure of public business, the matter has been forgotten. It is, therefore, necessary to put it in the view of the public; and it is hoped, that, in the next property act, the exemption will be inserted in the clearest and most comprehensive language.

2. When a Church Session has as much money, either by any legacy, or by savings in years of plenty, they sometimes purchase a rood or two, sometimes an acre of land, or a small house in any town in their neighbourhood, the rents of which are applied to the supplying of the enrolled poor of the parish. But owing to the Commissioners for Special Purposes, under the Property Tax Act, being Englishmen, who reside in London, and are unacquainted with the institu-

tions of Scotland, they have refused to exempt, from this tax, the rents of different small parcels of land, belonging to the Church Sessions of this county. The Writer of this Report, and other three of his brethren, claimed an exemption from the Property Tax, in favour of some small patches of land, belonging to the poor of their respective parishes, and situated in the burgh of Inverury. To obtain this exemption, the Reporter, and the Rev. ROBERT LESSEL, minister of Inverury, made separate affidavits, that the rents of these two small properties belonged to the poor of their parishes, and were applied solely for their support. When these affidavits were given in at Aberdeen, it was objected that they had been made in the country, before two Justices of the Peace, and not in that city, before the Commissioners of the Property Tax. To obviate this, the affidavits were made a second time before the Commissioners, as required, and according to a particular printed form, the blanks of which were filled up by the claimants, when they signed the affidavits. They were then delivered to Mr. WILLIAM DINGWALL FORDYCE, Clerk to the Commissioners, who behaved in the most obliging manner, and transmitted them to HENRY MCKENZIE, Esq. Comptroller of Taxes, Edinburgh. He sent them to the Special Commissioners under the Property Act in London, who alone have power to grant the exemption.

After waiting several months, a return came from the Special Commissioners, who, in direct opposition to the 62d section of the *Property Tax Act*, refused to grant the exemptions claimed. On a different application from the *Church Treasurer* of the parish of Belhelvie, whose lands belonging to the poor, are also situated in the burgh of Inverury, they also refused an exemption from the property tax; and on a second representation from that parish, they insisted, that the Church Treasurer should shew, that the rents of the burgh-roods

roods belonging to the poor, were not applied to the lessening of any poor's rate, or parochial assessment. Surely the *poor's funds* in Scotland are not *to be taxed*, because we have no *poor's rates*, or as an equivalent for the poor's rates of England. But if this ought to be so, the Commissioners for Special Purposes were not Legislators; they held their office under an act, which expressly exempted all charity-lands, whether in England or in Scotland, whose rents were applied to charitable purposes. If the Scotch Barons of Exchequer, or any set of Commissioners, who knew the institutions of Scotland, had been appointed to judge these claims, they would have sustained them immediately. And if the Special Commissioners in London had considered, that the *affidavits* were made according to *their own printed form*, they were bound to grant the exemption, unless they believed that the claimants were guilty of perjury.*

3. When, owing to the necessities of the poor of their parish, a Church Session disposes of any property belonging to them, this must be done by *public auction*, to prevent fraud, or remove all suspicion of improper conduct. Yet, in this case, both the auction duty, and the stamp duty, on the transfer of property, must be paid by those trustees for the poor in Scotland, though neither of these is paid by the church-wardens of England. The Writer of this Report, and the other members of his Church Session, owing to the decline of the stocking manufacture, which was the chief support of the women, who reside in cottages, were induced to sell the land in Inverury, on which they had illegally been obliged to pay the Property Tax. It was sold for L.151, (which was at the rate of L.100 per English, or L.127 for the
Scotch

* I was informed lately, that the *Special Commissioners* have agreed to refer this question to the *Crown Lawyers*.—G. S. K.

Scotch acre.) The auction duty amounted to *four pounds eight shillings and one penny*, and the tax on the stamped paper to *thirty shillings*; so that this small property belonging to the poor, paid 10 per cent. annually, while they possessed it, and nearly 4 per cent. of its total amount, when, from the decline of our staple manufacture, it was sold by auction.— The Writer of this Report applied to the Scotch Commissioners of Excise, to obtain redress as to the auction duty; but received for answer, that " no relief from duties could " be given, that the Board have no power of dispensing with " duties, that no exemption was given by law for sales for " the benefit of the poor, and that it was out of the power of " the Commissioners to give up the duty in this case, how- " ever much they regretted it."

The Reporter should not have had the approbation of his own mind if he had neglected to lay these statements before the public. If the matter had personally regarded himself, he should have thought it improper to have introduced it into this work; but as it respects the poor, not only of his parish, but of all the parishes in Scotland; and as in this section he was bound to state any facts that regarded the poor, he considers that his silence would have been criminal. The late eloquent EDMUND BURKE, in his Reflections on the French Revolution, has censured, in the most poignant language, the French National Assembly, for destroying the charitable institutions of France, which he emphatically terms *meddling with the poor's box*. Any interference of this kind is abhorrent to those principles of humanity which have ever distinguished the British Legislature; who certainly never intended that the poor's funds, or any charitable institution, in any part of the empire, should be taxed: And it will afford the most sincere pleasure to the Writer of this Report, if this work go to a second edition, to strike out the four last paragraphs,

graphs, and to insert in their place, in a single sentence, that, "by a mistake, or inattention of some persons, who "draw up the acts of Parliament, certain taxes had been im- "posed on the poor's funds; but that as soon as this was "known to the Legislature, *all the money* that had been receiv- "ed from the trustees of the poor *was appointed to be restored*, "and that all interests of money, rents of land, legacies, and "property of every kind, belonging to the poor, are ex- "empted from the Property Tax, the Auction Duty, the "Stamp Duties, and all taxes whatever."

In the mean time, he would state, that the sums annually given to the poor in the city and country parishes of Aberdeenshire, are very considerable; and that in calamitous seasons, very liberal subscriptions were made in all parts of the county for the benefit of the poor.

Besides the parochial supplies given to the poor by the Church Sessions, who are in place of the church-wardens of England, and that given by the incorporated trades in the city, and different charitable institutions, *box-clubs*, as they are termed in the plan of the corrected reports, or *friendly societies*, (whose object is the mutual supply of each other's wants,) are generally spread over the county, and are very beneficial. They preserve that independency of spirit which prevents many of the Scotch peasants from accepting charity, and without hurting the pride, or wounding the delicacy of those who have seen better days, effectually relieve their afflictions, which a reverse of fortune renders most poignant. It is one of the first duties of the rich to encourage those societies, to attend their meetings, and to see that their money be lent on proper security, and its annual rents be given to the indigent. And it is also incumbent on the Legislature to exempt all such property, and its yearly produce, from all taxes to Government.

SECT.

SECT. XII.—POPULATION.

The population of the county, in 1801, was found to be 123,082

In the Statistical Account of Scotland, from 1790 to 1798, it was - - - 122,949

In the account taken by Dr. Webster, about the year 1755, it was - - - 116,836

This would make the increase since 1755 only 6,113

But it is generally known that Dr. Webster's account was rather exaggerated. He made it up from the examination rolls of the clergy, and he added *two-ninth parts* of this number, for the children under five or six years old, who were not in these parish lists of examinable persons. To verify, or correct this proportion, the Writer of this Report in 1778, took a list of *the whole population* in his parish, when he found it was exactly 900; while that of children *not examinable*, was 151. According to Dr. Webster, it would have been 200. And as the Doctor wished to give a favourable representation to Government, it is probable that the population of the county, in 1755, did not exceed 112,000

And that it has increased nearly one-tenth since that period—or - - - 11,082

Whatever that increase be, it has taken place only in the towns; for in the country parishes, in general, there is a decrease of the population. In 1801, Aberdeen and Old Machar, that is the cities of Old and New Aberdeen, with the country district belonging to the former, contained 27,508

In 1755, they were said by Dr. Webster to amount to - - - - - 15,730

This would make the increase to be three-fourths of that in 1755—or - - - 11,778

In 1801, Peterhead, with its country parish was   4,491
In 1755, it was stated by Dr. WEBSTER to have been   2,487

This shews an increase of   -   -   -   2,004

In like manner, Huntly, in 1801, amounted to   2,863
In 1755 it is stated by Dr. WEBSTER at only   1,900

This shews an increase of   -   -   -   963

The increase in these three places amounts to   14,745

On the other hand, in the higher district of Marr, the population of the parishes of Crathie, Glenmuick, Aboyne, and Strathdon, in 1755, was   -   -   -   8,386
In 1801, it amounted only to   -   -   6,047

This shews a decrease of   -   -   2,339

In several of the parishes in the interior of the county, the population has been for some time on the decrease. The principal cause of this was the fear of the scarcity of fuel, when the peat-mosses began to be exhausted, after the introduction of carts. To prevent this scarcity, the farmers were improperly restricted in regard to the number of their subtenants, instead of being prohibited from giving subleases of large crofts, to persons who kept horses and carts, which enabled them to exhaust the peat-mosses. A cottager, with an acre of land for a cow, should never be turned out, if honest and industrious. A large crofter, with weak horses, and a bad plough, should never be tolerated. The taking off the duty on coals, (a most important benefit to the north of Scotland) has put an end to all that anxiety about fuel; and several of our landed proprietors are now attempting to raise villages, and wish to induce labourers to settle on their estates. With respect to such of them as are still resident in the country, they will probably succeed; but there is little chance of their bringing back those persons who were formerly obliged to

quit

quit their small possessions, and remove to Aberdeen. For most of these, on their removal, not only sold off their livestock, farm-implements, and crop of all kinds, but engaged in some new employment about the manufactories; and as the value of money is now so much reduced, they have not so much capital as is necessary now for building a cottage, and purchasing a cow, or two cows, if they were to return to the country. For it deserves to be considered, that the household furniture, and working implements of a manufacturer, are of much less value than the live-stock, and crops of corn and grass, of a cottager, who has only a few acres of land. The Writer of this Report wishes to impress upon the minds of all the landed proprietors, that they deeply injure their estates, or prevent their improvement, when they either remove, or prohibit their tenants from retaining an industrious labourer, who is a most useful member of society.

Tables of births, burials, and marriages, unless they were taken from every parish, and from registers, which were kept correctly, would not give any true idea of the increase of population. In the city of Aberdeen, the number of births does not much exceed that of deaths. In the Reporter's parish, at about 14 miles from that city, there are three births for every two deaths; and forty-one males are born for every forty females. The number of marriages cannot afford the least data for any correct calculation, excepting it had been laid down as a *general rule* in drawing up the *reports*, that this was to be estimated, by calling that *one* marriage in which both the man and the woman resided in the same parish, and by calling it *one-half* of a marriage, in which only one of them was resident,—or by reckoning those marriages only in which *the man* resided, in the lists of every parish.

The *increase* or *decrease of the population* has not depended solely on food, or even on food and the probability of getting em-

employment, but has depended on the easiness or difficulty of getting cottages and small patches of land, commonly called crofts.

The county of Aberdeen is *under peopled*. But its population is not affected by the *price* of wheat, nor even by that of oatmeal, which is the principal article of food. In 1783, when oatmeal sold at 1s. 6d. per peck, of 8 Dutch, or 8¼ English lbs. there was not one-fourth part of the usual number of marriages. In 1801, when the peck of meal was above 2s. 6d. (frequently 2s. 9d.) marriages were not checked nearly in the same proportion. The reason of the difference is, that in 1783, there was very little demand for black cattle, or for labour of any kind. In 1800, cattle sold at an advanced price, and with a steady demand; and there was in general abundance of employment for labourers of all descriptions, either at public works, such as turnpike roads, or in the service of the manufacturers or farmers. It was not with a captious view, to find fault with the question in the plan of the Corrected Reports, " *At what price of wheat* is the district " *over-peopled*, or *under-peopled*," but to shew, that the population of a county does not depend on *the money price of wheat*, or of any kind of corn, but on the facility of obtaining cottages, or places of residence, on the demand for labour, and *the real price of that labour* ; that is, how much food for man, whether from corn, vegetables, or butcher-meat, a labourer can earn by his industry. It is hoped, that this correction of the question, and the answer here given to it, may be useful to the country.

In regard to the *healthiness of the county of Aberdeen*, in general it is very healthy. But there no doubt are various gradations of salubrity of climate, between the air in some of the narrow lanes and low-lying streets, in some parts of the city of Aberdeen, and that of the spacious and elevated new streets ;—

streets;—and also in the country, between the atmosphere in the rooms of the Printfield Manufactory, at Woodside, (which is filled with labourers, and kept as pure as is possible, by ventilators,) and that of the pure and elastic atmosphere of Braemar, where the current of air sweeps along the mountain, and down the valley, and whither the consumptive patient flies in quest of health in the summer months.— On the whole, though our insular situation gives us more frequent changes of weather than are found on the continent of Europe, and though a great part of the county of Aberdeen, from extending as a large promontory into the German Ocean, is exposed to all the vicissitudes of climate, our people in general are robust and healthy, and many live to a very advanced age.

The food, and manner of living of the inhabitants has already been mentioned in general terms. But there is a great diversity in this respect between the country people, who live principally on meal and milk, (their porridge at breakfast, and their brose, or sowens at dinner or supper, being all different preparations of oatmeal) and the manufacturers in the towns, whose morning repast is imported partly from the East, and partly from the West Indies, whose dinner consists chiefly of butcher meat, and a few vegetables, with a small proportion of bread, and whose food in general unites a greater value of fermented or spirituous liquors to a less quantity of solid nourishment. Yet man is an accommodating animal, in respect to the mode of supporting his existence. And all these different persons may live comfortably, and enjoy good health, if they abstain from the immoderate use drink, and especially of ardent spirits.

CHAP.

CHAPTER XVII.

OBSTACLES TO IMPROVEMENT.

SECT. I.—RELATIVE TO CAPITAL.

It must be acknowledged, that the want of capital is a great obstacle to the improvement of a county. Only 360,000 Scotch acres of the surface of Aberdeenshire are as yet arable, and not 60,000 of these are brought to bear nearly the maximum produce of which they are capable. It would probably require three millions of pounds, or a sum equal to the whole agricultural capital belonging to the farmers, to improve the whole 360,000 acres of *arable* land, to the degree of which they are susceptible. It would require at least seven millions sterling, a sum nearly equal to the present value of all the lands of the county, or the whole capital which belongs to the landed proprietors, considered in that character, to improve the other 360,000 Scotch acres, partly by the plough, and partly by the spade and mattock, a species of improvement in which Aberdeenshire can give instruction to all the counties of Great Britain. Thus it would require the whole capital possessed both by the landholders and farmers, to establish the dominion of the plough over all that portion of the surface of the county, which can ever defray the expence of improving and cultivating, to such a degree as is both prudent and practicable. What then can be done, where the want of capital is so great? The Writer of this Report does not shrink from answering this question, but he hopes to be allowed to do so in his own manner. Where the spade and mattock must be used to make way for the plough, he hopes to be indulged in taking an allusion from the sciences.— When a mathematician cannot extract a surd root, or solve a complicated equation, by direct steps, he has recourse to substitution, and is contented with approximation. In like manner let our farmers begin by regular approximations to improve the barren lands, adjacent to the arable, by laying that manure upon the newly trenched or ploughed lands, (that have been lately added) which they would otherwise have laid on the old arable grounds; and let these last remain a year or two longer in grass; also when they are torn up, let them be gently cropped, so as to allow the greatest part of the manure on the farm to be applied to the newly improved fields. Let our merchants and manufacturers, acting as auxiliaries to the farmers, turn a part of their capital to a new channel, by investing it in the purchase of what lands the proprietors are willing to sell, and in the improvement of what they have either purchased, or taken in lease. But whenever they have improved these lands, let them be rented by farmers; for a merchant is seldom qualified to be a raiser of corn, though he can let his inclosures of grass to advantage by public auction. By this means, much may be done, in the way of approximation and substitution, to the improvement of our inarable land. And if the Legislature would, instead of giving subsidies to foreign princes, only allow the half of these *numeral co-efficients,* as premiums to the improvers of our barren acres, we would *extract roots from the earth,* where this at present is *impossible;* but which by approximation and substitution, could certainly be effected.— And instead of *maintaining the balance of power,* we would add at once to the *strength and wealth of the empire*, which would, (by the union of *agriculture, manufactures, commerce,* and *arms,*) bid defiance to the unprincipled ambition of Bonaparte, and remain unaffected by the convulsions and revolutions which have distracted the Continent of Europe.

SECT. II.—OBSTACLES TO IMPROVEMENT, RELATIVE TO PRICES.

The great number of public works in this county, within the last fifteen years, has greatly raised the price of labour; and no doubt there are local and temporary obstacles to improvement, arising from prices, when the money price of corn, or live-stock is low, and that of labour is high. But these things will in a little time find their own level. In some cases, the wise enactments of an enlightened legislature are highly beneficial. But in general, Parliament should not interfere without a very strong reason—" nisi dig-
" nus vindice nodus—inciderit."

SECT. III.—OBSTACLES TO IMPROVEMENT, RELATIVE TO EXPENCES,

Present Questions of difficulty that cannot always be solved by the farmer. By the assistance of the proprietor in advancing a certain proportion of the expence,—or the whole, on payment of interest, and by a long lease; with a rise of rent, at certain periods, (for a lease of 19 years is by far

too

too little when the spade and mattock are applied,) barren lands may be trenched, and wet lands under-drained, at certain seasons of the year; especially when wages are low, or when the earth is more easily penetrable than at other periods.

SECT. IV.—OBSTACLES TO IMPROVEMENT, FROM WANT OF POWER TO INCLOSE,

SELDOM occur in this county. Proprietors, whose estates are contiguous, rarely object to inclosing; and by the Law of Scotland, any one party can compel the other to enclose. There are sometimes cases in which there may be misunderstandings about the division of a moor, or of a hill; but in this cate it is the extent of the property, or the proportion belonging to each, that renders these disputes obstacles to improvement.

SECTS. V. AND VI.—OBSTACLES FROM TITHES AND POOR'S RATES,

UNKNOWN in this county.

SECT. VII.—OBSTACLES FROM THE WANT OF DISSEMINATED KNOWLEDGE.

THE Writer of this Report feels a pleasure in stating, that the want of disseminated knowledge is less felt in Scotland than in England. The establishment of parochial schools has
at

at a very small expence, (formerly not L.8000, and still not exceeding L.20,000 annually) tended very much to disseminate knowledge among all the peasants and lower classes in Scotland. Of the two kinds of knowledge, viz. agricultural libraries, and cheap publications, we do not feel any want. For,

1. As to agricultural libraries, such of our farmers as can afford to get a regular education, attend the classes of Natural History, Mathematics, and Chemistry. And with those who are contented with practical knowledge, the *Farmer's Magazine* is in very general circulation.

2. As to cheap publications. These are not always the most proper. One good book is worth a waggon load of trash. But one publication, which might, by the Board of Agriculture, be rendered both a cheap, and an useful one, is much wanted, not only in this country, but in every county in Great Britain. This is a book, neither voluminous, nor couched in learned phrases, which would point out concisely all the bad practices in the different districts, which ought to be abandoned and avoided, and all the good practices in husbandry, which deserve to be universally known, and generally imitated.

SECT. VIII.—ENEMIES.

The enemies of the farmer, as enumerated in the Corrected Plan of the Board, are,

1. The red, or wire-worm.—Ground that has been *newly broken up* out of grass, is frequently injured by these insects, which are often very destructive.

2. The slug, or short whitish snail, which is particularly destructive to the young plants of turnip, is very effectually destroyed by night-rolling. About one o'clock in the morn-

ing the slugs begin to feed on the tender plants; and by rolling at that hour, they are destroyed in myriads:

3. For killing rats various means have been devised. Both traps and poison are used by the inhabitants, and a professed rat-catcher annually travels through the county, who receives a stipulated yearly payment for killing rats. He is very successful while on this tour, and leaves a quantity of poison in every family where he is employed. It is remarkable, that the old Scotch rat was more easily poisoned than the large rats imported from the Baltic. Therefore in *the Select Transactions* * *of the oldest Agricultural Society in Scotland*, a receipt is given for destroying the Baltic and other rats, in the following words:—" Since the common poison given to other " rats will not destroy the Baltic rats, take the following in" gredients, mix them together, make them up into pills, " and lay them in their runs:—viz. *one ounce of oil of anise* " *seeds, half a pound of arsenic, two ounces of nux vomica, grat-* " *ed*, and *one pound of hogslard*. This poison the Baltic rats, " and others, will be sure to eat, and it will as sure kill " them."

For killing mice, a dose of *ipecacuanha* in a little meal or malt, and various kinds of traps, are used.

4. Sparrows, and other vermin, are troublesome, but they are also useful. It is well known, that *rooks*, commonly, but improperly, termed *crows*, are very beneficial in destroying worms. But it is not so generally known that the small insect, the ant, destroys the caterpillar. The late Mr. FAR-QUHARSON

---

* This society was called the Society of Improvers in the Knowledge of Agriculture in Scotland. It was instituted July 13th, 1723, and published its Transactions occasionally till January 29, 1743, when its Select Transactions were collected into a volume, a copy of which is in my custody. It consisted of 42 Peers, and 260 Commoners—of whom 4 Peers, and 25 Commoners belonged to Aberdeenshire.—G. S. K.

QUHARSON of Invercauld, one year had excellent gooseberries, when they generally failed in the other gardens in his neighbourhood. The gardener said, that this was owing to his digging up some ant-hills, and strewing the contents near the roots of his gooseberry bushes. The ants destroyed the young caterpillars, and preserved the leaves and the fruit.

5. Certain *vexatious circumstances* deserve to be mentioned as hostile to the farmer. By a gross blunder in the person who drew up the Act for the Local Militia, the time for purging the militia lists, or examining into the causes of exemption, is in the month of September, or in the middle of the harvest in this county. And the Deputy Lieutenants have been obliged to fine persons, who could not leave their harvest, and who had been drawn and balloted, or had relevant excuses. Also by not fixing the distance at which a local militiaman may be trained, some highlanders, from the highest lands in Braemar, have been obliged to travel to Ellon, a distance of nearly 60 miles. It is proper that these things should be known; as the Writer of this Report believes they need only be known, that they may be corrected.

Another vexatious circumstance merits attention. This is a tax upon horses, by persons who possess lands above L.10 of rent. This is not imposed upon the smallest farmer in England who does not pay L.20 of rent; and as the English farmer pays tithe, land-tax, and poor's rates, a farm of L.20 rent in England, is generally equal to one of L.35 in Scotland. When a farmer in England pays the property tax on one-half of his rent, the Scotch farmer, whose rent to the proprietor is much higher, ought seldom to pay above two-sevenths, in order to be exactly in the same situation with the English farmer. No favour is here asked—justice only is required by the Scotch farmers, especially by such of the small ones, as, in many places, are poorer than the day-labourers.

CHAPTER XVIII.

MISCELLANEOUS ARTICLES.

1.—AGRICULTURAL SOCIETIES ESTABLISHED.

Of these we have several in the different divisions, and even inferior districts. In the division of Marr, there is one in Aberdeen, one in Alford, and one in Kincardine. These meet frequently for the purpose of improving in agricultural knowledge, and for deciding ploughing matches; and in one of the inferior districts of Alford, there is a Friendly Society. In the divisions of Buchan and the Garioch, and in various other places, there are Societies for all the above purposes; and their effects are found to be very beneficial both in the improved cultivation of our lands, and for the support of the infirm, the widow, and the fatherless. But it is unnecessary to be more particular in this article.

2.—OTHERS WANTED.—*We stand in need* of some societies for offering premiums to the women, who will undertake to hoe turnips, and perform other neat and light farm work. A few guineas laid out on new gowns, or other premiums, would, in a few years, be worth as many hundred pounds in the hoeing of turnips; and by enabling the farmer to have a great proportion of this valuable green crop, would add as many thousand pounds annually to the total produce of the county. Enough is said to the landed proprietors

prietors and farmers of this respectable county, to induce them to excite the emulation and activity of the young women, (whose wages are by far too little) to handle the hoe, where there is little employment for the distaff, the spinning wheel, or the knitting of stockings.

3.—MILL MULTURES.—The thirlage to mills is, in many places, still found to be a grievance, notwithstanding the Act for converting this servitude into an annual payment of rent to the owner of the mill. The reason why this Act has been less beneficial, is, that a landed proprietor, whose estate is bound, or astricted, to the corn mill of a neighbouring landholder, must pay a rent not only for the mill multures, strictly so called, but also for the knaveship or services. For the first of these he ought to have made a full indemnification; but he should pay nothing as a recompence for the latter, where he does not require any services from the miller, but sells his corn unground to the corn-merchant, or grinds it on another mill.

CONCLUSION.

Means of Improvement, or the Measures calculated for that Purpose.

Though these have been incidentally mentioned in all parts of the Report, yet it may be useful to bring into one view, what should be attended to by the farmers, by the proprietors, and even by the Legislature.

To the farmers, the Writer of this Report would say, as a friend to the interests of agriculture, " Begin with attention to your dunghill, remembering that *muck is the mother of meal.* Let no part of the dung or urine of your cattle be lost. Collect all the peat-moss that you can; all the scrapings of ditches, or foundations of earthen fences, on your farm, and lay them in the bottom of your dunghill. Let a covering of horses dung, of earth, or peat-moss, and the dung of black cattle alternately, promote the fermentation of your dunghill and also prevent either the scorching heat of summer, or the chilling frosts of winter, from exhausting your dung. If you have any soil already so deep, that it requires no aid of compost, you may lay your manure unmixed on such lands. Plough your ridges straight; and set them off narrow where the ground is wet, and broad where it is dry, and when you are laying down grass seeds. Never take above two white crops in succession; and, as far as is possible, take but one crop of corn before your sow turnips, and let the next be laid down either with bear or oats. Sow early, and on a dry bed, that you may avoid a late and dangerous harvest. Let your corns be always ready for the sickle, but not too ripe before you cut them down; and be very cautious in taking them too early into the stack-yard. Drill all your turnips, and horse-hoe them carefully; but let not your drills be above 27 inches distant from each other, if you have a sufficient quantity of dung. You cannot raise a heavy crop if your drills be too wide; nor can you promise that it will be

clean,

clean, if you sow them in broad-cast. Whenever a few ridges are cleared of turnips, plough immediately for the sake of burying the rotted leaves or slimy matter; and plough either across or diagonally, that the dung in the drills may be equally mixed with the soil. Sow not the land that you lay out with grass seeds, with too great a quantity either of corn or of rye-grass: for too thick sowing checks the tender plants of clover. And let not the dread of a dull market for your bear or bigg, induce you to lay down above one-half of your grass seeds with oats. Remember that in a late harvest bear is your steady crop: and that in all seasons it is the best nurse for the young plants of clover. Always cut your grass intended for hay, before the juices are dried up in the stalk; and do not rashly turn, in moist weather, the new-mown hay in the swath, as it can resist the rains better in that position, than after it is turned over. Let it not be too much dried before it is put into the *cole;* and in a dubious season, remember the old proverb, *to make hay while the sun shines.*— In regard to the live-stock, be attentive to every thing that regards your breed of cattle; for if you lose the character of the breed, you must also lose that preference in the market, which gives you a high price and a ready sale. Feed from the calf to the ox; but attempt not to fatten when nature loses its relish for food. Be not dependent on other counties for your farm-horses; but raise as many as you can on your own farms, and from the best breeds which you can purchase. Endeavour, when you can, to get a few sheep of the South Down, or any valuable kind both for wool and mutton; and also try, by proper crosses, and good treatment, to raise the native breed, uniting the hardiness of this to the better qualities of the other. Never over-charge your farm with live-stock, but give a liberal allowance to every animal, that they may give you an ample return when carried to the market. Beware of scourging your land too far, to be avenged of your landlord, if you are leaving your farm, lest you hurt yourself in making this experiment; but keep it as much as you can in a progressive state of improvement. Be kind to all your servants, and let not the cottager's cow want grass, or his children milk: for he will repay it in attention both to your farm and to your live-stock. Let there be abundance of food to every creature, but no waste, except in making straw for the dunghill, when forage is abundant.— Lastly, do every thing in its season: and observe order in all your farming operations, never seeming to be negligent, but avoiding that appearance of bustle, which indicates that a man has more to do than he can well manage.—By industry and

and attention, endeavour to enlarge your quantity of arable land: and establish the dominion of the plough, by a commendable ambition, which destroys weeds, and raises food both for man and beast."

To the Landholders of this respectable county, the Writer of this Report would say, with the greatest deference—"The laws of this free country have given you the property of estates, some of which you hold in your own hands, and the rest you have let to farmers. In regard to what is in your natural possession, give the best example of good farming to your tenants. Procure always the best kinds of seed corn, of live-stock, and of farming implements; and let your dependents have a share of your best kinds of seed, at a moderate price, the use of your male animals kept for rearing livestock, and occasionally of any valuable implement of husbandry, that they may be induced to get another, on a less expence, made for themselves. In general, be contented with raising grass and green crops, and a small proportion of corn; remembering that a *tenant who rises early, and is his own bailiff, or farm-overseer, is best qualified to be a corn-farmer.* As to the lands which you let to such a person, remember that though the law gives you a *dilectus personæ*, or choice of the man who is to be your tenant, the preference should be given to the best farmer, and not merely to him who will promise to pay you the highest rent. It was an excellent observation of a most intelligent landed proprietor, who was perhaps also the best practical farmer in the north of Scotland—" That a good tenant was as well worth * L.500 to the " proprietor, as a good horse was worth L.25." Never admit a new farmer whose character is dubious; never reject an old tenant whose character for industry and integrity is clear, while he is willing to pay you a reasonable rent. Assist an active man, who is enterprising, with a little capital, in making great improvements. In a calamitous season, never exact too high a price for the rent payable in corn, which the farmers

---

* This speech was made to me by the late Mr. BARCLAY of Ury, in 1780. He was offered L.71 yearly for a small farm; and I heard him, with some surprise, offer it to another for L.60, and also to advance him, without interest, L.200 for building houses. He observed my astonishment, and said, " You think that I was either foolish or generous in my offer. I never was more selfish in my life. That man to whom I made this offer, is worth L.500 to get him on my estate. For a good tenant is as well worth L.500 to a proprietor, as a good horse is at present worth L.25."—G. S. K.

mers could not pay; as this *is adding to their loss*, and must diminish the *value of your land*. For they cannot *improve to advantage* when the *elements oppose*, and their landlord *fleeces them*. On the other hand, remember that a reasonably high rent, where the tenant does not fall behind, whets his industry, and tends to improve your property. Let not the restrictions in your leases be complicated; but let them be strictly enforced, except where a satisfactory reason is shewn for having occasionally departed from them. Give your farmers comfortable and durable houses; and let them not waste that capital on buildings which should be devoted to the cultivation of your lands. Never permit long arrears of rent, except where one crop has suffered by a calamitous season, and a second has been defective from the want of good seed; but never push a tenant for prompt payment at the precise day when his rent is due, if a sudden fall in the price of corn, or rather of cattle, shall render it difficult for him to pay his rent immediately. By delaying two or three months you can lose only *one* per cent., and he may save ten times as much. This delay will give him new zeal to improve your land.— Where an industrious farmer, by great exertions, breaks up a considerable extent of barren land, do you defray the expence of drains upon a moderate per centage; and where he lays on dung, let him have *lime* for two or three years without any interest, till the land is able to defray the expence incurred. Remember that you are lending this money to your own land, and that at the end of his lease it will repay you for that small advance of capital. Let your old tenants have the first offer of their leases, if they have behaved well; and let them always be taught to *respect themselves*. Then, on the best grounds they will *respect you*. Finally endeavour that both your tenants and your estate be in a progressive state of improvement.

To these admonitions to farmers and landholders, the Writer of this Report would, in conclusion, add, a humble address to the Legislators of the British Empire.

"Senators and Representatives of the people!—A man who has the warmest zeal for the prosperity of his country requests your indulgent attention. From the top of Ben Mac Douie, the second mountain in Great Britain, elevated above the contentions of party, standing nearly in the centre of Scotland, and alternately beholding the cultivated districts of the east, which are washed by the German Sea, and the mountainous regions on the west, where Ben Nevis, still higher than Ben Mac Douie, (a sublime image of British commerce,)

merce), lifts up its head, and where the prospect is bounded by the Atlantic Ocean—will you be pleased to take a comprehensive view of this province of the empire; of the means of its improvement, and the measures best calculated for this purpose. You cannot here expect to see the velvet lawn, or the Cotswold downs of England, or the elegant diction and attractive eloquence of St. Stephen's Chapel; but from the high elevation of this alpine district, you may obtain a comprehensive view of the principles of legislation; and among the rocks of granite ye may, with a little attention, discover a few of the topazes, or beryls, that are found in these mountains, or a few striking sentiments, expressed, though not in elegant, yet in perspicuous language.

The chief resources of every great nation must be drawn from the cultivation of the soil. But Great Britain, which formerly exported great quantities of corn, does not now supply itself with provisions. The county of Aberdeen, if properly cultivated, would maintain eight times its present population; or would support a million of souls, by the produce of a million of acres.—By offering high premiums, equal to one-third of the expence of trenching by the spade and mattock, (a mode of improvement for which this county is so much distinguished) a great proportion of our barren lands might be brought into cultivation; and instead of sending persons convicted of felony to Botany Bay, at a great expence, they might be doomed to the useful employment of improving the soil of their native country.—By granting a freehold qualification to every person holding of the crown, who possessed 104 Scotch acres of arable land, (the antient extent of a freehold, without any regard to its valued rent) many thousand acres of barren land would be rendered arable, without any other premium from Government, than this political advantage.—By offering various premiums for raising the *greatest quantity* of produce on our arable lands, a spirit of emulation may be excited, and a far greater value of food may be raised.—By giving to every proprietor of an entailed estate, the same powers in regard to granting leases, or paying for the expence of draining and enclosing, of building houses, and planting trees, as if he had held the property in fee simple, the improvement of the county would be carried on successfully on many estates, which are at present neglected.— By encouraging the raising of flax and hemp in our own fields, we may be rendered less dependent on other nations; but your first care should be to raise a sufficient quantity of human food, whether of corn or butcher meat.—In this case, be pleased to consider that improper corn laws, (enacted to appease the clamours of merchants, manufacturers, and other in-

ha-

habitants of the cities, towns, or villages,) would be very injurious to agriculture, if they prevented the money price of corn from keeping pace with that standard; and that the importation *of labour* of such rude produce, as we both can, and should raise in this country, would also be hurtful to the farmer. It should ever be remembered by you, Legislators, that by enacting wise and salutary laws, the wealth of the nation is found to consist in the *sum total* of our manufactures, commerce, and agriculture; but that the establishing of improper regulations, calculated to promote the interests, or appease the discontents of one class of the community, is a *diminution* of the *total amount*, in *proportion* to the errors of the law.

From our extensive colonies and commerce, the annual importation of sugar is now equal, in point of the intrinsic value of saccharine matter, to half the quantity of barley that is now raised in Great Britain and Ireland; and the last year's importation of cotton, viz. 560,000 bags, nearly *seventy thousand tons*, was probably more than half the weight of the whole wool that was raised within the United Kingdom.—The former, in years of scarcity, can supply you with both fermented liquor, and ardent spirits, leaving the barley crop entire for human food. The latter, when the Continent of Europe is open, affords you raw materials for the most valuable manufactures. While you give a decided preference to the rum, sugars, and cotton of our colonies, over the brandy, geneva, and flax of the Continent, let the British distillery never be supplied by the sugars of the West Indies, when you have barley at home in sufficient quantity; nor let even the cotton of the Bahamas, or of the East Indies, interfere with the wool of the United Kingdom.

Government must be supported, and can be maintained only by its revenues, arising either from direct or indirect taxes. The imposing of taxes is indeed a choice of difficulties; and the wisdom of a Statesman consists in his making the most judicious selection. Let never the existence, the real strength, or the improved agriculture of the kingdom, be endangered by an improper tax. Remember, that the *malt tax* is a premium to the grazier, and that it has, in this country, diminished the quantity of bear or bigg. The proportion of tax on our malt, compared to that of barley, is more injurious to our agriculture, than ten times the sum would be, if levied directly from the farmers or proprietors.—We live under the same government, and in the same island with our brethren in England, and in the south of Scotland. Nothing can save us in a calamitous season, except our crop of bigg; and no crop, in any season, is so proper as a nurse to grass seeds. The improvement of our agriculture has been much checked
by

by the unfair proportion of tax on that inferior species of barley, which has fallen, at least, 15 per cent. in its *relative* value, to both wheat and oats, since 1802. From your wisdom, and from your justice, an alteration of this proportion is expected.

Whatever taxes, from the exigencies of the state, you may impose on the nation at large, it is hoped that you will instantly repeal those, which certainly without any intention on your part, have been imposed on our poor's funds, friendly societies, and benevolent and charitable institutions; and that in the first property-tax act, you will appoint all sums, which have been levied from those classes of men, to be instantly repaid. This you owe to the generous character of the nation.

You have great honour in enacting laws for the improvement of the highways, and for the building of bridges in the highlands. But the forests of Marr ought to be rendered more accessible, as they would afford great supplies to the British navy. When the earldom of Marr was forfeited, the government gave away the *lands and forests,* retaining only the *patronage of the churches.* They ought to have retained the *forests,* which have produced to the proprietors more than thrice the sum which they paid for the lands. All that now can be done, is to get a rail-way, or some easy mode of carrying wood from Braemar to Aberdeen.

In the opposite corner of the county, a good harbour, and a Custom House at Peterhead, might counteract the schemes of a foreign enemy, as well as promote the interests of commerce.

Lastly, It would tend much to the advantage of this county, and of the whole kingdom, that the equalization of our weights and measures were committed to the Royal Society of London, whose venerable President is so zealous to promote the interests of science. Their report should then be ratified by Parliament."

These outlines of public laws or regulations are respectfully submitted to the Legislators of the United kingdom; and if any expression is deficient in elegance or urbanity, the Writer appeals from their ears to their understandings, from their taste to their generosity. He appeals to the liberality and magnanimity of the British Legislature, to which no man ever appealed in vain.

APPENDIX.

CORRECTIONS, ADDITIONS, AND EXPLANATIONS.

As the above contains the only Agricultural Survey ever made of the County of Aberdeen, (for Dr. ANDERSON's Original Report was written in Edinburgh) it was impossible to avoid committing errors, or to omit stating articles that deserved to be mentioned. To render these faults as few as possible, several copies of the printed sheets were handed through the different districts, as soon as they came from the press; and such defects as could be supplied in the subsequent parts of the work, were made up in the most proper places.—What could not be introduced in that manner, are now added, from information sent me by a number of Gentlemen, whose names I am not at liberty to mention. The Report itself was drawn up in the *third person*, to prevent the appearance of *egotism*; but the corrections are marked in the *first person*, because it is more candid and more manly to say, *I committed such an error*, than to say, *in arranging a number of articles, such an error has crept into the Report*. It is a fasle pride that prevents a man from frankly acknowledging where he is in the wrong; and it is because it is well known in the county of Aberdeen, that I have not spared either expence, time, or personal labour, and exertion;—that I have drawn up this paper, in order to render this work as correct as possible. For perfection, though it ought to be aimed at, can never be attained.—G. S. K.

Page 27. In the Meteorological Tables, I divided the four quarters of the year, according to the *legal division* that is established in *Scotland*. But if I had begun the spring quarter with the 1st of March, the summer quarter with the 1st of June, the autumn quarter with the 1st of September, and the winter quarter with the 1st of December, the seasons would have been more accurately discriminated; and both the heat of summer, and the cold of winter, would have been a little greater. What led me to adopt the Scotch division of the seasons, was, that the table of the heat of the thermometer, at Gilcomston, was kept in that way, and only for one year.

A Gentleman, to whom I am particularly indebted for many judicious observations on the climate of Aberdeenshire, remarks, that it is not in the morning and evening, but at noon, and after it, that, on the sea-coast particularly, there is the greatest deficiency of heat. As a proof of this, he mentions, that of 74 days, of summer, 1809 and 1810, during which

which he observed the heights of the thermometer, between 9 and 11 in the morning—and at noon, and from thence to 3 o'clock, there have been only 11, in which the latter period exceeds that of the former. The cause which he assigns appears to be a very just one, that a cool sea-breeze generally sets in, either at noon, or a little before mid-day.

P. 52. The same gentleman adds, that though the north side of the hills *is generally* the best, it is not uniformly so; for he has known a few exceptions. Another well informed friend assigns two reasons why the north side is generally the most fertile, viz. that the snow melts more quickly on the south, and runs off more rapidly; and also that a greater quantity of snow commonly lies on the south side of the mountains than on the north, being blown over by the north west winds that prevail in the winter.

P. 61. To the list of Minerals ought to be added a mine of black lead, in the neighbourhood of Huntly, upon the bank of the Deveron, between its junction with the Boggie and the old castle. It is not fit for pencils, but good enough for making crucibles; and several casks of it have been sent to London by commission.

It should also have been mentioned that manganese is found on both sides of the Don, and northwards as far as Uday.

P. 66. It has been already mentioned by me, that I did not mean to say what the actual rents of the salmon fishings amounted to. But I have not expressed myself accurately in p. 68. I believe that the *produce* of the two rivers may be nearly £.35,000; but the *rent*, (or *free profit*, where in the possession of the proprietors) is not above half that sum; and it is so variable, that I wish, as I mentioned in another place, to say nothing on the subject of rent.

P. 75. Hector Boethius' account of the Bishop of Aberdeen's resolution for fixing his residence at what is thence called the Bishop's Loch is "amœnitate loci, et arboribus captus"—a description that by no means corresponds with the situation of it for the last two centuries. Indeed a few trees have been lately planted around it. The *Corby Loch*, in its neighbourhood, is a much larger piece of water, but was neglected to be mentioned.

P. 86. It is believed that more than one-fourth of the landed property has been purchased by new proprietors since the union of the two kingdoms in 1707.

The observations respecting the obstruction to improvements, from entails, might have been more correctly stated. My meaning, which I wish to be fully understood, is that the heir in possession can neither grant a sufficiently long lease for great undertakings in improving barren lands, nor bind the next possessor, (in that capacity) to pay debts incurred by a tenant, and consequently due to him for making

im-

improvements. The Act, called the Lord Chief Baron's Act, goes to only a limited extent, and is clogged with restrictions. Nothing can be more clear to me, than that the possessor of an *entailed estate* ought to have the same powers for granting leases, and for making covenants, in leases, binding on the estate, as if he held it in fee simple. It is quite sufficient to gratify either family pride, or patriarchal affection, that security be given that an entailed estate shall not be alienated.— It is too much to allow, that it should either not be improved at all, or improved at the expence of others, who were deprived of their just claims, by the death of the heir in possession. A lease, though it were for 100 years, if no fine was paid, and if sold by auction, ought to be declared a good one. *Salus populi lex suprema.* But all long leases ought to have rises of rent, as stimulants to improvements.

P. 93. I am conscious that several of the descriptions of the gentlemens' seats are not so appropriate as I could have wished. But the truth is, it is both difficult to find appropriate expressions in matters of taste; and it is sometimes invidious to make those distinctions which a regard to truth demands.

It is proper, however, to mention, that when I wrote out a clean copy of my manuscript to the printer, I had no idea of giving any particular account of the *buildings* of proprietors, but had selected only a few of the most remarkable—that when I afterwards saw, by mere accident, the Plan of the Corrected Reports, which from some unaccountable cause, had not reached me, and saw there, that three subdivisions of that section were required to be filled up, (viz. 1. Those advantageously situated. 2. Those which were well planned for country gentlemen of moderate fortunes; and, 3. Those which were elegantly constructed.) I found it would be necessary to be more particular. And when I was favoured with a copy of Mr. ROBERTSON's Survey of the neighbouring county of Kincardine, and saw that this able and judicious Writer had devoted 17 pages to a description of the houses of proprietors, I thought that I should be accused of a culpable neglect, if I did not give a more particular account of those of Aberdeenshire, than what I had prepared for the press.— Therefore I drew up the article which is now printed in the Report from page 93 to page 129, and which was swelled from 6 to 36 pages. It was drawn up rather hastily, and when I was in bad health. I corrected one of the proof sheets when confined to bed, and I know that the gentlemen of this respectable county will not, after this circumstance is mentioned, be fastidious in criticising this part of the Survey, (which personal respect for them alone, induced me to draw up,) immediately before my first expedition to the mountains, when the press was ready to receive this section, and when my time was not in my own power.

After this acknowledgment, I beg leave to supply omissions and correct the errors in the Report.

1. I omitted *to publish any account* of the Castle of Craig of Auchindoir, belonging to JAMES GORDON, Esq. advocate in Edinburgh. It is the residence of an ancient family, who have possessed it for several centuries; and is distinguished by a fine romantic glen, filled with various kinds of wood.—The present proprietor has spared no expence in ornamenting his seat, or in improving his personal farm. Owing to a mistake in selecting the castles from the other houses, this castle, though mentioned in nearly the same terms in the manuscript, was omitted in printing.

2. The Castle of Braemar, formerly an occasional residence of the Earls of Marr, and afterwards a garrison for keeping the Highlanders in subjection, is now untenanted; and is the property of Mrs FARQUHARSON of Invercauld, who keeps it in repair. It also was omitted.

3. Among the villas near Aberdeen, one of the neatest is Woodhill, the property of GEORGE HOGARTH, Esq. present Dean of Guild of that city. It is ornamented with an excellent garden, well stocked with fruit trees, and has one of the best constructed hot-houses in the vicinity of Aberdeen. At some distance from it is another commodious house, with a garden attached to it, belonging to WILLIAM FORBES, Esq.

A third neat villa, called Glenburnie, the property of Mr CHARLES WALKER, has also been unintentionally omitted.

Some of the landed gentlemen probably account these villas unworthy of being noticed in this Survey. I am of a different opinion, because most of them are built on fields which, 100 years ago, had as many stones lying on their surface, as are now standing in a more regular order, than when they lay as shapeless masses of granite. Where the cultivation of the soil has been carried on at an expence unequalled in Europe, a merchant or manufacturer's villa, even without an elegant garden and hot-house, merited a place in this Survey.

To these omissions may be added the following errors committed:

P. 102. The modern house built by the Earl of ABOYNE is attached to the old castle.

P. 102. Lord FORBES has got a plan for a new house at Pitachy, which his absence from his country prevents from being presently carried into execution.

P. 124. Cairness is probably the best modern house in the county; it was built within these 20 years. PLAYFAIR was the architect. But the wood is not thriving—I saw it a few months ago, and it is declining. Indeed wood does not thrive,

thrive, except in dens, or hollow places, near the coast of Buchan.

Crimonmogate has no wood; but is distinguished by the best stone walls, round its inclosures, that are to be found in the county, on so large a scale.

Druminner, or Castle Forbes, is the residence of ROBERT GRANT, Esq.

Huntly Lodge, though possessed by the late Duchess of GORDON, was built by the Duke, as the residence of his factor, or principal manager, Mr HAMILTON.

Disblair is an estate which lies partly in the parish of Fintray, and partly in that of Newmachar; but the house is in the former parish.

There are *three* excellent *gardens* near Fraserfield; but the *cottage* which attracted my notice so particularly, is that which belonged to THOMAS LEYS, Esq. late Provost of the city of Aberdeen.

P. 134. A gentleman who is well acquainted with both the law and practice as to melioration of farmers houses, justly remarks that several removing tenants have left their houses in a worse state than when they entered them 19 years before; and yet, from the great decrease of the value of money, or the particular increase in the money price of wood, have drawn double the sum from their successors, or the proprietor, which they paid on their entry. This is a necessary consequence of the farmers being the principal owners of their farm buildings. Yet an English landholder, who pays a large sum annually for repairs, would have no objection to such a loss occasionally.

I am happy to find that this intelligent gentleman agrees with me in thinking that the entry to a farm should commence a little before Martinmas.

P. 254. From the same respectable and accurate observer, I quote with pleasure the following article on the cultivation of wheat.

" It would seem that there had not as yet been sufficient
" experience in raising wheat, to enable us to form a judg-
" ment of the expediency of extending its culture in this
" county. The deficiency of our native farmers, in skill, is
" probably a greater bar to its entrance than either soil or
" climate. There is a great deal of soil in the better parts
" of the county, (indeed in all the districts, except the gra-
" velly hills of Deeside), that would probably answer for
" wheat. And though neither our climate, nor perhaps that
" of any part of Scotland, will produce the finest grain of
" this species, I am persuaded, from what I have seen in
" very exposed situations more than 400 feet above the le-
" vel of the sea, in counties northward of the Lothians, that
" wheat will, in common years, ripen tolerably in the ave-

"rage climate of the arable part of the county. Wheat
"bears a great deal of stormy weather. That which was
"raised by the late ROBERT BARCLAY, Esq. of Ury, in 1788,
"stood the severity of that season better than any other crop.

"What is called Spring, or Siberian wheat, was raised in
"this county a good many years ago by a few persons, who
"abandoned the culture of it, from finding it an inferior and
"unmarketable grain. It has again been introduced within
"these few years; but from the same cause, those who have
"raised the greatest quantity seem disposed to give it up.—
"From the small trials I have made, only of an acre or two,
"I think myself entitled to recommend it for sowing grass
"seeds on light land. It stands better than bear; and is not
"much longer on the ground."

"July or August seem to be too early for sowing wheat.—
"The autumn months are those in which our climate will
"best stand a comparison with the climate of our northern
"neighbours; and in an ordinary season the growth before
"November would be too great. Perhaps, however, the
"first half of September would be the fittest period."

To these judicious observations on wheat, I would add that, in 1782, not only Mr BARCLAY of Ury's wheat stood the severity of the season better than the oats or two-rowed barley, but that several of the better sort of farmers, who sowed wheat in Kincardineshire in 1782, got it ground on corn mills, and used it for porridge to their families, as the most œconomical food, while they made use of oat meal and bear meal for baking into bread. But I must add, that the bear or bigg of 1782 was not only superior to the wheat, but of uncommonly fine quality, while the oats were generally bad; and that Mr BARCLAY's barley was of so extremely coarse a quality, that even in that calamitous season, it could with difficulty be used in a mixture with other meal.—A very decisive argument in favour of attempting to extend the cultivation of wheat, and of preferring the raising of bear or bigg to that of barley, in most parts of Aberdeenshire.

I am happy here to state that, in 1810, the quantity of wheat sown within the year has increased very much. By particular enquiries which I made, I find that it is nearly double of what it was only three years ago—probably 700 acres at least. In the small parish of Kinellar 56 Scotch, or above 71 English acres, were sown with wheat; and 36 Scotch, or 45 English acres, were sown with the same grain in the parish of Turriff. I saw a field of 5 or 6 acres of excellent wheat at Abergeldy, 46 miles from Aberdeen. But I find that the two-rowed barley is so precarious a crop, that last year there was not above 200 acres of it raised in the county; and except on the sea-coast, it will be generally given up.

In a most judicious Letter from Baron HEPBURN, to JAMES FERGUSON, Esq. of Pitfour, (which has been printed by him), sowing wheat deep, ploughing it in, and leaving the clods rough during winter, are strongly recommended; and if attended to, might prevent a great deal of the wheat from being thrown out in the spring.

P. 303. "*Tares,*" another well-informed gentleman observes, "have succeeded in some places where the climate "should not be superior to the best parts of this county."— But he adds,

P. 304. "That buck-wheat is certainly not worth cultivat"ing. In bad soils, oats will thrive much better; and, in "bad seasons, the former does not at all ripen. Some sown "at Scotstown, near Aberdeen, in 1799, continued to blos"som till winter; but produced scarcely any grain.

P. 310. "Carrots, however, have been used successfully "by him on peat-moss. The carrots were excellent; tho' "the crop was not so heavy as that reported in the printed "statement by Sir JOHN SINCLAIR. Also that,

P. 321. "Scorzonera is now not uncommon at our tables;" and that,

P. 331. "About 20 years ago, lucern was raised in small "quantities, by several people. It answered well in dry "soils and seasons, affording sometimes five cuttings in a "year; and was peculiarly valuable in a dry spring, where "all other grasses were stinted; but the expence of the cul"ture (constant hoeing) was so great, that it was, after a "few years trial, abandoned."

P. 346. He suggests an additional reason for early cutting of hay, in this county, that the chance of dry weather is greater. In most years, in the end of June, there is drought; in July, rains.

P. 365. He remarks, that the plants which thrive best in all land, whether rich or poor, are those that have been raised in rich (not newly dunged) soil. A gentleman in the Highlands, who has planted a great deal in hard wood and larches, says, he found the plants brought from the hot-beds between Leith and Edinburgh answer, upon the whole, best. Our nurserymen, in general, suffocate their plants, by setting them too close. Every gentleman who plants on a large scale, should have his own nursery, for which the soil should be light and rich. He adds,

P. 273. Beech, it may be suspected, is not a natural wood in Aberdeenshire. It is, according to Dr WALKER, an exotic; and few were planted in Scotland till after the revolution.

P. 275. Though what can be strictly called sea-spray cannot extend far from its margin, the effect of the sea air is observable

servable for several miles from the coast, and is extremely prejudicial to every root or tree in situations exposed to it.

P. 376. In general it will be found that, in the mosses near the coast, the trees are oak, birch, allar, and hazel, and no fir. In those which are 10 miles, or more, inland, fir prevails.

P. 377. The same well informed gentleman says, " After " 20 years experience in planting trees exposed to the sea " air, and in the course of that period frequently changing my " opinion as to the root of the fittest to be planted, I think " that persons on the sea coast, who have not the advantage " of glens, and other sheltered situations, and cannot spare " ground for large plantations, but can afford tolerable soils " for those they plant, should have recourse to the ash. This " useful, though not very ornamental tree, if it can be fed " with a tolerable depth and richness of soil, stands the sea " blast better than the plane or sycamore, which hitherto " has been so much extolled for that purpose. Our an- " cestors seem accordingly to have been partial to this tree, " it being the one most commonly found about farm-yards, " in avenues, &c. although, unless on this account, it would " be one of the worst, as it is a great robber of the soil, and " its leaves are thought prejudicial to grass. The laburnum " is a very low growing tree in any situation, and I cannot " say I have observed it to prosper in the sea breeze. In this " county, and in most parts of Scotland, unless in large " mountain plantations, the tree most planted of late, per- " haps in the proportion of 1000 to *one* of any other, is the " larch; and in general, when planted not singly, but in " clumps or belts, and in soil not very wet and stiff, its pro- " gress, unless in the maritime parts, has been prodigious for " the first 10 or 12 years. It is however, the opinion of a " very judicious and experienced planter (Lord ABOYNE), that " the larch is not a fit tree for poor hilly soil, where Scotch " firs thrive well; his Lordship having observed, that in such " situations, though it grows rapidly for the first dozen years, " it then begins to decay. The disease affecting the larch " has, within these ten years (about which time it first ap- " peared) been very prevalent, particularly in the eastern " district of the county, and in poor soils and exposed situ- " ations, in which the tree is stunted, and sometimes killed " by it. It is occasioned by an insect, which some who " have examined it, say is the common gnat."

In wet clay soils near the coast, the aller tree has been found to thrive; but this, and its sister tree the beech, are very liable to be blasted near Aberdeen.

Though one-tenth of the surface is, in some shape, covered with wood, yet the natural woods are so thinly stocked, that only 100 square miles of wood are to be considered as belonging to the county.

GENERAL TABLES

Of the POPULATION, &c. of the COUNTY of ABERDEEN.

Parish.	Population in 1755.	1796.	1801.	Ministers Stipends.	School-master's Salaries	Poor's annual Income
Aberdeen, Old & New, 6 minist.	15730	24227	27508	L. 1170	L. 260	
Aberdour	1397	1306	1304	130	20	L. 30
Aboyne	1695	1950	916	130	20	14
Alford	990	663	644	98	19	10
Auchindore	839	572	532	110	20	14
Auchterless	1264	1200	1129	126	20	46
Belhelvie	1471	1348	1428	147	22	30
Birse	1126	1253	1266	110	20	80
Bourty	525	450	445	122	20	15
Cabrach	960	700	684	83	20	8
Cairney	2690	2600	1561	160	20	10
Chapel of Garioch	1351	986	1224	200	23	42
Clatt	559	425	433	80	20	6
Cluny	994	885	821	130	20	24
Coldstone, Logie	1243	1182	861	120	20	10
Coull	751	465	679	103	20	5
Crathie, Braemar	2671	2251	1876	90	22	16
Crimond	765	917	862	127	38	35
Cruden	2549	2028	1934	170	22	28
Culsamond	810	618	730	118	20	37
Cushnie, and Leochell	1286	1001	888	120	20	20
Daviot	975	900	644	140	20	11
Deer, New	2513	2800	2984	176	22	46
Deer, Old	2813	3267	3552	170	22	70
Drumblade	1125	886	821	125	20	20
Drumoak	760	708	648	104	24	24
Dyce	383	352	347	130	19	19
Echt	1277	963	972	126	20	24
Ellon	2523	1830	2022	200	25	72
Fintray	905	920	886	147	20	24
Forbes and Kearn	456	370	413	67	20	4
Forgue	1802	1778	1768	120	20	53
Foveran	1981	1243	1391	180	22	38
Fraserburgh	1682	2200	2215	164	22	52
Fyvie	2528	2194	2391	170	22	41
Gartly	1328	1800	958	100	20	30
Glass	1093	776	793	160	20	16

Parish.	Population in 1755.	1796.	1801.	Ministers Stipends.	School-masters' Salaries	Poor's annual Income.
Glenbucket	430	449	420	L. 33	L. 19	L. 4
Glenmuik, &c.	2270	2117	1901	160	20	31
Huntly	1900	3600	2863	100	24	55
Insch	995	900	798	137	20	13
Inverury	730	712	783	170	20	23
Keig	499	475	379	120	20	10
Keith-hall	1111	858	853	140	22	39
Kemnay	643	611	583	110	20	37
Kildrummy	562	568	430	100	20	15
Kincard. O'Neil	1706	2075	1710	195	22	22
King Edward	1352	1577	1723	133	22	40
Kinellar	398	342	309	92	20	19
Kinethmont	791	850	784	140	20	23
Kintore	973	862	848	184	24	25
Leslie	319	392	367	128	20	12
Logie Buchan	575	538	539	132	22	11
Longside	1979	1792	1825	180	22	36
Lonmay	1674	1650	1607	168	20	33
Lumphanan	682	621	614	120	20	11
Machar, New	1191	1030	925	152	20	37
Meldrum, Old	1603	1490	1584	250	22	55
Methlick	1385	1035	1215	155	20	50
Midmar	979	945	803	160	20	20
Montquhitter	997	1470	1710	120	20	46
Monymusk	1005	1127	900	160	24	49
Newhills	959	1153	1305	165	20	32
Oyne	643	630	518	175	20	7
Peterculter	755	1002	871	145	22	36
Peterhead	2487	4200	4491	185	30	122
Pitsligo	1234	1300	1256	136	20	37
Premnay	448	450	486	200	20	9
Rathen	1527	1730	1558	155	20	40
Rayne	1131	1173	1228	160	20	23
Rhynie and Essie	836	681	676	100	20	14
Skene	1251	1233	1140	130	22	24
Slains	1288	1117	970	169	20	20
Strathdon	1750	1524	1354	200	20	14
Strichen	1158	1400	1520	110	20	47
Tarland	1300	1100	922	150	20	0
Tarves	2346	1690	1756	185	22	43
Tullynessle	335	396	350	160	20	6
Tough	570	560	629	102	20	4
Towie	656	550	528	73	20	10
Turriff	1897	2029	2090	170	28	61
Tyrie	596	864	1044	110	22	44
Udny	1322	1137	1242	192	22	32

APPENDIX.

OBSERVATIONS ON BRITISH GRASSES.

Although the number of native grasses found in this country is exceedingly great, yet the only one in cultivation is the ray grass, or lolium perenne; for clover and rib-grass are not ranked by botanists under that denomination. Ray grass is not only of excellent quality, and on the whole, a hardy species, but it is the only one, whose seed is procurable in large quantity of the seedsmen in this district. It is, on the other hand, short-lived on many soils, is far from productive, and it labours under a peculiar disadvantage, in the prevalence of the annual species, whose seed is often mixed with that of the perennial. It seems verging too, by artificial cultivation, long continued, and injudiciously conducted, into the condition of a biennial, or triennial; a course which red clover has already run[*], and which derives confirmation from the superiority of Mr. Pacey's ray-grass, which is nothing more than a restoration of the species, effected by collecting seed from the plant in its native state.

This deterioration of ray-grass seems to be imputable chiefly to two causes; to its being frequently sown on soils to which it is ill adapted, so that seed is saved from sickly plants, a practice too which has been of long continuance. But it is probably owing still more to that management of the hay-crop, which prevails all over Scotland, and which consists in allowing the seed to be completely formed, and nearly ripe, before the plant is cut down. By this means, ray-grass ripens its seed every year of the hay crop, and as it is then only seed is saved from it, it is never permitted to enjoy a single season to recruit its vigour, which it would do, were it either pastured during the first year, or cut previous to the formation of the seed when mown for hay. Annuals are evidently designed to ripen their seed every year; and though it be true, that perennials, such as the indigenous grasses, do the same in their native state,[†] yet, when to this is joined the stimulating effect of manure, whereby the growth is rendered large, and the seed weak, such a practice cannot be very long persevered in, without producing degeneracy in the species.

By delaying so long the hay season, a larger quantity of clover

---

[*] Mr. Davy has ascertained by chemical analysis, that clover contains an unusual proportion of gypsum, or sulphat of lime. As this crop is mostly carted off from the field, much of the gypsum is not restored to the soil, and in time a deficiency may ensue. This ingenious conjecture is supported by the great effects which had been before observed to attend dressing clover with gypsum.

[†] Even in their native state many grasses, e. g. the bristle leafed fescues, produce abundance of seed when young, but afterwards expend all their vigour in the formation of leaves, so that an old plant has often not a single seed stalk.

clover is indeed obtained as hay; but the total yield, taking in the whole season, is certainly no greater, than if it were cut at an earlier period:—it is probably less. Ray grass, like all the other species of gramina, attains to nearly its full size, before it flowers; after which time, while the seed is maturing, (a period of more than a month's duration) it scarcely adds any thing to its growth. It is, besides, more palatable to cattle, if cut at the flowering season; a fact which has long been familiar to the graziers in the midland counties of England, who employ no other in feeding calves which they wish to bring rapidly forward. If boiled into hay tea, it yields more extract than ripe hay, which is probably owing to a large additional quantity of silica, and other earths being taken up by the plant during the seed process; which gives to the stem greater rigidity, and also strength to support the weight of the head, or panicle. All the grasses contain a large proportion of earth. The hordeum pratense, or perennial field barley, and several of the annual bromes are, when young, as greedily eaten by live-stock as young wheat; but after the first week of June, when they begin to ripen, no animal, as Mr. Don observes, can be induced to touch them.*

According to the present system, hay is almost as much a robbing crop as oats or barley, since it is managed in a way nearly similar; whereas, if cut when in flower, (which in this latitude is about the middle of June), the field would be no more injured, than if its produce had been eaten by the teeth of cattle, instead of being mown by the scythe.

In the light-soiled districts of this county, nothing is more common than to find ray grass in a thin and sickly state, after the second and third years of the ley; owing, in some measure, to the imperfection of the previous fallow process, and to causes already insisted on, but which is in part attributable to the habits of the plant itself. When a field of middling fertility, especially if incumbent on a dry gravel, or wet tilly subsoil, has been laid down to grass, one or two tolerable hay crops are not unfrequently got from it; but after the third year, the clover and ray grass gradually disappear, and large blanks are then to be seen all over the surface. The holcus lanatus, anthoxanthum odoratum, agrostes, white gowan or bellis perennis, the crowfoots or ranunculi, and a long list of hardy plants, occupy the space deserted by the sown grasses; the vacancies begin to fill up, and in the course of the fifth or sixth year, the whole is co-
                                                                vered

* The coarse grasses, e. g. the Timothy and Cocksfoot, lose more in quality, by too long delayed cutting, than those of greater value. The two here mentioned are of very large growth, and they may attain to some consideration, should the proper hay season be attended on

vered with a thick, but coarse herbage, in which it would often be difficult to find a single blade of ray grass in a square ell.

In all such cases, there is a deterioration of the soil, which gradually consolidates, as is evinced by the flattening of the ridges, and by the firm texture of the sod, when turned up by the plough. This solidity is very different from the tenacity of clay, and resembles more that which is observable in flower-pots, whose earth has been kept too long unchanged, whereby it acquires a pale colour, a close, but friable texture, and a diminished bulk. In thin soils, incumbent on gravel, decomposition of vegetable matter is more rapid than its reproduction; and being, when decomposed, soluble in water, it is carried down through the porous subsoil along with the lime, animal manure, and whatever else water can hold in solution. For this reason, land of this description, having a south exposure, is generally more shallow and exhausted than when screened from the sun's heat, by an inclination to the north. With regard to soils having a cold retentive bottom, the case is somewhat different, for there the average heat of the summer months is insufficient to effect complete decomposition: hence, they frequently have the appearance of an imperfect peat. Decomposition of vegetables begins about 45° or 50° of FAHRENHEIT. Rich soils, where the decay of the year's growth is completely effected, and is also balanced by subsequent reproduction, may be kept under grass for an indefinite period of time; but in poor ones, when the above-mentioned processes have reached a certain length, ray grass is no longer able to maintain itself, and must give way to coarser, though hardier competitors. In the thin sterile outfields of this county, even the hardy agrostes, (or bents) are compelled to yield their place to the junci, carices airæ, nardi, hypni, &c.; the field is then provincially termed *reesque* At last heath, or whins, extirpate even these, and close the scene. Should, however, the ground be planted, the heath and whins die out in a few years, and are succeeded by the agrostes, airæ, anthoxanthum, &c. Such is the progression which every one may see in its various stages, by inspecting grass lands of these descriptions; and which seems to prove, that if such soils be worth cultivation at all, they should either be brought more frequently under the plough, and refreshed with manure, or sown down with some of the hardiest grasses, since they are quite unable to maintain themselves under ray-grass.

All attempts to arrest this natural progress of things, by top-dressing, will only end in disappointment and loss.— This is a matter susceptible of easy and distinct proof; but it is impossible to discuss a subject so extensive in this place. It may, however, be remarked, that no surfeit of lime and

earth,

earth, or of dung, has ever been known to kill the agrostes, ranunculi, bellis perennis, or the great variety of other weeds, which frequent old grass lands; and that if such a thing there be, as a clean top-dressed meadow, it derives an importance not only from its profit, but from its singularity.

What has been said respecting the tendency which ray grass has to die out on all soils, (rich loam and clay excepted) leads to a view of the merits of that species, compared with some others; a subject hitherto much neglected, but far too extensive to be compassed in an article already too diffuse.

Where ray grass has been repeatedly sown, and has as often disappeared, it is, on the present system, to be sown anew. Experience shews, that it is very far from being universally suitable, nor is this to be expected of any single species of grain or grass. If, indeed, it is in our power to raise the soil from poverty to richness, by means of lime and manure, this may be persevered in: but if otherwise, it is surely better to relinquish a system which is clearly unprofitable.

In the choice of a new species adapted to poor soils, no rule seems less exceptionable than that of examining the various indigenous grasses which have supplanted the lolium—of these, to select the best, and stock the ground with them in future. Were this method adopted, we should soon perceive, that of all the native species, inhabiting sterile soils, the holcus lanatus, or soft grass, and the anthoxanthum odoratum, or vernal grass, are the most common and most valuable. The former is easily recognizable by the woolly softness of its leaves and stalks; it is not found at very great altitudes, but it inhabits all soils, especially if moist, and appears of all others the best adapted for moss. It is a productive grass, but not well fitted for a hay crop, as when dried, it becomes soft and spungy, but, in a growing state, it is relished by every description of live-stock. It is, by Dr. Anderson, confounded with the holcus mollis, a very troublesome weed.

The vernal grass is easily known by the fragrant smell it imparts to hay. It is found on all soils and situations, from the summit of Ben Nevis, to the richest meadow. It is early, and palatable, but rather deficient in bulk.

The ample and valuable catalogue of fescues, furnishes several species, well adapted to inferior arable land. Such are the ovina, cambrica, duriuscula, and dumetorum, all of the best quality, all of great hardiness, as to soil and climate, as well as of large produce, (the ovina excepted!) The duriuscula would be proper for good arable, were it not that its root though not strictly repent, has a couchy habit, intermediate between a creeping and fibrous nature. The cambrica is, in all

all respects, as valuable, and is fibrous rooted. The festuca rubra would, from its superior size, be preferable to any of them, but it is a creeper.

There is a variety of a common weed, the avena elatior, now holcus avenaceus, or swine's arnut, which is not unworthy of notice. In hardiness, and in great produce, it has very few equals; so that, on some descriptions of poor soil, it may be found to yield the most profitable return; for farmers ought to keep in mind, that a good crop of an inferior grass, is almost always preferable to a bad crop of a better one; as a crop of oats, in many situations, yields more profit than one of wheat. This holcus is fibrous rooted, whereas the common arnut has a bulbose double pear-shaped root; which has induced some botanists to consider them as distinct species. It grows four feet high, and is eaten readily by cattle.

For arable, of 25s. to 30s. an acre, yearly value, no grasses seem more proper than the poas, nemoralis, glauca alba, and biflora, provided the soil be tolerably dry. Of these the two last are the earliest; in other respects the same description is applicable to them all. In quality they have no superior; they are a month earlier than ray grass, are hardier, and longer lived than it is. They will probably give two crops of hay in a season, each equal to a common hay crop, if cut at suming time. They are well adapted to the richest soils.

On land of the best quality, the meadow fescue, (festuca pratensis) will be found greatly superior to ray-grass. It is far more productive in stalk and leaf, is hardier, as to soil and climate, in earliness and quality much the same; and, like the other fescues, it stools out very much, gradually filling the whole surface.

The meadow foxtail, (alopecurus pratensis) is a month earlier than the preceding species, and is well calculated for rich soils, but not for poor ones. It is of great size, but has one disadvantage, in that it sends up few stems; it is, therefore, better fitted for pasture than for hay. It has no other fault, and is, perhaps the most eligible grass for rich meadows yet discovered, (unless the poa trivialis be an exception,) but the latter is of limited application, as it requires very rich soil in a warm climate.

The festuca elatior, triflora, and longifolia, may be classed together, being possessed of resembling qualifications.—They are, unquestionably, the most productive of all the British grasses, attaining to the average height of four feet and a half. The stems grow very close, and they abound in leaves. Mr. Don, (the best writer on this subject) Dr. Withering, and others, say, that the elatior is equal in quality to ray grass; but it seems inferior in this respect. On putting down a stone of hay from the elatior, before house-fed

cows

cows, and also a quantity of it before saddle horses, at hay and oats, I found that in both cases it was all eaten very readily. The elatior is, unfortunately, a hybrid, and does not therefore grow from seed, but it can be very easily and expeditiously propagated, by dividing and planting the roots, a practice possessing this advantage, that the plant being put into the ground in a state of maturity, has a decided superiority over weeds. But owing to this circumstance, it can never enter into a rotation of crops. The two others are lately discovered; the triflora does produce perfect seed, but many of them are abortive. It would undoubtedly be far easier to raise a thousand stones, or ten tons of hay, from an acre, under any of these fescues, than three hundred from one of ray grass; an assertion that will appear incredible to such as are unacquainted with them.

There are several other species deserving of notice, did this paper permit a longer detail, e. g. the poas, cœsia cristata, and flexuosa, the sesleria cœrulea, cynosurus cruciformis, melica cœrulea, &c. But it might seem a neglect, if, in discussing this subject, however superficially, the fiorin grass were entirely omitted. It is either the agrostis stolonifera, or a distinct species of agrostis; and its introduction is rather too recent to pronounce decisively on its merits. The stolonifera is indeed a common weed, but the fiorin may be a variety, and weeds are not therefore of no utility. Like other creepers, it is retentive of life in the highest degree, and thrives in very poor soil, as is proved by an experiment on it at Haddo-house, where it has been pretty extensively cultivated. On a thin gravelly heath it has made considerable shoots, and had it been planted at shorter distances, would have in time filled the ground. Should it be found to thrive on dry moss, it may serve to accomplish what is at present a great desideratum in agriculture; but the holcus lanatus bids fairer to effect this end. The expence of planting fiorin is about 10s. per acre.* It is very evident, that a creeper, which not two years fallow could extirpate, can never be advantageously introduced on arable land; but there are many soils, not of that description, where it may give a valuable return; and, at all events, any addition to the stock of grasses is desireable.

* The very great weights of hay said to be got from fiorin, may be explained by adverting to this circumstance, viz. that all plants which grow from buds, either in their roots or joints, retain their sap so long as they possess vegetable life, which (especially in a moist climate like that of Ireland) they will do for many weeks. If weighed during this period, the hay is one-fourth, or three-tenths weightier than when thoroughly dried. With us, fiorin does not grow during winter; and though palatable to cattle, when raised on good land, it is not relished by them when produced on inferior soils.

SHORT ACCOUNT OF TWO JOURNIES,

Undertaken with a view to ascertain the Elevation of the principal Mountains in the Division of Marr.

Though these are among the highest in Great Britain, their height had never been ascertained. One of the bounding mountains, viz. *Mont-Battock*, (on the top of which the counties of Aberdeen, Kincardine, and Forfar, meet,) had been measured by Mr. William Garden; and was stated by him to be 1155 yards, or 3465 feet above the level of the sea. As I knew that Mont-Battock was far inferior to those mountains, which were 40 miles farther west, and nearly in the middle between the Atlantic Ocean and German Sea, I presumed that the highest land in the kingdom was in this county.

Before I set out, on my first survey, I got a mountain-barometer, made by that excellent artist Mr. Thomas Jones, formerly of Mount-street, now of Kenton-street, London.—I also got a very good spirit-level, made for me by James Cassie, a very ingenious mechanic in my own parish, whose snuff boxes, and other neat trinkets, find a ready sale in all parts of the country, and even in the city of London. And what was of the utmost consequence, Mr. Professor Copland, of Marischal College, had, in the most friendly manner, given me his best advice for carrying on my operations on the mountains, and had very kindly undertaken to mark the heights of the barometer at his house at Fountainhall, by Aberdeen, which is 166 feet above the level of the sea, at half-flood. It is, by comparing the different heights of the barometer at two places, taken at the same hour, that the heights of mountains are calculated, after making allowance for the difference of the temperature of the air, and the expansion of the mercury, indicated by the heat of the attached and detached thermometers. In this paper I shall mark only the results of these calculations.

July 9, 1810.—Set out on my first survey of the mountains, and arrived at my friend, the Rev. James Gregory's at Banchory Ternan, where I staid all night. Found, by a medium of three observations taken this evening, and at two subsequent periods, that the surface of the Dee, at this time very low, was only 12 feet higher than Fountainhall, or 172 feet above the level of the sea.

July 10.—Called on Mrs. Farquharson of Finzean, and obtained a guide to Peterhill, which I reached by 4 o'clock, p. m. I found the elevation of its summit, above the said level,

vel, to be 1930 feet.—Returned to Finzean to dinner, at 6 o'clock, where I was most kindly treated for two days.

July 11.—In consequence of a great fall of rain, the rivulets were so much swelled, that Mont-Battock was inaccessible. That I might pay some attention to my hospitable landlady, I took the elevation of some of the neighbouring hills. Found the highest, viz. the Bughts of Glenferrick, was 1260 feet,—

The top of Ben-a-hard, from which there is an excellent prospect, 1080 feet,—

Corse-darder, where King Durdanus is said to be buried, 960 feet,—and

The Gallowhill, or place of executing criminals in the feudal ages, 940 feet.

July 12.—Set out for Mont-Battock; verified my former measure of Peterhill—laid my spirit level on its summit, and found that the highest part of Clochnaben, except the rock, was exactly on a level with the point on which I stood.— Therefore as the rock is not above 70 feet, Clochnaben cannot be above 2000 feet, though said to be 2370. Walked forward to Mont-Battock. Arrived at the Shank of Largo by 1 o'clock. Found it to be 1900 feet. Ascended the mountain, and staid an hour on the top of it. Quite disappointed in finding it only 2000 feet. This was verified by an observation made some months after. (I had formerly suspected that Mr. GARDEN, who was a most accurate man, had committed some error, when he measured this mountain; and I can account for so great an error no other way, than by supposing that he had either mistaken, or *put a wrong figure in his field book*, for the angle which marked the elevation of the Grampians above the flat land, called *the How of the Mearns*. Returned to Finzean at 6 o'clock. Dined, and took leave of my hospitable and intelligent landlady, and rode to the Manse of Birse, the parsonage house of my friend, the Rev. JOSEPH SMITH.

July 13.—Found the elevation of the Dee, at the intended bridge at Belwade, 310 feet.

Left my good friends after breakfast, and rode to Ballater-house, the residence of WILLIAM FARQUHARSON, Esq. of Monaltry. Found the elevation of the Dee, below the Bridge of Ballater, to be 790 feet.

The House of Ballater, 720 feet.

The top of the Craigs of Ballater, a romantic hill, close by the house, 1340 feet.

After dining, and spending three hours with the hospitable Mr. and Mrs. FARQUHARSON, my public spirited landlord accompanied me to the Manse of Crathie, where we arrived at a late hour, and were kindly received by Mr. and Mrs. McHARDY.

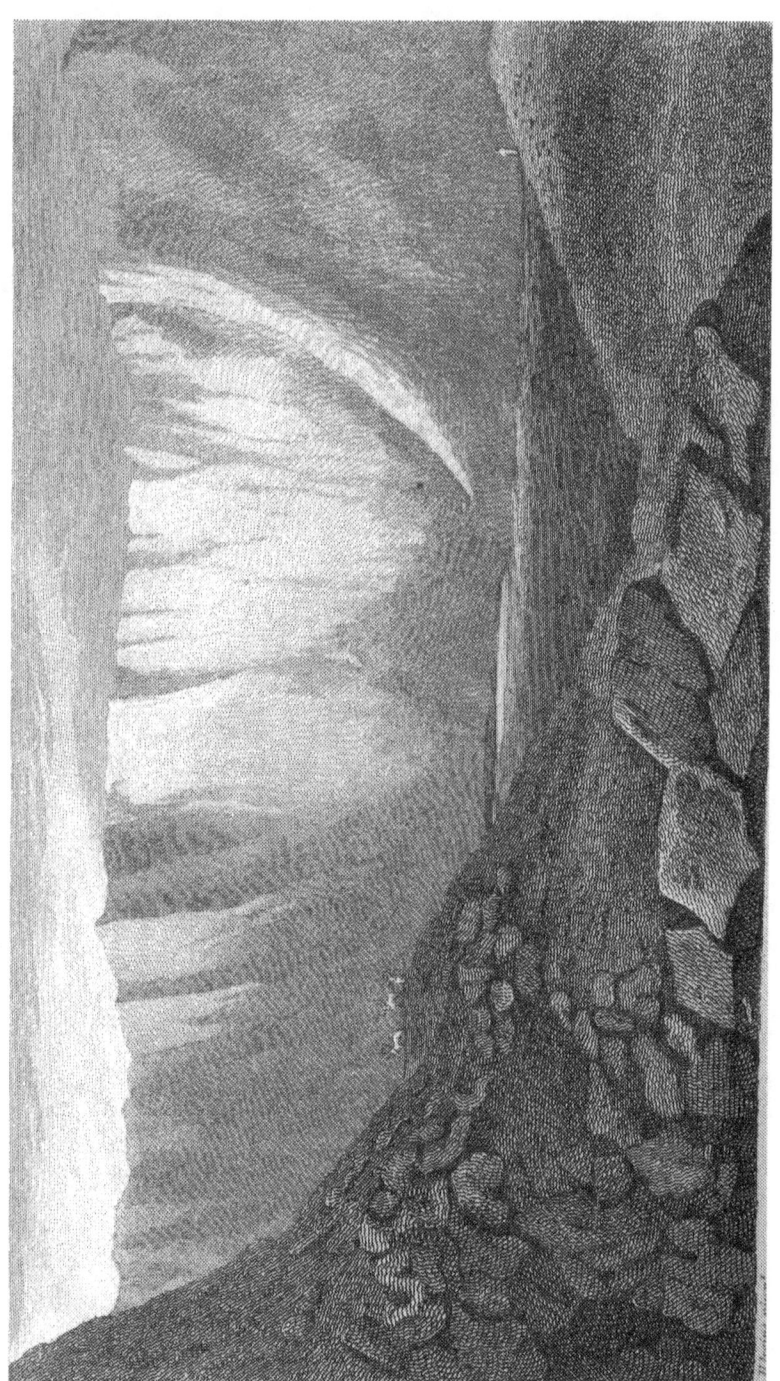

View of the Summit of Lochnagar

July 14.—Breakfasted at 7 o'clock, and set out with Mess. Farquharson and McHardy. On our way we were joined by Mr. Alexander Warren, a very ingenious man, who had measured the estate of Invercauld, consisting of nearly 100,000 Scotch acres, and to whom I was much indebted in my subsequent operations. Three guides took care of our horses, and conducted us in safety to the mountain top, which on the medium of three measures taken on this, and another visit on the 20th, was found to be elevated above the sea almost exactly 3800 feet. Laid my spirit level on the Ca Cairn, or highest top, in order to compare its elevation with the mountains near the source of the Dee. Found, that after making allowance for the curvature, they were considerably higher. Dined, and staid three hours on this interesting mountain, from whence the German Sea, from the Firth of Forth to the Murray Firth, and a great extent of country, was distinctly seen for about two hours; and where the varied scenery of woods, lakes, rivers, and mountains, and the height of the rocks of this mountain, varying from 950 to 1300 feet, and extending nearly two miles, would have detained us much longer, had not a thick fog, and every appearance of a great fall of rain, obliged us to descend as quickly as possible. Our descent was accompanied with a number of awkward tumbles, and one of the gentlemen was rolled nearly 100 feet; but no accident happened to any of us, except the loss of my spirit level. After 12 hours absence, we returned to the Manse of Crathie, completely drenched with rain; but we soon got dry clothes, and most hospitable entertainment.

July 15.—Preached at Crathie for my friend Mr. McHardy, but felt a considerable inflammation in the pleura from the great rains, and exertion of yesterday. Dr. Stuart took from me about eight ounces of blood, which gave me relief.

July 16.—Rose at 5 o'clock. Rode along with Mr. Warren to Mr. Stuart's, at Allanquoich, who is factor, or landsteward for Earl Fife. Here we breakfasted, and our hospitable landlord accompanied us to the mountains of Ben-a-bourd and Benavon. The former, viz. Ben-a-bourd, or Benavourd, (a Gaelic name, which signifies *the table hill*,) is an immense mass, without beauty or fertility, extending about three miles in length, and almost flat on the top; and was found to be elevated above the sea 3940 feet.

The latter was more interesting, having greater variety of surface. Its highest peak was 3920 feet, or 20 feet lower than Ben-a-bourd. But the mountain in general was 100 feet lower than the other. Several Scotch topazes and beryls are found in Lochnagar, and in both these mountains. We returned to Allanquoich at 8 o'clock at night, where we were entertained most hospitably, and had a sound sleep after the labours of the day.

July 17.—Rose at 5 o'clock; breakfasted before 6, and set out in about half an hour after. Called at Marr Lodge, the pleasant summer retreat of Earl FIFE, who had not as yet visited it. Took the elevation of the Dee at the Linn, which we found to be 1190 feet. (The Linn itself is a most attractive object; for the river Dee is confined within about 4 feet of breadth, between two rocks; and several fool-hardy people here leap across it; but it is a dangerous amusement, owing to the situation of the rocks on the south side, or right bank of the river.) Pursuing our course, we next took the elevation of the Dee at its confluence with the Geaully, a little above the Doubrach, (which is the highest arable, or inhabited land in the county,) and found it nearly 1300, or more exactly 1294 feet, above the level of the sea.

We then travelled in a direction due north, or at right angles to our former course, along the valley of Glen Dee, for about 5 or 6 miles, till we arrived at the junction of the Garchary and the Guisachan. Here the Dee loses its name; and its elevation above the sea, at the point where its two mountain streams unite, is 1640 feet. (The only human habitation is a shealing, belonging to some farmers in Badenoch, who rent these glens from Earl FIFE, and whose shepherds reside here a few months in summer.)

Following the banks of the Garchary, we travelled about 4 miles farther, to its confluence with the Larig, where the elevation of the united stream was found to be 1984 feet above the sea. Here we took a luncheon, and as neither my hospitable landlord, nor our guide had ever been at the source of the Dee, we were doubtful what course to pursue, when, fortunately, a man going with provisions from Badenoch to the shealing, came up to us, and pointed out the line that we should follow; for we had no road, nor even a foot-path.

At 2 o'clock, P. M. we set out to climb the mountain, still keeping in sight of the river. In a few minutes we came to the foot of a cataract, whose height we found to be above 1000 feet, and which contained about a fourth part of the water of which the Garchary was now composed. In about half an hour after, we perceived that this cataract came from a lake in the ridge of the mountain of Cairntoul, and that the summit of the mountain was another 1000 feet above the loch, which is called Loch-na-youn, or the *blue lake*. A short time after, we saw the Dee, (here called the Garchary, from this rocky bed, which signifies, in Gælic, *the rugged quarry*,) tumbling in great majesty over the mountain, down another cataract; or, as we afterwards found it, a chain of natural cascades, above 1300 feet high. It was in flood at this time, from the melting of the snow, and the late rains;

APPENDIX. 645

rains; and, what was most remarkable, an arch of snow covered the narrow glen from which it tumbled over the rocks. Here our landlord and our guide ascended the mountain by an easier, though more circuitous course; but I was determined not to lose sight of the river, and Mr WARREN kindly accompanied me. We approached so near to the cataract as to know that there was no other lake or stream; and then we had to climb among huge rocks, varying from one to ten tuns, and to catch hold of the stones or fragments that projected, while we ascended in an angle of 70 or 80 degrees. A little before 4 o'clock, we got to the top of the mountain, which, (by information given me, before I set out, by GEORGE SKENE, Esq. of Skene,) I knew to be *Breriach*, or the speckled mountain. Here we found the highest well, which we afterwards learned was called *Well Dee*, and other five copious fountains, which make a considerable stream, before they fall over the precipice. We sat down, completely exhausted, at 4 o'clock, P. M. and drank of the highest well, which we found to be 4060 feet above the level of the sea: and whose fountain was only 35 degrees of heat, on the 17th of July, or 3 degrees above the freezing point. We mixed some good whisky with this water, and recruited our strength. Then we poured, as a libation, into the fountain, a little of the excellent whisky which our landlord had brought along with him. After resting half an hour, we ascended to the top of Breriach, at 5 P. M. and found it to be 4280 feet above the level of the sea. We then descended amidst a thick fog, which suddenly overwhelmed us, and attempted next to get to the top of Cairntoul, on the other side of the Garchary. We could not see an object at above 100 yards distance; and at last ascended one of the inferior peaks, but afterwards climbed up the rocks to the highest summit of Cairntoul, which we found to be only 5 feet higher than Breriach, and that *apparent* difference was only occasioned *by the weight of the atmosphere*. On this summit the rain poured out in such torrents, and the wind battered us so much, that two gentlemen, holding umbrellas over my head, could not protect me while I marked the height of the barometer in my journal.— We were obliged to leave the index, then draw on the brass cover, and inverting the barometer, to descend the mountain. Unfortunately we had no pocket compass, and afraid of falling over the huge rocks of Poten Duon, which are nearly 1600 feet high, we turned too much to the right hand, and completely lost our way. It was 9 o'clock at night before we found, that a small river, whose course we happily followed, was the Guisachan, or the other source of the Dee. And it was half an hour past 9, when we arrived at the junction of the two streams, and the shealing which we passed at noon. We were now completely exhausted with hunger and exertion; and the shepherds had neither ale, milk, whisky, nor any thing, except oatmeal, and bannocks baken of oatmeal, and nearly two inches thick. But hunger gives a better relish for food than the best sauces can do. And the butter, which we had untouched, spread on these bannocks, appeared to me the best meat I had ever tasted; while the stream of the Dee allayed our thirst. Our horses joined us at 10 o'clock, and we mounted, retraced our steps homeward, and arrived at Allanquoich, about half an hour past 1 next morning. There we received the kindest treatment, and afterwards enjoyed a sound sleep, after nearly 19 hours of fatigue.

July 18.—After breakfasting at 10 o'clock, Mr. WARREN and I returned to Crathie. At setting out, we took the elevation of Allanquoich, which we found to be 1100 feet; and on our way home, that of the Castle of Braemar, which was 1070,—of the Dee, at the bridge on the military road, near Invercauld, which was 1030,—and lastly, that of Manse of Crathie, which was 860 feet above the level of the sea. I then dressed, and rode to Abergeldie to dinner, where the kindest reception from Captain and Mrs. GORDON, made me forget the fatigue of the two preceding days. The excellent birch wine appeared to me superior to the finest Champagne; and the transition from most fatiguing exertion, and hours of fasting, to social enjoyment, in this hospitable family, added much to the happiness of the wearied traveller.

July 19.—Rode, after breakfast, to the Manse of Crathie, and after doing some duty for Mr. McHARDY, as a mark of my gratitude, I returned to Abergeldy to dinner. Here my generous landlord, with the true spirit of a highland chieftain, caused his horses to be put to the carriage, and went, with his lady and neice, to Altguisach. This is a most commodious cottage belonging to Captain GORDON, in the neighbourhood of Loch Muick, and of the mountain of Lochnagar, which I wished again to examine particularly; and the ladies also proposed to accompany me. But,

July 20.—The weather was such as rendered this impracticable to them. It did not, however, detain me; and my hospitable landlord gave me a pocket compass to assist me in finding my way, in case of accidents. Two guides also accompanied me, and I took the elevation of the following places above the level of the sea.

1. Altguisach, this hospitable cottage,   1360 feet.
2. Loch Muick, at its junction with the Glassilt, 1280
3. Loch Dowlach,   -   -   -   2050
4. Craig of Dowlach,   -   -   3250
5. Quarry of Corbreach,   -   -   3450

My faithful companion, Mr. WARREN, joined me on the top

top of Lochnagar, whose elevation we verified repeatedly, and found to be very nearly, as before, 3800 feet.

And the Lake of Lochnagar, at the foot of the rock, 1300 feet less, or, - - 2500

The other Lakes, called Loch-an-yeans, or bird's lakes, from - - - 2450 to 2800

The second top, or border of the White Month, 3780

We returned about 7 at night to this *hut* of hospitality, and spent the evening most agreeably. Our only regret was, that owing to the weather, the ladies had not been able to visit the mountain.

July 21.—Breakfasted at 6 o'clock, and took leave of my hospitable landlord, and my kind assistant, Mr. WARREN, at 7.—A guide, with abundance of provisions, accompanied me to Mont Keen, where I arrived at 11; and leaving my horse with my guide, reached the summit of the mountain a little before noon. Staid there about half an hour, and found its elevation to be 3180 feet. Descended and joined my guide at 1; crossed the country on my way to Morven; and took my luncheon in the hills, beside a clear fountain, about 2 P. M. Arrived at the Bridge of Ballater at 4, dismissed my guide, called at Ballater house, and requested another guide to follow me to Morven. My request was granted, and my guide taking charge of my horse, I ascended the mountain, and a little after 6 o'clock P. M. I found the elevation of its highest top to be exactly 300 feet less than Mont Keen, or 2880 feet. After staying there half an hour, I descended the mountain, mounted my horse, and arrived at Ballater House a little before 9. There, a kind reception from Mr. and Mrs. Farquharson, recruited my strength, and the excellent birch wine, like what I had, in a similar case, tasted at Abergeldy, both exhilarated and cooked me, when parched with travelling above 14 hours in this mountainous district. A sound and refreshing sleep succeeded to an evening spent most agreeably.

July 21.—I preached and dined at Glenmuick, with Mr. and Mrs. BROWN, and returned to Ballater at night; where pleasant company, and another night's rest, completed my recovery from the fatigues of this expedition, and I proposed to return next day.

July 22.—Set out on my return; and in my way to Aberdeen, found the elevation of the Dee, at the proposed bridge at Potarch, to be 280 feet; that of the same river, at the Bridge of Banchory, 172; the top of the Tower of Drum, 460 feet; the Dee, at the Church of Drumoak, 90; and at the influx of the Burn of Culter, 60 feet.

This last measure, which I verified by going to the Quay at Aberdeen, at half-flood, and to Fountain-hall, where Professor COPLAND obligingly gave me all the measures of his barometer, (which compared with those taken in the course of my journey, enabled me to give the above calculations,) concluded my first expedition to the mountains of Marr; which had taken up two weeks, during which time I had travelled above 330 miles.—But though I had left Aberdeen in rather bad health, the pure air, and the kind hospitality which I met with every where, not only enabled me to bear the fatigue which I had undergone, but gave me a degree of health and strength, which I had not enjoyed for some months preceding.—At the same time I must remark, that the man must have good stamina, who expects to regain health by measuring mountains.

After returning home, and calculating the heights of Cairntoul, Breriach, and the other mountains of Marr, I thought it would be worth while to see whether these were the highest in this county; and also, when measured with the same barometer, whether they were equal or superior to Ben Nevis, on the confines of Inverness and Argyllshires, and to Cairngorum, in the head of Banffshire, which were reputed to be the highest mountains in the island. The thick fog, and excessive rains on the 17th of July, when I visited Breriach and Cairntoul, and the loss of my spirit-level at Lochnagar, rendered it impossible for me to know whether Ben MacDouie was higher than either of these. And though my barometer, made by JONES, was an excellent instrument, yet its having a moveable zero was a defect, which prevented me from depending upon it within twenty or thirty feet. I therefore wrote to him, returning my barometer, which had sustained a little injury, and desired him to repair it, and also to send me another with a fixed guage point. He did so, and I got notice of their having come to Aberdeen. But when I went to that city, on my second journey, I found that the Captain of the ship, from mere inattention, had carried them back to London. This was provoking; but I was determined not to lose the season, which I must have done had I waited their return from London.

Invention supplies the defects of learning; and *to be fertile in expedients*, though not equal to *true wisdom*, is the *next useful attainment*. I knew that Dr. DAUNEY, the Sheriff Substitute of Aberdeenshire, had one of Mr. JONES' barometers, and that GEORGE SKENE, Esq. of Skene, had another, made by the same instrument maker. Therefore, I applied to Dr. DAUNEY, who very readily lent me his barometer, which, fortunately, was an excellent one, and had a fixed guage point. And my son ALEXANDER went to Carieston, in the county of Forfar, to get from Mr. SKENE, the loan of his barometer. As Professor COPLAND was going on a visit to the Duke of GORDON, he kindly wrote to Mr. HOY, at Gordon Castle,

APPENDIX. 649

Castle, (whom I found to be a worthy man, and a correct mathematician) to take the heights of the barometer, at that place, while I was employed on the mountains.

On the 10th of September I left Aberdeen, and on the 12th I went to Abergeldy. On the 13th, I met my son, with a very ingenious friend of his, Mr. WILLIAM RAMSAY, son to the Rev. WILLIAM RAMSAY, minister of Cortachy.— They brought Mr SKENE's barometer; and after taking leave of the hospitable family at Abergeldie, we reached Allanquoich that night about 10 o'clock.

On the 14th we breakfasted at Marr Lodge, at 7 o'clock, where we were joined by our guides; and we checked our former measures of the elevation of the Dee, till we came to the summits of Cairntoul and Breriach. We found the snow was melted between the wells of Dee, and on the top of the rocks, though a considerable quantity of it remained on the north ridge of Cairntoul, which the sun never reaches. Only three wells out of five contained any water, and their united stream was not above one-third part of what it was on the 17th of July. We found that the heat of the water was 40 degrees of Fahrenheit, and that the peak of Cairntoul, and the top of Breriach, were exactly of the same elevation, and 60 feet lower than we found them, owing to the thick fog and rains, on the former survey. The spirit level was applied, as well as the barometer, to prove their comparative heights. And as the atmosphere was very clear this day, the distance of 60 feet of elevation might be called the *aberration of the barometer*, or the *difference* between the measure of a mountain above 4200 feet high, in a *clear and light atmosphere*, and that when it was *heavy or clouded*. The difference of 5 feet between the two mountains, when the atmosphere on the 17th of July was sensibly heavier at Cairntoul than it had been two hours before at Breriach, also attracted our notice, and shewed that the state of the weather should be always marked, when a mountain barometer is used, for marking accurate elevations.

Having adjusted these matters, we directed the spirit level to Cairngorum, distant about 6 miles, and found it was *considerably lower*. But when we directed it across the mountain torrent, from the Larig to the top of Ben MacDouie, we found it was considerably higher; and though not two miles distant, it was inaccessible that day; it being now past six o'clock, and Aviemore 12 miles distant. The barometer which my son had brought with him, evidently had too little mercury, and could not be trusted in an accurate experiment. We therefore set out for Aviemore, where we arrived a little after 10 at night, and found the people all in bed; but at last got admission, and staid there all night.

Sep-

September 15.—After breakfasting at this inn I sent off my young friends to Ben Nevis, and set out for Gordon Castle; but staid that night at Aberlour, in the hospitable manse of the Rev. Mr. WILSON.

September 16.—Preached at Fochabers for the Rev. JOHN ANDERSON, whose acquaintance proved to be of the greatest use to me, and whose kindness I can never forget.

September 17.—Breakfasted, and spent the forenoon with Mr. Hoy, with whom I settled the plan of my future operations. His Grace the Duke of GORDON arrived a little before dinner, and it was impossible for him, or any person, to treat me with more kindness than he did. He lent me a pocket compass for fear of more accidents on the mountains, conversed about my plans, as if he had an interest in them, and was quite at home when conversing on this subject. I never spent a day more pleasantly, or in which I experienced more kindness from any man.

September 18.—Examined the height of the barometer along with Mr. Hoy, at 7 in the morning. Took farewell of this worthy man a little before 8, and breakfasted at Fochabers. Mr. ANDERSON gave me letters, and a plan of my route, which were of the greatest service to me. I then rode to Grantown, after dining with the kind-hearted Mr. WILSON at Aberlour, and drinking tea at the Manse of Inveraven.

September 19.—Rode to Aviemore, and then to Inverdruie, in quest of GORDON the fox-hunter, whom Mr. ANDERSON recommended to be my guide. Most kindly treated by Captain and Mrs. GRANT at Inverdruie, and also by Mr. GRANT of Rothiemurcus. Rode to Pitmain Inn, where I staid all night.

September 20.—After breakfast sent off my letters from Mr. ANDERSON. In a little time Captain McBARNET called for me, and conducted me to his house, near to which we met my two young friends. They breakfasted at Captain McBARNET's, and were most kindly used, after walking 18 miles from Garviemore. Captain McBARNET sent a servant and a poney with us to the shepherds at Inverishie, where we were to stay all night.

We returned to the Inn at Pitmain; and after taking a luncheon, rode to the shepherds at Inverishie, where we arrived a little before night, and were kindly treated by those hospitable people; and where we found GORDON the Foxhunter, who agreed to be our guide. We got a clean bed on soft hay in the barn, and my young men slept soundly. But a heavy shower of rain, and the want of shelter to my poor poney, kept me from sleep; and at 4 next morning, I awoke my young friends, and we breakfasted before 5, anxious to get forward on our journey.

Sep-

September 21.—A most delightful day, after a heavy rain for two or three hours in the night time. Set out a little after 5 o'clock, the shepherd attending us himself, along with GORDON, till we came to the base of Breriach, when he went to Glenmore with my poney, while Gordon conducted me to the highest well of Dee, by 10 o'clock, A. M. My young men, on the way gave me an account of their hospitable reception at Far, by JAS. MCINTOSH, Esq. and of their having measured the elevation of Ben Nevis, which they found to be, on a medium of two observations, 4350 feet. Well Dee was now reduced to 4000 feet in this pure atmosphere; and the top of Breriach, to 4220, or at most, 4230. At 1 o'clock we descended 2200 feet, and after crossing the Larig, ascended 2280 feet, to the top of Ben Mac Douie. We reached this at half-past 2 o'clock, P. M. and found it the second mountain in the island, and inferior only to Ben Nevis, by nearly 50 feet, or 4300 feet high, at a medium of three observations. We dined on the top of Ben Mac Douie, the thermometer being 47°, and the water in the highest fountain at 40°, in one of the hottest days of this season.

by Roberts 4374.

After remaining there above an hour, we set out for Cairngorum, to the summit of which we mounted at 5 P. M. We found this mountain about 250 feet less than Ben Mac Douie, and also inferior both to Breriach and Cairntoul, by 170 feet, its height not exceeding 4050 feet. But we had the most extensive and delightful prospect from it, owing to the state of the atmosphere. On the west, Ben Nevis, at 70 miles distance, was seen in all its glory, and the waves of the Atlantic appeared, not blue, but of a whitish, or cotton colour. Toward the north, the Ord of Caithness was distinctly visible, as well as a great part of the sea-coast, from the north-east and south-east. The south and south-west only presented us with a view of the mountains of Marr. We descended to Lord Huntly's Well, about 60 feet from the top, where we drank his Lordship's health, in a mixture of whisky and water, and then went to the Duke's summer retreat, at Glenmore, where we found my poney, and from whence we travelled to Mr STUART's of Pituelish, whose house we reached a little before 10 o'clock at night. Here we were most kindly treated by his son-in-law, the rest of the family being gone to bed, and here we found refreshing sleep.

by Roberts 4095

Low or ith

September 22.—After taking the elevation both of the Lake of Pituelish, and of the river Spey, we took leave of this hospitable family at 11 o'clock; and after calling at Abernethy, and dining at Tomantoul, we reached Liniorn, the residence of the Rev. SAMUEL MASON, an old class-fellow of mine, who is minister of the Chapel of Curgarf, as a missionary in the highland district. Here we were received most hospitably, and staid all night.

September 23.—Dismissed my young friends with a letter to Andrew Farquharson, Esq. of Breda, requesting him to mark the heights of the barometer on the two following days. Remained myself with my old schoolmate, for whom I preached in the Chapel, and spent the rest of the day in talking over the adventures of our younger years.

September 24.—Mr. Mason accompanied me to the source of the Don, about 6 miles from his house. We found its elevation above the sea only about 1740 feet. It was difficult indeed to find the source, as it was in peat mossy soil, and contained very little water; and the small streamlets issuing out of the peat moss, had run nearly 2 miles, and fallen 340 feet before they formed so large a stream as that which tumbles over the rocks of Garchary, from the three wells of the Dee. It deserves to be mentioned, that the fountain of Alt Vichal, or Michael's Brook, is within a few hundred yards of the source of the Don, and runs either into the Avon, which falls into the Spey, or into the Don, which is here 200 feet above the Avon, according as a turf falling, on either side, happens to direct its stream. It is considered, however, as belonging to Banffshire. At the bridge at Curgarf, the elevation of the Don was only 1240 feet.

Ascertained the elevation of the hills of Scroulac, Carnweach, and the Lecht, very nearly the same, viz. 2700 feet.

That of the hills of Cairnmore and Ben New, to be also nearly equal—1800 feet.

Took leave of my old school-fellow, and rode to Forbes' Lodge at night, where I was received very kindly by Captain Forbes of Inverernan.

September 25.—Found the elevation of the hill of Lonack 1210 feet.

Called at Major Forbes's, and found the elevation
of the Don, opposite to his house, - 950
Travelling homeward, found this at the influx of
the Busket - - - - 690
That opposite to Kildrummy, on the banks of the
Don, - - - - 540
The elevation of the Don at Breda - 420
That of the hill of Callievar - - 1480

Arrived at Mr. Farquharson's of Breda by 6 o'clock at night, a good deal fatigued; but the kind reception of my worthy host and his lady, made me forget all my labours.

September 26.—Left this worthy couple, and spent a few hours at Haughton, with John Farquharson, Esq. a very intelligent gentleman, who has ornamented his property with most extensive plantations. Rode home in the evening, after 16 days absence; and wrote the worthy Mr. Hoy of my safe arrival. From his answer to my letter, I was enabled to calculate the above elevations, as he marked the heights of the ba-

barometer at Gordon Castle, while I was employed in this survey; during which I travelled 400, and my son, 440 miles.

The height of the principal mountains, and the elevations of the Dee and Don, in these various places, will render this county better known than it was formerly. And though it was not in my department, I am pleased that I made this second survey: For in matters of research, it should ever be remembered, that nothing is done while any thing remains undone.

I should be void of every feeling of gratitude, if I omitted to acknowledge the obligations I was under to all classes of men, from his Grace the Duke of Gordon, and the different landed proprietors, to JOHN GORDON, the fox-hunter, in Badenoch, and my other guides in this tour. To the honour of the latter I must add, that they were reluctant to accept of so much recompence as I thought myself bound to make them for their trouble.

I conclude, with recommending it to every man who undertakes a similar expedition, to use always a spirit-level, along with his mountain barometer.

SURVEY OF THE LANDS OF ABERDEENSHIRE,

Which are situated within two miles and a half of Aberdeen.

IN November, 1809, I applied to the Magistrates of Aberdeen, through the medium of WM. SHEPHERD, Esq. who had been for many years eldest baillie of that city, to get a correct account of the extent, different rents per acre, and the various modes of cropping of all these lands, in the vicinity of Aberdeen, which are situated within this county. (For those which are separated from Aberdeenshire by the river Dee, being in the county of Kincardine, did not fall within my province.) My request was readily granted by the Magistrates; and GEORGE TURREFF, the Dean of Guild's officer, a very accurate and intelligent man, was employed to make an actual survey of the district, according to the instructions which I gave him.

His Report was a full, satisfactory, and minute account of the names of the landed proprietors, the extent of each field, the tenants' names, rent per acre, in money or corn, the length of lease, and cropping for the four preceding years. The following is a concise summary of an authentic and useful paper, which it is not possible for me to insert at full length.

The

The lands in the vicinity of Aberdeen are situated partly in the parish of St. Nicholas, and partly in that of Old Machar, between the rivers Dee and Don.

The parish of St. Nicholas contains the city of Aberdeen, sometimes called New Aberdeen, and the arable lands which belong to it. The whole extent of this parish, (including houses, streets, roads, gardens, and both arable and barren lands, and those which are within the flood-mark; and are occasionally overflowed by the river or the tide) is a little more than 780 Scotch, or nearly 1000 English acres. Of this the arable land is nearly one-third part. The sand on the beach, between the rivers Dee and Don, and a few small adjacent hills, those near the harbour, which are overflowed by the tide, and the houses of the city, including the gardens, streets, and roads, constitute the other two-thirds.

The ancient arable lands belonging to St. Nicholas' parish, are in a high state of cultivation; and are chiefly situated on the N. E. and N. N. E. of Aberdeen. Their rent, per acre, is various, the lowest being L.5, and the highest L.18 per Scotch acre. The average is at present above L.10;— but if there were not a number of subsisting leases, would be considerably higher. They are distinguished as follows:

1. The most fertile, are those which are called the *Sandilands*, and which, before their improvement by manure, were chiefly a collection of sand. They are now a sandy loam, incumbent on a bottom of sand. The tenants are, in general, gardeners, or persons who raise garden stuffs, and vegetables, for the inhabitants of the city. They use different modes of cropping; but the most common practice is to plant potatoes in March or April; to dig the early kinds in July, the rest in August; and immediately to plant greens or coleworts, which are ready for sale in November and December. By this means they obtain two crops every year, for several years in succession. When they think the land requires a little rest, it is laid down with bear, (not barley) and grass seeds; and after lying two years in that state, is broken up generally for potatoes. For they are particularly partial to this root, as it pays best, and seldom fails to succeed in this loose and rich soil. They commonly allow 40 loads of dung to the Scotch acre, and the load is about 15 cwt.; and by frequently manuring their ground, they raise from ten to twelve tuns of potatoes, and about half that weight of coleworts per acre. Their practice clearly shews, that in a light loam, a much more valuable crop of potatoes and greens may be raised, with good culture, than can be raised of wheat, on heavy lands, to which the same quantity of dung is applied. The highest rent here is L.18. Many of the fields are L.13 and L.14 per acre, on leases from nine to nineteen years.

2. The

2. The lands, from the Bowl-road to the north-west corner of the Broadhill, are also a sandy loam, now much enriched by manure, and incumbent on a bed of gravel. These, too, are chiefly rented by gardeners, and cropped in the same manner. The only exceptions are a few acres, which are employed by nursery-men in raising the plants of young trees; and two acres of moist soil, which are planted with willows, for making baskets. They are all let on 19 year's leases, except three fields. Their highest rent is L.16; but owing to the subsisting leases, the average is only L.10.— The same quantity of manure, viz. 40 loads, or 30 tuns, is allowed to the acre. Very rich crops of bear and grass are raised occasionally; from 9 to 12 quarters of the former, and 100 tuns of green clover, cut for their cows. The grass is too valuable here to be made into hay.

3. The lands lying north from the back of Princes' Street, and towards the Porthill. These are rented indifferently by gardeners, cow-feeders, and nursery-men. The gardeners raise vegetables, and garden roots, such as onions, carrots, &c. The cow-feeders raise turnips, coleworts, bear, and the sown grasses. They lay down from 36 to 40 loads of manure, at from 3s. 6d. to 4s. per load. The highest rent is L.11. The average is above L.3; but the greater part is on leases from 12 to 19 years.

4. The lands lying N. N. W. of Aberdeen, from the Porthill to Broadford inclusive. The soil here is much heavier; and that part which is called the Lochlands, and was anciently a lake, though drained at a great expence, is still moist in many places. The quantity of manure laid on it is 40 loads, of the best kind of dung, from 6s. to 7s. per load. A considerable proportion is laid out in nursery. The rest is rented by gardeners, or cow-feeders. The soil, in general, is a light sandy loam, upon a bottom of sand or gravel; and in one place of peat-moss. The depth of the soil is from 15 to 18 inches.

II. The parish of Old Machar reaches from the Dee to the Don, and excepting on the E. and S. E. surrounds the parish of St. Nicholas. It includes the city of Old Aberdeen, and a large district of country, containing from 5000 to 6000 acres, part of which is on the north of the Don. The greater proportion of arable land is betwixt the Dee and Don, and consists of three divisions.

The first is towards Old Aberdeen, where the soil in general is a deep loam, excepting the high grounds of Spittal, which are of a thin sandy soil. The rent varies from L.4 to L.9. In one place it is L.17 to a nursery-man; but the average is only L.7 per acre. The lands here are rented by small farmers, gardeners, and cow-feeders. The crops are turnips, potatoes, oats, bear, and sown grass, which generally

rally continues for two, and sometimes, though rarely, for three years. The quantity of dung applied is from 40 to 48 loads, from 6s. to 7s. 6d. per load.

The second district comprehends the lands lying N. and N. W. and extending above two miles from the city of Aberdeen. It is in general a rich loam, with more or less of sand, and in a few places is a little moist. The rent varies from L.4 to L.8. At an average is L.6 10s. In a few places it is as low as L.3. But the manufacturers at the Printfield have raised the value of the lands in its neighbourhood; and PATRICK KILGOUR, Esq. draws frequently L.20 for an acre of potatoes. Wheat here has been raised with success, although the thin sandy soils cannot, without a great quantity of manure, raise good crops of wheat. Mr. KILGOUR uses the dung of the butcher market for manuring his wheat land.

The third district includes the lands lying to the N. W. W. and S. S. W. of Aberdeen. Those near the city are very highly rented. In one case as high as L.20 per Scotch acre. At a greater distance they are as low as L.3. The average is about L.6. The soil here has been more indebted to art than to nature. The greater part of the lands has been trenched at an enormous expence, some as high as L.100 per acre.— But the *granite* has frequently paid a third part, and in some cases one half of that expence. Part of this district, near Aberdeen, has been feued out, not for building, but for garden ground, at L.21 per acre. No where in the island, perhaps no where in Europe, have the spade and mattock been applied at an equal expence, or with more success; yet in the more distant parts, where the population is thin, and where there is no demand for milk, till it be carried to Aberdeen, the lands are lower rented than in any of the other districts at the same distance from that city. There is no printfield, or great manufactory in this corner.

On the whole, the lands near Aberdeen are now as highly rented, and though naturally unfertile, produce as good crops of vegetables, garden roots, and every kind of corn, except wheat and two-rowed barley, as are raised in any part of the island.

It is deserving of notice, that at five miles distance from Aberdeen, Mr. ALEXANDER PIRIE, on the farm of Watertown, adjoining to his paper mill, raised the weightiest crop of turnips last year, that probably was raised in the island viz. 59 tons on the Scotch acre, of globe turnips, at two feet ten inches distance between the drills; 57 tons of red topped turnips, at two feet ten inches between the rows; and 52 tons, at two feet three inches, of *red tops* also. An example of excellent husbandry, and a decisive proof of the advantage of narrow drills, when turnips are horse hoed, and the land is properly pulverized.

GENERAL RULES FOR EITHER PROPORTIONING OR EQUALIZING THE DIFFERENT TAXES ON CORN, MALT, AND SUGAR SPIRITS.

I shall, in this short paper, suggest a few leading principles, on a subject which I have impartially and attentively considered. I do this, as a person who has very little interest in the discussion, and with the greatest deference to the Legislature; and I request, that laying aside all prejudices and passions, and taking a *comprehensive view* of the *real interests of the nation*, the following General Rules, or Principles of Legislation, may be examined impartially.

1. When the price of corn in general, and of barley in particular, is *so very high*, as that importation of foreign corn is allowed on paying the *lowest duties*, (which, in this state ought to be *merely nominal,*) all distillation of spirits from either corn or malt ought to be prohibited, not by a temporary act of the legislature, or an occasional exertion of power by the King's Ministers, but, by a general regulation in an act of Parliament; which may at all times be known to his Majesty's subjects. In this situation of the country, the distillation of sugar spirits should both be permitted and encouraged; and the duties both on rum and on spirits distilled from sugar, ought to be precisely the same as are imposed upon the corn or malt spirits of the united kingdom. Nay, in this state of our supply, and very high price of corn, *every facility* should be given to the use of sugar in the *brewery,* by granting a deduction of *at least four shillings,* on every hundred cwt. of sugar, which, at the sight of the Officers of Excise, is put into the brewer's mash tun.

2 When the price of barley is *so high* as that importation is permitted on paying the *first low duties,* (which should always be *moderate* when corn is high priced) both corn and malt spirits should be allowed a fair competition with rum and sugar spirits; and, in this state of the prices of barley, should pay no more tax than what is paid by corn or malt spirits, when the price of barley is moderate.—In short, rum from our West India Colonies, sugar, corn, and malt spirits distilled in Britain, should pay the same real tax to Government, including the customs-duty on sugar, and the malt-duty on corn or malt spirits, which had been previously paid to the Excise. This regulation would prevent our manufacturers from suffering great pressure from the high price of corn; and might even prevent that price from rising higher.

3. When the price of corn falls so low, as that barley cannot be imported except upon payment of high duties, the tax on corn or malt spirits should be only two-thirds of what is imposed on rum from our West India colonies, or sugar spirits distilled in Britain. In this state of the price of barley, both sugar and corn should be allowed to be distilled; but all rum

and sugar spirits, including the customs duty of 27s. per cwt. on sugar imported, ought to pay *three shillings* of tax, where corn or malt spirits pay *two shillings*, including the malt tax for what malt is used in the British distillery. The *equalizing of the taxes in the former case* was *necessary for the manufacturer;*—the *proportioning* these taxes now becomes *necessary* to the *farmer*, and is *beneficial to the nation at large*, by encouraging *the raising of corn* in Britain.

4. When the price of barley is so low as that it may be exported, either with or without a bounty, all sugar spirits ought to pay double the tax, that is paid on spirits distilled from corn or malt. In this state of the country, it is absolutely necessary to support British Agriculture, by encouraging the consumpt of British malt or corn spirits, *in preference* to expending money on bounties for exportation.

5. The taxes on brandy and geneva (now that Holland is annexed to France) ought to be the same; and, in my opinion, ought to be thrice as much as is imposed upon malt or corn spirits, and double of the tax on rum and sugar spirits, in the third mentioned case, or when the price of corn is moderate.

In applying these General Rules either to the equalization, or proportioning of the taxes on corn, malt, or sugar spirits, it should be remembered, that besides the wash and spirit duties, the malt used in the distillation of corn spirits, pays a tax in England equal to 4s. and four-pence per bushel, which at the average proportion of malt used by the English distillers, is equal to 7 pence half-penny on every gallon of corn spirits; and that the customs tax on sugar paid on importation, is equal to two shillings and three pence on the gallon of sugar spirits. These three different items therefore, must be subtracted from the sum, whatever it be, that is fixed upon as the gross amount of the taxes or spirits, before we can either proportion or equalize the duties on the different kinds of spirit distilled in England.

The following calculations are made on the supposition, that 100 gallons of wash from sugar, yield 21 gallons, and 100 gallons of wash from corn or malt, yield 19 gallons of spirits, and that 1 cwt. of sugar produces 12 gallons, and a quarter of corn 18 gallons of spirits, of the strength of one to ten above hydrometer proof.

1. The tax on wash from malt, or corn, is now proposed to be one shilling and threepence halfpenny per gallon; and, consequently, on one hundred gallons is six pounds nine shillings and twopence, or fifteen hundred and fifty pence. As that quantity contains twenty-one gallons of spirits, the twenty-first part of this sum is the tax on every gallon of sugar-spirits paid by the distiller. This is a little more than six shillings, one penny, and three farthings—exactly six shillings, one penny, and seventeen-twenty-firsts of a penny.—
- Add

Add to this two shillings and threepence for the duty on importation, and the whole tax on sugar-spirits is a little more than eight shillings and fourpence three farthings.

2. The tax on corn-wash is now proposed to be one shilling and ninepence per gallon; and, consequently, on one hundred gallons is eight pounds fifteen shillings, or twenty-one hundred pence. This quantity must contain nineteen gallons of spirits, and, therefore, the nineteenth part of this sum, viz. nine shillings, twopence, and a little more than a halfpenny, (exactly ten-nineteenths of a penny) is the proposed duty on every gallon of spirits distilled from corn or malt. By two clauses in the bill, this is called nine shillings and fivepence halfpenny; and in the several cases mentioned, this error adds threepence more to the duty on corn-spirits. But, independently of this, the malt-tax, on the proportion of barley malted for this purpose, is sevenpence halfpenny: so that the whole tax on corn-spirits is above nine shillings and tenpence per gallon, or seventeen per cent. more than that imposed on sugar-spirits: and where nine shillings and fivepence halfpenny is charged on distillation only, it is twenty per cent. higher than the other. Whatever this may be termed, it is *not an equalization* of the duties on corn and sugar-spirits.—But,

Since this sheet went to press, I am informed that Mr. Perceval's intention was *not* to equalize the duty to be obtained from sugar and from malt spirits, but only to propose a rate of duty, which, supposing barley to be 38s. per quarter, and sugar 70s. per cwt. should render the distillation of sugar and corn spirits equally advantageous to the distiller.— This is incomparably more exceptionable than I had supposed. It *fixes* a proportion of price between sugar and barley, (which fluctuates every day, and not according to that proportion), and gives a decided preference to sugar spirits, by imposing less duty on strong wash from sugar, than is proposed to be made on weaker wash from corn.

Besides, if barley fall in price in England, as bear has rapidly fallen in this county, all spirits distilled from sugar ought to pay not *seventeen per cent. less*, but *fifty per cent. more* tax than is paid on corn-spirits.

The paper which is subjoined to this will shew, that I wish to give every encouragement to sugar, where it does not interfere with British agriculture. But, as an honest man, I must maintain, that wherever it is used in the distillery, it should not pay less tax on its wash than what is paid on corn-wash for the same real value of saccharine matter.

The *money-price* either of sugar or of barley has *nothing to do with this question.* When sugar is compared to barley, *the only fair ratio* of taxation is the *quantity* of *saccharine* or *fermentable matter* that can be obtained from *each*. *Every other ratio is both false and foreign.*

The following paper was drawn up two years ago; and was communicated to a Gentleman in office. It is now inserted by authority from the President of the Board of Agriculture; and some additions have been made to it, in consequence of Mr. PERCEVAL's resolutions for equalizing the duties on distillation of sugar, malt, and corn spirits. It is drawn up in the form of questions and answers, as it was intended to be sent to the Distillery Committee of the House of Commons, if they had entered upon the discussion in 1809. But they merely met for correcting a mistake in the former act, for inserting the words *Great Britain* instead of *England*, occasioned by an error in copying the bill.

Question 1.—What is the proportion between the value, in point of saccharine matter, of all the sugar now annually imported into Great Britain, and that of the whole malt used in the kingdom, for the purpose of making ale and spirits?

Answer.—The value of the whole sugars imported, or the quantity of saccharine matter contained in them, (including what is brought from the ceded islands,) is, for the above purpose, equal to *one-half* of the malt, that is now annually made in Great Britain and Ireland; or in other words, would make one-half of all the ale and spirits which are produced from malt.

Quest. 2.—In what way do you compute the proportional or relative value of sugar and malt, when you find that the saccharine matter, in all the malt, is only double of that contained in the sugars?

Answ.—I consider 1 cwt. 3 qrs. (or 196 lbs. neat) of Muscovado sugar to be equal to a quarter of malt, at an average of the different qualities of malt and sugar. I have extracted spirits both from malt and from sugar, have examined the gravity of the worts, and find that this is the nearest general proportion which I can give of their comparative values, since the deterioration of the quality of our sugars, from the introduction of the Bourbon cane.

Quest. 3.—What do you assume in your calculations, as the total amount of the importation of sugars, and of the quantity of malt annually made in the United Kingdom?

Answ.—I find, from the public accounts presented to the House of Commons, that the total quantity of malt, made in seasons of moderate fruitfulness, is, in round numbers, four mil-

---

Mr. PERCEVAL has found that the distillers make 12 gallons of spirits from the cwt. of sugar; and as 18 gallons are the greatest average of spirits extracted from a quarter of malt, this would make 6 cwt. of sugar equal to 2 qrs. or 1 cwt. 2 qrs. (or 168 lbs. neat) equal to a quarter of malt. But this is owing to the distiller's using only the best or strongest sugars; and from the experiment which I made in distilling sugar of an inferior quality, I cannot suppose less than 196 lbs. to be equal to a quarter of malt, in point of saccharine matter, or real value.

millions of quarters, or 32 millions of bushels. I also find that the quantity of sugar now annually imported is three millions and a half of cwts. which quantity is equal to two millions of quarters of malt. Some years, indeed, it is below this quantity; but if the unfruitful seasons of 1796, 1799, and 1800, be taken into the account, the quantity of malt is, at an average, as much less than four millions of quarters, as that of sugar, since we got the ceded islands, is less than three and one-half millions of cwts.

Quest. 4.—Is malt valuable for its saccharine matter only? Or are there any other circumstances that ought to be taken into the account, when sugar is thus compared with malt?

Answ.—Malt is *valued chiefly* for its saccharine matter; and it is *commonly valued*, in proportion to the quantity of extract which it is supposed to yield; or in the brewer's language, for the saccharine matter which it contains. But the grains of malt, and more especially of a mixture of malt and barley, after being treated with water as far as it is thought proper to search them, are valuable as food for cattle, both from the mucilage and hulls of barley, and also from the coarser saccharine matter which the brewer does not extract. And even the spent wash of corn spirits, as it never can be so thoroughly attenuated as wash from sugar, is more valuable as food for cattle. On the other hand, the extract of malt, or what is called the saccharine matter of malt or corn wash, is not so pure as that of sugar dissolved in water; and in fact will not yield so much spirits of a given strength from the same gravity of worts, when as well fermented as can be done by the distiller. And it also deserves to be mentioned, that the expence of brewing from malt, or a mixture of malt and barley, is considerable, while that of mixing sugar and water, is very trifling. Taking all these things into the account, the value of all the sugars, now annually imported from the West Indies, is equal in point of real value, not money price, to one-half of the whole malt now used in the United Kingdom. I might add, that the rum imported into Great Britain and Ireland, is equal to one-half of the barley that is used in the distilleries; so that the produce of the sugar cane imported into the mother country from our colonies, (since the large importation from the ceded islands,) is now equal to one-half both of our brewery and distillery. When that additional importation, and the diminution of the usual quantity of malt in 1797, 1800, and 1801, are attentively considered, the produce of the sugar cane is fully equal to one-half of the barley that is either brewed into ale, or distilled into spirits.

Quest. 5.—For what purposes do you consider this great importation of sugar to be valuable to Great Britain and Ireland, as well as to the West India planters and merchants?

Answ.—I cen-

Answ.—I consider every production of British, Irish, or colonial labour, to be beneficial to the nation at large; more especially when both the raw materials, and the manufacture, or the labour bestowed in fitting the rude produce for the market, are all belonging to this empire. In regard to sugar, it is advantageous in a way not commonly considered. Since the year 1767, owing to the great increase both of our luxury and of our population, Britain, which formerly exported at a low rate, has imported a great quantity of corn, at a high money-price. But sugar is a most nourishing article of food, even in years of plenty, or moderate fruitfulness; while in years of scarcity, if used by our brewers and distillers, it would enable us to make our barley into meal, and would supply us with both ale and spirits. In a calamitous season, if the taxes on sugar were reduced from 27s. to 20s. per cwt. (which is very nearly equal to the tax on malt, at 4s. 4d. per bushel,) and if both the brewery and distillery were limited to the use of sugar, the great quantity of barley which, even in an unfruitful year, would be made either into meal or pot barley, or mixed with wheaten flour or oatmeal, would go a considerable way to supply our want of wheat and oats; and would not only prevent the danger of a famine, but would keep down the very high prices of foreign grain.— Even when we do now import foreign corn, we are enabled by exporting great quantities of sugar, to barter one article of food for another. I have examined attentively the state of our exports and imports of grain since 1697; and have occasionally seen, for many years, the accounts of the exports of sugars to the continent of Europe. And on the whole I find that we have exported sugars equal in point of nourishment, or of pure saccharine matter, to all the corn that we have imported for food, under the designations of wheat, rye, barley, and oats. So that in every view, the great importation of sugar is an advantage to the nation, although, from a variety of causes, it has of late been unprofitable to the grower.

Quest. 6.—What are the causes to which you ascribe the present distress of the West India planters?

Answ.—They are various; but may be reduced to three classes. 1. Those, which it is impossible for either them or us to remove. 2. Those which may be removed, or rendered less hurtful to the planters. 3. Those which are occasioned by our laws and regulations—but which may be alleviated, if not altogether removed, by an alteration, or repeal of those laws.

Quest. 7.—Can you state concisely, the causes of the distress of the West India planters, which cannot be removed by either us or them?

Answ.—I con-

APPENDIX. 663

Answ.—I consider that they are, 1. A great rise in the money price of every thing that the West India planter needs to purchase or employ ; such as provisions, lumber, machinery, and implements; also shipping charges, freight, and insurance, whether he exports or imports any articles.— 2. A great fall in the price of sugar, and of whatever he has to sell, occasioned by what is termed a glut in the market, from the quantity that is now raised both in the old plantations of our ancient colonies, in the new plantations that have been made in these, and in all the plantations belonging to the ceded islands, (which have greatly increased our total importation of sugar,) and also from the want of demand for our sugars on the continent of Europe, or in places to which the produce of our colonies was formerly carried.

Quest. 8.—What are the causes of the distress, under which the West India planter labours, which he himself might remove, or alleviate? You may state these pretty fully.

Answ.—I apprehend, that these relate to the raising of provisions—to the curing of his sugars—and to the distillation of his rum. On all these important articles errors may be corrected, and defects supplied.

1. He has too little of his land in provision grounds, and often consumes more than is necessary. He should always raise as much provisions as he possibly can. Nothing renders either an individual, a colony, or a nation, more dependent than the want of provisions. The planter, therefore, should raise those articles, which he at present imports, as far as he can; and when he cannot raise the same article of the first necessity, he should endeavour to procure substitutes for it, as far as the climate will permit, that he may be less dependent on importation. In 1810, when oats cost L.2 10s. per quarter, in London, and at least, L.3 in St. Christophers, the planter who ordered 400 quarters to be bought, or L.1200 to be laid out for feeding his horses, (as appears in evidence before the Sugar Committee) certainly wanted economy. If he wished to keep a number of saddle-horses, he should have delayed till he returned home to Britain, as the small island of St. Kitt's was not extensive enough for his amusement. If he wished merely to employ horses in cultivating his plantation, the tops, or juices of the small canes, and the offals of the sugar mixed with other food, might have been used as substitutes for oats, at one-half of their price in general, and one-fifth part of that price in 1800. The planter, for the purpose of *human food*, must also learn to increase the extent of his provision grounds.

2. The West India planter does not cure his sugar properly; in many cases, at any rate, he neglects to do this.— Hence he sometimes loses *an eighth part*, or 12½ per cent. by leakage, before his sugars are landed in Europe, owing to

their being ill cured. Before he is ready to load a cargo, a ship appears in the offing, or is known to sail in a few days for Britain. All his people are incessantly employed to get as much as is possible of the crop, no matter how ill cured, put on board expeditiously, and transferred from the overseers to the merchant, or ship's captain. The introduction of the Bourbon cane, which is more productive as to quantity, demands that more attention should be paid, than hitherto has generally been done, to the curing of sugar. Yet in fact, less attention is now paid to it than formerly. The planter should clay all his sugars, where it is possible for him to do this, and should cure them well if he cannot clay them. With a glut in the market, and a great increase of freight, he should endeavour to have no leakage, but fine sugars in as compact a shape as may be done. Though the quantity exported were less, the prices and profits would be greater.

3. He pays too little attention to the distillation of his rum. The climate has done much for him in promoting the fermentation of his liquors, and the attenuation of his wash. Therefore he does nothing for himself. A saccharometer is scarcely known by name; and the use of this valuable instrument is not at all understood. In some of the islands, single rum, in its impure and bulky state, is sent to North America for lumber, and bartered to great disadvantage. In many places, the strongest of the low wines, or of this coarse, because only once distilled, spirit, is mixed with the double rum, or what has been twice distilled; and this compound mass is sent to Britain, of a certain strength, but with too little regard to quality. In the West India distillery, the fermentation of the liquor is left to the climate; and a voyage across the Atlantic supplies the place of a rectifier.—This mixture of first and second distillations may be allowed in England, where ardent spirits are rectified after being distilled. But it is intolerable to mix, in this way, the rum of the West Indies, (which is meant to be drank without rectification,) and it is folly to set up this impure spirit as a rival to both brandy and geneva. The West India planter ought to study distillation as a science; to be provided with a saccharometer for ascertaining the gravity, and attenuation of his liquor, and proportioning its strength to the heat of the season, should distil slowly, mixing water with his low wines, if too strong, and avoiding all mixtures of the essential oils, or of the feints with the pure spirits. By attending to these things, he will be enabled to procure both a greater quantity and a better quality of rum.

Quest. 9.—What are the causes of the distresses of the West India planters, which may be either removed or alleviated, by repelling or altering the British laws or regulations, which affect our colonies. You may here be both minute

nute and particular in stating both these causes, and what remedies, in your opinion, would tend to lessen or remove those distresses.

Answ.—There are various causes of the present distresses of our West India planters, which certainly may be removed, or at any rate much alleviated by an alteration of some of our laws, and a repeal of others. I consider that the following are the principal causes of the evils which belong to this class of legal distresses or difficulties.

1. The West India colonist is obliged to import every thing except lumber and provisions, from the mother country, and is prohibited from exporting any thing except rum and molasses, in exchange for those most necessary articles.

2. He is not allowed to refine his sugars, except for domestic use; and must send them to Britain to be refined for sale.

3. The tax on all Muscovado sugars is the same, without any distinction of quality; and on coarse sugars is very high, being 27s. per cwt.

4. The duty on clayed sugar is too high, being one-sixth part more than on Muscovado, or 31s. 6d. per cwt.

5. The duty on rum is also the same in all cases, and is not so much inferior to that of brandy and geneva, as it ought to be, in order to encourage the produce of our colonies, as opposed to that of our enemies.

6. The duty on rum is nearly 23 per cent. greater than what was formerly imposed on sugar-spirits, and is 34 per cent. more than what was formerly laid on these.

I shall now state the nature, extent, and remedies of these evils, with the greatest deference to the legislature, after remarking that they are occasioned, in a great measure, by our former connexion with the American States, and by the situation of our West India islands, in the neighbourhood of these States, which were also formerly British colonies.

1. The West India planters are obliged to import every thing that they want, except lumber and provisions, from the mother country, and are prohibited from exporting any thing except rum and molasses, in exchange for these necessary articles. Here it will not be denied, that whatever Britain can send to the Colonies, ought to be got from the mother country. It may even be admitted, that whatever the West India islands need from the north of Europe, should come to America in British bottoms; but it ought to be conceded, that those things, which the West Indies require as articles of the first necessity, either from the United States, or any other place in that neighbourhood, ought to be allowed to be imported into these islands, in either British, American, or in Colonial ships; also, that in return for such articles, the West India planter should be permitted to send refined sugar, and double rum, rather than molasses and rum

once

once distilled, which he sends at present. His imports and exports too, as far as is possible, ought to be carried either in British, Irish, or Colonial shipping; and he should export his produce in exchange for necessary articles, not in its rude and coarse state, but in the most perfect condition of manufacture. Nothing can be more improper, than the present practice of permitting wheaten flour, (the finished manufacture of the American agriculturist,) to be imported into the West Indies in American ships, and of returning by the same vessels at a great discount, (because sugar is not allowed to be bartered) single rum and molasses, the imperfect manufacture of the West India planter, which are to become the materials of the distillery of New England. The remedy here is obvious, namely to allow our planters to export sugar in exchange for lumber and provisions; and as much as possible in British, Irish, or Colonial shipping.

2. The West India planter is not allowed to refine his own sugar; except for domestic use; but must send it to Britain to be refined for sale. Whether the British Legislature should permit the colonies to refine the sugars which are to be sent to the mother country, for our internal consumpt, may be doubtful. But whether they should allow them to refine their sugar for bartering with the American States, or other nations, in exchange for lumber, provisions, and articles which they cannot want, will admit of no doubt with any impartial man, who has studied the subject. That can never be a wise regulation, of which the reverse appears to be founded in prudence or expediency. The sending of sugar, when refined, in exchange for these necessaries, is giving a manufactured article in exchange for raw materials, rude produce, and commodities of the first necessity, than which, nothing can be more politic or more expedient. The reverse, therefore, can never be a matter of prudence. Hence I conclude, that it would be very proper to allow the colonies to barter refined sugar and double rum, rather than send molasses and single rum to America. But this, as far as is possible, should be done in British or Colonial shipping; for nothing can be more improper, than the carrying on this trade in ships belonging to the United States. This is equally hurtful both to the wealth and the strength of the nation.

3. The tax on all Muscovado sugar is the same; and on coarse sugars is very high, being 27s. per cwt.—good Jamaica sugar at 60s. per cwt. can more easily afford to pay 27s. of duty, than coarse sugar from the Bourbon cane can pay 18s. when it is sold at 40s. I would here observe, that since the introduction of this species of sugar cane from Otaheite, (although it be a valuable kind, not only from yielding productive crops on good land, but also for growing on soils that would either raise the native cane of the West Indies very

ry imperfectly, or could not raise it at all,) the quality of sugar has certainly been much deteriorated of late years, and is much inferior to what it was formerly. In this situation, a duty *ad valorem*, or in proportion to the price of sugar, would be the fairest mode of imposing the tax on sugars of widely different quality. But as this might lead to frauds which would be injurious to the revenue, the next best method would be to encourage the claying of sugars, especially if the West India planter is not allowed to send refined sugars to Britain.

4. The duty on clayed sugars is too high, being one-sixth part more than that on Muscovado, as above mentioned.— It is owing to this that the planter is induced to send his sugars to Britain without being clayed; for the clayed sugar does not contain a sixth part more of saccharine matter, in a given weight, than the Muscovado contains; and the molasses, or refuse of the clayed sugar, is made into rum, which pays a much higher duty than is imposed on the sugar. Instead of a high duty on clayed sugars, the planter should pay only a reasonable one, in order to induce him to clay all his sugars. By this means there will be no leakage in the passage to Europe, nor any unproductive labour in carrying watery and heavy, instead of dry, and comparatively lighter sugar. And more rum will be produced from clayed, than from imperfectly cured sugar. Therefore I would suggest, that the duty on clayed sugar should be only 28s. per cwt. or 1s. more than Muscovado sugar pays. Both the planter and the revenue would gain by this alteration. The planter would save what he loses by leakage, and would have better sugar and more rum to export; and the revenue would gain the additional shilling on every cwt. of sugar, and all the duty on the additional quantity of rum, imported into Britain. The cargo also would be lighter. For rum is one-fourth part lighter than the sugar from which it is extracted.

5. The duty on rum is the same in all cases; and is not so much inferior to that on brandy and geneva, as it ought to be, in order to encourage the produce of our colonies, in opposition to that of our enemies. This duty at present, including the rates of the Excise and Customs, is 11s. 3d. and nearly one-third, exactly 11-30ths of a penny per gallon, while that on brandy is 16s. 7d. and 1-6th part of a penny, exactly 19-120ths of a penny; and that on geneva is 16s. 5d. and 4-5ths of a penny per gallon. I would not recommend the prohibiting the two last; but in the present state both of Europe and of our colonies, I think that the tax on foreign spirits should be double of that on rum. This duty should not be the same in all cases; but when the distillation of corn and malt is prohibited in this country, the tax on rum, and also on spirits distilled from sugar, should be no more than

than what is at other times paid by the distiller for corn or malt spirits. For the revenue loses nothing if it get as much duty for rum as for corn spirits; and in this state of the country, all the corn that can be spared, should be used for human food. When corn is very cheap, the tax on rum ought to be one-half more than on malt or corn spirits, to give a decided preference to British agriculture. But when it becomes expedient to prohibit their distillation, the tax on rum and on sugar should be for a time reduced to the lowest duty which is paid on corn-spirits.

6. The duty on rum is nearly twenty-three per cent. more than was lately imposed on sugar-spirits distilled in England; although it is obvious, that whatever be the duty on rum, or on the spirit made from the offals of sugar in the West Indies, ought also to be the duty on all spirits made from sugar in Britain. But, by the proposed law, the duty on sugar-spirits distilled in England is to be thirty-four per cent. or above two shillings and tenpence halfpenny less than that on rum, and seventeen per cent. less than that on corn-spirits. Nothing more preposterous can be imagined. It is saying to the West India planter—" You distil rum from the offals of sugar, there-
" fore you must pay a high tax." To the British distiller of sugar—" You distil from the best Muscovado sugar, there-
" fore you must pay a low tax;" and to the British farmer—
" Because the money price of barley has not risen with that
" of labour, and because the malt tax has been more than
" tripled within these few years, you must pay such a tax
" on corn-wash, as shall be a virtual prohibition of the distil-
" lation of corn spirits."

Quest. 10.—But will it not give relief to the West India *Planter*, if a low duty is imposed on sugar-wash that is distilled in Britain?

Answ.—It will be very convenient for the West India *Merchant* who has a quantity of sugar on hand. But the best method of granting relief to the *Planter*, or *grower of sugar*, is to impose a lower duty (such as twenty-eight shillings per cwt.) on all his clayed, or imperfectly refined sugar; and a reasonable tax on rum, (especially when corn is at a high price) and at all times a much less tax than on brandy and geneva : so that he may be induced to distil a greater quantity of rum, and be enabled to draw more money for a less weight and bulk, but a better quality of sugar. To save him twenty per cent. of shipping charges; ten per cent. of leakage, occasioned by exporting coarse moist sugars ; and ten per cent. of taxes, would be a permanent relief to the Planter, and would be no improper interference with the interests of British agriculture. But he ought *on no account to pay less tax on his sugar or rum*, than is *paid on malt and corn-spirits* : for this is acting

ting very *unfairly* by the farmers of the mother country, and, in the end, will be found injurious to the nation at large.

Quest. 11.—Can you give, in a few sentences, a general idea of the weight, bulk, and best method of packing the produce of the West India Planter?

Answ.—Rum is, in general, only seven-eighths of the specific gravity of water. Sugar, when in lumps, is sixty per cent. weightier than water; but when broken, it is only about one-third heavier. If the Planter was allowed to send it home in lumps, with the best of his Muscovado or clayed sugar beat hard between the lumps, there would be no leakage, and the least space would be occupied. And by sending a much greater quantity of rum, the cargo, on the whole, would contain no useless weight, which all moist sugar does. To every indulgence of this nature he is fairly entitled; and no bad effect could arise from it, except that the price of molasses would be higher, and that the sugar-bakers would have less employment.

Quest. 12.—Could sugar be advantageously introduced into the brewery, when the price of corn is extremely high?

Answ.—I have both brewed and distilled sugar. And I know that it makes excellent fermented liquor; and that where its wort is made as *strong* as that from malt, it will keep for an equal length of time, if properly managed. Some brewers from ignorance, and others from parsimony, made their worts from sugar too weak; and, therefore, they did not keep.—But though I wish the brewery to be supplied from *malt* and not from *sugar*, I must candidly declare, that sugar makes excellent ale, though its flavour is different from that of malt liquor. In 1800, when the price of corn was so very high, the reduction of the sugar duty, from 27s. to 20s. (or even lower, when put at sight of the excise-officer into the brewer's mash tun,) would have saved our barley, and diminished the quantity of corn imported, as well as the extremely high prices paid for foreign corn.

Quest. 13.—Would a mixture of sugar and malt be proper in such cases.

Answ.—It would be proper only because it would yield a liquor with a malt flavour. But as sugar ferments more easily than malt does, the malt should be mashed by itself; and after being fermented for 24 hours at least, any quantity of boiled sugar and water, that was desired, might be added. This is more necessary in the distillery than in the brewery. For the spirit is apt to assume a cloudy appearance, when it goes to the rectifier, owing to the extract of sugar being in the acetous state before that of the malt is completely fermented.

Quest. 14.—What is the price of barley and bigg in Aberdeenshire?

Answ.

Answ.—We do not raise much barley in this county, and what we do raise is generally of an inferior quality. The fiar prices of barley, struck a few weeks ago by the Sheriff and a Jury, were L.1 3s. per Aberdeenshire boll, or L.1 7s. 7d. 1-5th per English quarter. Of bear or bigg, we raise a considerable quantity, though less than we did ten years ago, owing to the want of licensed distilleries, and to the high proportion of malt tax. The fiar prices of bear, struck at the same time with barley, were L.1 1s for the boll, county measure, or L.1 5s. 2d. 2-5ths for the Winchester quarter. But at present there is no demand, even at this low price; and the brewers will not purchase bear except of the finest quality. Five bolls of my own bear, which weighed only an oz. less than 45 lbs. per bushel, *were rejected* a few weeks ago; although I have had bear formerly accepted when only 42 or 43 lbs. per bushel. But bear at that time was in demand, and both the malt and ale duties were much less than at present.

Quest. 15.—Is there much illicit distillation carried on in Aberdeenshire; and by whom is it carried on?

Answ.—There is a considerable quantity of spirits made by illicit distillers. But they are generally the poorer peasants only, who engage in this illicit traffic. They carry it on in the night, and during the winter season, and in very small quantities, seldom more than two gallons of spirits being distilled at one time. They use no proper stills, but a wooden head and two copper pipes; their whole apparatus amounting only to a few shillings. Yet by distilling slowly, and from malt, with a small proportion of potatoes, they make a good spirit; though they seldom get a gallon of spirits from a bushel of malt. A legal distiller, from the same quantity of materials, makes at least 50 per cent. of more spirits. And the only remedy for curing this illicit distillation, is to grant licences to small stills in all parts of the county. This would bring a considerable revenue to government, and render the labour of the people more productive. It is because I have no interest in this matter, and because I know that a *reasonable tax* would be paid to government, without costing *a farthing to the legal distiller, merely by extracting more spirits from the bushel of malt*, than are extracted by the poor ignorant country people, that I state this particularly.

Quest. 16.—What duty would you think it fair and equitable that the Highland Distiller should pay, in order that Smuggling may be checked, and that the Government may obtain all the revenue that can be had from that part of the kingdom?

Answ.—It is a matter of considerable delicacy to answer this question; but I shall do it candidly. Whenever the English

English Distiller pays seven shillings, and the Lowland Scotch Distiller pays five shillings, (which is the present proportion) the Highland Distiller should pay four shillings, including the duties on the still, the wash, the spirits, and the malt. And to prevent any injury to the Lowland Distiller from smuggling, all spirits sent into the Lowlands by permit, should pay other two shillings of duty. I suspect I shall be blamed for proposing so much; but I consider all ardent spirits as a fair subject of taxation; and I feel myself bound to give a conscientious opinion, though it should displease all parties.

Quest. 17.—What are the Duties which, by the bill lately brought into Parliament, are imposed upon Spirits distilled in the Lowlands, including the duty on the contents of the still, the duty on wash and spirits, and both the sugar-duty on importation, and the malt-duty for the proportion of malt used?

Answ.—The duties on corn-spirits in the Lowlands, per gallon, are - - - - - £0 7 8¼
On sugar-spirits, reckoning every thing, they are only - - - - - - 0 5 10½

Or less than on corn by 32 per cent.; or - £0 1 10¼

Quest. 18.—What are the Duties which, by the same bill, are proposed to be laid on the Highland Distiller, on Corn and on Sugar-Spirits?

Answ.—On sugar-spirits they amount to nearly five shillings; or - - - - - £0 4 11$\frac{12}{14}$
On corn-spirits, if 100 gallons of wash yield 10 of spirits, exactly - - - 0 8 0
On spirits from pure malt, as distilled in this county, - - - - - 0 10 0

No Highland Distiller can take out a licence on such terms: but smuggling will prevail; the Government will lose its revenues; and the morals of the people will be corrupted.

On the Method of ascertaining the specific Gravity of different Roots.

From observing fifty tuns of White Globe, and only thirty-six of Red-topped, in the same field, I was led to examine the specific gravity of Ruta Baga, Carrots, Potatoes, and of different kinds of Turnips.

As to potatoes, ruta baga, and carrots, the process was very easy. I first weighed a plant of ruta baga in air, after cutting off the leaves and roots; then hung it in water, tied to a piece of strong thread, fixed to one scale of a beam which was very sensible, or turned easily. Into the other scale I put as many ounces and grains as exactly balanced the ruta baga plant while

while it hang in water: the weight necessary to balance this indicated the specific gravity of the ruta baga. Thus a plant which weighed one hundred ounces in air, was balanced by two and a half ounces appended to the opposite scale, while the ruta baga was hung in water. The specific gravity of ruta baga, therefore, was 102.5. That of potatoes and carrots was found in the same manner.

But all turnips are lighter than water. Therefore I took a piece of pack-thread and tied round a turnip, and weighed it in air: then I affixed as many weights to the pack-thread as sunk the turnip in a large cask of water. The weights were of a mixture of brass and copper, and eight times as weighty as water. Therefore, I subtracted the eighth part from the whole weight, and then I had the proportion which the turnip wanted of being equal to the specific gravity of water. Thus a turnip of 12¼ pounds required 16 ounces weight of metal weights to sink it in water. The eighth part of this is 2 ounces: therefore, deducting those 2 ounces for the bulk or space occupied by the weights, I said—If 12¼ pounds, or 200 ounces want 14 ounces to sink it, what will 100 pounds want?— The answer was 93. Therefore, the specific gravity of the turnip was 93.

The two largest turnips which I found were at Pitfour.— The one was a white globe, weighing 32 pounds 12 ounces, and its specific gravity was 87. The other a red-topped, weighed 28 pounds 12¼ ounces, and its specific gravity was 88.

Mr Whyte, at Bridgend of Longside, had 48 tons 8 cwt. on the Scotch acre of white globe, which had been manured with one part of dung and three parts of peat-moss. This fact deserves attention.

P. S.—Since this Appendix went to press, there have been three Oxen killed at Aberdeen. The two first were fed by Mr WALKER at Wester Fintray, and weighed, of Averdupoise, neat pounds—

	Beef.	Tallow.	Hide.	Total.
The First,	1592	243	143	1978
The Second,	1546	292	138	1976
The Third,	1612	172	177	1961

This last was a twin ox, reared by F. GARDEN CAMPBELL, Esq. of Troup, and had been five years a bull.—The above weights are exclusive of head, feet, tongue, hearts, and kidneys— which, in the smallest ox, were exactly 112 pounds, or 1 cwt.

THE END.

D. Chalmers & Co.
Printers, Aberdeen.

www.ingramcontent.com/pod-product-compliance
Lightning Source LLC
Chambersburg PA
CBHW021348290426
44108CB00010B/149